MEMBRANE IN CANCER CELLS

ANNALS OF THE NEW YORK ACADEMY OF SCIENCES
Volume 551

MEMBRANE IN CANCER CELLS

Edited by Tommaso Galeotti, Achille Cittadini,
Giovanni Neri, and Antonio Scarpa

The New York Academy of Sciences
New York, New York
1988

⊗ The paper used in this publication meets the minimum requirements of American National Standard for Information Sciences—Permanence of Paper for Printed Library Materials, ANSI Z39.48-1984.

The photograph on the cover of the paperbound edition of this book shows a ceramic ex voto (5 × 8 inches, ca. 1675) from the sanctuary of the Church of the Madonna dei Bagni in the Umbrian region of Italy. Ex voto tiles are votive offerings that were made to obtain protection against disease and other calamities.

Library of Congress Cataloging-in-Publication Data

Membrane in cancer cells.

(Annals of the New York Academy of Sciences, ISSN 0077-8923; v. 551)
Result of a conference held June 13–16, 1988 in Perugia, Italy; co-sponsored by the New York Academy of Sciences and the Università cattolica del Sacro Cuore.
Includes bibliographies and index.
1. Carcinogenesis—Congresses. 2. Cancer cells—Growth—Regulation—Congresses. 3. Cell membranes—Abnormalities—Congresses. I. Galeotti, T. (Tommaso) II. New York Academy of Sciences. III. Università cattolica del Sacro Cuore. IV. Series. [DNLM: 1. Cell Membrane—congresses. 2. Cell Transformation, Neoplastic—congresses. 3. Neoplasms—etiology—congresses. W1 AN626YL v.551 / QZ 202 M5335 1988]
Q11.N5 vol. 551   500 s   89-2999
[RC268.5]   [616.99′4071]
ISBN 0-89766-490-6 (alk. paper)
ISBN 0-89766-491-4 (pbk.)

SP
Printed in the United States of America
ISBN 0-89766-490-6 (cloth)
ISBN 0-89766-491-4 (paper)
ISSN 0077-8923

ANNALS OF THE NEW YORK ACADEMY OF SCIENCES

Volume 551
December 30, 1988

MEMBRANE IN CANCER CELLS[a]

Editors and Conference Organizers
TOMMASO GALEOTTI, ACHILLE CITTADINI,
GIOVANNI NERI, and ANTONIO SCARPA

CONTENTS

---

[a]This volume is the result of a conference entitled Biological Membranes in Cancer Cells co-sponsored by the New York Academy of Sciences and the Università Cattolica del Sacro Cuore and held in Perugia, Italy on June 13–16, 1988.

Part II. Ionic Signals

Part III. Growth Factors and Oncogene Product

Part IV. Therapeutic Strategies

Financial assistance was received from:
- AZIENDE CHIMICHE RIUNITE F. ANGELINI
- BECKMAN ANALYTICAL S.P.A.
- CASSA RISPARMIO DI ROMA
- CATHOLIC UNIVERSITY SCHOOL OF MEDICINE, ROME
- THE COUNCIL FOR TOBACCO RESEARCH—U.S.A., INC.
- FARMADES S.P.A.
- FARMITALIA CARLO ERBA
- FLOW LABORATORIES
- MERCK SHARP & DOHME RESEARCH LABORATORIES
- MINISTRY OF EDUCATION OF ITALY
- PFIZER CENTRAL RESEARCH
- PRISMA
- SERONO
- SIGMA-TAU FARMACEUTICI
- S.L.A.L. S.R.L.
- SQUIBB
- TECHNOCHIMICA MODERNA
- THE UPJOHN COMPANY

BRITTON CHANCE

A Tribute to Britton Chance

This conference is in honor of Dr. Britton Chance and to commemorate his 75th birthday.

Dr. Britton Chance is a key figure in the development of modern biochemistry and biophysics. A unique feature of Dr. Chance's scientific achievements is the significance of his research to several fields of science, ranging from mathematics through biochemistry and electronics to clinical medicine. In all of these fields, Dr. Chance has made fundamental and original contributions, documented by an impressive and possible unprecedented record of publications, now approaching 1000 in number. His research has encompassed studies at the cellular, molecular, and more recently, at the whole-body level. Perhaps even more important is the way in which Dr. Chance has opened new avenues with his novel approaches. He has continually stimulated other scientists and other laboratories to continue and expand studies which he pioneered, often developing new technological approaches. Dr. Chance may be looked upon as the "father" of quantitative spectrophotometry, an area of research that began as part of his interest in fast-reactions methods, leading to the first feedback-stabilized optical system with unprecedented stability. The invention of the double-beam or dual-wavelength spectrophotometer, during the years 1947–1951, made it possible to measure enzyme action in isolated organelles and tissues with reliability and precision. This instrument further led to many key developments in biophysics, biochemistry, and molecular biology in different laboratories throughout the world.

Dr. Chance's academic career has evolved completely at the University of Pennsylvania, where he was nominated professor of Biophysics and Physical Biochemistry, and Director of the Johnson Foundation of the late 1940s. In Philadelphia, Dr. Chance has created a point of attraction for many scientists from different countries and of different ages, and there he mastered new developments to investigate quantitatively cellular metabolism. In 1951, Dr. Chance catalyzed the development of the Johnson Foundation Analogue Computer, which led to the computer solution of a variety of problems in enzyme kinetics. A classical example of this is the seminal work on the intermediates in the reaction of peroxidase with H_2O_2, the first direct demonstration of the existence of a Michaelis-Menten complex. Dr. Chance led the way to a better understanding of the function of myoglobin and cytochrome in the tissue, and of the effects of cardiac ischemia by the invention of a time-sharing spectrophotometer. The use of this instrument in a pioneering study on the perfused heart linked this basic approach to important medical applications and opened a new method for studying cardiac structure and function *in vivo*. Along the sames lines, Dr. Chance's intense commitments to the quantitative investigations of metabolism led to the invention of a laser-activated fluorometer especially designed for use during human neurosurgery. This line of research is being recently developed and extended by Chance himself, using picosecond spectroscopy for noninvasive mapping of muscle and brain metabolism in health and disease.

An exciting expansion of Dr. Chance's lifetime work on noninvasive techniques for metabolic studies has been his role in the development and utilization of NMR spectroscopy for biochemical studies in humans. In this case, the great strength of the method is the combination of the electro-optical and the electromagnetic techniques for the determination of key variables related to physiology and pathology.

During his outstanding career, Dr. Chance has received many awards and honors. He is a member of many scientific societies including the National Academy of Sciences in the United States, and the Royal Society of the UK and the Academia Nazionale dei Lincei of Italy abroad. Additionally, he played an important role in the inception and growth of international scientific unions, and among others, has been President of the International Union of Pure and Applied Biophysics (IUPAB).

Dr. Chance's passion for instrumentation extends beyond biochemistry and biophysics. Among his important contributions in other fields is the development of radar with other colleagues at MIT, the invention of a "computer bomber" employed during World War II, and the design in the late 1940s of an automatic steering device for vessels. His enthusiasm for sailing and his ability as a skipper are legendary. Britt's competitive spirit, courage and technical skills made him a pioneer of modern technological sailing, and a gold-medal Olympic winner in the 5.5-m International Class in Helsinki. His delight in sailing has been transmitted to his children, and one of them, Britton C. Jr., is an outstanding naval architect, and designer of very original boats competing successfully in the America's Cup.

Although one could continue to enumerate his impressive individual contributions to the body of scientific knowledge, it is far more important to stress that his outstanding quality has been the transfer of concepts and techniques from one discipline to another, and the dissemination of these methods among his many co-workers. His original contributions span the entire scientific spectrum from electronics to chemistry and biology. In this era of fragmentation and specialization of knowledge, his work and scientific contributions represent an outstanding exception.

—THE EDITORS

Introductory Remarks

T. GALEOTTI, A. CITTADINI, AND G. NERI

Department of General Pathology
Catholic University
School of Medicine
Rome, Italy

A. SCARPA

Department of Physiology and Biophysics
Case Western Reserve University
Cleveland, Ohio 44106

This conference is the third of a series that began in Rome in 1982. The goal then seemed very ambitious: to discuss and build a consensus on the role(s) of biological membranes on the process of carcinogenesis. Having cell membranes as a focus for distant and diverse experimental approaches, in this and previous conferences, we brought together scientists from different disciplines to look at the evidence of primary and secondary alterations of cell membrane in malignancy.

Our original goal now appears less ambitious and utopian than it did six years ago, when we started the conference series. This conference has shown that a unifying view is finally taking shape, although still fuzzy in many details, and that the experimental evidence of links and connections among phenomena occurring from plasma membranes to intracellular compartments and the nucleus is becoming increasingly convincing.

At the same time, as knowledge develops, established observations and interpretations are reexamined critically and new evidence is placed in proper perspective. Membrane lipid damage by oxy-radicals, and its implications in cellular differentiation and tumor promotion, is one of the established evidences that may need some reconsideration. A number of articles in this *Annals* indicate that there are some conflicting reports about the concept that oxidative stress is a crucial mechanism in the control of cell proliferation.

Ion flux across cell membranes is another critical process that is becoming more clearly defined as we learn about new systems that integrate our understanding of cell calcium and H^+ homeostasis. At the same time it is becoming well established that the relationship between ion fluxes and cell growth regulation is mediated by oncogene products.

Oncogenes are clearly central to carcinogenesis as coding genes for growth factors and growth factor receptors, and although we all understand the relevance of these gene products for the process of neoplastic transformation, it is important to be aware of the distinction between genes that are necessary for cell growth and genes that act as regulators of cell growth. All this will become clearer as we proceed in our understanding of the molecular mechanisms through which these factors act, and of the mutations that are responsible for the deregulation of their action.

At the bottom of these new and fascinating insights there continues to be the tantalizing prospect of their possible therapeutic applications. The role of cell surface proteins and glycoproteins in cell aggregation and in the spreading of metastases cannot be overemphasized, and it is quite obvious that a better knowledge of their structure and function could be exploited to devise specific means of blocking tumor invasion.

Many of the above points are summarized and discussed critically in the comprehensive article by Britton Chance, to whom this conference has been dedicated. After illustrating a brilliant example of new instrumental approaches to the study of the energy metabolism of neoplastic tissues *in vivo,* and relating diagnostic implications, Dr. Chance makes one very provocative point: that the prevailing approach to cancer treatment by oxidative stress, based on what is known about the energy metabolism of cancer cells, may have to be reshaped in favor of a "reducing" stress. This is clearly a challenge that will have to be met in the future.

Optical and Nuclear Magnetic Resonance Studies of Hypoxia in Human Tissue and Tumors[a]

B. CHANCE, E. BORER, A. EVANS, G. HOLTOM,
J. KENT, M. MARIS, K. McCULLY,
J. NORTHROP, AND M. SHINKWIN

Department of Biochemistry and Biophysics
University of Pennsylvania
Philadelphia, Pennsylvania 19104

I am deeply appreciative for the opportunity to present my current work on NMR studies of tumor cells to an audience of friends and colleagues who are most experienced in this study area, and at the same time to find myself in this historic theater in the beautiful medieval town of Gubbio. My lecture attempts to bridge the gap in understanding normal and abnormal biochemistry of tissues through the use of two noninvasive biochemical techniques—optical and nuclear magnetic resonance spectroscopy. But let me pause a moment before starting and pay tribute to one of the founding members of this group, Al Lehninger, and to note his impact upon Italian biochemistry.[1]

Oxygen delivery and cell energetics of solid tumors are very complex processes and are not fully understood in spite of years of study. The tumor cells live by the stimulation of oxygen delivery to the tumor mass and die by virtue of the failure of adequate alimentation. It is now possible to better quantitate the tumor oxygenation noninvasively by nuclear magnetic resonance spectroscopy of the cell energy metabolism[2,3] and by optical spectrophotometry of the deoxygenation of hemoglobin. Furthermore, the possibility of optical ranging of the tumor mass by picosecond pulse light spectroscopy will be presented.

The application of magnetic resonance spectroscopy to tissues is one of the simplest forms of this sophisticated technique where absorption of radiofrequency waves by the magnetically aligned nuclear spins[4] is the central aspect of the technology (FIG. 1).[4] An ingenious Fourier-transform spectrometer allows the simultaneous measurement of the absorption of nuclei of the same kind in different environments over a small band of frequencies. For the different phosphate compounds of muscle, the nuclei absorb, over a limited range, 100 ppm.[5] A large and constant magnet of 1.5–4.7 tesla is essential and is probably best known because of the danger of the fringe field which may extend to intensities of 5 gauss, at 10–15 meters from large high-field superconducting magnets. Smaller magnets were designed for *in vivo* spectroscopy. When George Radda and I joined forces in 1975[6] to apply NMR to the study for signals in the human body, we

[a]This work was supported by Grants HL-31934, NS-22881, RR-02305 and HL 18708 from the National Institutes of Health.

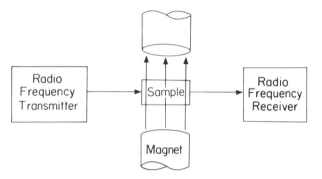

FIGURE 1. Schematic diagram of the NMR apparatus.

acquired at that time a 7-inch bore magnet for arm and leg studies which we are still using for neonate studies. The key feature of an NMR magnet is that the object to be examined has to be put inside the magnet (FIG. 2).[7] No exteriorized field has been obtained. Furthermore, the homogeneous field volume is not only located near the middle of the magnet, but it also occupies only a small percentage of the length of the magnet. These restrictions favor the use of either animal models or arms and legs of adult humans for magnets that do not permit entry of the thorax. Obviously, neonates can be accommodated in small-magnet units as in the study of infant neuroblastoma described herein.[2,8]

Phosphorus nuclear magnetic resonance spectroscopy (^{31}PMRS) affords a quantitative assay of cell ATP. Cells devoid of ATP will soon suffer ionic disequilibration, osmotic stress, membrane rupture, and loss of cell contents. Thus, ATP can correctly be said to be essential to cell function and failure to recuperate ATP spells cell death

FIGURE 2. The homogeneous field volume is within the magnet and only objects that can be inserted into the magnet can be studied.

(FIG. 3).[9,10] In fact, one of the clear-cut delineations of free-radical damage to tissues is indeed the loss of ATP after oxygen reflow to previously ischemic tissues. No more quantitative assay of free radical damage is available.

Normally, ATP is maintained constant through the cooperation of oxidative phosphorylation and the creatine kinase equilibrium. Phosphocreatine (PCr) itself is responsive to the rate of cell energy metabolism, with the result that the sum of PCr plus inorganic phosphate (P_i) is maintained constant and P_i itself is maintained at much lower levels than expected from the biochemical analysis of total cell phosphate. Such determinations indicate that the P_i that is involved in metabolic control is only 10 percent of the total phosphate pool.[11] Thus, much P_i is bound and inaccessible to the control of respiration. This phenomenon is also observable in liver and in those lines of tumor cells that lack creatine kinase and a creatine pool.

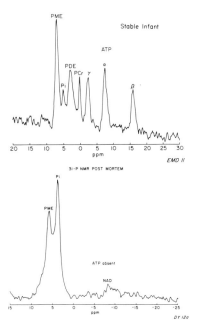

FIGURE 3. NMR spectra of live and dead neonate brain illustrating the large significance of the changes of phosphocreatine, inorganic phosphate, and ATP.

The predominant signals of phosphorus metabolism originate in the cytosolic space, where the creatine kinase system approaches equilibrium and maintains a nearly constant cytosolic ATP concentration by variation of the ADP (and P_i) concentrations in response to the needs of respiration. The signals that originate in the mitochondrial space are more difficult to observe than those originating from the cytosol (FIG. 4).[12] The actual concentration of ADP is too small to be measured, but can be readily calculated from the creatine kinase equilibrium to be between 5 and 50 μM, depending upon the level of respiratory activation expected from an *in vitro* K_m of approximately 20 μM.[13] P_i is found to be ~1 mM and is about equal to its regulatory K_m. Functional stresses readily raise phosphate to the 5–10 mM region, well beyond its

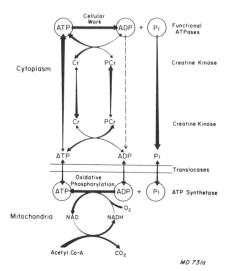

FIGURE 4. Schematic diagram of the cytosolic-mitochondrial relationships appropriate to NMR study.

regulatory level in cells that regulate metabolism through the ADP concentration. In tumor cells and liver that lack creatine kinase, P_i may be regulated at 0.5–5 mM.[14]

ADP will be expected to rise and fall with increasing and decreasing activity. Indeed, as ADP goes to zero and flux goes to zero, the phosphate potential, ATP/ADP $\times$ P_i, will rise to a maximum. Since ATP is constant and P_i is in excess, the phosphate potential is best represented by the ADP value, which in turn is $P_i/PCr \times 33$ for pH 7, making some significant approximations.[12,15] Under physiological metabolic stress (for example, rapid growth of tumor cells, ion transport in the kidney, or exercise in muscle), the P_i/PCr rises. While this algorithm is applicable to the tumor cell as well as to the human muscle, the muscle provides exemplary studies of the relationship between ADP concentrations and metabolic stress, which are readily observed because oxidative metabolism is controlled by musclar work.[16,17] No such protocol is available in tumor cells, although the stress of ion transport is afforded by cooling ascites cells to 0° C and raising the temperature to 20°C so that effused ions are transported.[18]

Such perturbations are readily explained by a Michaelis-Menten formulation for the velocity of oxidative metabolism, V, in relation to its maximum V_M:

$$\frac{V}{V_M} = \frac{1}{1 + \dfrac{K_{ADP}}{ADP}} \tag{1}$$

$$\frac{V}{V_m} = \frac{1}{1 + \dfrac{K_s}{s}} = \frac{1}{1 + \dfrac{K_{ADP}}{ADP}} \tag{2}$$

where s is ADP and K_s is taken to be the 20 μM *in vitro* value,[13] because mitochondria are one of the few organelles that can be extracted from the tissue without disruption of their activities other than that of depletion of substrates.

Hypoxia is believed generally to be a characteristic state of tumors in rapid growth or necrosis and affords the basis for a long-standing theory of Otto Warburg[19] and Dean Burk.[20] However, the precise measurement of hypoxia in cell and tissues is a difficult task; detailed studies have been made by a number of investigators using oxygen microelectrodes, which indicate wide ranges of tissue oxygen levels. However, these techniques are invasive and difficult to apply to human subjects. What does PMRS tell us about hypoxia? The general equation for oxidative metabolism,

$$3\,ADP + 3\,P_i + \tfrac{1}{2}\,O_2 + NADH + H^+ \rightarrow 3\,ATP + NAD^+ + H_2O \qquad (3)$$

indicates that not only can ADP and P_i be regulatory, but also oxygen itself can be regulatory if the tissue oxygen falls to the level of its K_m. This, of course, applies to other components of the equation and indeed to regulatory factors that indirectly control NADH delivery. However, we concentrate our attention upon tissue oxygen, and represent simultaneous ADP and oxygen control. The V/V_m equation now becomes only slightly more complex[15]:

$$\frac{V}{V_m} = \frac{1}{1 + \dfrac{K_{ADP}}{ADP} + \dfrac{K_{O_2}}{O_2}} \qquad (4)$$

A very important aspect of this relationship is the response of P_i/PCr to oxygen. Whenever the oxygen decreases into the range of its affinity constant, there can be a compensatory rise of ADP and P_i/PCr in order to maintain V/V_m constant. However, as ADP rises, it in turn will not only reach its K_m, but will also rise to a point where it is ineffective in increasing respiration further. Under these conditions, V_m falls and V_m approaches V. At this point the system is no longer under metabolic control by respiratory activity and the only remaining regulatory function is that of glycolysis, which operates at a very high ADP concentration and high (P_i/PCr).

There are many applications of PMRS to hypoxia and in line with my presentation on tumor hypoxia, it is appropriate to give two examples: one of the peripheral weakness of patients with congestive heart failure is faulty oxygen delivery. As FIGURE 5 shows, for a given work load, they need almost twice the ADP (P_i/PCr) to maintain the work load as compared to age-matched controls.[21]

Another example is afforded by peripheral vascular disease. The stacked plot of spectra (FIG. 6) taken from a diseased limb show that one to two minutes' exercise nearly completely depletes the PCr pool and raises the P_i near its maximum; this is the condition cited earlier where ADP (P_i/PCr) rises in response to an oxygen deficiency.[22] The recovery from exercise is prolonged since the rate of ATP resynthesis is limited by

FIGURE 5. Effect upon P_i/PCr of peripheral weakness in patients with congestive heart failure ($\times$) compared with age-matched control (O).

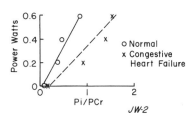

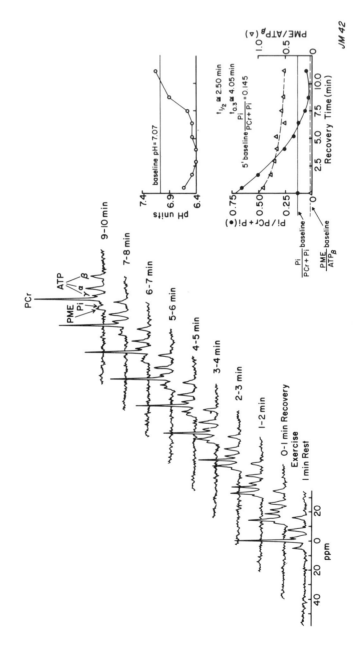

FIGURE 6. Stacked plot of the time course of post-exercise recovery changes in the limb of a patient with peripheral vascular disease.

oxygen delivery and occurs over an interval of nearly 10 minutes, whereas normal limbs recover in under a minute.[16]

PMRS is a very sensitive oxygen indicator, one that comes "into register" when the tissue oxygen level approaches the oxygen affinity of cytochrome oxidase ($10^{-6}10^{-7}$M). It is often desirable to have an "early warning" of imminent tissue hypoxia, particularly in a heterogeneous tissue volume such as a tumor where the hypoxia may be significantly regionalized. In order to accomplish this, we have taken advantage of some principles introduced in the 1940s by Glenn Millikan[23] for tissue hemoglobinometry using visible and near-red light. Millikan used red and green filters, although it is obvious that the green filter transmitted light in a near-red region appropriate for tissue study as well as in the green region, as indeed Jobsis' later studies clarified.[24] The spectrum of hemoglobin in the deoxy and oxy states is indicated by

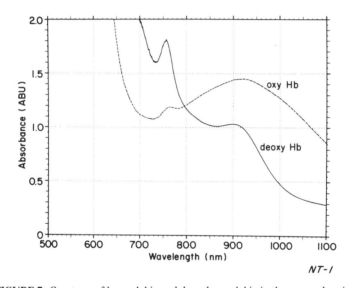

NT-1

FIGURE 7. Spectrum of hemoglobin and deoxyhemoglobin in the near-red region.

FIGURE 7 (S. Nioka, personal communication), and it is clear that the peak at 760 nm of the deoxy species can be measured with respect to the isosbestic point at 800 nm. The total HbO_2 + Hb is measured at the later wavelength, and the difference of absorption is measured at 760–800 nm, corresponding to the increase of Hb concentration. The diagram of a simple dual wavelength spectrophotometer attached to the human leg is shown in FIGURE 8, where the two light sources and the two detectors at 760 and 800 nm are in contact with the leg and are appropriately spaced to permit an average penetration of 2–3 cm.[25] The typical readout from such an instrument on ischemia of the calf is indicated in FIGURE 9, which shows that the hemoglobin (and myoglobin in this case) respond sensitively to occlusion of the circulation, as would indeed be expected from a tumor in which the growth had exceeded the oxygen supply.

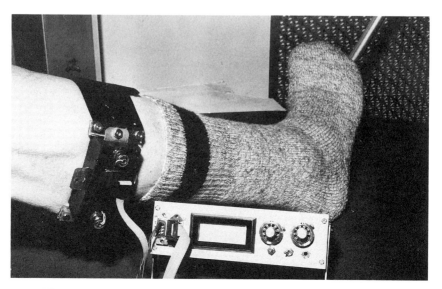

FIGURE 8. Application of a portable spectrophotometer to a human subject's leg.

One of the great problems of continuous illumination of tissue is that the detector receives light rays having travelled all possible paths, short, medium, and long, which are then averaged indiscriminately. We have recently demonstrated that picosecond light pulses can be used selectively to optically range the tissue and thus enable a selection of the path lengths over which photons have migrated in their progress toward the detector. Thus this system has the great advantage that the optical path length is defined and quantitative estimates of absolute concentration changes are indeed possible.[26]

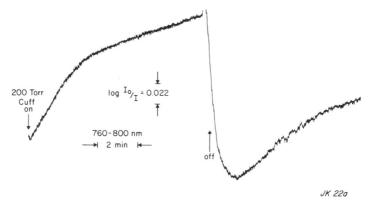

JK 22a

FIGURE 9. Recordings of the response of the spectrophotometer traces to occlusion of the human limb. Note the biphasic character of the deoxygenation.

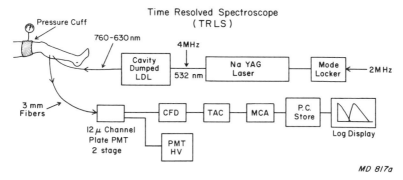

FIGURE 10. Block diagram of the picosecond technique as applied to the human leg.

The pulse responses that are obtained from this technique are shown in FIGURE 10 and it is seen that in spite of a very short imput pulse (<100 psec), the photons continue to emerge from the tissue in an exponential fashion over several nanoseconds. They are observed to have migrated over distances up to one meter within the calf.[25]

If the supply of oxygen to the leg is occluded, as is done with the continuous light method (FIG. 9), the light absorption is increased, and the migration is impaired (FIG. 11). The photon exit is less effective, in proportion to the absorption of light, here measured per centimeter of optical path; the time scale of the abscissa is converted to optical path length by the known velocity of light in tissue (23 cm per nsec). Thus the value of the absorption per centimeter is the slope of the lines of FIGURE 11, and the

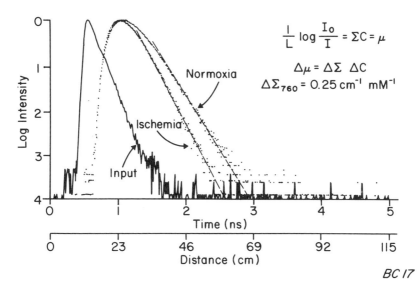

FIGURE 11. The effect of ischemia on the logarithmic exponential decay of photons exiting from the human arm.

incremental slope (cm^{-1}) can be used to calculate on an absolute basis the increase of deoxy hemoglobin concentration (Δc mM) using the known value of increment of extinction ($\Delta\epsilon$ cm^{-1} × mM^{-1}) and applying the Beer-Lambert relationship.[26]

$$\Delta c = \frac{\Delta\mu}{\Delta\epsilon} \qquad (5)$$

The application of this principle to the study of tumor oxygenation is illustrated below.

STUDIES OF TUMORS

In order to introduce the study of tumors, we will refer to our published work with Evans and D'Angio on malignant (stage IV) and spontaneously regressing (stage IV-S) tumors.[2] I am, however, reproducing the data for facets of the experiment that

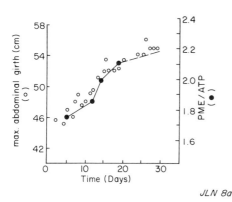

FIGURE 12. Correlation of tumor diameter and PME/ATP in growing human neuroblastoma of liver prior to effective therapy.

JLN 8a

were poorly understood at that time. FIGURE 12 illustrates the time course of phosphoethanolamine increase, which is correlated with the increase of tumor mass for the type IV-S tumors, which were in a phase of rapid growth. Other studies show a similar correlation. The time course of PME/ATP changes was followed over 32 weeks in our published study[2] (FIG. 13) showing rises of PME prior to effective therapy, which fell thereafter, thus correlating well with the tumor volume as indicated by FIG. 12. It was our conclusion at that time that high phosphoethanolamine levels were consistent with a rapid rate of tumor growth, supported by Kano's observation of phosphoethanolamine as a growth factor for tumors *in vitro*.[27] Furthermore, the close correspondence of neuroblastoma levels of PME with those observed in infant brains during rapid development is supportive of this viewpoint.

Although we did not perceive the value of continuous monitoring of phosphoethanolamine levels during tumor therapy in the first cases presented to us, we had an opportunity to later examine the kinetics of phosphoethanolamine changes during therapy of a similar neuroblastoma. Here three rounds of therapy were followed and in

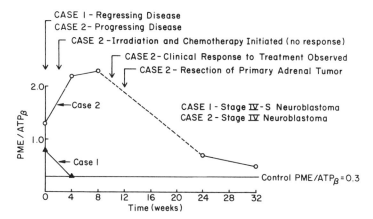

FIGURE 13. Phosphomonoester/ATPβ as a function of time in type IV and type IV-S neuroblastoma.

the first two, precipitous decreases of phosphoethanolamine followed within 4 hours of the therapeutic procedure (cisplatin) (FIG. 14).[8] These kinetics were confirmatory of the results previously obtained, but they also suggested a much closer correlation of therapeutic response and PME level to the point that clinical outcome could be predicted prior to the observation of the regression of the tumor volume (dashed traces).

In order to comprehend the rather complex kinetics of pre- and post-therapy energy metabolism we have combined the data from PME/ATP with that of PCr/P_i in a longitudinal study over 70 days (FIG. 15), involving three rounds of cisplatin therapy. The PME/ATP trace is identical to that of the previous figure. The alterations of PCr/P_i, which we stated earlier to be inversely related to the tissue oxygen level, between the first and second observations and prior to therapy showed that the PCr/P_i

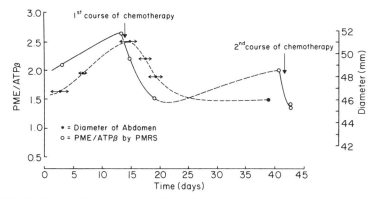

FIGURE 14. Time profile of responses of PME/ATP and tumor volume to therapy in a second neonatal neuroblastoma.

falls, suggesting that the tumor volume has reached a size where oxygen regulates the rate of metabolism of the tumor. After therapy, there is an abrupt rise of PCr/P_i testifying to better oxygenation of the tumor, and on the fortieth day, it even exceeds the level observed on day zero. The second round of therapy caused the opposite response to that of the first round, namely, a loss of PCr/P_i while by 70 days the value has recovered to slightly above the level on zero day.

An explanation based upon oxygen delivery to the tumor suggests that in the first 12 days the tumor is growing to the point where hypoxia is observed. Therapy, which decreases the growth rate and the tumor diameter by the 20th day, favors increased delivery of oxygen to the tumor and increased the oxidative metabolism of the tumor. This is mirrored to some extent by the increase of PME/ATP, which again indicates an

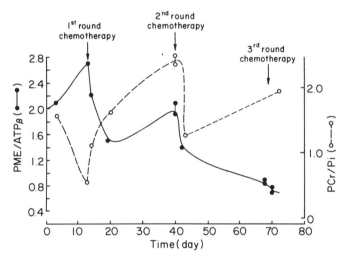

FIGURE 15. The time course of PCr/P_i and PME changes in neuroblastoma during therapy.

increase in growth rate. Thus, the first round of therapy establishes substrate and oxygen delivery conditions that are obviously favorable to tumor growth.

The second round of therapy, however, differs dramatically from the first. The energy state of the tumor significantly deteriorated and is accompanied by a fall of PME/ATP. One explanation for the decrease of PCr/P_i is the response of the cells to the therapy itself rather than the effect of the cell killing upon substrate delivery to the survivors. Whatever that effect may be, it is a transient one and by the 70th day, the energy level has recovered somewhat. It should be noted that the liver itself has no PCr and the residual signal in the surviving tumor cells following the third round of therapy may well be an authentic tumor signal. However, the thin layer of overlying musculature in this neonate could be responsible, and depth-resolved spectroscopy, not available at the time of these studies, would have been the preferable technique for these studies.

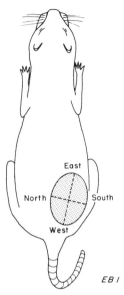

FIGURE 16. Location of hepatoma in flank of rat.

OPTICAL RANGING OF HEPATOMAS

Pulsed light spectroscopy of animal tumors to detect hemoglobin deoxygenation and eventually to localize it is exemplified by FIGURES 16 and 17, which show photon migration characteristics of an exteriorized rat hepatoma 3–4 cm in diameter (FIG. 16). The response to pulsed light is similar to that of the human calf shown in FIGURE 8, that is, the photon migration decays exponentially (FIG. 17). A displacement of the receiving probe to a portion of the tumor that contained a liquified cyst (FIG. 18) shows that photon migration through the cyst is much more effective (small slope) than that in the cellular part of the tumor, suggesting that localized structural features can be identified by the pulsed light technique.

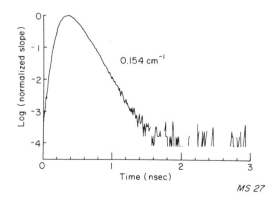

FIGURE 17. Photon migration in homogenous portion of slow-growing exteriorized hepatoma 4 cm in diameter in rat in FIGURE 16.

Hypoxia of the tumor tissue is expected to cause the same changes in the photon migration kinetics as were shown in calf muscle. Localization of zones of the tumor containing deoxyhemoglobin seems possible.[28] The growth rate, energy state, and oxygenation are obvious and crucial parameters in the monitoring of therapeutic procedures in tumors. The possibility that the intensity of the phosphoethanolamine peak in the PMRS spectrum determined on an absolute basis or relative to ATP will index the growth rate is currently limited to a handful of studies in neuroblastoma in infants. However, others have been using this criterion on adult tumors (nonbrain) with similar positive correlations with the effect of therapy.[29]

SUMMARY

Correlations of energy state with response to therapy are more difficult to analyze because of the large effect of tumor clearing and oxygenation upon the tumor energy

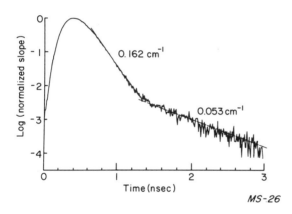

FIGURE 18. Photon migration in heterogeneous (liquified) portion of slow-growing exteriorized Morris hepatoma measuring 4 cm in diameter.

state as detected by PMRS alone. The combination of time-resolved hemoglobinometry using picosecond laser technology and localized PMRS seems appropriate to unravel the complexities of therapeutic intervention, tumor energetics, and oxygenation.

REFERENCES

1. PAPA, S., B. CHANCE & L. ERNSTER, Eds. 1987. Cytochrome Systems: Molecular Biology and Bioenergetics. Plenum. New York.
2. MARIS, J. M., A. E. EVANS, A. C. MCLAUGHLIN, G. J. D'ANGIO, L. BOLINGER, H. MANOS & B. CHANCE. 1985. [31]P Nuclear magnetic resonance spectroscopic investigation of human neuroblastoma *in situ*. New Engl. J. Med. **312:** 1500–1505.
3. EVANOCHKO, W. T., T. C. NG & J. D. GLICKSON. 1984. Application of in vivo NMR spectroscopy of cancer. J. Mag. Reson. Med. **1:** 508–534.
4. LEIGH, J. S., JR. Personal communication.
5. GADIAN, D. G., G. K. RADDA, M. J. DAWSON & D. R. WILKIE. 1982. pH$_i$ measurements of

cardiac and skeletal muscle using ^{31}P NMR. *In* Intracellular pH: Its Measurement, Regulation and Utilization in Cellular Functions. : 61–77.

6. GADIAN, D. G., D. I. HOULT, G. K. RADDA, P. J. SEELEY, B. CHANCE & C. BARLOW. 1976. Phosphorus magnetic resonance studies on normoxic and ischemic cardiac tissues. Proc. Natl. Acad. Sci. USA **73**: 446–448.
7. CHANCE, B., S. ELEFF & J. S. LEIGH. 1980. Noninvasive, nondestructive approaches to cell bioenergetics. Proc. Natl. Acad. Sci. USA **77**: 7430–7434.
8. CHANCE, B. & J. NORTHROP. 1985. How MR spectroscopy is deployed depends upon intended goals. Diagnostic Imaging (Nov): 311–320; 360.
9. YOUNKIN, D. P., M. DELIVORIA-PAPADOPOULOS, H. SUBRAMANIAN, *et al.* 1984. Studies of cortical metabolites in postasphyxiated infants. Pediat. Res. **18**: 357.
10. DELIVORIA-PAPADOPOULOS, M. & B. CHANCE. 1987. ^{31}P NMR spectroscopy in the newborn. *In* Neonatal Intensive Care. R. D. Gurhrie, Ed.: 153–179. Churchill Livingstone. New York.
11. CHANCE, B. & P. K. MAITRA. 1963. Determination of the intracellular phosphate potential of ascites cells by reversed electron transfer. *In* Control Mechanisms in Respiration and Fermentation. B. Wright, Ed. : 307–332. Ronald Press. New York.
12. CHANCE, B., J. S. LEIGH, JR, B. J. CLARK, J. MARIS, J. KENT, S. NIOKA & D. SMITH. 1985. Control of oxidative metabolism and oxygen delivery in human skeletal muscle: A steady-state analysis of the work/energy cost transfer function. Proc. Natl. Acad. Sci. USA **82**: 8384–8388.
13. CHANCE, B. & G. R. WILLIAMS. 1955. Respiratory enzymes in oxidative phosphorylation. I. Kinetics of oxygen utilization. J. Biol. Chem. **217**: 383–393.
14. TANAKA, A., B. QUISTORFF & B. CHANCE. 1988. Inorganic phosphate as regulator of oxidative phosphorylation in gluceneogenesis and urea synthesis as studied by phosphorus magnetic resonance in perfused rat liver. Proceedings of the 7th Annual Meeting of the Society for Magnetic Resonance in Medicine, San Francisco. p. 32.
15. CHANCE, B., J. S. LEIGH, JR., J. KENT, K. MCCULLY, S. NIOKA, B. J. CLARK, J. M. MARIS & T. GRAHAM. 1986. Multiple controls of oxidative metabolism as studied by phosphorus magnetic resonance. Proc. Natl Acad. Sci. **83**: 9458–9462.
16. PARK, J. H., R. L. BROWN, C. R. PARK, K. MCCULLY, M. COHN, J. HASELGROVE & B. CHANCE. 1987. Functional pools of oxidative and glycolytic fibers in human muscle observed by ^{31}P magnetic resonance spectroscopy during exercise. Proc. Natl. Acad. Sci. USA **84**: 8976–8980.
17. RADDA, G. K. & D. J. TAYLOR 1985. Applications of nuclear magnetic resonance spectroscopy in pathology. Int. Rev. Exp. Pathol. **27**: 1–58.
18. AULL, F. & H. G. HEMPLING. 1963. Sodium fluxes in the Ehrlich mouse ascites tumor cell. Am. J. Physiol. **204**: 789–794.
19. WARBURG, O. 1931. Metabolism of Tumors. R. R. Smith. New York.
20. BURK D. (1939). A Colloquial consideration of the Pasteur and non-Pasteur effects. Cold Spring Harbor Symp. Quant. Biol. **7**: 420–423.
21. WIENER, D. H., L. I. FINK, J. MARIS, R. A. JONES, B. CHANCE & J. R. WILSON. 1986. Abnormal skeletal muscle bioenergetics during exercise in patients with heart failure: Role of reduced muscle blood blow. Congestive Heart Failure **73**: 1127–1136.
22. ZATINA, M. A., H. D. BERKOWITZ, G. M. GROSS, J. M. MARIS & B. CHANCE. 1986. [31]P nuclear magnetic resonance spectroscopy: Noninvasive biochemical analysis of the ischemic extremity. J. Vasc. Surg. **3**: 411–420.
23. MILLIKAN, G. A. 1937. Proc. Roy. Soc. London B. **129**: 218–223.
24. JOBSIS, F. F., J. H. KEIZER, J. C. LAMANNA & M. ROSENTHAL. 1977. Reflectance spectrophotometry of cytochrome a,a₃ in vivo. J. Appl. Physiol. **113**: 858–871.
25. CHANCE, B., S. NIOKA, J. KENT, K. MCCULLY, M. FOUNTAIN, R. GREENFELD & G. HOLTOM. 1988. Time-resolved spectroscopy of hemoglobin and myoglobin in resting and ischemic muscle. Anal. Biochem. **174**: 698–707.
26. CHANCE, B., J. S. LEIGH, H. MIYAKE, D. S. SMITH, S. NIOKA, R. GREENFELD, M. FINANDER, K. KAUFMANN, W. LEVY, M. YOUNG, P. COHEN, H. YOSHIOKA & R. BORETSKY. 1988. Comparison of time-resolved and -unresolved measurements of deoxy-hemoglobin in brain. Proc. Natl. Acad. Sci. USA **85**: 4971–4975.

27. KANO-SUEOKA, T. & J. E. ERRICK. 1981. Effects of phosphoethanolamine and ethanola-
    mine on growth of mammary carcinoma cells in culture. Exp. Cell Res. **136:** 137–145.
28. PATTERSON, M. S., B. CHANCE & B. C. WILSON. 1988. Time resolved reflectance for the
    noninvasive measurement of tissue optical. J. Appl. Optics. Submitted for publication.
29. KARCZMAR, G. S., J. POOLE, M. D. BOSKA, D. J. MEYERHOFF, B. HUBESCH, K. ROTH, G. B.
    MATSON, J. ARBEIT, S. VALINE & M. W. WEINER. 1988. ^{31}P MRS study of response of
    human tumors to therapy. Proceedings of the Seventh Annual Meeting of the Society for
    Magnetic Resonance in Medicine, San Francisco.

Membrane Antioxidants

NORMAN I. KRINSKY

Department of Biochemistry
Tufts University Health Sciences Campus
Boston, Massachusetts 02111

INTRODUCTION

The concept of membrane antioxidants usually brings to mind those compounds specifically able to inhibit lipid peroxidation. In particular, α-tocopherol (FIG. 1; I) has been identified as the major lipid-soluble antioxidant in biological systems,[1,2] and we have some understanding of how it functions in this role.[3] However, there are many other compounds that function as membrane antioxidants, both directly and indirectly, and this chapter will review some of the evidence regarding these compounds. In addition, there are several enzyme systems that also function to protect membranes against oxidant damage.

The oxidation reactions that involve membrane lipids are shown in FIGURE 2. The initiation step consists of a radical species ($R\cdot$) that can abstract a hydrogen atom from a lipid substrate (LH), resulting in the formation of an alkyl species ($L\cdot$). The propagation step of lipid oxidation involves the very rapid reaction between the alkyl radical and dioxygen to form the lipid peroxyl radical ($LOO\cdot$). The peroxyl radical in turn can react with another lipid substrate to generate a new alkyl radical and the hydroperoxide (LOOH). Termination of the process can occur by a variety of procedures including the interaction of two peroxyl radicals to form stable products, or the reaction of the peroxyl radical with an inhibitor (IH) to form a new hydroperoxide and the relatively stable radical inhibitor.

The process *in vivo* is probably more complicated, because reinitiation reactions can occur when the hydroperoxide, LOOH, is used as a substrate in the presence of transition metals. Under these circumstances, the hydroperoxide can undergo a Fenton-type reaction in the presence of reduced transition metals, such as Fe^{2+}, to form an alkoxyl radical ($LO\cdot$) or can form another peroxyl radical in the presence of oxidized transition metal, such as Fe^{3+}. Both of these radical species can then reinitiate lipid peroxidation by interacting with lipid substrates to form a new $L\cdot$.[4]

Membrane oxidation is not limited to the lipid components of the membrane, but can also involve the membrane proteins, which can undergo oxidation in terms of forming new disulfide bridges or nonreducible covalent linkages when exposed to oxidant stress.[5,6]

CHEMICAL ANTIOXIDANTS

One way to think about the compounds that act as membrane antioxidants is to divide them into a lipid-soluble group and a water-soluble group, as shown in TABLE 1.

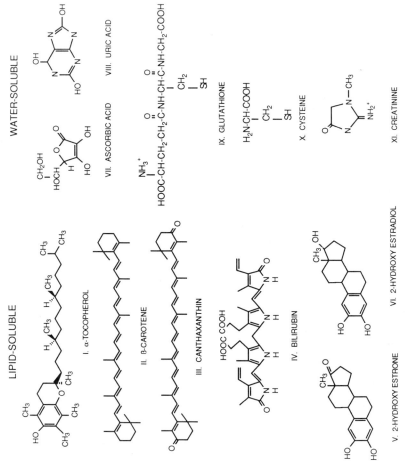

FIGURE 1. The structure of some representative lipid-soluble and water-soluble membrane antioxidants.

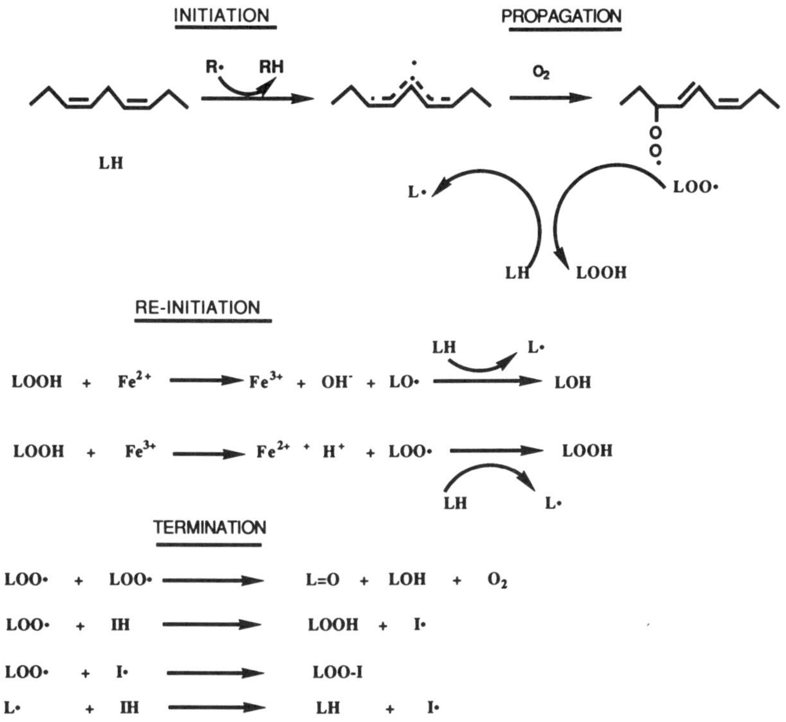

FIGURE 2. The reactions of lipid peroxidation. Fatty acids, such as linoleic acid, are represented by LH; alkykl radicals by L·, peroxyl radicals by LOO·, alkoxyl radicals by LO·, inhibitors by IH, and radical initiators by R·.

This table is not complete; occasionally there are single reports of antioxidant action which require further verification.

Lipid Antioxidants

Lipid antioxidants have been the compounds most extensively studied with respect to their ability to act as membrane antioxidants. To a large extent, this reflects the

TABLE 1. Chemical Antioxidants

Lipid-Soluble	Water-Soluble
(1) Tocopherols	(1) Ascorbic Acid
(2) Carotenoids	(2) Uric Acid
(3) Bilirubin	(3) Glutathione
(4) Catechol Estrogens	(4) Cysteine
(5) Ubiquinone	(5) Histidine Peptides
	(6) Creatinine

interest of investigators in lipid peroxidation and the fact that there are relatively straightforward techniques for evaluating this process in biological membranes.[7] It also probably reflects the availability of artificial membrane models consisting of liposomes that would only reflect oxidative damage to lipid components of membranes.[8] The best known members of this family of lipid-soluble antioxidants are the tocopherols.

Tocopherols

Tocopherols, such as α-tocopherol (vitamin E), have been studied for many years and have proven to be extremely effective lipid-soluble inhibitors of the propagation

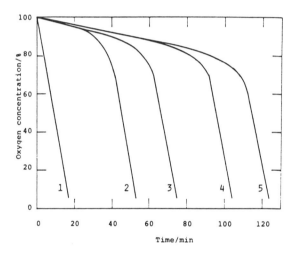

FIGURE 3. The inhibition of the oxidation of methyl linoleate induced by 2,2'-azobis(2,4-dimethylvaleronitrile) by various μM concentrations of α-tocopherol ($1 = 0$, $2 = 22$, $3 = 45$, $4 = 67$, $5 = 112$). (From Niki et al.[11] Reprinted by permission.)

step of lipid peroxidation.[1,3,9] The reactions involved in inhibiting this step in lipid peroxidation are depicted below:

$$\alpha\text{-T} + \text{LOO}\cdot \rightarrow \alpha\text{-T}\cdot + \text{LOOH}$$

$$\alpha\text{-T}\cdot + \text{LOO}\cdot \rightarrow \text{LOO-}\alpha\text{-T}$$

The first product is the α-tocopheroxyl radical (α-T$\cdot$), which is a resonance-stabilized oxygen-centered radical. The α-tocopheroxyl radical can react with another peroxyl radical to form a stable adduct, which has been isolated.[10] The reactions depicted above clearly indicate that α-tocopherol has the capacity to neutralize two propagating peroxyl radicals.

The net effect of the addition of α-tocopherol to a membrane system is an inhibition of the propagating step of lipid peroxidation. To evaluate this process, one can measure oxygen uptake in either a homogeneous solution or in a natural or artificial membrane in the presence or absence of an antioxidant. The effects on oxygen consumption of α-tocopherol are shown in FIGURE 3.[11] The rate of oxygen consumption proceeds linearly in the absence of the antioxidant until all of the oxygen is consumed. In the presence of the antioxidant, there is an extended lag period before the rate of oxygen consumption corresponds to that found in the control preparation. This lag period is attributed to the inhibition of the propagating steps by the presence of the antioxidant. A similar type of analysis can be carried out for any antioxidant that inhibits the propagating steps of lipid peroxidation.

By means of this type of analysis, the effectiveness of α-tocopherol as a lipid antioxidant has been studied in homogeneous soution, in soybean phosphatidylcholine liposomes, and in red blood cell ghost membranes. In the liposome and membrane preparation, the effectiveness of α-tocopherol is only 1–2% of that in homogeneous solutions. This finding has been attributed to a lower mobility of α-tocopherol in the membranes and to the greater probability of chain propagation in the more tightly structured environment of the membranes.[12]

Under some conditions, α-tocopherol can also act as a pro-oxidant with respect to lipid peroxidation. Yamamoto and Niki[13] reported that α-tocopherol inhibited lipid peroxidation in liposomes treated with Fe^{2+}, but when the liposomes were treated with Fe^{3+}, α-tocopherol acted as a pro-oxidant. They suggested that the α-tocopherol was reducing Fe^{3+} to Fe^{2+}, which is much more reactive in the decomposition of LOOH to radical species. In addition, Fukuzawa et al.[14] reported a similar pro-oxidant effect of α-tocopherol in sodium dodecyl sulfate micelles containing LOOH and Fe^{2+}. In this case, α-tocopherol enhanced lipid peroxidation by regenerating Fe^{2+}.

It has been suggested that tocopherols are the only important lipid inhibitors of the propagation steps of lipid peroxidation that are present in either human red blood cells or rat liver.[1,2] However, some of the remaining lipid-soluble compounds may also play an important physiological role in inhibiting lipid peroxidation.

Carotenoids

For many years, carotenoids such as β-carotene (FIG. 1; II) have been considered to be antioxidants on the basis of their ability to quench photochemical reactions leading to lipid peroxidation.[15,16] β-carotene in particular has the ability to quench singlet oxygen.[17,18] Because carotenoids are bleached when exposed to radicals such as those that arise during lipid peroxidation,[19–21] these pigments must also intercept active oxygen species. This is not surprising because their long conjugated double-bond system would make them excellent substrates for radical attack. In fact, when carotenoids are exposed to the trichloromethylperoxyl radical ($CCl_3OO\cdot$) generated during the pulse radiolysis of chloroform in the presence of oxygen, the carotenoid molecule is very rapidly bleached[22] (FIG. 4). In fact, the bleaching of carotenoids when they interact with radicals has been used very successfully to measure the kinetic parameters of a variety of oxygen radical species.[23–25]

Carotenoids have also been shown to act as antioxidants in vivo. When ascorbic-acid-deficient guinea pigs were exposed to small amounts of carbon tetrachloride, the extent of lipid peroxidation, as measured by pentane and ethane production, was decreased in those animals that received injections of β-carotene.[26] In addition,

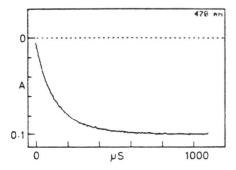

FIGURE 4. The bleaching of β-carotene by the trichloromethylperoxyl radical formed by pulse radiolysis of chloroform in the presence of oxygen. (From Packer *et al.*[22] Reprinted by permission.)

mitochondria from rats that received dietary supplements of canthaxanthin (FIG. 1; III; 4,4'-diketo-β-carotene) showed a lower level of lipid peroxidation than did rats maintained on a chow diet.[27]

The first attempt at quantitatively evaluating the effectiveness of β-carotene as an antioxidant at physiological oxygen tensions was carried out by Burton and Ingold,[28] who reported that β-carotene was a better antioxidant at 15 torr (2% oxygen) than at 150 torr (20% oxygen). They also reported that β-carotene acted as a pro-oxidant at 760 torr. Presumably, once we understand the mechanism whereby carotenoids act as antioxidants, we will be able to explain this observation.[29] Burton and Ingold[28] suggested that β-carotene might react directly with the peroxyl radical to form a resonance-stabilized carbon-centered radical, as shown in FIGURE 5. If that were the case, it is conceivable that carotenoids should be able to quench two peroxyl radicals

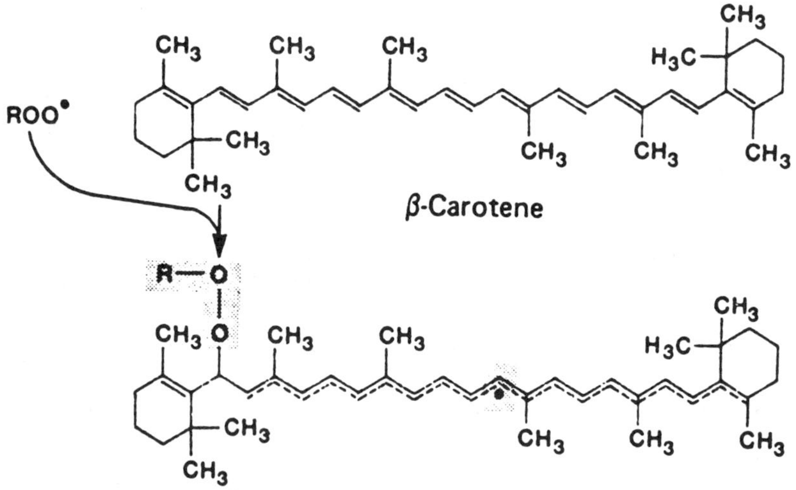

FIGURE 5. The hypothetical formation of a resonance-stabilized carbon-centered radical species formed when β-carotene reacts with a lipid peroxyl radical (LOO·). (From Burton and Ingold.[28] © 1984 by the AAAS. Reprinted by permission.)

(LOO·), as described above for α-tocopherol, in the mechanism shown below:

$$\text{LOO·} + \text{CAR} \xrightarrow{} \text{LOO-CAR·} \xrightarrow{\text{LOO·}} \text{LOO-CAR-OOL}$$

Unfortunately, none of these potential intermediate forms has been isolated and we know very little about the chemistry of the reaction of carotenoids with radicals.[21–22]

We have demonstrated, as have many others, that carotenoids are very effective quenchers of singlet oxygen.[17,30] In addition, we were able to demonstrate that carotenoids such as β-carotene and canthaxanthin were able to inhibit the formation of malondialdehyde (MDA) in a liposome system treated with ferrous chloride (FIG 6).[31] The prolonged lag phase before lipid peroxidation was observed may well be related to the reports that carotenoids may function as dietary anticarcinogenic compounds.[32–34]

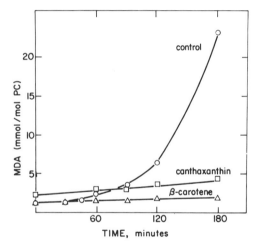

FIGURE 6. β-Carotene and canthaxanthin protection of Fe^{2+}-induced lipid peroxidation in egg phosphatidylcholine liposomes. (From Krinsky and Deneke.[31] Reprinted by permission.)

Bilirubin

The concept that waste products of metabolism may play additional roles in the body has been developed by Ames and his associates.[35,36] They have looked at a number of waste products, and have presented very convincing evidence that bilirubin (FIG. 1; IV), the degradation product of all of our heme proteins, is a very effective lipid antioxidant.[36,37] They incorporated bilirubin into homogenous solutions of linoleic acid (LOOH) or into multilamellar liposomes prepared from purified soybean phosphatidylcholine and exposed both of these preparations to a lipid-soluble radical generator, 2,2'-azobis(2,4-dimethylvaleronitrile) (AMVN). Under these circumstances, micromolar concentrations of bilirubin effectively scavenged peroxyl radicals and acted as a chain-breaking antioxidant. Stocker et al.[36] observed some differences between the effects of bilirubin in LOOH solutions as opposed to the liposome preparation with respect to the effects of either 2% oxygen or air. In the LOOH solution, α-tocopherol was a better antioxidant than bilirubin in both air or in 2% oxygen, but in liposomes, bilirubin was almost as good as α-tocopherol in air, and appeared to be a slightly better antioxidant than α-tocopherol under 2% oxygen. These results (FIG. 7) indicate strongly that bilirubin, if associated with lipid membranes, can serve as an effective biological antioxidant.

Catechol Estrogens

Nakano and his associates[38,39] have presented evidence that estrogens, and in particular catechol estrogens such as 2-hydroxy estrone (FIG. 1; V) and 2-hydroxy estradiol (FIG. 1; VI), are potent inhibitors of lipid peroxidation induced by the

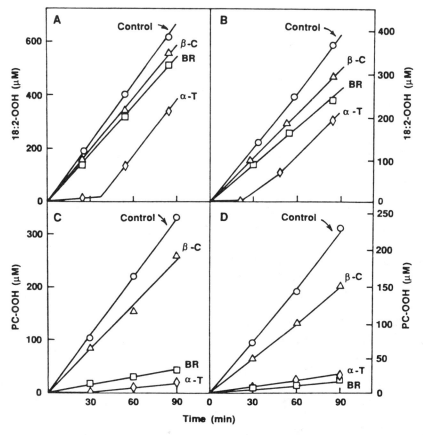

FIGURE 7. Comparison of the rate of oxidation when either linoleic acid (**A,B**) or soybean phospholipid liposomes (**C,D**) are peroxidized by exposure to the lipid-soluble radical generator, AMVN, in the presence of either air (**A,C**) or 2% oxygen (**B,D**). α-Tocopherol (α-T), bilirubin (BR), and β-carotene (β-C) were all used at 10 μM, and lipid peroxidation was assayed by detecting either the linoleic acid hydroperoxide (18:2-OOH) or the phosphatidylcholine hydroperoxide (PC-OOH). (From Stocker et al.[36] © 1987 by the AAAS. Reprinted by permission.)

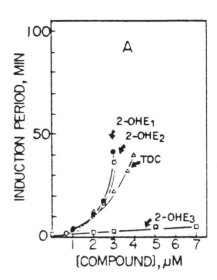

FIGURE 8. The effect of varying the concentration of catechol estrogens on the induction period of phospholipid liposomes treated with Fe^{3+}-ADP-adriamycin, in comparison to α-tocopherol (TOC). The compounds used are 2-hydroxy estrone (2-OHE$_1$), 2-hydroxy estradiol (2-OHE$_2$) and 2-hydroxy estratriol (2-OHE$_3$). (From Nakano *et al.*[39] Reprinted by permission.)

addition of Fe^{3+}-ADP-adriamycin to liposomes. They measured both elongation of the induction period as well as inhibition of oxygen consumption, and, as seen in FIGURE 8, both of these catechol estrogens proved to be better inhibitors than α-tocopherol. It is interesting that both of these catechol estrogens are among the principal metabolic products of the parent estrogens in man,[40] and may represent another example of the usefulness of metabolic waste products.

Ubiquinone

There have been a number of reports that ubiquinone, in both the reduced and the oxidized form, can protect membrane phospholipids against peroxidative reactions.[41,42] Although the mechanism of this effect is not at all understood, some of the recent work has suggested[43] that it might scavenge singlet oxygen (although there are still no documented reports of singlet oxygen formation during lipid peroxidation in biological membranes) or perturb the lipid membrane and thus inhibit the decomposition of hydroperoxides.

Water-Soluble Antioxidants

TABLE 1 lists not only the lipid-soluble antioxidants just discussed, but also includes the large number of water-soluble radical scavengers that have been reported to protect cells against oxidative damage. In many cases, these compounds act as preventive antioxidants that inactivate free-radical precursors and thereby decrease the initiation of the radical processes that initiate lipid peroxidation. In other cases, these water-soluble compounds can act as chain-breaking antioxidants that react directly with oxygen-centered radicals. Our understanding of the mechanism of action of water-soluble antioxidants was clarified by the work of Barclay, Locke, and MacNeil[44] who demonstrated for the first time the synergistic effect of ascorbic acid (FIG. 1; VII) and α-tocopherol in preventing lipid peroxidation in a micellar preparation of linoleic acid or methyl linoleate. This synergistic mechanism may well explain the inhibition of membrane lipid peroxidation by many other of the water-soluble antioxidant compounds.

Ascorbic Acid

After the report of Barclay *et al.*[44] appeared, several other groups investigated the synergism between ascorbic acid and α-tocopherol. The experimental system utilized

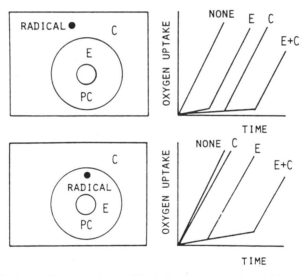

FIGURE 9. Oxidation of soybean phosphatidylcholine liposomes induced either by water-soluble or lipid-soluble radical generators (R·). When the R· is in the aqueous environment, the effects of α-tocopherol (E) and ascorbic acid (C) are additive, whereas when the R· is in the lipid environment, the effects are synergistic. (From Niki *et al.*[7] Reprinted by permission.)

both lipid-soluble and water-soluble radical initiators, allowing the investigators to select the environment and the rate at which radicals were generated.[11,45–47] The effect is seen most clearly in FIGURE 9, which is taken from a recent review by Niki.[7] When α-tocopherol and ascorbic acid were tested with a water-soluble initiator, an *additive* effect was observed, whereas the use of a lipid-soluble initiator allowed the demonstration of a *synergistic* effect between ascorbic acid and α-tocopherol. Since the ascorbic acid is consumed in these reactions before the α-tocopherol, Niki *et al.*[11] proposed the following reactions sequence to account for this effect, where α-T represents α-tocopherol, C represents ascorbic acid, and LOO· represents the lipid peroxyl radical:

$$\begin{array}{ccccc} \text{LOO·} & \text{LOOH} & \text{C} & \text{C·} & \longrightarrow \text{dehydroascorbate} \\ \alpha\text{-T} & \longrightarrow & \alpha\text{-T·} & \longrightarrow \alpha\text{-T} & \end{array}$$

In this experiment, α-tocopherol remained almost unchanged until all of the ascorbate was consumed, and only then was the α-tocopherol used up.

Uric Acid

Uric acid is another waste product that has been shown to have strong antioxidant activity at a physiological concentration.[35,48] It is very effective when lipid peroxidation is initiated with a water-soluble radical source, but when tested against a lipid-soluble radical generator, it proved to be ineffective.[49]

Glutathione, Cysteine, Histidine Peptides, and Creatinine

There are several other water-soluble antioxidants that have been reported to block radical reactions, therefore having the capacity to inhibit membrane oxidation. Among these are glutathione,[50] cysteine,[50] histidine peptides,[51] and creatinine,[52] which all have clear antioxidant effects, but would only be effective.against water-soluble radicals or might interfere with radical-generating conditions. As such, they, and the other water-soluble antioxidants, must be considered as secondary lines of defense in inhibiting membrane oxidation.

ANTIOXIDANT ENZYMES

The ability of various enzymes, such as those listed below, to prevent membrane damage, has been very well documented.[53-54] These antioxidant enzymes include:
1. Se-Glutathione peroxidase
2. Phospholipid hydroperoxide glutathione peroxidase
3. Glutathione S-transferase
4. Other antioxidant enzymes such as:
   catalase
   Cu-Zu superoxide dismutase
However, in only a few cases can one detect a direct effect on the membrane components that can lead to an inhibition or reduction in the level of membrane oxidation.

Se-Glutathione Peroxidase

Although twenty years have elapsed since it was first reported that glutathione peroxidase (GSH-Px) was able to reduce fatty acid hydroperoxides in cell membranes,[55] the field has remained controversial because some systems were described in which this enzyme was ineffective.[56,57] However, the suggestion that phospholipase A_2 (PLaseA$_2$) might play a role in this process by liberating the hydroperoxy fatty acid from membrane phospholipids[58] has opened up the field for a new series of investigations.

The observation that PLaseA$_2$ activity is increased when liposome membranes are peroxidized and that the enzyme removes a large portion of the peroxidized fatty acids[59] and that PLaseA$_2$ is required for the detoxification of phospholipid hydroperoxides[60,61] has led to the following proposal for the combined action of PLaseA$_2$ and GSH-Px in protecting membrane phospholipids (PLH) from lipid peroxidative damage[62]:

$$\text{PLH} \xrightarrow{\ \ R\cdot\quad RH\ \ } \text{PL}\cdot \xrightarrow{\ \ O_2\ \ } \text{PLOO}\cdot \xrightarrow{\ \ LH\quad L\cdot\ \ }$$

$$\text{PLOOH} \xrightarrow{\ \ \text{PLaseA}_2\ \ } \text{P + LOOH} \xrightarrow{\ \ \text{GSH-Px}\ \ } \text{LOH}$$

When the phospholipid undergoes radical attack and forms the hydroperoxide, the lipid bilayer is disrupted and the oxidized fatty acid becomes available as a substrate for PLaseA$_2$. Once released, this fatty acid hydroperoxide is reduced by GSH-Px in the presence of glutathione to form the hydroxy fatty acid, and the membrane phospholipid is repaired by a re-acylation reaction. Thus, GSH-Px, acting in concert with PLaseA$_2$, can effectively prevent additional peroxidative damage initiated by the phospholipid hydroperoxide by removing it and reducing it to the harmless hydroxy fatty acid.

The process of releasing lipid hydroperoxides might play an important regulatory role in prostaglandin metabolism, for such a release would increase the "peroxide tone," which in turn stimulates additional PLaseA$_2$ to release substrates for cyclooxygenase.[63]

Phospholipid Hydroperoxide Glutathione Peroxidase

Ursini and his associates[64] have isolated a protein from liver, originally called peroxidation-inhibiting protein (PIP), which acted by reducing membrane phospholipids and was distinct from GSH-Px or glutathione S-transferase. As their work has continued, they now report that it is a seleno-protein, renamed phospholipid hydroperoxide glutathione peroxidase (PL-OOH GSH-Px) and that the major difference between it and Se-GSH-Px is that PL-OOH GSH-Px acts directly on membrane phospholipid hydroperoxides.[65] Thus, this enzyme reduces the hydroperoxy fatty acids directly without the necessity of invoking PLaseA$_2$ activity.

Glutathione-S-Transferase

These enzymes, sometimes referred to as selenium-independent GSH peroxidases, are capable of inhibiting membrane lipid peroxidation, but only in the presence of

PLaseA$_2$.[66] Other enzyme preparations have been described that require glutathione for inhibition of lipid peroxidation, but they do not appear to be glutathione-S-transferase-type enzymes.[67]

Other Antioxidant Enzymes

It would be remiss to talk about antioxidant enzymes and not mention either catalase or Cu-Zn superoxide dismutase. Both of these enzymes carry out their principal function by removing sources of radicals (i.e., hydrogen peroxide and superoxide, respectively) and may not play a direct role in membrane antioxidant action. However, a recent report reviewing the unique properties of tumor membranes suggests that part of the antioxidant defenses of these membranes may be due to changes in their contents of both prooxidant and antioxidant enzymes.[68]

SUMMARY

In this chapter, I have discussed both lipid-soluble and water-soluble antioxidants that exert protective action with respect to inhibiting lipid, and in some cases, protein oxidation in both natural and artificial membranes. In addition, recent work has begun to clarify exactly how antioxidant enzymes can protect membranes against peroxidative damage. Although we have some understanding of the mechanisms of several of these antioxidants, much work remains to be done before we can begin making dietary recommendations that may have profound implications with respect to aging, cancer, and the many other human diseases that have been associated with radical-induced damage.

REFERENCES

1. BURTON, G. W., A. JOYCE & K. U. INGOLD. 1983. Is vitamin E the only lipid-soluble chain-breaking antioxidant in human blood plasma and erythrocyte membranes? Arch. Biochem. Biophys. **221:** 281–290.
2. INGOLD, K. U., A. C. WEBB, D. WITTER, G. W. BURTON, T. A. METCALFE & D. P. R. MULLER. 1987. Vitamin E remains the major lipid-soluble, chain-breaking antioxidant in human plasma even in individuals suffering severe vitamin E deficiency. Arch. Biochem. Biophys. **259:** 224–225.
3. BURTON, G. W. & K. U. INGOLD. 1981. Autoxidation of biological molecules. 1. The antioxidant activity of vitamin E and related chain-breaking phenolic antioxidants *in vitro*. J. Am. Chem. Soc. **103:** 6472–6477.
4. GIROTTI, A. W. 1985. Mechanisms of lipid peroxidation. J. Free Rad. Biol. Med. **1:** 87–95.
5. YAMAMOTO, Y., E. NIKI, J. EGUCHI, Y. KAMIYA & H. SHIMASAKI. 1985. Oxidation of biological membranes and its inhibition. Free radical chain oxidation of erythrocyte ghost membranes by oxygen. Biochim. Biophys. Acta **819:** 24–36.
6. DEAN, R. T. & K. H. CHEESEMAN. 1987. Vitamin E protects proteins against free radical damage in lipid environments. Biochem. Biophys. Res. Commun. **148:** 1277–1282.
7. NIKI, E. 1987. Antioxidants in relation to lipid peroxidation. Chem. Phys. Lip. **44:** 227–253.
8. CHATTERJEE, S. N. & S. AGARWAL. 1988. Liposomes as membrane model for study of lipid peroxidation. Free Rad. Biol. Med. **4:** 51–72.
9. MIKI, M., H. TAMAI, M. MINO, Y. YAMAMOTO & E. NIKI. 1987. Free radical chain

oxidation of rat red blood cells by molecular oxygen and its inhibition by α-tocopherol. Arch. Biochem. Biophys. **258**: 373–380.

10. MATSUMOTO, S., M. MATSUO, Y. IITAKA & E. NIKI. 1986. Oxidation of a vitamin E model compound 2,2,5,7,8-pentamethylchroman-6-ol, with the tert-butylperoxyl radical. J. Chem. Soc. Chem. Commun.: 1076–1077.

11. NIKI, E., T. SAITO, A. KAWAKAMI & Y. KAMIYA. 1984. Inhibition of oxidation of methyl linoleate in solution by vitamin E and vitamin C. J. Biol. Chem. **259**: 4177–4182.

12. NIKI, E. 1987. Lipid antioxidants: How they may act in biological systems. Br. J. Cancer **55**(Suppl. VIII): 153–157.

13. YAMAMOTO, K. & E. NIKI. 1988. Interaction of α-tocopherol with iron: antioxidant and pro-oxidant effects of α-tocopherol in the oxidation of lipids in aqueous dispersions in the presence of iron. Biochim. Biophys. Acta **958**: 19–23.

14. FUKUZAWA, K., K. KISHIKAWA, T. TADOKORO, A. TOKUMURA, H. TSUKATANI & J. M. GEBICKI. 1988. The effects of α-tocopherol on site-specific lipid peroxidation induced by iron in charged micelles. Arch. Biochem. Biophys. **260**: 153–160.

15. KRINSKY, N. I. 1979. Carotenoid protection against oxidation. Pure Appl. Chem. **51**: 649–660.

16. KRINSKY, N. I. 1982. Photobiology of carotenoid protection. *In* The Science of Photomedicine. J. D. Regan & J. A. Parrish, Eds.: 397–403. Plenum. New York.

17. FOOTE, C. S. & R. W. DENNY. 1968. Chemistry of singlet oxygen. VIII. Quenching by β-carotene. J. Am. Chem. Soc. **90**: 6233–6235.

18. KRINSKY, N. I. 1979. Biological roles of singlet oxygen. *In* Singlet Oxygen. H. H. Wasserman & R. W. Murray, Eds.: 597–641. Academic Press. New York.

19. FRIEND, J. 1958. The coupled oxidation of β-carotene by a linoleate-lipoxidase system and by autoxidizing linoleate. Chem. Ind. (London): 597–598.

20. BEN AZIZ, A., S. GROSSMAN, I. ASCARELLI & P. BUDOWSKI. 1971. Carotene-bleaching activities of lipoxygenase and heme proteins as studied by a direct spectrophotometric method. Phytochemistry **10**: 1445–1452.

21. KLEIN, B. P., D. KING & S. GROSSMAN. 1985. Cooxidation reactions of lipoxygenase of plant systems. Adv. Free Rad. Biol. Med. **1**: 309–343.

22. PACKER, J. E., J. S. MAHOOD, V. O. MORA-ARELLANO, T. F. SLATER, R. L. WILLSON & B. S. WOLFENDEN. 1981. Free radicals and singlet oxygen scavengers: reaction of a peroxy-radical with β-carotene, diphenyl furan and 1,4-diazobicyclo(2,2,2)-octane. Biochem. Biophys. Res. Commun. **98**: 901–906.

23. SARAN, M., C. MICHEL & W. BORS. 1980. The bleaching of crocin by oxygen radicals. *In* Chemical and Biochemical Aspects of Superoxide and Superoxide Dismutase. J. V. Bannister & H. O. A. Hill, Eds.: 38–44. Elsevier/North-Holland. Amsterdam.

24. BORS, W., C. MICHEL & M. SARAN. 1984. Inhibition of the bleaching of the carotenoid crocin. A rapid test for quantifying antioxidant activity. Biochim. Biophys. Acta **706**: 312–319.

25. BORS, W., C. MICHEL & M. SARAN. 1985. Determination of kinetic parameters of oxygen radicals by competition studies. *In* Handbook of Methods for Oxygen Radical Research. R. A. Greeenwald, Ed.: 181–188. CRC Press. Boca Raton, FL.

26. KUNERT, K.-J. & A. L. TAPPEL. 1983. The effect of vitamin C on *in vivo* lipid peroxidation in guinea pigs as measured by pentane and ethane production. Lipids **18**: 271–274.

27. MAYNE, S. T. & R. S. PARKER. 1987. Dietary canthaxanthin as a protective agent against lipid peroxidation in biomembranes. Fed. Proc. **46**: 1189.

28. BURTON, G. W. & K. U. INGOLD. 1984. β-Carotene: An unusual type of lipid antioxidant. Science **224**: 569–573.

29. KRINSKY, N. I. 1988. Mechanisms of inactivation of oxygen species by carotenoids. *In* Anticarcinogenesis and Radiation Protection. M. G. Simic & O. Nygaard, Eds.: 41–46. Plenum. New York.

30. ANDERSON, S. M. & N. I. KRINSKY. 1973. Protective action of carotenoid pigments against photodynamic damage to liposomes. Photochem. Photobiol. **18**: 403–408.

31. KRINSKY, N. I. & S. M. DENEKE. 1982. The interaction of oxygen and oxy-radicals with carotenoids. JNCI **69**: 205–210.

32. PETO, R., R. DOLL, J. D. BUCKLEY & M. B. SPORN. 1981. Can dietary β-carotene materially reduce human cancer rates? Nature **290**: 201–208.

33. KRINSKY, N. I. 1988. Evidence for the role of carotenes in preventive health. Clin. Nutr. **7:** 107–114.
34. KRINSKY, N. I. 1988. Carotenoids and cancer in animal models. J. Nutrition. In press.
35. AMES, B. N., R. CATHCART, E. SCHWIERS & P. HOCHSTEIN. 1981. Uric acid provides an antioxidant defense in humans against oxidant- and radical-caused aging and cancer: A hypothesis. Proc. Natl. Acad. Sci. USA **78:** 6858–6862.
36. STOCKER, R., Y. YAMAMOTO, A. F. MCDONAGH, A. N. GLAZER & B. N. AMES. 1987. Bilirubin is an antioxidant of possible physiological importance. Science **235:** 1043–1046.
37. STOCKER, R. & B. N. AMES. 1987. Potential role of conjugated bilirubin and copper in the metabolism of lipid peroxides in bile. Proc. Natl. Acad. Sci. USA **84:** 8130–8134.
38. SUGIOKA, K., Y. SHIMOSEGAWA & M. NAKANO. 1987. Estrogens as natural antioxidants of membrane phospholipid peroxidation. FEBS Lett. **210:** 37–39.
39. NAKANO, M., K. SUGIOKA, I. NAITO, S. TAKEKOSHI & E. NIKI. 1987. Novel and potent biological antioxidants on membrane phospholipid peroxidation: 2-hydroxy estrone and 2-hydroxy estradiol. Biochem. Biophys. Res. Commun. **142:** 919–924.
40. GELBKE, H. P., P. BALL & R. KNUPPEN. 1977. 2-Hydroxyoestrogens. Chemistry biogenesis, metabolism and physiological significance. Adv. Ster. Biochem. Pharmacol. **6:** 81–154.
41. MELLORS, A. & A. L. TAPPEL. 1966. The inhibition of mitochondrial peroxidation by ubiquinone and ubiquinol. J. Biol. Chem. **241:** 4353–4356.
42. TAKAYANAGI, R., K. TAKESHIGE & S. MINAKAMI. 1980. NADH- and NADPH-dependent lipid peroxidation in bovine heart submitochondrial particles. Biochem. J. **192:** 853–860.
43. CABRINI, L., P. PASQUALI, B. TADOLINE, A. M. SECHI & L. LANDI. 1986. Antioxidant behavior of ubiquinone and β-carotene incorporated in model membranes. Free Rad. Res. Comms. **2:** 85–92.
44. BARCLAY, L. R. C., S. J. LOCKE & J. M. MACNEIL. 1983. The autoxidation of unsaturated lipids in micelles. Synergism of inhibitors vitamins C and E. Can. J. Chem. **61:** 1288–1290.
45. BARCLAY, L. R. C., S. J. LOCKE, J. M. MACNEIL, J. VAN KESSEL, G. W. BURTON & K. U. INGOLD. 1984. Autoxidation of micelles and model membranes. Quantitative kinetic measurements can be made by using either water soluble or lipid-soluble initiators with water-soluble or lipid soluble chain-breaking antioxidants. J. Am. Chem. Soc. **106:** 2479–2480.
46. NIKI, E., A. KAWAKAMI, Y. YAMAMOTO & Y. KAMIYA. 1985. Synergistic inhibition of oxidation of phosphatidylcholine liposomes in aqueous dispersions by vitamin E and vitamin C. Bull. Chem. Soc. Jpn. **58:** 1971–1975.
47. DOBA, T., G. W. BURTON & K. U. INGOLD. 1985. Antioxidant and co-antioxidant activity of vitamin C. The effect of vitamin C, either alone or in the presence of vitamin E or a water-soluble vitamin E analogue, upon the peroxidation of aqueous multilamellar phospholipid liposomes. Biochim. Biophys. Acta **835:** 298–303.
48. MATSUSHITA, S., F. IBUKI & A. OKI. 1963. Chemical reactivity of the nucleic acid bases. I. Antioxidative ability of the nucleic acids and their related substances on the oxidation of unsaturated fatty acids. Arch. Biochem. Biophys. **102:** 446–451.
49. NIKI, E., M. SAITO, Y. YOSHIKAWA, Y. YAMAMOTO & Y. KAMIYA. 1986. Oxidation of lipids. XII. Inhibition of oxidation of soybean phosphatidylcholine and methyl linoleate in aqueous dispersions by uric acid. Bull. Chem. Soc. Jpn. **59:** 471–477.
50. TERAO, K. & E. NIKI. 1987. Damage to biological tissue induced by radical initiator 2,2'-azobis (2-amidinopropane) dihydrochloride and its inhibition by chain-breaking antioxidants. J. Free. Rad. Biol. Med. **2:** 193–201.
51. DAHL, T. A., W. R. MIDDEN & P. E. HARTMAN. 1988. Some prevalent biomolecules as defenses against singlet oxygen damage. Photochem. Photobiol. **47:** 357–362.
52. GLAZER, A. N. 1988. Fluorescence-based assay for reactive oxygen species: A protective role for creatinine. FASEB J. **2:** 2487–2491.
53. FRIDOVICH, I. 1986. Superoxide dismutases. Adv. Enzymol. **58:** 61–97.
54. CHANCE, B., H. SIES & A. BOVARIS. 1979. Hydroperoxide metabolism in mammalian organs. Physiol. Rev. **59:** 527–605.
55. CHRISTOPHERSEN, B. O. 1968. The inhibitory effect of reduced glutathione on the lipid peroxidation of the microsomal fraction and mitochondria. Biochem. J. **106:** 515–522.

56. GIBSON, D. D., K. R. HORNBROOK & P. McCAY. 1980. Glutathione-dependent inhibition of lipid peroxidation by a soluble, heat-labile factor in animal tissues. Biochim. Biophys. Acta **620**: 572–582.
57. BELL, J. G., C. B. COWEY & A. YOUNGSON. 1984. Rainbow trout liver microsomal lipid peroxidation. The effect of purified glutathione peroxidase, glutathione S-transferase and other factors. Biochim. Biophys. Acta **795**: 91–99.
58. FLOHE, L. 1982. Glutathione peroxidase brought into focus. *In* Free Radicals in Biology. W. A. Pryor, Ed. Vol. V: 223–254. Academic Press. New York.
59. SEVANIAN, A., S. F. MUAKKASSEH-KELLY & S. MONTESTRUQUE. 1983. The influence of phospholipase A$_2$ and glutathione peroxidase on the elimination of membrane lipid peroxides. Arch. Biochem. Biophys. **223**: 441–452.
60. GROSSMANN, A. & A. WENDEL. 1983. Non-reactivity of the selenoenzyme glutathione peroxidase with enzymatically hydroperoxidized phospholipids. Eur. J. Biochem. **135**: 549–552.
61. SEVANIAN, A. & E. KIM. 1985. Phospholipase A$_2$ dependent release of fatty acids from peroxidized membranes. J. Free Rad. Biol. Med. **1**: 263–271.
62. VAN KUIJK, F. J. G. M., A. SEVANIAN, G. J. HANDELMAN & E. A. DRATZ. 1987. A new role for phospholipase A$_2$: protection of membranes from lipid peroxidation damage. TIBS **12**: 31–34.
63. HEMLER, M. E. & W. E. M. LANDS. 1980. Evidence for a peroxide-initiated free radical mechanism of prostaglandin biosynthesis. J. Biol. Chem. **255**: 6253–6261.
64. URSINI, F., M. MAIORINO, M. VALENTE, L. FERRI & C. GREGOLIN. 1982. Purification from pig liver of a protein which protects liposomes and biomembranes from peroxidative degradation and exhibits glutathione peroxidase activity on phosphatidylcholine hydroperoxides. Biochim. Biophys. Acta **710**: 197–211.
65. URSINI, F. & A. BINDOLI. 1987. The role of selenium peroxidases in the protection against oxidative damage of membranes. Chem. Phys. Lip. **44**: 255–276.
66. TAN, K. H., D. J. MEYER, J. BELIN & B. KETTERER. 1984. Inhibition of microsomal lipid peroxidation by glutathione and glutathione transferases B and AA. Biochem. J. **220**: 243–252.
67. GIBSON, D. D., J. HAWRYLKO & P. B. McCAY. 1985. GSH-Dependent inhibition of lipid peroxidation: properties of a potent cytosolic system which protects cell membranes. Lipids **20**: 704–711.
68. MASOTTI, L., F. CASALI & T. GALEOTTI. 1988. Lipid peroxidation in tumor cells. Free Rad. Biol. Med. **4**: 377–386.

DISCUSSION

A. RILEY (*University College, London*): I would like to comment on the regeneration of α-tocopherol from the tocopheryl radical by ascorbate that you mentioned in connection with the elegant work by Niki and coworkers; this reaction has been demonstrated by pulse radiolysis experiments carried out by the Brunel group (Packer *et al.,* Nature, 1984). I would also ask a question concerning the clear difference in mechanism that is implied by the fact that β-carotene is the only lipophilic antioxidant that you mentioned in your comprehensive review which is effective solely at low O_2 tensions. Why is β-carotene apparently unable to act as a membrane antioxidant at O_2 tensions above 2 percent?

N. I. KRINSKY: The lack of effect of β-carotene at high O_2 tension suggests that there is a competitive reaction between β-carotene and O_2. For example, the alkyl radical, L·, may either react with β-carotene or with O_2. At high pO_2, the latter reaction would be favored.

G. GUIDOTTI (*Università di Patologia Generale, Parma, Italy*): You mentioned bilirubin as a physiologic antioxidant. I wonder what is going on in pathologic situations such as the Krigler-Najjar syndrome, in which the bilirubin concentration in plasma increases between 5 and 50 times. Is peroxidative stress lower in these patients?

KRINSKY: I don't know of any studies where a clear antioxidant effect was measured under conditions of hyperbilirubinemia.

R. S. SOHAL (*Southern Methodist University, Dallas, Texas*): I think that the potential of oxidants to cause damage under physiological conditions may be exaggerated, and one-sided emphasis is often placed on the deleterious role of oxygen radicals. There is some evidence (to be mentioned in my paper) that oxidative stress may, in fact, play a useful role in cell physiology, especially in relation to the induction of cellular differentiation. Of course, I do not intend to imply that antioxidants do not play a useful role in aerobic cells; my concern is that the prevailing view, that the minimization of oxidants would be to the benefit of the cell, may be oversimplified.

KRINSKY: What is probably important is a proper balance in terms of antioxidants that can regulate radical reactions and radical processes that are necessary for normal biological functions.

L. ERNSTER (*University of Stockholm, Stockholm, Sweden*): You mentioned ubiquinone as an example of a hydrophobic antioxidant. The evidence accumulated over the years indicates that it is the reduced form of ubiquinone that acts as an antioxidant. (For a recent review, see Beyer *et al.*, Chem. Scr. Vol. 27, 1987).

KRINSKY: Thank you for bringing this to my attention.

Metals and Membrane Lipid Damage by Oxy-Radicals

GIORGIO MINOTTI

Institute of General Pathology
Catholic University
School of Medicine
00168 Rome, Italy

INTRODUCTION

The reaction of molecular oxygen with membrane unsaturated lipids (LH) to form lipid hydroperoxides (LOOH) is popularly referred to as lipid peroxidation:

$$LH + O_2 \rightarrow LOOH \tag{1}$$

Written as such, lipid peroxidation would appear a straightforward uncatalyzed reaction. But lipid peroxidation instead occurs as a rather complex multi-step process in which the formation of LOOH is preceded by formation of both lipid and nonlipid radicals. The complexity of lipid peroxidation and its contingency on free-radical intermediates stem from the electronic configuration of oxygen and lipids, which is that of triplet versus singlet multiplicity, respectively.[1] Uncatalyzed lipid peroxidation is, therefore, a spin-forbidden reaction.

According to a schematic view[2] the overall machinery of lipid peroxidation can be suddenly triggered by a reactive species (R·, the so-called "initiator"[3]) that overcomes the dissociation energy of an allylic bond, thereby causing hydrogen abstraction and formation of a lipid alkyl radical (L·):

$$LH + R^{\cdot} \rightarrow L^{\cdot} + RH \tag{2}$$

Unlike ground-state lipids the corresponding alkyl radicals will rapidly add molecular oxygen to form lipid peroxyl radicals (LOO·), which eventually liberate LOOH via hydrogen abstraction from a neighboring allylic bond:

$$L^{\cdot} + O_2 \rightarrow LOO^{\cdot} \tag{3}$$

$$LOO^{\cdot} + LH \rightarrow LOOH + L^{\cdot} \tag{4}$$

The sequence of Reactions (2)–(4) clearly indicates that the formation of a first generation of LOOH is invariably paralleled by the formation of a second generation of lipid alkyl radicals. Therefore, lipid peroxidation must be thought of as a process that requires externally generated oxidant(s) for the initiating event, yet it retains the unique feature of self-maintaining and rapidly expanding the oxidative attack from one allylic bond to another one.

The realization that lipid peroxidation may represent a common pathway in the onset of apparently unrelated diseases as well in the onset of drug therapeutic or toxic effects[4,5] has prompted impetus in the identification of the initiator(s). Both metals and

34

oxygen radicals have been implicated, although with two very different motivations[5]: (1) lipid peroxidation is initiated by the hydroxyl radical, ($\cdot$OH), a powerful oxidant ($E° = 1.83$ V)[6] that is formed via metal-catalyzed reactions between superoxide (O_2^-) and hydrogen peroxide (H_2O_2); or, (2) lipid peroxidation is initiated by metal-oxygen complexes that stem from the redox cycling of metals themselves with oxygen and the O_2^-/H_2O_2 couple. In other words, it has been suggested that lipid peroxidation can be initiated either by an oxygen-centered radical that is formed via metal-catalyzed reactions, or by a metal-centered radical that is formed via oxygen- and oxy-radical-catalyzed reactions.

This article focuses on the biochemical aspects and biomedical implications of the two proposed mechanisms, special emphasis being given to the role of iron.

PROPOSED INITIATORS OF LIPID PEROXIDATION

The Hydroxyl Radical

Mechanisms of Metal-Catalyzed Hydroxyl Radical Formation

For several years it was thought that the hydroxyl radical could be formed via the so-called Haber-Weiss reaction, which consists of one electron transfer from O_2^- to H_2O_2:

$$O_2^- + H_2O_2 \rightarrow O_2 + OH^- + \cdot OH \qquad (5)$$

This reaction was originally proposed as metal-catalyzed,[7] yet the requirement for such catalysis was systematically ignored until the mid 1970s, when both chemical and physical considerations led many authors to re-evaluate the role of copper, iron, and other transition metals in O_2^-- and H_2O_2-dependent reactions. The Haber-Weiss reaction ought, therefore, be rewritten as it follows, wherein M is indicative of any redox active metal which first picks up one electron from O_2^- and subsequently loses one electron to reductively cleave H_2O_2.

$$O_2^- + M^{n+1} \rightarrow O_2 + M^n \qquad (6)$$

$$H_2O_2 + M^n \rightarrow OH^- + \cdot OH + M^{n+1} \qquad (7)$$

The final step of the Haber-Weiss reaction (i.e., the metal-catalyzed reductive cleavage of H_2O_2) is sometime referred to as Fenton's reaction.[8]

The metal-catalyzed Haber-Weiss reaction has attracted consensus as a mechanism by which naturally occurring metals (such as iron and copper) and ubiquitous by-products of oxygen utilization (O_2^- and H_2O_2) may subtly threaten the cell by forming a powerful oxidant such as $\cdot$OH. Most importantly, the metal-catalyzed Haber-Weiss reaction would rationalize the existence and physiological significance of enzymes like superoxide dismutase (SOD), which speeds up the dismutation of O_2^- to H_2O_2, or like glutathione peroxidase (GSH-Px) and catalase, which in turn reduce H_2O_2 to water or water and molecular oxygen, respectively.[9] As a matter of fact, neither O_2^- nor H_2O_2 is of major concern for reaction with most biomolecules, lipids included. Viewed in this context, the effort of aerobic cells to evolve enzymatic systems aimed at lowering the concentration of O_2^- and H_2O_2 would appear unmotivated.

Conversely, if one realizes that $O_2^{\bar{\cdot}}$ is required for metal reduction and that reduced metals would in turn cleave H_2O_2 to $^{\cdot}OH$, then the availability of SOD (to inhibit metal reduction) and that of catalase or GSH-Px (to prevent H_2O_2 accumulation) both turn out to be crucial to protect the cell from deleterious fluxes of $^{\cdot}OH$.

The avidity of $O_2^{\bar{\cdot}}$ for metals is worthy of further considerations. Superoxide can certainly serve as a one-electron reductant for M^{n+1}, yet it may also serve as a one-electron oxidant for M^n, according to the following schema:

$$O_2^{\bar{\cdot}} + M^n \xrightarrow{2H^+} H_2O_2 + M^{n+1} \tag{8}$$

With certain metals (e.g., iron and copper), the rate constant for the $O_2^{\bar{\cdot}}$-dependent M^n oxidation may actually exceed the rate constant for the H_2O_2-dependent M^n oxidation.[10,11] Keeping this in mind, one would conclude that reaction of $O_2^{\bar{\cdot}}$ with ferric iron or cupric copper sparks a sequence of reactions culminating in H_2O_2 accumulation rather than in H_2O_2 breakdown and $^{\cdot}OH$ liberation. How and when, therefore, can the $^{\cdot}OH$ formation occur? It is my personal experience that the potential for $O_2^{\bar{\cdot}}$ to behave as a versatile reductant/oxidant for metals is dominated by the presence and nature of chelators. For example, one can find that ADP-chelated iron is first reduced by $O_2^{\bar{\cdot}}$ and then oxidized by H_2O_2, whereas citrate-chelated iron is first reduced by $O_2^{\bar{\cdot}}$ and then reoxidized by $O_2^{\bar{\cdot}}$ itself.[12] These observations underscore that naturally occurring ligands, such as ADP or citrate, may shift metals toward pathways culminating in H_2O_2 breakdown, $^{\cdot}OH$ formation, and cell damage; or toward pathways culminating in H_2O_2 accumulation and harmless activation of the catalase/GSH-Px route. In brief, the reaction of $O_2^{\bar{\cdot}}$ with metals should not be considered as an invariably deleterious event.

Regardless of the above critical considerations, *in vivo* formation of $^{\cdot}OH$ remains contingent on the availability of low molecular weight metal complexes. With specific regard to iron much attention must be given to ferritin, the most important intracellular iron storage protein. Ferritin is a multi-subunit shell-shaped protein with a central core in which iron can be stored in its ferric form[13] and from which iron can be released upon reduction to the ferrous form.[14,15] Owing to its small size and natural propensity to function as ferric reductant, $O_2^{\bar{\cdot}}$ can easily enter the narrow channels leading to the core of ferritin, thereby promoting the reductive mobilization of the iron that is stored therein. This picture would reinforce the idea that the first step of *in vivo* Haber-Weiss reactions (i.e., the reduction of Fe^{3+} to Fe^{2+}) does occur via $O_2^{\bar{\cdot}}$-dependent mechanisms and, therefore, is inhibited by SOD, a "protective" enzyme. There are, however, numerous compounds of either toxicological or pharmacological relevance which can readily mobilize ferritin iron via $O_2^{\bar{\cdot}}$-independent, and hence SOD-insensitive, mechanisms. This is the case, for example, of the monocation radical of paraquat, a dication herbicide possessing a devastating lung toxicity; and it is also the case of the semiquinone radical or adriamycin, an anthracycline co-characterized by broad-spectrum antitumour activity and severe heart toxicity. Both paraquat and adriamycin undergo reduction by the microsomal cytochrome P-450 reductase and the corresponding one-electron-enriched forms may then reduce redox-active sites of ferritin from which electrons are finally shuttled to the central core to release iron.[16,17] The monocation radical of paraquat or the semiquinone radical of adriamycin can also reoxidize at the expense of oxygen to form $O_2^{\bar{\cdot}}$; however, the so-formed $O_2^{\bar{\cdot}}$ does not compete with the drug radicals to mobilize ferritin iron. In this context, the effect of SOD would be that of enhancing the dismutation of O_2^- to H_2O_2, which in turn reacts with the released iron to form $\cdot OH$. In other words, SOD would unexpectedly exaggerate $\cdot OH$ formation, somewhat behaving as "pro-oxidant" enzyme rather than as "antioxidant" or "protective" enzyme.

Evidence for or Evidence against the Initiation of Lipid Peroxidation by the Hydroxyl Radical

A role for $\cdot OH$ as initiator of lipid peroxidation has been envisioned by several investigators.[18–22] The conclusion is almost entirely based on the repeated finding that chemicals which intercept $\cdot OH$ (the so-called "$\cdot OH$ scavengers" or "$\cdot OH$ traps") prevent, to variable extent, lipid peroxidation. In this respect, a nonpreconceived overview of the literature would actually suggest that "pure" $\cdot OH$ scavengers do not exist and that the inhibitory effect of these chemicals on lipid peroxidation is probably indicative of nonspecific side effects. Thus, it has been shown that reaction of Fe^{2+} with H_2O_2 is greatly facilitated by benzoate[23] and formate,[24] whereas it is inhibited, to a variable extent, by mannitol[23] and chloroacetate.[24] All these scavengers will, therefore, affect lipid peroxidation by targeting the redox cycling of iron rather than by preventing reaction of $\cdot OH$ with the lipid substrate. Other scavengers, such as Tris, are rather good chelators and may affect lipid peroxidation by sequestering iron in a redox inactive form.[25] Finally, there are scavengers that do not chelate iron or interfere with its redox cycling, yet they may inhibit enzymatic sources of the O_2^- and H_2O_2 required for $\cdot OH$ formation. This is the case of thiourea, which is a non-negligible inhibitor of xanthine oxidase.[3] Therefore, thiourea should not be used in lipid peroxidation studies that rely on xanthine oxidase as a source of superoxide and hydrogen peroxide.

Earlier considerations about the multiple nonspecific effects of $\cdot OH$ scavengers would point to SOD or catalase as more reliable and specific tools to establish the intermediacy of $\cdot OH$ in lipid damage. Unfortunately, the effects of the two enzymes on different in vitro lipid peroxidation systems are multifaceted and do not provide any unifying interpretation of the role of O_2^- or H_2O_2 as forerunners of $\cdot OH$ and lipid damage. For example, it has been shown that lipid peroxidation is inhibited, unaffected, or even stimulated by SOD.[26–28] Superoxide can be formed during the microsomal oxidation of NADPH or during the oxidation of xanthine by the flavoprotein xanthine oxidase. In the first case O_2^- will most likely be formed at the site of cytochrome P-450, with minor contribution by the P-450 reductase[28]; in the second case O_2^- will be formed via the so-called "univalent flux" of electrons from xanthine to molecular oxygen via xanthine oxidase.[29] Morehouse et al.[28] have shown that when the concentration of NADPH and that of xanthine oxidase are adjusted to values that yield comparable fluxes of O_2^-, the peroxidation induced by NADPH is not inhibited by SOD, whereas that induced by xanthine oxidase is greatly, if not completely, inhibited by SOD. In this study ADP-Fe^{3+} was used to catalyze $\cdot OH$ formation and lipid damage; fully comparable results can be obtained by replacing ADP-Fe^{3+} with ferritin as a physiological donor of iron.[30]

The diverging effects of SOD in the two different peroxidation systems could theoretically be reconciled by assuming that: (1) in an NADPH-dependent reaction, O_2^- is liberated at sites that cannot be sterically entered by SOD (the "active site" of

P-450 reductase or that of cytochrome P-450); and (2) in a xanthine oxidase-dependent reaction, $O_2^{\cdot-}$ is liberated in the aqueous phase of incubations, where it can be easily recognized and intercepted by SOD. Although intriguing in principle, the veracity of this interpretation is weakened by the finding that in an NADPH-dependent, Fe^{2+}-catalyzed system the addition of SOD increases the yield of $\cdot OH$.[26] This result clearly indicates that the $O_2^{\cdot-}$ formed during the NADPH-supported microsomal electron transport *can* be intercepted and enzymatically converted by SOD to H_2O_2, which eventually fuels Fenton's reaction. In their comparative study on NADPH- versus xanthine oxidase-dependent reactions, Morehouse *et al.*[28] also showed that the purified P-450 reductase couples the oxidation of NADPH with the reduction of several ferric chelates, both aerobically and anaerobically, suggesting that the one-electron-enriched form of this enzyme may represent an additional example of an organic radical capable of direct metal reduction. Very recent studies[27] indicate that in a reconstituted mixed-function oxidase system the cytochrome P-450 itself can shuttle electrons from the reductase to ADP-Fe^{3+}, although it is not clear whether the juxtaposed lipid milieu has some role in the reaction. Oxygen- and $O_2^{\cdot-}$-independent reduction of ADP-Fe^{3+} has also been observed by Vegh *et al.* using NADPH and intact microsomes.[31]

Keeping in mind this body of results and information, I would suggest that the effects of SOD on NADPH- and xanthine oxidase-dependent lipid peroxidation can be tentatively reconciled as follows: (1) in an NADPH-dependent, Fe^{3+}-catalyzed system, the first step of the Haber-Weiss reaction (i.e., the Fe^{3+} to Fe^{2+} reduction) is mediated by the P-450 reductase or by the P-450 itself via $O_2^{\cdot-}$-independent mechanisms, and hence SOD will not inhibit Fe^{2+} formation and initiation; and (2) in a xanthine oxidase-dependent, Fe^{3+}-catalyzed system, iron reduction is mediated by $O_2^{\cdot-}$, and SOD will inhibit both Fe^{2+} formation and lipid damage. In light of this subtle, yet dramatic difference between NADPH- and xanthine oxidase-driven systems, it is rather difficult to conceive a role for $O_2^{\cdot-}$ as unambiguous forerunner of $\cdot OH$ and lipid damage, or a role for SOD as an invariably "protective" or "antioxidant" enzyme.

As a final tool to establish the intermediacy of $\cdot OH$ in metal-catalyzed lipid damage one can use catalase, the rationale being that scavenging the H_2O_2 required for Fenton's reaction would inhibit $\cdot OH$ formation and initiation. The data in the literature are in dramatic contrast to the validity of this rationale. Indeed, it has been clearly shown that NADPH-dependent ADP-Fe^{3+}-catalyzed lipid peroxidation is unaffected by the addition of large "pharmacological" amounts of catalase.[32] Conversely, and perhaps more importantly, it has been shown that the addition of H_2O_2 is paralleled by impressive inhibition of the lipid damage.[32] These results would suggest that the H_2O_2 formed during the NADPH-supported microsomal electron transport is not required for iron-catalyzed lipid peroxidation. It would also appear that, starting with Fe^{3+}, lipid peroxidation is contingent on the formation of some Fe^{2+} and that excessive reoxidation of such Fe^{2+}, while generating more $\cdot OH$, will prevent initiation. In agreement with these conclusions Saito *et al.*[33] have shown that in an NADPH- and paraquat-dependent system (i.e., in a system in which the monocation radical of paraquat reduces both ferritin iron and molecular oxygen) lipid peroxidation is inhibited by SOD and stimulated by catalase. These results may appear extremely confusing, yet they can be "easily" interpreted by keeping in mind that lipid peroxidation requires the mobilization of ferrous iron from ferritin and that such ferrous iron must be "protected" from excessive H_2O_2-dependent reoxidation. Thus, it is not surprising that lipid damage is inhibited by SOD, which accumulates H_2O_2 from O_2^{-}, and is exaggerated by catalase, which instead prevents H_2O_2 accumulation. Viewed from another perspective, these results indicate that lipid peroxidation is minimal under conditions that favor $^{\cdot}OH$ formation (reaction of Fe^{2+} with H_2O_2) and maximal under conditions that prevent $^{\cdot}OH$ formation (decomposition of H_2O_2 before reaction with Fe^{2+}).

Similar results have been obtained in xanthine oxidase-dependent systems.[16,34]

Metal-Oxygen Complexes

The Perferryl Ion

If not the hydroxyl radical, "who" is responsible for metal-dependent lipid damage? A role for a metal-centered radical was proposed some 25 years ago by Hochstein and Ernster.[35] These authors showed that in the presence of ADP the microsomal oxidation of NADPH was accompanied by the formation of malondialde-hyde (MDA), a well-known byproduct of the peroxidative decomposition of unsaturated lipids. In a subsequent communication[36] the same authors pointed out that ADP was contaminated by iron and that such contamination was essential for lipid peroxidation to occur. It was concluded that microsomal electron transport causes reduction of ADP-Fe^{3+} to a perferryl ion which is capable of abstracting hydrogen from the allylic bond of unsaturated lipids:

$$ADP\text{-}Fe^{3+} + O_2 + reductase^- \rightarrow ADP\text{-}Fe^{2+} - O_2 + reductase \qquad (9)$$

$$ADP\text{-}Fe^{2+} - O_2 + LH \rightarrow ADP\text{-}Fe^{2+} - O_2 - H^{+1} + L^{\cdot} \qquad (10)$$

The hypothesis that lipid peroxidation is promoted by a perferryl ion encircles a number of both practical and experimental considerations. First, the mechanism of perferryl ion formation (Fe^{3+} to Fe^{2+} reduction and subsequent binding of Fe^{2+} to oxygen) helps us to understand why in both NADPH- and xanthine oxidase-dependent systems only ferric reduction is required for initiation, no matter how such reduction is achieved (i.e., via O_2^{-}-independent or O_2^{-}-dependent mechanisms, respectively). Furthermore, the structure of the perferryl ion implies that subsequent encounters of this metal-oxygen complex with H_2O_2 will regenerate Fe^{3+}, thereby inhibiting lipid peroxidation and providing a basis for the stimulatory effect of catalase. As a final yet unobvious consideration one should take into consideration that the formal charge of iron in the perferryl ion is +6, which may account for a rather unique electron avidity.[1]

Microsomes supplemented with ADP-Fe^{2+} rapidly undergo quite extensive peroxidation irrespective of the presence of enzymatic sources of O_2^{-} or H_2O_2.[37] This simple yet clear-cut experiment would substantiate the "perferryl ion model" of lipid peroxidation, that is, it would confirm that binding of Fe^{2+} to molecular oxygen to form a perferryl ion is essential for the lipid damage to occur.

The Ferrous-Dioxygen-Ferric Complex

Partial evidence for the intermediacy of a perferryl ion in the initiation of lipid peroxidation is seen in observations by Bucher et al.[38] These investigators found that, in agreement with the "perferryl ion model" of lipid peroxidation, the addition of ADP-Fe^{2+} to incubations containing microsomal phospholipid liposomes resulted in SOD- and catalase-insensitive MDA formation. However, it was observed that ADP-Fe^{2+}-dependent peroxidation of the purified lipids was preceded by a rather extensive lag phase which was not observed during incubation of ADP-Fe^{2+} with intact microsomes. The message from these experiments was that the nature of the lipid substrate (extracted lipids or protein-associated lipids) may dominate formation and/or reactivity of the initiators. More specifically, it was evident that in the presence of protein-free lipids the reaction of ADP-Fe^{2+} with oxygen to form a perferryl ion cannot account for the overall mechanism of initiation. The finding that the lag phase observed during reaction of ADP-Fe^{2+} with phospholipid liposomes was completely abolished by simultaneous addition of ADP-Fe^{3+} and the realization that microsomes cause extensive autoxidation of ferrous iron prompted the authors to conclude that: (1) lipid peroxidation is initiated by a complex of Fe^{2+} with Fe^{3+}, and perhaps with oxygen to form a ferrous-dioxygen-ferric complex; (2) in incubations containing extracted lipids the lag phase is indicative of the time required for ADP-Fe^{2+} to slowly autoxidize to ADP-Fe^{3+} and to combine with it to initiate peroxidation; and (3) in incubations containing microsomes the autoxidation of ADP-Fe^{2+} is so rapid that the Fe^{2+}-Fe^{3+} complex is formed almost instantaneously with a concomitant burst of peroxidation.[38,39]

The "ferrous-dioxygen-ferric model" of lipid peroxidation is a matter of both positive and negative considerations. Major negative consideration is that chemical structure, stability constant, and paramagnetic features of this hypothetical complex have not yet been described. On the other hand, the assumption that lipid peroxidation is promoted by a ferrous-dioxygen-ferric complex rationalizes a rather subtle observation that is not encompassed by the "perferryl ion model," the observation being that maximal rates of lipid peroxidation are linked to $Fe^{2+}:Fe^{3+}$ ratios approaching unity. Evidence that the $Fe^{2+}:Fe^{3+}$ ratio dominates the onset and extent of lipid peroxidation has been obtained by Braughler et al.[40–42] and by Minotti and Aust[3,23,25,43] in nonenzymatic systems in which the oxidation of ferrous iron (by H_2O_2 or lipid substrates) or the autoxidation of ferrous chelates were monitored in conjunction with the formation of MDA.

The observation that ferritin-catalyzed enzymatic peroxidation is stimulated by catalase does not argue against the proposed requirement for both ferrous and ferric iron in the initiation step. In this situation the ferric iron required for initiation will be formed via the autoxidation of the ferrous iron released from ferritin, the autoxidation being favored by chelators and only scarcely affected by catalase.[25] The latter will instead exaggerate lipid damage by diverting H_2O_2 from the complete oxidation of the residual Fe^{2+}.[25]

BIOMEDICAL IMPLICATIONS AND PERSPECTIVES

The repeated finding that ferritin supplies the iron required for *in vitro* lipid peroxidation does not necessarily imply that it will also supply the iron required for *in vivo* lipid peroxidation. Ferritin is a predominantly cytosolic protein, and sterical distance from the sites at which $O_2^{\bar{}}$ or other ferric reductant are being formed (microsomes, mitochondria) may lessen its importance as a source of readily available iron. In this respect it is crucial to point out that biological membranes contain significant amounts of ferritin. Mitochondria exhibit two distinct binding sites for ferritin[44,45]; critical amounts of ferritin remain tightly bound to microsomes as well,[46-48] although no specific anchorage site has been so far described.

Very intriguingly, microsomes have also been found to contain a small pool of nonheme iron which is not accounted for by ferritin, and hence is tentatively referred to as nonheme-nonferritin. The existence of this elusive nonheme-nonferritin iron was first described by Montgomery et al.[46] and subsequently confirmed by Thomas and Aust.[47] More recently I have shown that *in vivo* treatment with phenobarbital causes simultaneous induction of cytochrome P-450 and depletion of nonheme-nonferritin iron, somewhat indicating that this peculiar form of nonheme iron may serve as precursor of the heme iron associated with inducible cytochrome P-450 isozymes.[48] The finding that methemalbumin, an inhibitor of heme synthesis, prevents both the PB-induced increase in cytochrome P-450 and the PB-induced depletion of nonheme-

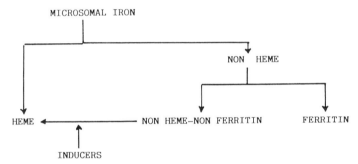

FIGURE 1. Schema for microsomal heme and nonheme iron and for the effect of heme synthesis inducers.

nonferritin iron would validate this interpretation.[48] A schematic diagram for microsomal heme versus nonheme iron is given in FIGURE 1.

On the basis of these observations I realized that microsomes may actually contain two forms of nonheme iron both susceptible, at least in principle, to reductive mobilization. In order to evaluate the distinct contribution of these two pools of iron in reductive mobilization processes I also developed an experimental system aimed at removing nonheme-nonferritin iron, by means of *in vivo* treatment with phenobarbital; or ferritin, by means of *in vitro* manipulations of the isolated microsomes (e.g., chromatography on Sepharose CL-2B). By combining the two procedures one can eventually obtain four different types of membranes: native microsomes, which contain both ferritin and nonheme-nonferritin iron; native-chromatographed microsomes, which contain only nonheme-nonferritin iron; phenobarbital-induced microsomes, which contain only ferritin; and phenobarbital-induced/chromatographed microsomes, which do not contain either ferritin or nonheme-nonferritin iron (FIG. 2).

The results obtained with these membranes indicate that adriamycin couples the

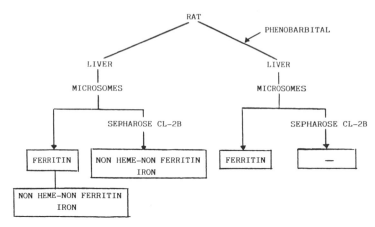

FIGURE 2. "Flow chart" for the isolation and preparation of microsomes with different nonheme iron composition.

microsomal oxidation of NADPH with the reductive mobilization of iron (TABLE 1) and that the reaction is considerably more effective in native microsomes than in chromatographed or phenobarbital-induced microsomes (TABLE 2). Overall, it would appear that both ferritin and nonheme-nonferritin iron are released during reaction of NADPH with microsomes and adriamycin. Consistently, phenobarbital-induced/ chromatographed microsomes do not release detectable iron, in obvious agreement with the fact that they do not contain either ferritin or nonheme-nonferritin iron (see also TABLE 2).

I would not overburden this section by showing that NADPH- and adriamycin-dependent release of iron is SOD-insensitive (indicating that the iron is reduced and released by the semiquinone radical of the drug[49]), or by showing that adriamycin-dependent lipid peroxidation is inhibited by SOD and stimulated by catalase (indicating that, in analogy to earlier observations in paraquat-driven systems, both ferrous and ferric iron are required for initiation[50]). Rather, I would conclusively emphasize that biological membranes are naturally loaded with the iron required for oxidative

TABLE 1. NADPH- and Adriamycin-Dependent Release of Fe^{2+} from Microsomal Membranes

Addition	Fe^{2+} Release (nmol mg protein^{-1} min^{-1})
None	—
NADPH	0.04
Adriamycin	—
NADPH and adriamycin	0.89

NOTE: Incubations (1 ml final volume) contained rat liver microsomes (0.5 mg protein/ml) and bathophenanthroline (0.5 mM) in 50 mM NaCl, pH 7.0, at 37°C. Where indicated, NADPH (0.25 mM) and adriamycin (0.25 mM) were included. Iron release was monitored at 530–560 nm.

damage and that the iron will be released only in the presence of a drug (e.g., adriamycin) which may impose both therapeutic and toxic effects. Major implication of this concept is that the beneficial or pathologic effects of xenobiotics on different tissues will depend not only, and perhaps not much, on the presence of "protective enzymes," but rather on the presence and relative distribution of membrane iron liable to reductive mobilization and subsequent redox cycling.

CONCLUSIONS

In this article mechanisms of metal-catalyzed lipid peroxidation have been presented and discussed. Although *in fieri* and still a matter of critical re-evaluation, the concept that lipid peroxidation is promoted by metal-oxygen complexes appears to motivate new and less conventional fields of investigation, especially if one considers that "protective" enzymes like SOD or catalase may not prevent or may even facilitate the formation and reactivity of these initiators. Most importantly, the proposal that

TABLE 2. NADPH- and Adriamycin-Dependent Fe^{2+} Release from Different Microsomal Membranes

Microsomes	Fe^{2+} Release (nmol mg protein^{-1} min^{-1})
Native	0.89
Native, chromatographed	0.42
PB-induced	0.34
PB-induced, chromatographed	—

NOTE: Incubations (1 ml final volume) contained rat liver microsomes (0.5 mg protein/ml), adriamycin (0.25 mM), and bathophenanthroline (0.5 mM) in 50 mM NaCl, pH 7.0, at 37°C. Reactions were started by the addition of NADPH (0.25 mM) and iron release was monitored at 530–560 nm.

biological membranes are abundantly loaded with the iron required to form the perferryl ion or a ferrous-dioxygen-ferric complex prompts new interest in therapeutic strategies aimed at switching on or off *in vivo* lipid peroxidation.

ACKNOWLEDGMENTS

I would like to thank Dr. Tommaso Galeotti (Catholic University, Rome, Italy) and Dr. Steven D. Aust (Utah State University, Logan, Utah) for generous help and support throughout this work.

REFERENCES

1. AUST, S. D. & B. A. SVINGEN. *In* Free Radicals in Biology. W. H. Pryor, Ed. **Vol. 5:** 1–28. Academic Press. Orlando, FL.
2. NIKI, E. 1987. Chem. Phys. Lipids **44:** 227–254.

3. MINOTTI, G. & S. D. AUST. 1988. Chem. Biol. Interactions. In press.
4. SIES, H., Ed. 1985. Oxidative Stress. Academic Press. London.
5. AUST, S. D., L. A. MOREHOUSE, & C. E. THOMAS. 1985. J. Free Rad. Biol. Med. 1: 3–25.
6. CZAPSKI, G. In Methods in Enzymology. L. Packer, Ed., Vol. 105: 209–215.
7. HABER, F. & J. J. WEISS. 1934. Proc. R. Soc. London Ser. A 147: 332–351.
8. FENTON, H. J. H. 1894. J. Chem. Soc. 65: 899–910.
9. FRIDOVICH, I. 1986. Arch. Biochem. Biophys. 247: 1–11.
10. RUSH, J. D. & B. J. BIELSKI. 1985. J. Phys. Chem. 89: 5062–5066.
11. CZAPSKI, G. & S. GOLDSTEIN. 1986. Free Rad. Res. Comms. 1: 157–161.
12. MINOTTI, G. & S. D. AUST. 1987. Arch. Biochem. Biophys. 253: 257–267.
13. HARRISON, P. M. 1977. Semin. Hematol. 14: 55–70.
14. THOMAS, C. E., L. A. MOREHOUSE & S. D. AUST. 1985. J. Biol. Chem. 260: 3275–3280.
15. BOLANN, B. J. & R. J. ULVIK, 1987. Biochem. J. 1987. 243: 55–59.
16. THOMAS, C. E. & S. D. AUST. 1986. J. Biol. Chem. 261: 13064–13070.
17. THOMAS, C. E. & S. D. AUST. 1986. Arch. Biochem. Biophys. 248: 684–689.
18. LAI, C. & L. H. PIETTE, 1977. Biochem. Biophys. Res. Commun. 78: 51–59.
19. LAI, C. & L. H. PIETTE. 1978. Arch. Biochem. Biophys. 190: 27–38.
20. GUTTERIDGE, J. M. C. 1984. Biochem. J. 224: 697–701.
21. GIROTTI, A. W. & J. B. THOMAS. 1984. J. Biol. Chem. 259: 1744–1752.
22. GIROTTI, A. W. & J. B. THOMAS. 1984. Biochem. Biophys. Res. Commun. 118: 474–480.
23. MINOTTI, G. & S. D. AUST. 1987. J. Biol. Chem. 262: 1098–1104.
24. MINOTTI, G. & J. SCHUBERT. Unpublished observations.
25. MINOTTI, G. & S. D. AUST. 1987. Chem. Phys. Lipids. 44: 191–208.
26. FONG, K. L., P. B. MCCAY, J. L. POYER, B. B. KEELE & H. MISRA. 1973. J. Biol. Chem. 248: 7792–7797.
27. MOREHOUSE, L. A. & S. D. AUST. 1988. Free Rad. Biol. Med 4: 267–277.
28. MOREHOUSE, L. A., C. E. THOMAS & S. D. AUST. 1984. Arch. Biochem. Biophys. 232: 366–377.
29. NAGANO, T. & I. FRIDOVICH. 1985. J. Free Rad. Biol. Med. 1: 39–42.
30. KOSTER, J. F. & R. G. SLEE, 1986. FEBS Lett. 199: 85–88.
31. VEGH, M., A. MARTON & I. HORVATH. 1988. Biochim. Biophys. Acta 964: 146–150.
32. MOREHOUSE, L. A., M. TIEN, J. R. BUCHER & S. D. AUST. 1983. Biochem. Pharmacol. 32: 123–127.
33. SAITO, M., C. E. THOMAS & S. D. AUST. 1985. J. Free Rad. Biol. Med. 1: 179–185.
34. TIEN, M., B. A. SVINGEN & S. D. AUST. 1982. Arch. Biochem. Biophys. 216: 142–151.
35. HOCHSTEIN, P. & L. ERNSTER. 1963. Biochem. Biophys. Res. Commun. 12: 388–394.
36. HOCHSTEIN, P., K. NORDENBRAND & L. ERNSTER. 1964. Biochem. Biophys. Res. Commun. 14: 233–238.
37. SVINGEN, B. A., F. O. O'NEAL, & S. D. AUST. 1978. Photochem. Photobiol. 28: 803–810.
38. BUCHER, J. R., M. TIEN & S. D. AUST. 1983. Biochem. Biophys. Res. Commun. 111: 777–784.
39. AUST, S. D., J. R. BUCHER & M. TIEN. 1984. In Oxygen Radicals in Biology and Medicine. W. Bors, M. Saran & D. Tait, Eds.: 147–154. De Gruyter. German Democratic Republic.
40. BRAUGHLER, J. M., L. A. DUNCAN & R. L. CHASE. 1986. J. Biol. Chem. 261: 10282–10289.
41. BRAUGHLER, J. M., J. F. PREZENGER, R. L. CHASE, L. A. DUNCAN, E. J. JACOBSEN & J. M. MCCALL. 1987. J. Biol. Chem. 262: 10438–10440.
42. BRAUGHLER, J. M., R. L. CHASE & J. F. PREZENGER. 1987. Biochim. Biophys. Acta 921: 457–464.
43. MINOTTI, G. & S. D. AUST. 1987. Free Rad. Biol. Med. 3: 379–387.
44. ULVIK, R. J. 1982. Biochim. Biophys. Acta 715: 42–51.
45. ULVIK, R. J. & I. ROMSLO. 1981. Biochim. Biophys. Acta 635: 457–469.
46. MONTGOMERY, M. R., C. CLARK & J. L. HOLTZMAN. 1974. Arch. Biochem. Biophys. 160: 113–118.
47. THOMAS. C. E. & S. D. AUST. 1985. J. Free Rad. Biol. Med. 1: 293–300.
48. MINOTTI, G. 1988. Arch. Biochem. Biophys. Submitted for publication.
49. MINOTTI, G. 1988. Arch. Biochem. Biophys. In press.
50. MINOTTI, G. 1988. Arch. Biochem. Biophys. Submitted for publication.

DISCUSSION

H. C. BIRNBOIM (*Ottawa Regional Cancer Centre, Ottawa, Ontario, Canada*): Can the ferrous iron released by adriamycin from ferritin or nonferritin iron be recaptured readily by ferritin?

G. MINOTTI: The basic rule of iron "movements" is that mobilization from storages implies Fe^{3+} to Fe^{2+} reduction, whereas accumulation inside the storages implies Fe^{2+} to Fe^{3+} oxidation. Once released from microsomes, Fe^{2+} can be oxidized to Fe^{3+} and readily re-stored inside the membrane storage sites. In the specific case of adriamycin, this would appear not to be the case, since adriamycin is an outstanding Fe^{3+} chelator and competes with ferritin, for example, to maintain the iron outside the membrane.

L. ERNSTER (*University of Stockholm, Stockholm, Sweden*): How do you explain that if you use ADP-chelated iron to initiate microsomal lipid peroxidation you do not need both ferrous and ferric iron for maximal activity?

MINOTTI: Initiation of microsomal lipid peroxidation by ADP-Fe^{2+} is greatly facilitated by the ferroxidase activity of the microsomes. Rapid microsome-induced ADP-Fe^{2+} autoxidation will therefore yield the ferric iron required for initiation.

G. FISKUM (*George Washington University, Washington, DC*): What is the relationship between the pool size of nonheme-nonferritin iron and adriamycin resistance in adriamycin-resistant tumor cells?

MINOTTI: Work is now in progress to answer this question. As a matter of fact we are studying heme metabolism and nonheme iron-heme iron functional relationships in Morris hepatomas with differing degrees of cytochrome expression and nonheme iron disposition. These tumors might provide us with a very simple model to focus on a positive relationship between adriamycin-induced iron mobilization and adriamycin-induced cytoxicity.

N. I. KRINSKY (*Tufts University, Boston, Mass*): What are the kinetics of O_2^- production and Fe^{2+} release from NADPH- and adriamycin-treated microsomes?

MINOTTI: Superoxide generation and reductive mobilization of microsomal iron are simultaneous processes. However, I would point out that the iron is reduced and released by the adriamycin semiquinone radical, which is also formed in conjunction with superoxide. The semiquinone radical of adriamycin stems from the NADPH-supported, P-450 reductase-catalyzed one-electron reduction of the drug.

P. A. RILEY (*University College, London, U.K.*): Could I ask whether your observations on iron availability for reductive mobilization by the adriamycin semiquinone would illuminate the organ-specific toxicity of the drug?

MINOTTI: I would predict that it can. We are now studying iron content and composition in tissues, such as the heart, that are selectively and dramatically sensitive to adriamycin toxicity.

G. DELICONSTANTINOS (*University of Athens, Athens, Greece*): We have reported (Biochem. Pharmacol. **36**: 1153–1161) a direct effect of adriamycin on the membrane fluidity (increase of fluidity) in neural cells; is there a similar correlation with the Fe^{2+} flux system?

MINOTTI: Definitely. Binding of adriamycin to microsomal membranes dominates the redox cycling of the released ferrous iron as well as the effects of SOD and catalase therein. I don't exactly know whether modifications in membrane fluidity have some role in this effect, yet the potential for membranes to regulate the redox behavior of the adriamycin-Fe^{2+} complex is very clear.

C. FRANCESCHI (*Institute of General Pathology, Modena, Italy*): Do you have data on pathologic conditions of iron overloading such as hemochromatosis?

MINOTTI: At this time we have no data on iron overload situations. As shown by Sassa (J. Biol. Chem. **254:** 729–735) some mice strains spontaneously develop hemochromatosis. However, we would rather prefer "moderate" iron overload schedules in which ^{59}Fe can be monitored during its circulation through ferritin or nonheme-nonferritin pools and eventually heme pools, in the presence of adequate inducers. This system might provide help in "looking" at iron during its multifaceted utilization and reductive mobilization processes.

Lipid Peroxidation in Cancer Cells: Chemical and Physical Studies[a]

LANFRANCO MASOTTI, EMANUELA CASALI,
NICOLA GESMUNDO, AND GIORGIO SARTOR

Institute of Biological Chemistry
School of Medicine
University of Parma
43100 Parma, Italy

TOMMASO GALEOTTI AND SILVIA BORRELLO

Institute of General Pathology
School of Medicine
Catholic University
00168 Rome, Italy

MARCO V. PIRETTI AND GIAMPIERO PAGLIUCA

Department of Biochemistry
University of Bologna
40126 Bologna, Italy

INTRODUCTION

Cellular metabolism generates oxygen radicals and related free-radical species as normal byproducts whose steady-state concentration, compatible with cell viability,[1-4] is maintained by an integrated defense system. This comprises proteins and small molecules located in the cytosol and membranes[5] (ceruloplasmin, transferrin, glutathione, ascorbate, vitamin E, β-carotene, coenzyme Q, uric acid, and bilirubin), scavenger enzymes[6] (superoxide dismutase, glutathione peroxidase, and catalase), several redox enzymatic systems (e.g., glutathione reductase, DT-diaphorase), and DNA repair enzymes. Intracellular cytotoxic levels of oxy-radicals that lead to damage of DNA, proteins, lipids and carbohydrates can be reached under conditions of pro-oxidant/antioxidant imbalance. In tumors there seems to be an abnormal oxygen metabolism where oxygen toxicity is attained, chiefly owing to a deficiency of antioxidant defenses.[7]

In recent years, lipid peroxidation in tumors has been the subject of thorough investigation by several research groups. *In vitro* studies have been mainly carried out on cellular membranes isolated from solid as well as systemic tumors, and have established the following facts: (1) Tumor subcellular organelles exhibit a very low degree of "peroxidizability" that has been shown to be related to the growth rate of the tumor.[8-16] (2) Associated with such a low susceptibility to peroxidation are (*a*) changed

[a]This work was supported by grants from AIRC 1987 (L. M. and T. G.) and C. N. R. Special Project "Oncology", Grant No. 87.01346.44 (L. M.) and Grant No. 87.01295.44 (T. G.).

lipid composition of cellular membranes, whose content in polyunsaturated fatty acid (PUFA) is markedly decreased,[17–23] and (*b*) changed static and dynamic properties of the membrane.[24–26] (3) Cellular oxy-radical scavenging enzymes are markedly reduced.[22,27–34]

Many such results have been obtained by employing total microsomal fractions, and the question has been raised as to whether the data could be indicative of the membrane characteristics because of the heterogeneity of the subcellular fraction.[7] In this paper an answer will be provided to this question. Furthermore, the use of microsomal membranes has allowed us to draw interesting conclusions on the changes in composition and susceptibility to lipid peroxidation of a hepatoma that spontaneously increased its growth rate.

All this has brought additional experimental evidence in support of a possible role of the abnormal oxy-radical metabolism in the control of tumor growth: lipid peroxidation products, which are supposed to be greatly diminished in tumors, would become unable to exert the normal control on cell division.[16,22,35,36]

In order to provide a firmer ground for such a role, we then studied the extent of endogenous peroxidation and the chemical nature of its products in hepatoma membranes.

COMPARISON BETWEEN TOTAL, SMOOTH, AND ROUGH ENDOPLASMIC RETICULUM

As pointed out previously, the information on the chemical composition, the structural organization, and the peroxidizability of tumor membranes, as obtained mainly by employing total microsomal fractions, needed to be confirmed on purified, homogenous membrane preparations because microsomes contain Golgi membranes, endoplasmic reticulum, and plasmalemmas.

Thus we compared total, rough, and smooth microsomal membranes isolated from

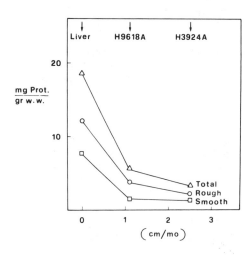

FIGURE 1. Content of microsomal membranes in rat liver and Morris hepatomas as a function of tumor growth rate.

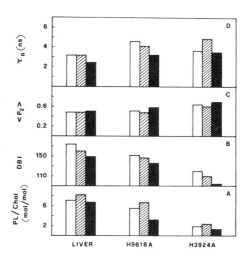

FIGURE 2. Phospholipid/cholesterol ratio (A), double-bond index (B), order parameter (C), and rotational correlation time (D) of DPH in rat liver and Morris hepatoma microsomes. □, total microsomes; ▨, rough endoplasmic reticulum; ■, smooth endoplasmic reticulum.

rat liver and two Morris hepatomas with different growth rates, the fast-growing 3924A and the slow-growing 9618A.[37]

FIGURE 1 shows the microsomal membrane content of rat liver (ACI/T strain) and the two hepatomas. Clearly such content diminishes for all the fractions with increasing growth rate of the tumor. This trend seems of particular relevance in view of a hypothetical role that membrane lipid byproducts, resulting from the oxy-radical attack, might play in the control of tumor proliferating activity.

FIGURE 2 provides a picture of the lipid composition and structural organization of the same three microsomal fractions. The content of phospholipids (PL) decreases with increasing growth rate of the tumor along with the degree of fatty acid unsaturation.

The decreasing trend of the phospholipid content in the total microsomal fraction has been confirmed by gas-liquid chromatography (GLC) (FIG. 3). If we assume that the cholesterol content[38] is constant, the PL/cholesterol ratio decreases from 4.3 in rat liver, to 3 in hepatoma 9618A, and to 1.1 in hepatoma 3924A.

Consistently, the data reported in FIGURE 2 show that the membranes become more ordered, as indicated by the increase of the value of the order parameter of 1,6-diphenyl-1,3,5-hexatriene (DPH), and less fluid, as shown by the increase of the rotational correlation time of the probe. The above data on the purified microsomal fractions therefore support the validity of the conclusions previously drawn from the experiments carried out employing total microsomes.

The alterations in lipid composition and physical properties of the bilayer, described above, could well be the result of a peroxidative insult, particularly effective in the cells of the most undifferentiated tumor (the 3924A). Indeed FIGURE 4 shows that when normal liver microsomal membranes are subjected to a peroxidative stress, their degree of fatty acid unsaturation and molecular order become close to those of the hepatoma 3924A.

The decrease in the enzymatic antioxidant defenses, characteristic of transformed cells, could be the basis for the loss in intracellular membranes, the decrease of the total phospholipid content, and the decrease in the content in PUFA. This chain of

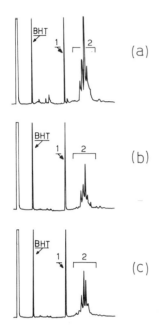

FIGURE 3. GLC profile of microsomal total lipids. The first intense peak after the solvent is BHT (2,6-di-*tert*-butyl-*p*-cresol); peak 1 is cholesterol and the system of peaks, 2, corresponds to diglycerides. Experimental conditions: gas capillary column 9 m coated with SE52 (0.3 mm i.d./ 0.15 μm d.f.); the injection was made on the column at 40°C. The column temperature was programmed at 350°C at a rate of 20°C per min. The carrier gas was H_2, used at a flow rate of 14 ml/min. (**a**) Rat liver; (**b**) H9618A; (**c**) H3924A.

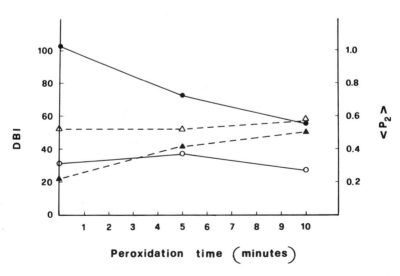

FIGURE 4. Degree of unsaturation of the fatty acid residues and order parameter of total microsomes subjected to peroxidation by xanthine + xanthine oxidase. (●) DBI and (▲) $\langle P_2 \rangle$ of rat liver microsomes; (○) DBI and (△) $\langle P_2 \rangle$ of microsomes from 3924A.

events would be the cause of the very low production of peroxidation products exhibited by cancer cells. If one considers that such byproducts do inhibit division in proliferating cells,[39] the hypothesis that the deficiency of these metabolites causes the tumor cell to lose at least partially the ability to control its growth rate seems to be reasonably supported. This seems in agreement with recent findings that the spontaneous increase in growth rate of the originally very slow-growing Morris hepatoma 9618A results in the loss of microsomal cytochrome P-450, an important intramembranous propagator of lipid peroxidation, and arachidonic as well as docosaesenoic acids. Consequently the hepatoma loses its ability to form peroxidation products (FIG. 5).

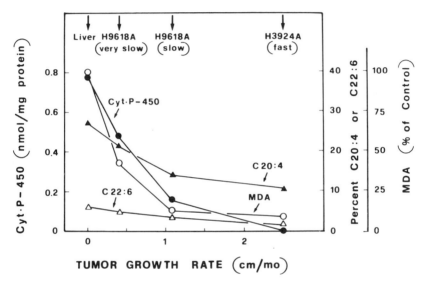

FIGURE 5. Some biochemical parameters of hepatoma microsomes as a function of the growth rate of the tumor. MDA is the measure of lipid peroxidation induced by *tert*-butyl-hydroperoxide.

The working hypothesis has been based on results obtained from studies on completely transformed and malignant cells subjected to exogenous oxidative stress. It is therefore necessary to investigate whether membranes per se contain products of the peroxidative attack, to determine their chemical nature and quantity, and to define their role in interfering and controlling the mechanisms of cell proliferation. Then the investigation would extend to cell populations in the course of proliferation in order to establish a correlation between the degree of deviation of the tumor and the formation of peroxidation byproducts; this would definitely answer the question of whether peroxidation plays a role in tumor growth. We have therefore focused our attention first on the determination of endogenous nonvolatile peroxidation products of PUFA in microsomes of rat liver and Morris hepatomas 9618A and 3924A.

GAS-LIQUID CHROMATOGRAPHY AND GAS-MASS (GC-MS) ANALYSIS
OF ENDOGENOUS LIPID PEROXIDATION PRODUCTS

The method that we have employed[40,41] is the only one so far available that performs the reduction of hydroperoxyl groups and the transmethylation of the polar lipids at the same time and that differs totally from the others by operating in a reducing medium. Thus not only are two reactions carried out in one step, but also at the same time the oxidation of polyunsaturated residues is avoided, an event that takes place at a high rate in the presence of air, especially at drastic operating conditions. It is to be stressed that this method does not produce artifacts during the manipulation of the sample as, to different extents, the other methods do.

After subjecting the sample to alkaline-reducing transmethylation, the fatty acid composition was directly determined by GLC; the results are reported in TABLE 1. The decrease in PUFA on going from the control to the two hepatomas is very evident and is in agreement with the data reported in FIGURE 2.

TABLE 1. Fatty Acid Composition of Total Microsomes from the TLC Band with Rf = 0.58

Fatty Acid	Composition (%)		
	Liver	H9618A	H3924A
16:0	19.60	27.28	24.20
16:1	0.41	4.87	5.47
18:0	26.32	12.82	19.65
18:1Δ9	4.40	20.61	24.04
18:1Δ11	1.69	10.28	4.89
18:2	13.62	6.22	8.91
20:4	26.87	14.08	10.55
24:0	1.33	0.40	0.77
22:6	5.74	3.43	1.52

The material recovered after alkaline-reducing transmethylation was fractionated by thin-layer chromatography (TLC) using a solvent with low eluting power in order to obtain a good resolution of the bands containing the oxidation products. The chromatogram was resolved in four bands, of which the second (RF = 0.11), was analyzed by GLC to identify the nonvolatile oxidation products containing alcohols formed by the reduction of esters during alkaline-reducing transmethylation[41] and hydroxyesters produced by the oxidation of the unsaturated residues. The GLC profile is shown in FIGURE 6. In the chromatogram peak 2 and peak 3 are hydroxymethylesters, whereas the remaining peaks, collectively indicated as 1, correspond to fatty alcohols formed by reduction of the corresponding fatty residues of the diacylglycerides. These last compounds together amount to about 2% of the total methylesters.[40] Peak 3, which is well evident only in liver microsomes, is located in a zone in which the C20:3-OH methylesters are eluted, whereas peak 2 is found in the region of C18:1-OH methylesters. It has to be noted that the C20:3-OH methylesters' band nearly disappears, whereas the C18:1-OH methylesters' band is greatly reduced in the two hepatomas.

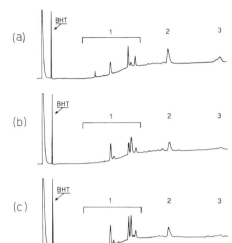

FIGURE 6. GLC profile of the material recovered from the TLC band with Rf = 0.11. Refer to the text for details. (**a**) Rat liver; (**b**) H9618A; (**c**) H3924A.

GC-MS was then employed to identify the chemical structure of such compounds. The analysis has been carried out only in rat liver microsomes because the material from the hepatoma microsomes was not sufficient for the analysis (FIG. 7). The very low amount of the single constituents of the sample allowed only a tentative interpretation of the corresponding mass spectra. Indeed, the fragments M-15, M-31, M-15-32 and M-90 present in spectrum (a) precisely identify the molecular weight of

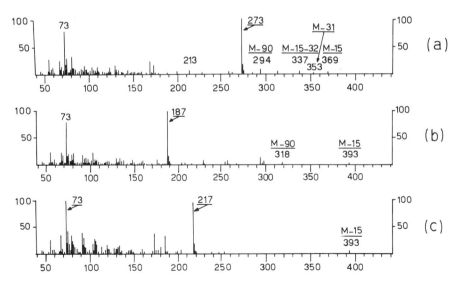

FIGURE 7. Mass spectra of the oxidation products from rat liver microsomes. The mass spectra have been recorded using a Finnigan 1020 GC-MS combination. Peak 2 (**a**) and peak 3 (**b** and **c**) of FIGURE 6.

the compound (384), corresponding to a C18:1-OH; and the intense fragment at $m/z = 273$ indicates that the O-trimethylsilylether (O-TMS) group is bound to C10 (FIG. 8). In the case of spectrum (b) and spectrum (c) the molecular weight can be determined with less certainty, because of a lower number of fragments identifiable in the spectra. Nevertheless the molecular weight can be identified as being equal to 408, corresponding to C20:3-O-TMS. The two intense peaks at $m/z = 187$ and $m/z = 217$ indicate that the O-TMS is bound to C14 in the second compound and to C6 in the third one (FIG. 8). From these data it seems therefore that oxidation occurs specifically in selected positions. We do not know at present the mechanism of oxidation, it being possible that the oxy-derivative is formed either by formal addition of H_2O, or by direct

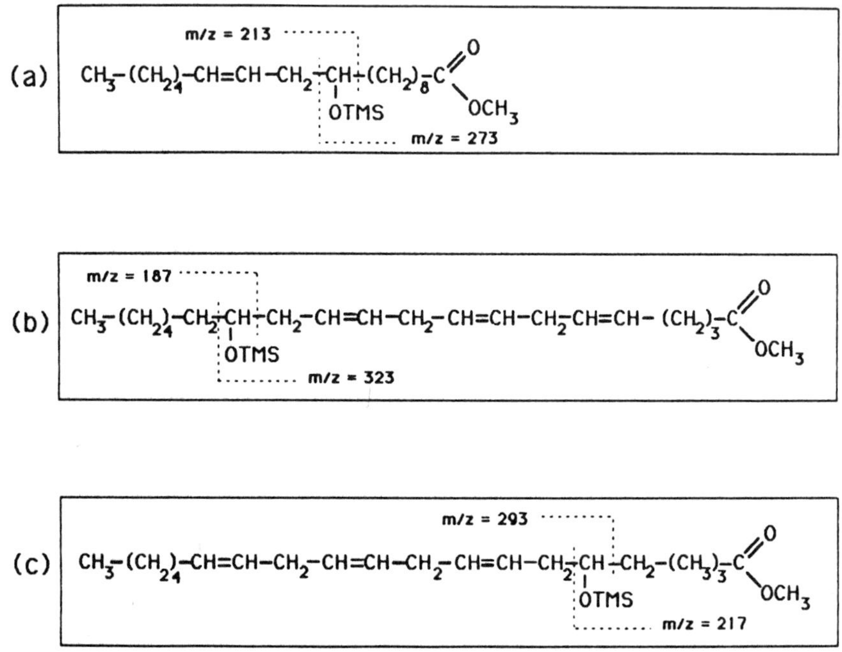

FIGURE 8. Fragmentation scheme of the mass spectra reported in FIGURE 7.

attack of OH·, or by the reduction, under our analytic conditions, of an hydroperoxyl group specifically formed. Moreover we have failed, at the present stage of our investigation, to find evidence of hydroxyl derivatives of low molecular weight aldehydes.

SUMMARY

Our studies on the biochemical composition and the structural organization of smooth and rough endoplasmic reticulum isolated from Morris hepatomas 9618A and

3924A confirm the results obtained employing the total microsomal fraction. We have definitely established the following facts: (1) Tumor subcellular organelles exhibit the very low degree of peroxidizability that has been shown to be related to the growth rate of the tumor. (2) Associated with such a low susceptibility to peroxidation are (*a*) changed lipid composition of cellular membranes, whose content in polyunsaturated fatty acid is markedly decreased, and (*b*) changed static and dynamic properties of the membrane. Previously it was also found that cellular oxy-radical scavenging enzymes are markedly reduced.

From these data, it is possible to infer that tumor membranes are altered structurally and functionally in part as the result of an oxy-radical-induced damage that occurs *in vivo* under conditions of oxygen toxicity. This seems to be supported by recent findings that the spontaneous increase in growth rate of the originally very slow-growing Morris hepatoma 9618A results also in the loss of cytochrome P-450 (an important intramembraneous propagator of lipid peroxidation) as well as of C20:4 and C22:6. Studies performed by GLC and GC-MS on the fatty acid residues of phospholipids of rat liver microsomes show the presence of C20:3-OH and C18:1-OH, but no hydroxyl derivatives of low molecular weight aldehydes. The hyroxyl derivatives of arachidonic acid and linoleic acid are present in much smaller amounts in the microsomes isolated from H9618A and H3924A.

REFERENCES

1. OSHINO, N., B. CHANCE, H. SIES & T. BUCHER. 1973. The role of H_2O_2 generation in perfused rat liver and the reaction of catalase compound I and hydrogen donors. Arch. Biochem. Biophys. **154**: 117–131.
2. TYLER, D. D. 1975. Polarographic assay and intracellular distribution of superoxide dismutase in rat liver. Biochem. J. **147**: 493–504.
3. BOVERIS, A., E. CADENAS & B. CHANCE. 1981. Ultraweak chemiluminescence: A sensitive assay for oxidative radical reactions. Fed. Proc. **40**: 195–198.
4. PRYOR, W. A. 1986. Oxy-radicals and related species: Their formation, lifetimes and reactions. Annu. Rev. Physiol. **48**: 657–667.
5. SIES, H. 1986. Biochemistry of oxidative stress. Angew. Chem. Int. Ed. Engl. **25**: 1058–1071.
6. FREEMAN, A. B. & J. D. CRAPO. 1982. Biology of disease: Free radicals and tissue injury. Lab. Invest. **47**: 412–426.
7. GALEOTTI, T., S. BORRELLO, G. MINOTTI & L. MASOTTI. 1986. Membrane alterations in cancer cells: The role of oxy radicals. Ann. N. Y. Acad. Sci. **488**: 468–480.
8. LASH, E. D. 1966. The antioxidant and prooxidant activity in ascites tumors. Arch. Biochem. Biophys. **115**: 332–336.
9. BURLAKOVA, E. B. 1975. Bioantioxidants and synthetic inhibitors of radical processes. Russ. Chem. Rev. **44**: 871–880.
10. PLAYER, T. J., D. J. MILLS & A. A. HORTON. 1977. NADPH-dependent lipid peroxidation in mitochondria from livers of young and old rats and from rat hepatoma D30. Biochem. Soc. Trans. **5**: 1506–1508.
11. PLAYER, T. J., D. J. MILLS & A. S. HORTON. 1979. Lipid peroxidation of the microsomal fraction and extracted microsomal lipids from the DAB-induced hepatomas. Br. J. Cancer **39**: 773–778.
12. BURLAKOVA, E. B., E. M. MOLOCHKINA & N. P. PAL'MINA. 1980. Role of membrane lipid oxidation in control of enzymatic activity in normal and cancer cells. Adv. Enzymol. Regul. **18**: 163–179.
13. DIANZANI, M. U., R. A. CANUTO, M. A. ROSSI, G. POLI, R. GARCEA, M. E. BIOCCA, G. CECCHINI, F. BIASI, M. FERRO & A. M. BASSI. 1984. Further experiments on lipid peroxidation in transplanted and experimental hepatomas. Toxicol. Pathol. **12**: 189–199.

14. SHARMA, S. C., R. J. SCHAUR, H. M. TILLIAN & E. SCHAUENSTEIN. 1984. Low inducibility of lipid peroxidation by ferrous ions in two rat ascites hepatoma lines. IRCS Med Sci. Pharmacol **12:** 236–237.
15. CHEESEMAN, K. H., M. COLLINS, K. PROUDFOOT, T. F. SLATER, G. W. BURTON, A. C. WEBB & K. U. INGOLD. 1986. Studies on lipid peroxidation in normal and tumour tissues. Biochem. J. **235:** 507–514.
16. BARTOLI, G. M. & T. GALEOTTI. 1979. Growth-related lipid peroxidation in tumour microsomal membranes and mitochondria. Biochim. Biophys. Acta **574:** 537–541.
17. FEO, F., R. A. CANUTO, G. BERTONE, R. GARCEA & P. PANI. 1973. Cholesterol and phospholipid composition of mitochondria and microsomes isolated from Morris hepatoma 5123 and rat liver. FEBS Lett. **33:** 229–232.
18. HOSTETLER, K. Y., B. D. ZENNER & H. P. MORRIS. 1976. Abnormal membrane phospholipid content in subcellular fractions from the Morris 7777 hepatoma. Biochim. Biophys. Acta **441:** 231–238.
19. REITZ, R. C., J. A. THOMPSON & H. P. MORRIS. 1977. Mitochondrial and microsomal phospholipids of Morris hepatoma 7777. Cancer Res. **37:** 561–567.
20. MORTON, R., M. WAITE, J. W. HARTZ, C. CUNNINGHAM & H. P. MORRIS. 1978. The composition and metabolism of microsomal and mitochondrial membrane lipids in the Morris 7777 hepatoma. *In* Advances in Experimental Medicine and Biology. H. P. Morris & W. E. Criss, Eds. **92:** 381–403. Plenum. New York.
21. HOSTETLER, K. Y., B. D. ZENNER & H. P. MORRIS. 1979. Phospholipid content of mitochondrial and microsomal membranes from Morris hepatomas of varying growth rates. Cancer Res. **39:** 2978–2983.
22. BARTOLI, G. M., S. BARTOLI, T. GALEOTTI & E. BERTOLI. 1980. Superoxide dismutase content and microsomal lipid composition of tumors with different growth rates. Biochim. Biophys. Acta **620:** 205–211.
23. BORRELLO, S., G. MINOTTI, G. PALOMBINI, A. GRATTAGLIANO & T. GALEOTTI. 1985. Superoxide-dependent lipid peroxidation and vitamin E content of microsomes from hepatomas with different growth rates. Arch. Biochem. Biophys. **238:** 588–595.
24. GALEOTTI, T., S. BORRELLO, G. PALOMBINI, L. MASOTTI, M. B. FERRARI, P. CAVATORTA, A. ARCIONI, C. STREMMENOS & C. ZANNONI. 1984. Lipid peroxidation and fluidity of plasma membranes from rat liver and Morris hepatoma 3924A. FEBS Lett. **169:** 169–173.
25. CAVATORTA, P. L. MASOTTI, G. SARTOR, M. B. FERRARI, E. CASALI, S. BORRELLO, G. MINOTTI & T. GALEOTTI. 1985. Lipid peroxidation of microsomes from Morris hepatoma 3924A: Order parameter and dynamic properties determined by fluorescence anisotropy decay. *In* Cell Membranes and Cancer. T. Galeotti, A. Cittadini, G. Neri, S. Papa & L. A. Smets, Eds.: 269–274. Elsevier. Amsterdam.
26. MASOTTI, L., P. CAVATORTA, E. CASALI, A. ARCIONI, C. ZANNONI, S. BORRELLO, G. MINOTTI & T. GALEOTTI. 1986. O_2^--dependent lipid peroxidation does not affect the molecular order in hepatoma microsomes. FEBS Lett. **198:** 301–306.
27. YAMANAKA, N. Y. & D. DEAMER. 1974. Superoxide dismutase activity in WI-38 cell cultures. Effects of age, trypsinization, and SW-40 transformation. Physiol. Chem. Phys. **6:** 95–106.
28. DIONISI, O., T. GALEOTTI, T. TERRANOVA & A. AZZI. 1975. Superoxide radicals and hydrogen peroxide formation in mitochondria from normal and neoplastic tissues. Biochim. Biophys. Acta **403:** 292–300.
29. BOZZI, A., I. MAVELLI, A. FINAZZI-AGRO', R. STROM, A. M. WOLF, B. MONDOVI' & G. ROTILIO. 1976. Enzyme defense against reactive oxygen derivatives. II. Erythrocytes and tumour cells. Mol. Cell. Biochem. **10:** 11–16.
30. PESKIN, A. V., I. B. ZBARSKI & A. A. KOSTANTINOV. 1976. An examination of the superoxide dismutase activity in tumour tissue. Dokl. Acad. Nauk SSSR **229:** 751–754.
31. PESKIN, A. V., Y. M. KOEN, I. B. ZBARSKI & A. A. KONSTANTINOV. 1977. Superoxide dismutase and glutathione peroxidase activities in tumours. FEBS Lett. **78:** 41–45.
32. BIZE, I. B., L. W. OBERLEY & H. P. MORRIS. 1980. Superoxide dismutase and superoxide radical in Morris hepatomas. Cancer Res. **40:** 3686–3693.
33. BARTOLI, G. M., T. GALEOTTI, S. BORRELLO & G. MINOTTI. 1982. Loss of defensive enzymes and oxygen-mediated damage of microsomal membranes in liver and hepato-

mas. *In* Membranes in Tumour Growth. T. Galeotti, A. Cittadini, G. Neri & S. Papa, Eds.: 461–470. Elsevier. Amsterdam.

34. MOCHIZUCHI, Y., Z. HRUBAN, H. P. MORRIS, A. SLESERS & E. L. VIGIL. 1971. Microbodies of Morris hepatomas. Cancer Res. **31:** 763–773.

35. BARTOLI, G. M., G. PALOMBINI & T. GALEOTTI. 1978. Lipid peroxidation in tumour microsomal membranes. 12th FEBS Meeting Dresden, Abstract 2424.

36. MASOTTI, L., E. CASALI & T. GALEOTTI. 1988. Lipid peroxidation in tumour cells. Free Rad. Biol. Med. **4:** 377–386.

37. MORRIS, H. P. & B. P. WAGNER. 1968. Induction and transplantation of rat hepatomas with different growth rate (including "minimal deviation" hepatomas). *In* Methods in Cancer Research. H. Busch, Ed. **4:** 125–152. Academic Press. New York.

38. MASOTTI, L., P. CAVATORTA, G. SARTOR, A. ARCIONI, C. ZANNONI, G. M. BARTOLI & T. GALEOTTI. 1982. A fluorescence depolarization investigation of membranes from tumour cells with different growth rate. *In* Membranes in Tumour Growth. T. Galeotti, A. Cittadini, G. Neri & S. Papa, Eds.: 39–50. Elsevier. Amsterdam.

39. CORNWELL. D. G. & N. MARISAKI. 1984. Fatty acid paradoxes in the control of cell proliferation: prostaglandins, lipid peroxides and co-oxidation reactions. *In* Free Radicals in Biology. W. A. Pryor, Ed. **6:** 95–148. Academic Press. New York.

40. PIRETTI, M. V., G. PAGLIUCA & M. VASINA. 1987. Proposal of an analytical method for the study of the oxidation products of membrane lipids. Anal. Biochem. **167:** 358–361.

41. PIRETTI, M. V., G. PAGLIUCA & M. VASINA. 1988. Transmethylation of neutral and polar lipids with $NaBH_4$ in the presence of NaOH. Chem. Phys. Lipids **47:** 149–153.

DISCUSSION

B. SZWERGOLD (*Fox Chase Cancer Center, Philadelphia, Pa.*): Were the levels of arachidonate comparable in the several cells you looked at?

L. MASOTTI (*University of Parma, Parma, Italy*): No; in fact it may be that the lowered levels of the oxidation products of arachidonate may simply reflect a lower concentration of the substrate.

C. FRANCESCHI (*University of Modena, Modena, Italy*): Are you confident, possibly on the basis of other data in the literature, that your observations may be generalized to other tumors? Did you make similar studies on liver from aged animals?

MASOTTI: No, I would not generalize, at this stage of research, our observations to tumors other than hepatomas. As for the second question we did not study liver in aging, but I believe that our method of lipid analysis would be useful to extend the knowledge on membrane lipid composition of aged cells.

L. ERNSTER (*University of Stockholm, Stockholm, Sweden*): Have you measured the cytochrome b_5-dependent fatty acid desaturase activity of the tumor microsomes? Could the low PUFA content of the tumor microsomes be due to a diminished fatty acid desaturase activity?

MASOTTI: No, it has not been measured. Among the other factors mentioned so far, a diminished desaturase activity could contribute to the decrease in PUFA.

H. WOHLRAB (*Boston Biomedical Research Institute, Boston, Mass.*): Do you have an explanation for the increased growth rate of the very slow hepatoma to an intermediate growth rate with time?

MASOTTI: This type of evolution occurs often in highly differentiated hepatomas, such as 9618A, because they tend to dedifferentiate spontaneously.

N. KRINSKY (*Tufts, University, Boston, Mass.*): Could you comment on the reports of Mountfond and Wright on the presence of neutral lipids in tumor membranes?

MASOTTI: We did not find neutral lipids in microsomal membranes. They have been recently reported as being present in plasma membranes of activated and transformed cells in a sort of "enclave" made, if I recall it correctly, by phospholipids that would provide the wall for the structure. I find this report quite interesting even if I think that the structural and functional significance of their presence has to be established. Conceivably they might be important, for example, in metastases.

G. DELICONSTANTINOS (*University of Athens, Athens, Greece*): Are there any other tumor cell types which express increased rigidity other than hepatomas and Hodgkin T cell?

MASOTTI: All solid tumors investigated so far show an increased molecular order and a decreased fluidity of cellular membranes. In membranes of leukemia cells fluidity increases.

Oxidative Stress and Cellular Differentiation

R. S. SOHAL,[a,b] R. G. ALLEN,[c] AND C. NATIONS[a]

[a]Department of Biological Sciences
Southern Methodist University
Dallas, Texas 75275

[c]Laboratory of Investigative Dermatology
The Rockefeller University
New York, New York 10021

INTRODUCTION

It is axiomatic that the transformation of cells from one phenotype to another depends, ultimately, on the expression of specific genes. Several types of extracellular and intracellular inductive signals that modulate gene expression during embryogenesis and carcinogenesis have been recognized.[1] This article describes the results of our studies on the differentiation of microplasmodia of the syncytial slime mold *Physarum polycephalum* into spherules. Our findings implicate the involvement of oxidative stress as a causal factor in differentiation. Examples from the literature are provided to support our view that oxidative stress may be a general rather than an isolated phenomenon during cellular differentiation.

LIFE CYCLE OF *PHYSARUM*

The life cycle of *Physarum polycephalum,* shown in FIGURE 1, alternates between haploid and diploid forms, depending upon environmental conditions.[2] The plasmodium, which is a multinucleated syncytium, can differentiate into sporangia that undergo meiosis to form spores. The haploid spores germinate to form amebae, which can also reversibly transform into flagellated swarmers. Pairs of amebae or swarmers can fuse to form a diploid zygote. The zygote then develops into a plasmodium. The transformation of plasmodium into sporangia occurs in response to starvation, in the presence of light. Plasmodia can also undergo a diploid form of differentiation in response to starvation but in the absence of light. During this type of differentiation, referred to as spherulation, the plasmodium cleaves into clusters of hard-walled spherules. Each cluster is a sclerotium. In a shake-flask culture, a plasmodium breaks up into numerous microplasmodia, which spherulate to form microsclerotia upon transfer to a non-nutrient salts-only medium. Spherules germinate to form microplasmodia in shaker cultures containing nutrient medium.

[b]The research work of R.S.S. is supported by Grant RO1 AG7657 from the N.I.H.-N.I.A.

PHYSARUM POLYCEPHALUM AS A MODEL OF DIFFERENTIATION

The transformation of microplasmodia into spherules (FIG. 2) in a salts-only starvation medium has been widely used as a model for the study of events related to cellular differentiation.[3,4] This type of differentiation occurs in nature under environmentally stressful conditions akin to those in mammalian cell cultures with an inadequate supply of serum. In the laboratory, spherulation is usually induced by the transfer of plasmodia from a nutritionally balanced medium to the salts-only medium.

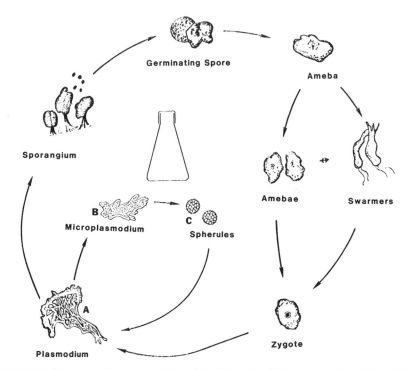

FIGURE 1. Diagrammatic representation of the life cycle of *Physarum polycephalum*. Plasmodia (**A**) grown in shake-flasks break up into microplasmodia (**B**), which upon transfer to salts-only medium differentiate into spherules (**C**).

During spherulation, cell walls are secreted as the synthesis of a variety of spherule-specific proteins from newly transcribed mRNA proceeds.[5,6] The transition of microplasmodia to spherules can be detected microscopically by the appearance of cell walls; this allows easy quantification of the rate of differentiation. The transition is synchronous; therefore all biochemical events involved in this form of differentiation are amplified. Another advantage of *Physarum* as a model of differentiation is that a control strain (LU887 × LU897, "white" or W strain) which fails to differentiate in the salts-only medium has been characterized.[8,9] Furthermore, it is also possible to

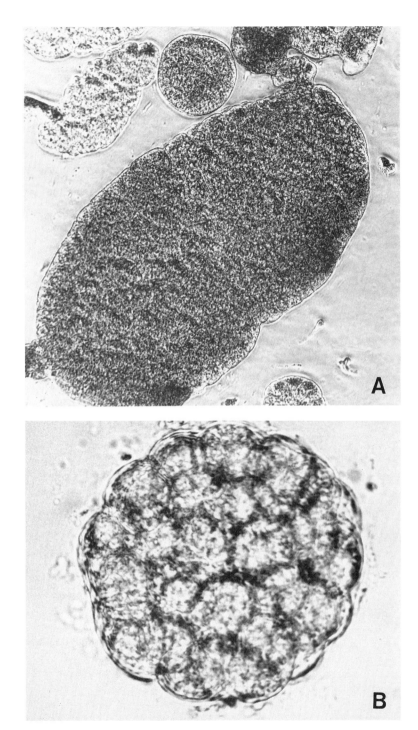

FIGURE 2. Phase-contrast photomicrograph of (A) microplasmodium (original magnification ×750) and (B) microsclerotium (×1200). (FIGURE 2B is from Henney.[77] Reproduced by permission.)

create heterokaryons by fusing nondifferentiating and differentiating strains, without the use of chemical agents, thus aiding in the identification of specific factors affecting the process of differentiation.[10,11]

FIGURE 3 compares the rates of differentiation in different stains of *Physarum* after transfer from the nutritive growth medium to the non-nutrient salts medium. The rate of differentiation is highest in the Y (M_3cVII) strain, whereas the W strain does not differentiate under these conditions and is not depicted in the figure. The heterokaryon (H) formed by the fusion by Fy (LU887 × LU836), a differentiating strain, and W, the nondifferentiating strain, showed a relatively slow rate of differentiation.[12]

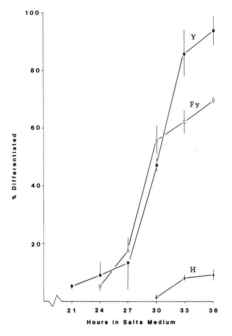

FIGURE 3. Time course of differentiation of various *Physarum* strains after transfer from growth to salts medium. Y and Fy are differentiating strains, whereas the W strain is a nondifferentiation strain and is thus not shown here. The heterokaryon (H) was obtained by the fusion of Fy and W strain. The code name of the Y strain is M_3cVII, the W strain is LU887xLU897, and the Fy strain is LU887xLU863. (From Allen *et al.*[12] Reproduced by permission.)

INDICATIONS OF OXIDATIVE STRESS AS A CORRELATE OF SPHERULATION

A variety of co-ordinated changes, described below, occur during the spherulation of *Physarum* which, taken together, indicate that the level of oxidative stress increases during this process. The rate of oxygen consumption decreases during the starvation-induced differentiation of the Y and Fy strains. But the rate of cyanide-insensitive oxygen consumption increases during spherulation[12]; cyanide-insensitive respiration is an indirect indicator of the generation of partially reduced oxygen species.[13] The W strain does not show an increase in cyanide-insensitive respiration. The activity of cyanide-resistant superoxide dismutase (SOD) activity increased up to 46-fold during spherulation (FIG. 4).[12] This is probably in response to an upsurge in O_2^- generation.

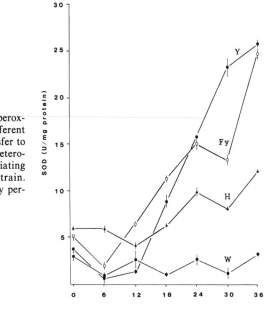

FIGURE 4. Cyanide-insensitive superoxide dismutase (SOD) activity in different strains of *Physarum* following transfer to salts-only starvation medium. H = heterokaryon formed by fusion of differentiating Fy and nondifferentiating W strain. (From Allen *et al.*[12] Reproduced by permission.)

The concentration of reduced glutathione (GSH) decreases by 90% during spherulation (FIG. 5). The concentrations of H_2O_2 and lipid peroxides increase significantly.[12,14] The rate of differentiation of the Y and the Fy strains parallels the rate of changes in their SOD activities and correlates inversely with their GSH content.[12,14] The nondifferentiating W strain does not exhibit any of these biochemical changes following its transfer to the salts-only medium.[12,14,15] Also, the heterokaryon, formed by

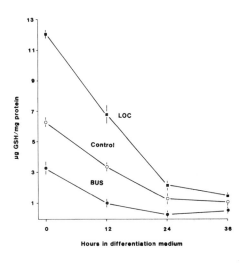

FIGURE 5. Changes in GSH concentration in cells of the Y strain treated with 4 mM L-2-oxothiazolidine-4-carboxylate (LOC) or 4 mM buthionine sulfoximine (BSO). (From Allen *et al.*[14] Reproduced by permission.)

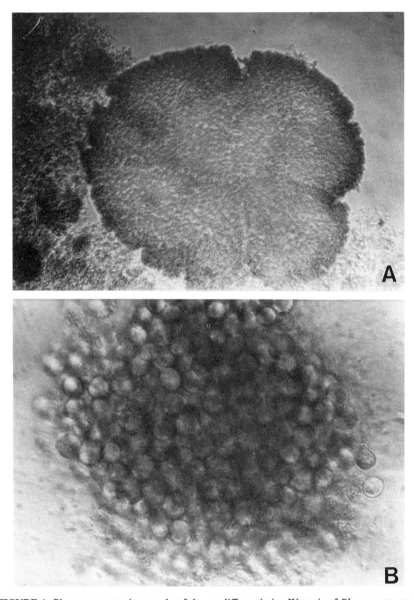

FIGURE 6. Phase-contrast micrographs of the nondifferentiating W strain of *Physarum* treated with (**A**) inactive SOD and cultured in salts medium for 11 days (magnification ×200), (**B**) SOD-containing liposomes and cultured in salts medium 8 days (magnification ×700), and (**C**) D-amino acid oxidase-containing liposomes and cultured 8 days in salts medium (magnification ×700). (From Allen *et al.* Reproduced by permission.)

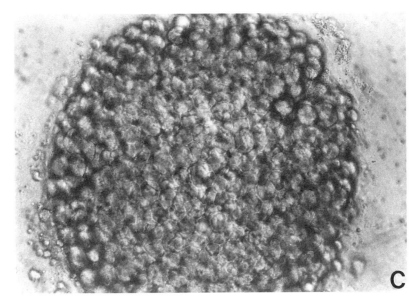

FIGURE 6. (continued)

the fusion of the differentiating Fy and the nondifferentiating W strain, exhibits a relatively slow rate of differentiation and a correspondingly retarded pace of changes in SOD, GSH, and peroxides.[12] These biochemical changes are interpreted by us to be manifestations of an increase in the level of oxidative stress during spherulation.

EFFECTS OF EXPERIMENTAL VARIATIONS IN OXIDATIVE STRESS ON SPHERULATION

To determine whether spherulation is linked to elevated oxidative stress, the levels of the latter were experimentally varied in the Y strain by altering the endogenous concentration of GSH and by exposure of microplasmodia to paraquat, a known generator of O_2^-. To decrease the level of GSH, which would enhance the level of oxidative stress, L-buthionine-SR-sulfoximine (BSO), an inhibitor of γ-glutamylcysteine synthetase, was added to the growth as well as to the salts-only differentiation medium. To increase GSH concentration and thus reduce the level of oxidative stress, L-2-oxothiazolidine-4-carboxylate (LOC), which stimulates GSH synthesis, was added to both the growth and salts medium. At the time of transfer to the salts-only differentiation medium (0 hr), the GSH concentration in LOC-treated cultures was twice the GSH level of controls, whereas BSO-treated cultures had only about half the amount present in the controls (FIG. 5). GSH concentration decreased 80 to 90% in all cultures, including untreated cultures as well as those exposed to BSO or LOC, following transfer from growth to the salts-only medium. As compared to the controls, the rate of spherulation was faster in BSO-treated cells and slower in LOC-exposed cells. SOD activity during spherulation increased more rapidly in the BSO-treated cultures and more slowly in LOC-treated cultures than in the controls. Thus, the rate at which SOD activity increased paralleled the rate of spherulation and the rate of spherulation was inversely related to the concentration of GSH.[14]

The herbicide paraquat (1,1-dimethyl-4,4-dipyridium dichloride) is widely believed to exert its toxic effects by the generation of O_2^-. Effects of paraquat exposure on spherulation were studied in the Y strain and the nondifferentiating W strain. Paraquat-treated Y cultures exhibited higher rates of spherulation and a greater elevation of SOD activity than did the controls. The inductive potency of paraquat was highest 8 hours after the transfer of plasmodia to the differentiation medium, which suggested the existence of a temporal phase of sensitivity to oxidative stress. Paraquat-treated W strain microplasmodia exhibited some increase in SOD activity, but no differentiation. However, if the paraquat treatment of the W strain was preceded by exposure to BSO in the growth medium, the plasmodia transformed into apparently immature spherules; mature hard-walled spherules were never observed.[15]

INDUCTION OF SPHERULATION IN A PREVIOUSLY NONDIFFERENTIATING STRAIN

To further investigate the role of oxidative stress and SOD as causal factors in differentiation, Cu-Zn SOD was introduced into the previously nondifferentiating W strain of *Physarum* via liposomes. The treatment with liposomes resulted in a 4.5-fold increase in SOD activity as compared to control cultures exposed to empty liposomes. After 8 days in the salts-only medium, none of the control cultures survived, whereas a proportion of the cells in SOD-treated cultures had begun to differentiate into spherules (FIG. 6).[16] If the cultures had been previously treated with 1 mM BSO throughout the growth phase, which caused an approximate 70% decrease in GSH, and then treated with SOD-containing liposomes in the salts medium, a greater percentage of cells exhibited differentiation as compared to cultures with no prior exposure to BSO. BSO alone did not induce differentiation.[16] Because of the protective role of SOD in oxygen toxicity, we recognized that SOD-treatment of cultures might result in either an increased or a decreased level of oxidative stress; the mechanism by which SOD acts in the induction of spherulation in *Physarum* might involve either the elimination of O_2^- or an increased generation of H_2O_2. Direct addition of H_2O_2 to salts medium was found to be ineffective for exploring this problem because the medium rapidly degraded H_2O_2 because of the presence of metallic ions; however, the liposomal addition of D-amino acid oxidase to the microplasmodia, with a subsequent transfer to a medium containing D-methionine, increased the intracellular concentration of H_2O_2 more than 5-fold and also induced the differentiation of about 10% of the W cells. Pretreatment with BSO approximately doubled the yield of differentiated material. No differentiation was induced by liposomes that contained inactive D-amino acid oxidase or by salts medium that contained no D-methionine.[16]

Strong oxidants such as plumbagin and cumene hydroperoxide (200 μM) also induced differentiation in 30–50% of the cells of the W strain of *Physarum*. In contrast, none of the antioxidants, such as tocopherol, β-carotene, butylated hydroxytoluene, ascorbate or dithiothreitol could induce differentiation.[16] It seems that augmentation in SOD activity during differentiation contributes to the increase in oxidative stress by generating H_2O_2. It has also been reported by Elroy-Stein *et al.*[17] that overproduction of SOD, achieved by the insertion of additional SOD genes, resulted in an increase in the level of oxidative stress.

RELATIONSHIP BETWEEN OXIDATIVE STRESS, CALCIUM, AND DIFFERENTIATION

Increase in oxidative stress has been shown to cause the release of calcium into the cytosol and the concentration of free cytosolic calcium.[18–21] In several cell types, calcium has been identified as an inducer of differentiation[22,23]; therefore, we hypothe-

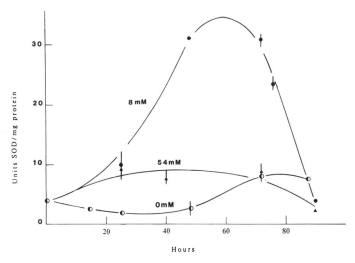

FIGURE 7. Effect of $CaCl_2$ on the activity of cyanide-resistant SOD activity in Y strain *Physarum* microplasmodia in salts-only differentiation medium containing 0 mM, 8 mM and 54 mM $CaCl_2$. (From Nations *et al.* Reproduced by permission.)

sized that oxidative stress may be affecting cellular differentiation by inducing an increase in the concentration of cytosolic calcium or by altering its intracellular distribution. As also mentioned above, spherulation of the Y strain is normally induced by transfer from the growth medium, which contains no added $CaCl_2$, to the salts-only starvation medium, which contains 8 mM $CaCl_2$. We have found that differentiation in the Y strain can also be induced in the growth medium by the addition of high concentrations of $CaCl_2$ (54 mM), although spherulation proceeds at a slower rate than in the salts medium. It was also found that microplasmodia transferred to salts medium that contained no $CaCl_2$ failed to spherulate. Deletion of other salts, including $MgCl_2$, $MnCl_2$ and $ZnCl_2$, had no effect on differentiation. The rate of differentiation was quite similar in salts media containing 1.6 and 5.5 mM $CaCl_2$, but was much slower in the medium containing 54 mM $CaCl_2$.

A surprising finding was that addition of 54 mM $CaCl_2$ to either growth or salts medium induced differentiation in the W strain, which had not previously been observed to differentiate in the normal salts-only medium, which contains 8 mM $CaCl_2$. Most notably, the striking increase in SOD activity and decrease in GSH concentration that accompany differentiation do not occur in salts media containing 54 mM $CaCl_2$ (FIG. 7). If microplasmodia of Y or W strains are cultured in a nutrient medium containing 8 mM $CaCl_2$ they will differentiate in the normal salts-only medium, but the increase in SOD activity accompanying differentiation does not occur.[24] We have routinely used the simplified nutrient medium of Brewer and Pryor.[25]

The results of these experiments can be summarized as follows: oxidative stress induces spherulation only if adequate amounts of calcium are available. Differentiation can be induced by calcium alone, albeit in high concentrations, without a concomitant increase in oxidative stress. These results suggest that an increase in oxidative stress during differentiation is a physiological phenomenon that functions to mobilize calcium, which in turn apparently acts as the underlying signal for spherulation. The striking ability of *Physarum* to accumulate calcium has been demonstrated by Achenbach *et al.*[26]

OTHER EXAMPLES OF ALTERATIONS IN OXIDATIVE STRESS DURING DIFFERENTIATION AND DEVELOPMENT

Changes in enzymic and nonenzymic antioxidant defenses have been noted in a variety of disparate systems during cell state transitions.[27] Tumor cells, which are believed to have regressed from a fully differentiated state, exhibit lower SOD activity than do normal cells of similar tissue origin.[28-31] SOD activity increases markedly during the development of acellular slime molds,[32] insects,[33,34] amphibians,[35] and mammals.[36-40]

It is known that elevation of ambient oxygen tension can stimulate the activity of SOD in certain tissues of adult animals,[41] and it has been hypothesized that removal of developing organisms from their *in utero* environment, at birth, into a relatively high-oxygen atmosphere may be the cause of subsequent increases in SOD activity.[42,43] Consistent with this hypothesis is the observation that the activity of SOD is much more readily induced by oxygen in the tissues of developing organisms and newborns than in adults.[42-47] However, birth and the consequent sudden change in concentration of available oxygen is a phenomenon occurring only in higher organisms; development-associated changes in SOD activity are also observed in simpler organisms that are not subjected to any sudden increase in the concentration of ambient oxygen in their life cycle. Catalase activity has also been reported to increase during development in insects,[48] amphibians,[49,50] and mammals.[39,40,51,52] The activity of glutathione peroxidase, which eliminates H_2O_2, has also been shown to increase during development in several different mammalian organs.[53-56]

GSH concentration has been generally found to be relatively high in rapidly proliferating cells, regenerating tissues, and tumor cells, and it often decreases in cells exhibiting a decline in mitotic activity. For example, GSH concentration declines by 60% in houseflies during metamorphosis[57] and 50% in developing rat lung.[58] A loss of

GSH has also been reported in both mouse myoblasts and in preadipocytes induced to differentiate in culture.[59] GSH concentration increases during the mitotic phase of regeneration and decreases during redifferentiation of regenerating tissues in amphibians.[60] Regeneration also leads to an increased chemical reducing capacity in mammalian tissues.[61,62] Cells that differentiate into tissues that retain a high regenerative capacity, such as liver tissue, exhibit no loss of GSH during development; in fact, GSH concentration can increase in some cases.[62] GSH concentration increases sharply during the development of meiotic tissues.[63]

Several studies have indicated that experimental variations in oxidative stress can modulate the process of differentiation and morphogenesis. For example, during development the phenotype can be alternately switched in the fungus *Mucor racemosus*,[64] chick bone and cartilage,[65] and mouse neuroblastema cells[66] by varying ambient oxygen concentration. Cultured mouse neuroblastema cells undergo morphologic and biochemical differentiation following 3 days of exposure to 80% oxygen. Differentiation is not blocked by KCN or lipid-soluble antioxidants such as vitamin E and butylated hydroxytoluene.[66]

Insects are remarkable for their tolerance of hypoxic conditions. The rate of (resting) oxygen consumption remains unchanged until ambient oxygen concentration is lowered to 5% or even below. Flour beetles, *Tenebrio moliter,* raised at 10.5% ambient oxygen concentration showed an order of magnitude higher rate of abnormal metamorphosis as compared to those kept at 21% oxygen concentration. Under hypoxic conditions up to one-third of the pupae formed pupae-adult intermediates, which possess a mixture of pupae and adult features.[67] It has been shown that the rate of O_2-production of cells is proportional to ambient oxygen concentration.[68] The developmental abnormality observed in *Tenebrio* is apparently not related to rate of oxygen utilization since hypoxia has no effect on oxygen uptake.[67] A decrease in the level of O_2-production and, inferentially, of oxidative stress may have been a factor in the inability of many *Tenebrio* pupae to undergo complete metamorphosis.

Agents such as X-irradiation, benzo [*a*] pyrene, and phorbol ester, which generate oxygen free radicals, have been shown to effectively induce transdifferentiation of cultured cells.[69,70] Buthioninine sulfoximine, an inhibitor of glutathione synthesis, enhances the rate of embryogenesis in carrot parenchyma cultures.[71] This BSO effect is similar to that observed in spherulating *Physarum.*[14] Conversely, antioxidants such as dimethylsulfoxide and dihydroxybenzoic acid have been found to delay the rate of differentiation in mouse neuroblastema cells[72] and to delay development in *Drosophila,*[73] respectively.

SPECULATIONS AND CONCLUSIONS

On the basis of our studies on *Physarum* and those conducted in other models it is evident that oxidative stress and calcium are causal factors in cellular differentiation in several systems. The primary question which arises from these studies is the nature of the mechanism by which these factors influence the pattern of gene expression. It has been demonstrated in mouse epidermal cells that active oxygen species, produced extracellularly by xanthine/xanthine oxidase, cause a significant increase in the rate of transcription of *c*-fos and β-actin genes.[74] The expression of the *c*-fos oncogene has also been shown to be inducible by calcium influx.[75] Calcium has been recognized as a regulator of a large number of cellular processes and also as a signal in cellular differentiation and proliferation. Compelling evidence exists to indicate that an increase in oxidative stress causes a corresponding increase in the concentration of cytosolic free calcium by inducing release of mitochondrial stores of calcium and by causing influx of extracellular calcium.[76]

On the basis of the evidence presented above, we propose a mechanism involving oxidative stress and calcium fluxes acting as permissive and obligatory factors, respectively, during cellular differentiation. It is postulated that induction of cellular differentiation is preceded by an upsurge in the rate of production of O_2^- and H_2O_2, which induces an increase in SOD activity and a decrease in GSH concentration apparently by its oxidation. As a result, the intracellular environment becomes progressively more pro-oxidizing, causing an increase in the concentration of cytosolic free calcium. In our view, calcium rather than oxidative stress acts as the specific intracellular signal for subsequent events associated with differentiation, because differentiation can also be induced by calcium without concomitant oxidative stress, but the reverse is not true: in the absence of calcium, oxidative stress cannot induce differentiation. This indicates that calcium and not oxidative stress is obligatory to the process of differentation. Oxidative stress acts as a physiological means for causing the increase in free calcium availability and is thus permissive to this process.

In summary, the evidence presented here indicates that oxidative oxygen free radicals play a role in cellular differentiation. It is hypothesized that the mechanism by which oxidative stress influences gene expression is due to its effect on the concentration of cytosolic free calcium, which acts as the specific or obligatory signal for differentiation. A survey of the literature suggests that cells of disparate organisms employ elevation of oxidative stress to induce differentiation and to mobilize calcium.

REFERENCES

1. DARNELL, J., H. LODISH & D. BALTIMORE. 1986. Molecular Cell Biology. Scientific American Books. New York.
2. RUSCH, H. P. 1969. Some biochemical events in the growth cycle of *Physarum polycephalum*. Fed. Proc. **28**: 1761-1770.
3. RUSCH, H. P. 1980. The search. *In* Growth and Differentiation in *Physarum polycephalum*. W. F. Dove & H. P. Rusch, Eds.: 1–6. Princeton University Press. Princeton, NJ.
4. SAUER, H. W. & G. PIERRON. 1983. Morphogenesis and differentiation in *Physarum*. *In* Fungal Differentiation. J. R. Smith, Ed.: 73–106. Marcel Dekker. New York and Basel.
5. BERNIER, F., V. L. SELIGY, D. PALOTTA & G. LEMIEUX. 1986. Gene expression during spherulation in *Physarum polycephalum*. Biochem. Cell Biol. **64**: 337–343.
6. SCHRECKENBACH, T. & C. VERFUERTH. 1982. Blue light influences gene expression and mobility in starving microplasmodia of *Physarum polycephalum*. Eur. J. Cell Biol. **28**: 12–19.
7. RAUB, T. J. & H. C. ALDRICH. 1982. Sporangia, spherules and microcysts. *In* Cell Biology of *Physarum* and *Didymium*. H. C. Aldrich & J. W. Daniel, Eds. II: 21–75. Academic Press. New York.
8. NATIONS, C. & J. L. MCCARTHY. 1984. Growth of white microplasmodia of *Physarum polycephalum*. Comp. Biochem. Physiol. **78A**: 459–462.
9. NATIONS, C., R. G. ALLEN & J. L. MCCARTHY. 1984. Nonhistone proteins, free radical defenses and acceleration of spherulation in *Physarum*. *In* Growth, Cancer and Cell Cycle. P. Skehan & S. J. Friedman, Eds.: 71–78. Humana Press. Clifton, NJ.
10. DEE, J. 1982. Genetic control of the life cycle and of plasmodium development. *In* Cell

Biology of *Physarum* and *Didymium*. H. C. Aldrich & J. W. Daniel, Eds. **2**: 212–251. Academic Press. New York.

11. COLLINS, O. R. 1982. *Didymium iridis* in past and future research. *In* Cell Biology of *Physarum* and *Didymium*. H. C. Aldrich & J. W. Daniel, Eds.: 25–57. Academic Press. New York.

12. ALLEN, R. G., R. K. NEWTON, R. S. SOHAL, G. L. SHIPLEY & C. NATIONS. 1985. Alterations in superoxide dismutase, glutathione, and peroxides in the plasmodial slime mold *Physarum polycephalum* during differentiation. J. Cell. Physiol. **125**: 413–419.

13. HASSAN, H. M. & I. FRIDOVICH. 1979. Intracellular production of superoxide radical and hydrogen peroxide by redox active compounds. Arch. Biochem. Biophys. **196**: 385–395.

14. ALLEN, R. G., K. J. FARMER, P. L. TOY, R. K. NEWTON, R. S. SOHAL & C. NATIONS. 1985. Involvement of glutathione in the differentiation of the slime mold *Physarum polycephalum*. Devel. Growth Differ. **27**: 615–620.

15. ALLEN, R. G., R. K. NEWTON, K. J. FARMER & C. NATIONS. 1985. Effect of the free radical generator paraquat on differentiation, superoxide dismutase, glutathione and inorganic peroxides in microplasmodia of *Physarum polycephalum*. Cell Tissue Kinet. **18**: 623–630.

16. ALLEN, R. G., A. K. BALIN, R. J. REIMER, R. S. SOHAL & C. NATIONS. 1988. Superoxide dismutase induces differentiation in the slime mold, *Physarum polycephalum*. Arch. Biochem. Biophys. **261**: 205–211.

17. ELROY-STEIN, O., Y. BERSTEIN & Y. GRONER. 1986. Overproduction of human Cu/Zn superoxide dismutase in transfected cells: Extenuation of paraquat-mediated cytotoxicity and enhancement of lipid peroxidation. EMBO J. **5**: 615–622.

18. JEWELL, S. A., G. BELLOMO, H. THOR, S. ORRENIUS & M. T. SMITH. 1982. Bleb formation in hepatocytes during drug metabolism is caused by disturbances in thiol and calcium ion homeostasis. Science **217**: 1257–1258.

19. JONES, D. P., H. THOR, M. T. SMITH, S. A. JEWELL & S. ORRENIUS. 1983. Inhibition of ATP-dependent microsomal Ca^{2+} sequestration during oxidative stress and its prevention by glutathione. J. Biol. Chem. **258**: 6390–6393.

20. ORRENIUS, S., S. A. JEWELL, H. THOR, G. BELLOMO, L. EKLOW & M. T. SMITH. 1983. Drug-induced alterations in the surface morphology of isolated hepatocytes. *In* Isolation, Characterization, and Use of Hepatocytes. R. A. Harris & N. W. Cornell, Eds.: 333–340. Elsevier/North Holland Amsterdam.

21. SMITH, M. T., H. THOR, S. A. JEWELL, G. BELLOMO, M. S. SANDY & S. ORRENIUS. 1984. Free radical-induced changes in the surface morphology of isolated hepatocytes. *In* Free Radicals in Molecular Biology, Aging and Disease. D. Armstrong, R. S. Sohal, R. G. Cutler & T. F. Slater, Eds.: 103–118. Raven Press. New York.

22. GILKEY, J. C., L. F. JAFFE, E. B. RIDGWAY & G. T. REYNOLDS. 1978. A free calcium wave traverses the activating egg of the medeka, *Oryzias latipes*. J. Cell Biol. **76**: 448–466.

23. KRETSINGER, R. J. 1980. Mechanisms of selective signalling by calcium. Neurosci. Res. Prog. Bull. **19**: 211–328.

24. NATIONS, C., R. G. ALLEN, A. K. BALIN, R. J. REIMER & R. S. SOHAL. 1987. Superoxide dismutase activity and glutathione concentration during the calcium-induced differentiation of *Physarum polycephalum* microplasmodia. J. Cell. Physiol. **133**: 181–186.

25. BREWER, E. N. & A. PRYOR. 1976. A simplified medium for *Physarum polycephalum*. Physarum Newsletter **8**: 45.

26. ACHENBACH, F., U. ACHENBACH & K. WOHLFARTH-BOTTEMAN. 1981. Calcium accumulation in plasmodia of *Physarum polycephalum*. Cell Calcium **2**: 587–599.

27. SOHAL, R. S., R. G. ALLEN & C. NATIONS. 1986. Oxygen free radicals play a role in cellular differentiation: An hypothesis. J. Free Rad. Biol. Med **2**: 175–181.

28. DIONISI, O., T. GALEOTTI, T. TERRANOVA & A. AZZI. 1975. Superoxide radicals and hydrogen peroxide formation in mitochondria from normal and neoplastic tissues. Biochem Biophys. Acta **403**: 292–300.

29. OBERLEY, L. W., T. D. OBERLEY & G. R. BUETTNER. 1980. Cell differentiation and cancer: The possible role of superoxide and superoxide dismutases. Med. Hypotheses **6**: 249–268.

30. OBERLEY, L. W. 1982. Superoxide dismutases in cancer. *In* Superoxide Dismutase. L. W. Oberley, Ed. **2**: 127–165. CRC Press. Boca Raton, FL.

31. OBERLEY, L. W. 1984. The role of superoxide dismutase and gene amplification in carcinogenesis. J. Theor. Biol. **106:** 403–422.
32. LOTT, T., S. GORMAN & J. CLARK. 1981. Superoxide dismutase in *Didymium iridis:* Characterization of changes in activity during senescence and sporulation. Mech. Ageing Dev. **17:** 119–130.
33. FERNANDEZ-SOUZA, J. M. & A. M. MICHELSON. 1976. Variation of the superoxide dismutases during the development of the fruitfly, *Ceratitis capitata.* Biochem Biophys. Res. Commun. **73:** 217–223.
34. MASSIE, H. R., V. R. AIELLO & T. R. WILLIAMS. 1980. Changes in superoxide dismutase activity and copper during development and ageing in the fruit fly *Drosophila melanogaster.* Mech. Ageing Dev. **12:** 279–286.
35. BARJA QUIROGA, G. & P. GUTIERREZ. 1984. Superoxide dismutase during the development of two amphibian species and its role in hyperoxia tolerance. Mol. Physiol. **6:** 221–232.
36. VAN HIEN, P., K. KOVACS & B. MATKOVICS. 1974. Properties of enzymes. I. Study of superoxide dismutase activity changes in human placenta of different ages. Enzymes **18:** 341–347.
37. RUSSANOV, E. M., M. D. KIRKOVA, M. S. SETCHENSKA & H. R. V. ARNSTEIN. 1981. Enzymes of oxygen metabolism during erythrocyte differentiation. Biosci. Rep. **1:** 927–931.
38. MAVELLI, I., A. RIGO, R. FEDERICO, M. R. CIRIOLO & G. ROTILIO. 1982. Superoxide dismutase, glutathione peroxidase and catalase in developing rat brain. Biochem. J. **204:** 535–540.
39. TANSWELL, A. K. & B. A. FREEMAN. 1984. Pulmonary antioxidant enzyme maturation in the fetal and neonatal rat. I. Development profiles. Pediatr. Res. **18:** 584–587.
40. GERDIN, E., O. TYDEN & U. J. ERIKSSON. 1985. The development of antioxidant enzymatic defense in the perinatal rat lung: activities of superoxide dismutase, glutathione peroxidase, and catalase. Pediatr. Res. **19:** 687–691.
41. CRAPO, J. D. & D. F. TURNEY. 1974. Superoxide dismutase and pulmonary oxygen toxicity. Amer. J. Pathol. **226:** 1401–1407.
42. AUTOR, A. P., L. FRANK & R. J. ROBERTS. 1976. Developmental characteristics of pulmonary superoxide dismutase: relationship to idiopathic respiratory distress syndrome. Pediatr. Res. **10:** 154–158.
43. FRANK, L., D. L. WOOD & R. J. ROBERTS. 1978. Effect of diethyldithiocarbamate on oxygen toxicity and lung enzyme activity in immature and adult rats. Biochem. Pharmacol. **27:** 251–254.
44. FRANK, L., J. R. BUCHER & R. J. ROBERTS. 1978. Oxygen toxicity in neonatal and adult animals of various species. J. Appl. Physiol. Resp. Environ. Exercise Physiol. **45:** 699–704.
45. YAM, J., L. FRANK & R. J. ROBERTS. 1978. Oxygen toxicity: comparison of lung biochemical responses in neonatal and adult rats. Pediatr. Res. **12:** 115–119.
46. BUCHER, J. R. & R. J. ROBERTS. 1981. The development of the newborn rat lung in hyperoxia: a dose-response study of lung growth, maturation, and changes in antioxidant enzyme activities. Pediatr. Res. **15:** 99–1008.
47. BOCHNER, B. R., P. C. LEE, S. W. WILSON, C. W. CUTLER & B. N. AMES. 1984. ApppA and related adenylylated neuclotides are synthesized as a consequence of oxidative stress. Cell **37:** 225–232.
48. NICKLA, H., J. ANDERSON & T. PALZKILL. 1983. Enzymes involved in oxygen detoxification during development of *Drosophila melanogaster.* Experientia **39:** 610–612.
49. CRABTREE, H. G. 1945. Influence of unsaturated dibasic acids of the induction of skin tumors by chemical carcinogens. Cancer Res. **5:** 346–351.
50. GIL, P., M. ALONSO-BEDATE & G. BARJA DE QUIROGA. 1987. Different levels of hyperoxia reversibly induce catalase activity in amphibian tadpoles. Free Rad. Biol. Med. **3:** 137–146.
51. FRANK, L., A. P. AUTOR & R. J. ROBERTS. 1977. Oxygen therapy and hyaline membrane disease: the effect of hyperoxia on pulmonary superoxide dismutase activity and mediating role of plasma or serum. J. Pediatr. **90:** 105–110.
52. FOREMAN, H. J. & A. B. FISCHER. 1981. Antioxidant defences. *In* Oxygen and Living Processes. D. L. Gilbert, Ed.: 65–90. Springer-Verlag. New York.

53. CRABTREE, H. G. 1944. Influence of bromobenzene on the induction of skin tumors by 3,4-benzpyrene. Cancer Res. **4:** 688–693.
54. DI ILIO, C., G. D. BOCCIO, E. CASALONE, A. ACETO & P. SACCHETTA. 1986. Activities of enzymes associated with metabolism of glutathione in fetal rat liver and placenta. Biol. Neonate **49:** 96–101.
55. FRANK, L. & E. E. GROSECLOSE. 1984. Preparation for birth into an O_2-rich environment: the antioxidant enzymes in developing rabbit lung. Pediatr. Res. **18:** 240–244.
56. PROHASKA, J. R. & H. E. GANTHER. 1976. Selenium and glutathione peroxidase in developing rat brain. J. Neurochem. **27:** 1379–1387.
57. ALLEN, R. G. & R. S. SOHAL. 1986. Role of glutathione in aging and development of insects. *In* Insect Aging. K. G. Collatz & R. S. Sohal, Eds.: 168–181. Springer-Verlag. Heidelberg.
58. WARSHAW, J. B., C. W. WILSON, K. SAITO & R. A. PROUGH. 1985. The response of glutathione and antioxidant enzymes to hyperoxia in developing lung. Pediatr. Res. **19:** 819–823.
59. TAKAHASHI, S. & M. ZEYDEL. 1981. γ-Glutamyl transpeptidase and glutathione in aging IMR-90 fibroblasts and in differentiating 3T3 L1 preadipocytes. Arch. Biochem. Biophys. **214:** 260–267.
60. BALINSKY, B. I. 1970. Embryology, 3rd ed.: 638–668. Saunders. Philadelphia.
61. FRASER, L. B. & D. B. CARTER. 1967. Variation of acid-soluble sulfhydryl groups during liver regeneration. Brit. J. Cancer **21:** 235–241.
62. HARMAN, A. W. & C. A. HENRY. 1987. Differences in glutathione-S-transferase activities in hepatocytes from postnatal and adult mice. Biochem. Pharmacol. **36:** 177–179.
63. CALVIN, H. I. & S. I. TURNER. 1982. High levels of glutathione attained during postnatal development of rat testis. J. Exp. Zool. **219:** 389–393.
64. PHILIPS, G. J. & P. T. BORGIA. 1985. Effects of oxygen on morphogenesis and polypeptide expression by *Mucor racemosus*. J. Bacteriol. **164:** 1039–1048.
65. PAWELEK, J. M. 1959. Effects of thyroxine and low oxygen tension on chrondrogenic expression in cell culture. Dev. Biol. **19:** 52–72.
66. ERKELL, L. J. 1980. Differentiation of mouse neuroblastoma under increased oxygen tension. Exp. Cell **48:** 374–380.
67. LOUDON, C. 1988. Development of *Tenebrio molitor* in low oxygen levels. J. Insect Physiol. **34:** 97–103.
68. TURRENS, J. F., B. A. FREEMAN, J. G. LEVITT & J. D. CRAPO. 1982. The effect of hyperoxia on superoxide production by lung submitochondrial particles. Arch. Biochem. Biophys. **217:** 401–410.
69. GUERNSEY, D. L., S. W. C. LEUTHAUSER & N. J. KOEBBE. 1985. Induction of cytodifferentiation in C3H/10T mouse embryo cells by X-irradiation and benzo [a] pyrene. Cell Different. **16:** 147–151.
70. LOMBARDI, T., R. MONTESANO & L. ORCI. 1987. Phorbol ester induces diaphragmed fenestrae in large vessle endothelium *in vitro*. Eur. J. Cell Biol. **44:** 86–89.
71. EARNSHAW, B. A. & M. A. JOHNSON. 1985. The effect of glutathione on development in wild carrot suspension cultures. Biochem. Biophys. Res. Commun. **133:** 988–993.
72. FORMANSKI, P. & M. LUBIN. 1972. Effects of dimethylsulfoxide on expression of differentiated functions in mouse neuroblastoma. J. Nat. Cancer Inst. **48:** 1355–1361.
73. LEVENGOOD, W. C. & R. DAMRAUER. 1969. Developmental inhibition in *Drosophila* using dihydroxybenzoic acid isomers. J. Insect Physiol. **15:** 633–641.
74. CRAWFORD, D., I. ZBINDEN, P. AMSTAD & P. CERUTTI. Oxidant stress induces the protooncogenes *c*-fos and *c*-myc in mouse epidermal cells. Oncogene. In press.
75. MORGAN, J. & T. CURRAN. 1986. Role of ion flux in the control of *c*-fos expression. Nature **322:** 552–555.
76. BELLOMO, G. & S. ORRENIUS. 1985. Altered thiol and calcium homeostasis in oxidative hepatocellular injury. Hepatology **5:** 876–882.
77. HENNEY, H. R., JR. 1982. *In* Cell Biology of *Physarum* and *Didymium*. H. C. Aldrich & J. W. Daniel, Eds.: 131–151. Academic Press. New York.

DISCUSSION

H. C. BIRNBOIM (*Ottawa Regional Cancer Centre, Ottawa, Ontario, Canada*): Is the heat-shock or stress protein reaction commonly seen as a result of a variety of treatments responsible for the differentiation of *Physarum* by starvation, and so on?

R. S. SOHAL (*Southern Methodist University, Dallas, Texas*): We have not investigated this interesting possibility.

M. MIRANDA (*University of Aquila, Italy*): Since differentiated cells have a less attachable genome conformation, might the active oxygen-induced differentiation of slime molds be a defense mechanism?

SOHAL: Your question is about philosophy of interpretation. Many people have said in the past that cells differentiate because of external stress. Differentiation in the slime mold could be thought of in terms of a defense mechanism.

G. ROTILIO (*University of Rome, Rome, Italy*): First, when you observe an increase of SOD activity in differentiating strains, is that total SOD activity or rather a specific isoenzyme (Mn or Cu-Zn)? Second, do you have any evidence that your liposomal SOD actually crosses the cell membrane?

SOHAL: In answer to your first question: *Physarum* exhibits relatively little cyanide-sensitive SOD activity. Regarding your second question, we believe that SOD gets into cells as the enzyme activity significantly increases after exposure to SOD-containing liposomes.

R. LOTAN (*University of Texas M.D. Anderson Cancer Center, Houston, Texas*): John Hickman has suggested that cytotoxic agents can induce differentiation of HL-60 promyelocytic leukemia by inducing stress because the dose-response curves for differentiation induction are slightly shifted to lower doses than are dose-response curves for cytotoxicity. Hence, in cancer cells the induction of sublethal stress can lead to differentiation.

SOHAL: I tend to agree with you.

Physiological and Pathologic Effects of Oxidants in Mouse Epidermal Cells[a]

PETER CERUTTI, GEORG KRUPITZA, ROGER LARSSON,
DOMINIQUE MUEHLEMATTER, DANA CRAWFORD,
AND PAUL AMSTAD

Department of Carcinogenesis
Swiss Institute for Experimental Cancer Research
Lausanne, Switzerland

An overwhelming amount of experimental evidence and some epidemiologic data indicate that carcinogenesis is a multi-step process. The simplest model distinguishes initiation and promotion and was originally recognized in mouse skin using croton oil as promoter.[1,2] Convincing evidence has now accrued for multistep processes in many organ systems and for the existence of many different types of promoters.[3] According to the general consensus, initiators induce genotypic changes and promotion results in the clonal expansion of "initiated" cells. The genotypic changes are often the consequence of the processing of carcinogen-induced DNA damage resulting in mutations, gene rearrangements, or gene amplication. The functional consequence is a *response modification* of the "initiated" cell to extra- or intercellular growth and differentiation signals or cytotoxic agents. Response modification has created promotability. Activation of certain proto-oncogenes, such as *erb*A, *ras, src, tck,* and *abl,* with membrane or cytoplasmic functions, and of *myc, myb,* and *fos* with nuclear functions can represent response modifications which manifest themselves upon exposure to an appropriate promotional signal.[4]

In our work on multi-stage carcinogenesis we have concentrated on the mechanisms of action of oxidant tumor promoters. The general message is that oxidants, besides being toxic, induce a multitude of (patho) physiological reactions in cells.[5–7] As demonstrated for the promotable JB6 clone 41 of mouse epidermal cells, exposure to active oxygen can cause growth stimulation rather than inhibition.[8] The final biological effect depends on a subtle balance. It is evident that the induction of reactions that are likely to be on the pathway to growth stimulation, such as S6-phosphorylation[9] and c-*fos* and c-*myc* expression,[10] can only come to fruition in a background of tolerable toxicity. Since the cellular antioxidant defense attenuates the effects of oxidants, it is bound to play a decisive role in this regard. In the context of oxidant tumor promotion, the elevated antioxidant defense of promotable JB6 clone 41 may be considered as a "response modification" to a cytostatic agent relative to nonpromotable clone 30.[4]

[a]This work was supported by the Swiss National Science Foundation, the Swiss Association of Cigarette Manufacturers, and the Association for International Cancer Research.

75

MECHANISMS OF ACTION OF OXIDANT TUMOR PROMOTERS

The recognition that oxidants can act as tumor promoters has stimulated experimental work in many laboratories. Unlike the phorbol esters and other experimental drugs, oxidants are ubiquitous and suspected to represent "natural promoters."[11] We have concentrated on the effects of active oxygen (AO) produced extracellularly by xanthine/xanthine oxidase (X/XO) on mouse epidermal JB6 cells. JB6 cells represent a useful model for the study of tumor promotion because clones have been derived from a common parent line which are susceptible or resistant to growth stimulation by a variety of promoters.[12] We found that the promotable JB6 clone 41 can be stimulated to grow in soft agar[12] and in monolayers upon treatment with several oxidants, whereas the nonpromotable clone 30 is slightly inhibited under the same conditions.[8] As described below we have studied the effects of AO from the cell envelope to the nucleus on reactions that are likely to be on the pathway of growth stimulation, such as changes in intracellular Ca^{2+} concentration, phosphophorylation of the S6-ribosomal protein,[9] DNA strand breakage and, as a consequence, poly ADP-ribosylation of chromosomal proteins,[8,13] and finally the modulation of the expression of the competence-related proto-oncogenes c-*fos* and c-*myc*.[10] These reactions were observed in both promotable (i.e., clone 41) and nonpromotable (i.e., clone 30) JB6 cells but there were significant quantitative differences. A possible reason for some of these differences was discovered when we found 2–3 fold-higher constitutive levels of the antioxidant enzymes Cu,Zn-superoxide dismutase (SOD) and catalase (CAT) in the promotable JB6 clone 41. We propose that a subtle balance between the dose of AO and the cellular antioxidant defenses determines whether toxic or pathophysiologic effects predominate (Ref. 14 and unpublished results).

Oxidant-Induced Phosphorylation of Ribosomal Proteins S6

Phosphorylation of the major ribosomal protein S6 occurs in response to the receptor-binding of hormones and growth factors and has been speculated to be required for the stimulation of protein synthesis in the acquisition of competence by quiescent cells.[15–21] It can involve several types of kinases, among them $Ca^{2+}/$ calmodulin-dependent enzymes[22] and $Ca^{2+}/$phospholipid-dependent protein kinase C (pKC).[23,24] Since we had observed that low concentrations of extracellular AO produced by X/XO stimulated the growth of JB6 clone 41 cells and caused a rapid increase in $[Ca^{2+}]_i$, it was of interest to study S6-phosphorylation. We exposed monolayer cultures of JB6 clone 41 cells to the following sources of AO: H_2O_2 produced by glucose/glucose oxidase, a mixture of O_2^- plus H_2O_2 generated by X/XO, and intracellular O_2^- induced by menadione. S6-phosphorylation was rapidly stimulated by all these sources of AO. The reaction was inhibitable by the addition of CAT but not SOD, the Ca^{2+} complexers quin 2 and EGTA, and the $Ca^{2+}/$calmodulin antagonist, trifluoperazine. We conclude that S6-phosphorylation of JB6 cells is triggered by H_2O_2 rather than O_2^- in a reaction that is mostly accomplished by a $Ca^{2+}/$calmodulin-dependent kinase. Further insight into the mechanism of oxidant-induced S6-phosphorylation derives from experiments with diamide, which oxidizes cellular SH functions.[25,26] Diamide strongly stimulated the reaction and we propose that the oxidant-induced increase in $[Ca^{2+}]_i$, which is required for S6-phosphorylation, is the consequence of the oxidative inactivation of a plasma membrane Ca^{2+} ATPase. This mechanism has been implicated in menadione- and cystamine-induced cytotoxicity. Our results indicate that oxidants can activate pathways of signal transduction which are used by receptor-mediated agonists. These pathophysiologic reactions of oxidants may be of particular importance in wound healing and in tumor promotion and progression.

Induction of Poly ADP-ribosylation of Chromosomal Proteins by Oxidants

Poly ADP-ribosylation of chromosomal proteins is a metabolic consequence of DNA strand breakage.[27] As expected, extracellularly generated AO induced poly ADP-ribosylation of JB6 cells and the reaction was more extensive in clone 30, which contained larger amounts of DNA breaks.[8] We are identifying the major poly ADP-ribose acceptor proteins in intact JB6 cells according to the following experimental design. Poly ADP-ribosylated proteins from nuclear extracts of AO-treated cells are purified by affinity chromatography on a boronate resin and characterized on immunoblots with antibodies against poly ADPR chains and specific acceptor proteins. Our results indicate that histone H3 serves as major acceptor and histones H2A and H2B as minor acceptors. ADP-ribose transferase was the most important nonhistone acceptor. The reaction was partially suppressed by the transferase inhibitor, benzamide. Significant but lower poly ADPR substitution was detected on topoisomerase I and on the protein encoded by the proto-oncogene c-*fos* (Ref. 13 and unpublished results). We can only speculate on the functional consequences of the poly ADP-ribosylation of these nonhistone proteins. Poly ADP-ribosylation, unlike phosphorylation, invariably inactivates enzymes, and it is conceivable that the automodification of ADPR-tranferase represents a feedback loop with the function to avoid excessive poly ADP-ribosylation and consecutive NAD and ATP depletion. Temporary inactivation of topoisomerase I by poly ADP-ribosyl substitution may transiently shut down DNA replication, giving the cell more time for repair before the damage is fixed by the replication machinery. In the context of the differential effect of AO on the growth of the promotable clone 41 and the nonpromotable clone 30 of JB6 cells, it should be reiterated that both DNA strand breakage and poly ADP-ribosylation were more extensive in the latter.[8]

Oxidants Induce the Proto-oncogenes c-Fos and c-Myc and β-Actin

For the recruitment of quiescent cells to enter the cell cycle, the reprogramming of the expression of numerous genes is required. Among them are the proto-oncogenes, c-*fos* and c-*myc*, which are induced by serum factors, growth factors, and the phorbol ester promoter PMA.[28–30] While their function is not fully elucidated, there is convincing evidence that c-*fos* serves as a transcriptional regulator,[28,29,31] while c-*myc* encodes a DNA replication function.[30,32,33] Since AO was capable of stimulating the growth of JB6 clone 41 it was of interest to study its effect on these genes. We found that an extracellular burst of AO produced by X/XO resulted in a transitory increase in the steady-state mRNA concentrations of c-*fos* within 15–30 min and of c-*myc* after 1–2 hr. Nuclear run-off experiments indicated that at least part of the increase in c-*fos* message, but not of c-*myc* message, was due to the stimulation of the rate of transcription. The responsiveness of these genes to AO was much more pronounced in the nonpromotable clone 30.[10,14] At first sight this result may be unexpected. However, no simple correlation between the magnitude of the induction of these genes and promotability needs to exist. Another important difference between the two JB6 clones discovered in our work lies in the high persistent c-*myc* expression in the promotable clone 41. Analysis by Southern blots indicated that the persistent c-*myc* mRNA levels in clone 41 were not due to gene amplification nor to rearrangements. Indications that high persistent *myc* expression might be associated with a preneoplastic state can be found in the literature. We speculate that high persistent c-*myc* expression may be

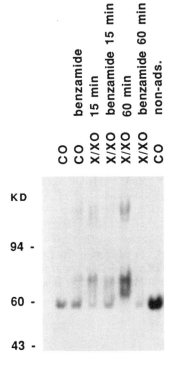

FIGURE 1. Immunoblot of nuclear extracts of active oxygen-treated JB6 cells with anti-c-*fos* antibody. Nuclear extracts of JB6 (clone 41) cells which had been treated with AO from X/XO (50 μg/ml X, 5 μg/ul XO) were applied on phenylboronate PBA 30 columns. The nonadsorbed flow-through and the eluted adsorbed material (for conditions see Adamietz and Rudolph[34]) were applied on 7% polyacrylamide gels, which were developed according to the method of Holtlund *et al.*[35] The proteins were electrotransferred to nitrocellulose according to standard procedures and reacted with monclonal antibody raised against a synthetic c-*fos* peptide (residues 359–378) and finally with ^{125}I-labeled sheep anti-mouse IgG.

incompatible with terminal differentiation. It may in part explain the promotability of JB6 clone 41.

Induction of c-*fos* transcription by AO is followed by an increase in *fos* protein from high constitutive levels within 60 min according to immunoblot analysis. As discussed in the previous section, AO causes DNA strand breakage and as a consequence stimulates the poly ADP-ribosylation of chromosomal proteins. Using a combination of phenyl boronate affinity chromatography and immunoblots, we found that *fos* protein served as a poly ADP-ribose acceptor in JB6 cells. Exposure to alkali of the nuclear proteins removed the poly ADP-ribose chains from the *fos* protein, and, as shown in FIGURE 1, pretreatment of cells with ADP-ribose tranferase inhibitor benzamide suppressed the reaction. Benzamide also strongly suppressed the induction of c-*fos* transcription in nuclear run-on experiments, suggesting a role for poly ADP-ribosylation in the modulation of c-*fos* expression by AO.[36]

Promotable Mouse Epidermal Cells JB6 Clone 41 Possess a Superior Antioxidant Defense

Several of the differences between the two JB6 clones just discussed suggested a superior antioxidant defense for the promotable clone 41. Therefore, we compared the activities and mRNA concentrations of the three major antioxidant enzymes, SOD, CAT and glutathione peroxidase (GPx) between the two clones. In support of our hypothesis we found that the activities of SOD and CAT were approximately twice as high in clone 41 and that the steady-state mRNA concentrations for these proteins were also considerably higher in this clone. This conclusion was further supported by immunoblot analysis, which indicated higher CAT-protein levels in clone 41 than in clone 30. These findings are remarkable because the two antioxidant enzymes SOD and CAT are increased coordinately in clone 41. Since the product of the action of SOD is H_2O_2, an increase in its activity is only beneficial to the cell if it is counterbalanced by a sufficient capacity for the destruction of H_2O_2. In contrast, the third major antioxidant enzyme, GPx, was present at comparable levels in both clones. The observed differences appear to be at the level of gene expression since both clones contained the same complement of SOD- and CAT-genes according to Southern-blot analysis (Ref. 14 and unpublished results). We conclude that the superior antioxidant defense protects it from excessive cytostatic effects of AO. Only when the cytostatic effects are moderate can the induced growth-related genes exert their functions.

REFERENCES

1. BERENBLUM, I. 1941. The carcinogenic action of croton resin. Cancer Res. **1**: 214.
2. BERENBLUM, I. & P. SHUBIK. 1941. A new quantitative approach to the study of stages of chemical carcinogenesis in mouse's skin. Br. J. Cancer **1**: 383.
3. SLAGA, T, Ed. 1983. Mechanisms of Tumor Promotion, Vol. **I–IV**. CRC Press. Boca Raton, FL.
4. CERUTTI, P. 1988. Response modification creates promotability in multistage carcinogenesis. Carcinogenesis **9**: 519–526.
5. CERUTTI, P. 1987. Genotoxic oxidant tumor promoters. *In* Banbury Report 25: Nongenotoxic Mechanisms in Carcinogenesis. T. Slaga and B. Butterworth, Eds.: 325–335. Cold Spring Harbor Press. Cold Spring Harbor, NY.
6. CERUTTI, P. 1987. Tumor Promotion by Oxidants. Theories of Carcinogenesis. O. H. Iversen. Ed.: 221–230. Hemisphere.
7. CERUTTI, P. 1988. Oxidant tumor promoters. *In* Growth Factors, Tumor Promoters and Cancer Genes. Nancy H. Colburn, Harold Moses & Eric J. Stanbridge, Eds.: 239–247. Alan R. Liss. New York.
8. MUEHLEMATTER, D., R. LARSSON & P. CERUTTI. 1988. Active oxygen induced DNA strand breakage and poly ADP-ribosylation in promotable and non-promotable JB6 mouse epidermal cells. Carcinogenesis **9**: 239–245.
9. LARSSON, R. & P. CERUTTI. 1988. Oxidants induce phosphorylation of ribosomal protein S6. J. Biol. Chem. In press.
10. CRAWFORD, D., I. ZBINDEN, P. AMSTAD & P. CERUTTI. Oxidant stress induces protooncogenes c-fos and c-myc in mouse epidermal cells. Oncogene. In press.

11. CERUTTI, P. 1985. Prooxidant states and tumor promotion. Science 227:375–381.
12. GINDHART, T., Y. NAKAMURA, L. STEVENS, G. HEGAMEYER, M. WEST, B. SMITH & N. COLBURN. 1985. *In* Cancer of the Respiratory Tract: Carcinogenesis—a Comprehensive Survey, Vol. 8. M. Mass *et al.*, Eds.: 341–367. Raven Press. New York.
13. CERUTTI, P., G. KRUPITZA & D. MUEHLEMATTER. Poly ADP-ribosylation of nuclear proteins by oxidant tumor promoters. *In* Proceedings of the 8th International Symposium on Niacin, Nutrition, ADP-Ribosylation and Cancer. Elaine and Myron Jacobson, Eds. In press.
14. CRAWFORD, D. & P. CERUTTI. 1988. Expression of oxidant stress-related genes in tumor promotion of mouse epidermal cells JB6. *In* Anticarcinogenesis and Radiation Protection. P. Cerutti, O. Nygaard, and M. Simic, Eds.: 183–190. Plenum Press. New York.
15. NOVAK-HOFER, I. & G. THOMAS. 1984. An activated S6 kinase in extracts from serum- and epidermal growth factor-simulated Swiss 3T3 cells. J. Biol. Chem. **259:** 5995–6000.
16. SMITH, C. J., C. S. RUPIN & O. M. ROSEN. 1980. Insulin-treated 3T3-L1 adipocytes and cell-free extracts derived from the incorporate ^{32}P into ribosomal protein S6. Proc. Natl. Acad. Sci. USA **77:** 2641–2645.
17. NISHIMURA, J. & T. F. DEUEL. 1983. Platelet-derived growth factor stimulates the phosphorylation of ribosomal protein S6. FEBS Lett. **156:** 130–134.
18. ERIKSON, E. & J. L. MALLER. 1985. A protein kinase from *Xenopus* eggs specific for ribosomal protein S6. Proc. Natl. Acad. Sci. USA **82:** 742–746.
19. PIERRE, M., D. TORU-DELBAUFFEM, J. M. GAVARET, M. POMERANCE & C. JAQUEMIN. 1986. Activation of S6 kinase activity in astrocytes by insulin, somatomedin C and TPA. FEBS Lett **206:** 162–166.
20. MATSUDA, Y. & G. GUROFF. 1987. Purification and mechanism of activation of nerve-growth factor-sensitive S6 kinase from PC12 cells J. Biol. Chem. **262:** 2832–2844.
21. EVANS, S. W. & W. F. FARRAR. 1987. Interleukin 2 and diacylglycerol stimulated phosphorylation of 40S ribosomal S6 protein. J. Biol. Chem. **262:** 4624–4630.
22. SCHULMAN, K. H., J. KURET, A. B. JEFFERSON, P. S. NOSE & K. H. SPITZER. 1985. Ca^{2+}/calmodulin-dependent microtubule-associated protein 2 kinase. Broad substrate specificity and multifunctional potential in diverse tissues. Biochemistry **24:** 5320–5327.
23. LE PEUCH, C. J., R. BALLESTER & O. M. ROSEN. 1983. Purified rat brain calcium- and phospholipid-dependent protein kinase phosphorylates ribosomal protein S6. Proc. Natl. Acad. Sci. USA **80:** 6858–6862.
24. PARKER, P. J., M. KATAN, M. D. WATERFIELD & D. P. LEADER. 1985. The phosphorylation of eukaryotic ribosomal protein S6 by protein kinase C. Eur. J. Biochem. **148:** 579–586.
25. KOSOWER, N. S., E. M. KOSOWER, B. WERTHEIM & W. S. CORREA. 1974. Diamide, a new reagent for the intracellular oxidation of glutathione to the disulfide. Biochem. Biophys. Res. Commun. **37:** 593–596.
26. GRAF, E., A. K. VERMA, J. P. GORSKI, G. LIPASCHUK, V. NIGGLI, M. ZURINA, E. CARAFOLI & J. T. PENNISTON. 1982. Molecular properties of calcium-pumping ATPase from human erythrocytes. Biochemistry **21:** 4511–4516.
27. ALTHAUS, F., H. HILZ & S. SHALL, Eds. 1986. ADP-ribosylation of proteins. Springer Verlag. Heidelberg.
28. VERMA, I. 1986. Proto-oncogene fos: a multifaceted gene. Trends in Genetics **2:** 93–96.
29. MÜLLER, R. 1986. Cellular and viral fos genes: Structure, regulation of expression and biological properties of their encoded products. Biochim. Biophys. Acta **823:** 207–225.
30. COLE, M. 1986. The myc oncogene: Its role in transformation and differentiation. Ann. Rev. Genet. **20:** 361–384.
31. DISTEL, R., H. S. RO, B. ROSEN, D. GROVES & B. SPIEGELMAN. 1987. Nucleoprotein complexes that regulate gene expression in adipocyte differentiation: Direct participation of c-fos. Cell **49:** 835–844.
32. KACZMAREK, L., J. HYLAND, R. WATT, M. ROSENBERG & R. BASERGA. 1985. Microinjected c-myc as a competence factor. Science **228:** 1313–1314.
33. HEIKKILA, R., G. SCHWAB, E. WICKSTROM, E., S. LOONG LOKE, D. H. PLUZNIK, R. WATT & L. NECKERS. 1987. A c-myc antisense oligodeoxynucleotide inhibits entry into S phase but not progress from G_0 to G_1. Nature **328:** 445–449.
34. ADAMIETZ, P. & R. RUDOLPH. 1984. ADP-ribosylation of nuclear proteins in vivo.

Identification of histone H2B as a major acceptor for mono- and poly(ADP-ribose) in dimethyl-sulfate-treated hepatoma AH 7974 cells. J. Biol. Chem. **259:** 6841–6846.
35. HOLTLUND, J., R. JEMTLAND & T. KRISTENSEN. 1983. Two proteolytic degradation products of calf thymus poly(ADP-ribose)-polymerase are efficient ADP-ribose acceptors. Europ. J. Biochem. **130:** 309–314.
36. KRUPITZA, G. & P. CERUTTI. Unpublished material.

DISCUSSION

T. GALEOTTI (*Catholic University, Rome, Italy*): The difference in enzymatic antioxidant defences between promotable and nonpromotable cells appears to be modest. Did you measure the fatty acid composition and vitamin E content of the two clones?

P. A. CERUTTI: In answer to your first remark: A fine balance of tolerable cytotoxicity and *inducibility* of growth-related genes (e.g., *fos* and *myc*) is required for promotability. Preliminary results show that transfection of nonpromotable (cl 30) cells with catalase expression vector leading to a 3-fold increase in CAT rendered them promotable.

We did not measure fatty acids and vitamin E, but it should be done!

H. C. BIRNBOIM (*Ottawa Regional Cancer Centre, Ottawa, Ontario, Canada*): Does benzamide inhibit *fos* induction by nonoxyradical mechanisms, such as growth factors?

CERUTTI: This has not been tested, but serum factors and TPA do not induce poly ADPR of c-*fos* in mouse epidermal cells.

C. BOREK (*Columbia Unversity, New York, N.Y.*): Response modification of cells to cytotoxic agents and transformation may vary with cell type and tissue. There have been data from our lab and others showing that human cells, both epithelial and fibroblasts, when transformed *in vitro* to malignancy do not exhibit obvious chromosomal ablation. This may be a much later phenomenon. A second point is that when we transfect cells with one oncogene, their response to radiation cytotoxicity is increased, and, when transfected with two oncogenes, there is a further enhanced response to cell killing by radiation, as compared to nontransfected cells.

R. S. SOHAL (*Southern Methodist University, Dallas, Texas*): Apparently, promotable cells, by virtue of their high antioxidant defenses, do not permit the development of optimal oxidative stress. Your finding is thus compatible with our finding and interpretation that increased oxidative stress is a causal factor in cellular differentiation.

CERUTTI: This still allows sufficient signal to modulate gene expression and the induction of competence-related genes (*fos, myc*), but cytotoxicity is tolerable.

R. LOTAN (*University of Texas M.D. Anderson Cancer Center, Houston, Texas*): Retinoic acid is an inhibitor of promotion of JB6 cells. In some cells (e.g., F-9 embryonal carcinoma, HL-60 leukemia) retinoic acid suppresses c-*myc* expression. You have shown that high or constitutive c-*myc* expression is related to promotion. Did you study whether retinoic acid decreases promotion by oxygen radicals? Does such inhibition involve c-*myc* suppression?

CERUTTI: We have *not* studied the effect of retinoic acid on persistent nor active oxygen-induced c-*myc* expression, nor whether retinoic acid inhibits active oxygen-induced growth stimulation of JB6 (cl 41) mouse epidermal cells.

C. FRANCESCHI (*University of Florence, Italy*): Can you comment on the intriguing relationship between aging and cancer? In aging there is an increased promotability, but the antioxidant and cellular defense systems appear to be reduced.

CERUTTI: The dramatic increase in cancer incidence with aging may not be a "promotional effect," but rather an accumulation of "interesting" response modifications (see the April 1988 issue of *Carcinogenesis*). Also keep in mind that a *fine balance* between protection from excessive cytotoxic effects and the inducibility of oxidant-inducible genes is required for promotion. It is difficult to predict in a particular tissue the exact balance that results in optimal promotability.

G. GUIDOTTI (*Università di Patologia Generale, Parma, Italy*): You mention the possibility that the *trans*-activator AP-1 that initiates the TPA induction of oncogenesis may be involved in the effect of oxidants. AP-1 could be the product of the *jun* oncogene. Is the *jun* oncogene activated in your system?

CERUTTI: My speculation was that poly ADPR of c-*fos* might alter its binding to AP-1 and affect its transcription regulatory activity for c-*fos* expression. No work was carried out on the *jun* oncogene.

Superoxide Anion May Trigger DNA Strand Breaks in Human Granulocytes by Acting at a Membrane Target[a]

H. C. BIRNBOIM[b]

Department of Experimental Oncology
Ottawa Regional Cancer Centre, and
Department of Microbiology and Immunology
University of Ottawa
Ottawa, Ontario, Canada

INTRODUCTION

Active oxygen species such as superoxide anion (O_2^-) and hydrogen peroxide (H_2O_2) occur widely in humans and animals during the course of normal cellular functions as well as after treatment with xenobiotics and ionizing radiation. There has been considerable interest in the possibility that such free-radical and related species may be causally involved in conditions such as cancer, aging, and ischemia-reperfusion injury as well as other diseases.[1] Because these are reactive and short-lived species which can in turn give rise to other reactive species, identifying precise reaction mechanisms can be exceedingly complex, especially when studied in intact animals, in tissues, or even in cells. This has led to the use of a number of model systems, some discussed in this volume, in an attempt to piece together the total picture of the true impact of active oxygen species on living organisms.

In the context of this presentation, I have distinguished two major classes of events that need to be considered in evaluating the biological significance of active oxygen species. These are illustrated in FIGURE 1. On the right limb of this scheme are *toxic events* having short-term consequences, such as death of cells, damage to tissues, and disruption of organ function. I propose that this limb may be particularly important in ischemia/reperfusion injury and relatively less important for cancer and aging. The other limb of the scheme is shown on the left. I propose that this limb contributes primarily to the *development of cancer*. It is axiomatic that a lethal cellular event precludes a malignant transformation event. If oxyradical and related species are important for cancer, either for initiation, promotion or progression, it is quite clear that they cannot be cytotoxic for the target cell. I suggest that the primary contribution of oxyradicals to cancer is at the stage of tumor progression, with lesser contributions to initiation or promotion.

It is now well known that multiple events are often required for the evolution of fully developed cancer. For example, even in cases of hereditary bilateral retinoblasto-

[a]This work was supported by a grant from the Medical Research Council of Canada.
[b]Address for correspondence: Dr. H. C. Birnboim, Department of Experimental Oncology, Ottawa Regional Cancer Centre, 501 Smyth Road, Ottawa, Ontario, Canada K1H 8L6.

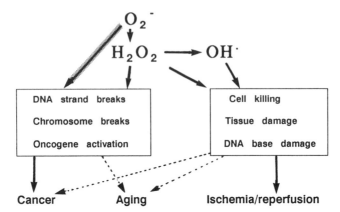

FIGURE 1. Dual effects of oxyradicals on biological systems. The *left limb* indicates a pathway postulated to be of greatest importance for carcinogenesis and possibly for aging. The *right limb* is the cytotoxic pathway, which may be of greatest importance for ischemia/reperfusion injury and also for aging.

ma, in which the initiating event is thought to occur in the germ line, a second event is thought to be required (reviewed briefly in Ref. 2). In transgenic animals in which an oncogene is introduced directly into the genome, only a fraction of the animals develop tumors, again indicating the need for some as yet undefined "spontaneous" event for full expression of the tumor phenotype.[3,4] I propose that DNA strand breaks caused by O_2^- and H_2O_2 are related to these "second events" because they promote chromosome breaks and rearrangements, leading to activation of oncogenes and inactivation of suppressor genes.

Hydroxyl radical (OH·) can be derived from these species in the presence of transition metals such as iron. I suggest that OH· is particularly important for the cytotoxic/tissue-damaging limb where its action may proceed primarily through lipid damage although it is so reactive that many other cell targets are possible. H_2O_2, at high concentrations, may be cytotoxic by causing depletion of ATP and NAD.[5] These are both massive disruptions which can lead to cell death. O_2^- and low concentrations of H_2O_2 may act more subtly since cell viability is important if the target cells are to become malignantly transformed. Most of our studies have concentrated on the effect of O_2^- and the mechanism by which extracellular O_2^- can cause strand breaks in nuclear DNA under conditions in which cell viability is not affected.

THE MODEL SYSTEM

A significant source of oxyradical species in intact animals and man is the "respiratory burst" of phagocytic cells such as granulocytes and macrophages. These cells use a membrane-bound NADPH oxidase complex to release relatively large amounts of O_2^- extracellularly in response to a variety of stimuli, including the tumor promoter, phorbol 12-myristate 13-acetate (PMA).[6–9] Most of the O_2^- which is formed is converted spontaneously or enzymatically to H_2O_2. We and others have shown that H_2O_2 added to intact cells can induce strand breaks in nuclear DNA. It has been much more difficult to show that extracellular O_2^- can also cause DNA strand breaks *separate from the effects of H_2O_2*. Our strategy for identifying O_2^--specific effects separate from H_2O_2-specific effects is shown schematically in FIGURE 2. When human granulocytes are stimulated with PMA, a "burst" of strand breaks in nuclear DNA lasting 30–40 min occurs shortly after the onset of O_2^- generation by these cells.[10,11] Addition of a small amount of catalase prevents the appearance of 50% of the strand breaks. One hundred times higher concentration of catalase has no further effect. However, the addition of superoxide dismutase (SOD) (in the presence of catalase) inhibits most of the remaining breaks. Since SOD is a protein that enters cells slowly, if at all, this observation allows us to define a population of DNA strand breaks that are specifically caused by extracellular O_2^-. Further evidence that extracellular O_2^- is indeed responsible for the breaks includes the demonstration that a negatively charged catechol compound (2,3-dihydroxybenzoic acid) can also effectively prevent the induction of breaks (unpublished experiments). Low concentrations of lipophilic SOD-mimetic compounds and catechol compounds are also very effective.[11,12] We have also shown that no breaks occur in granulocytes from patients with chronic granulomatous disease, a genetic defect in the NADPH oxidase that produces O_2^-.[13] Thus, we have accumulated considerable evidence that extracellular O_2^- can indeed cause strand breaks in the nuclear DNA of granulocytes in the absence of extracellular H_2O_2.

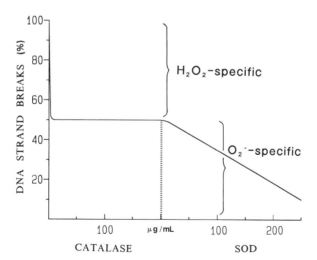

FIGURE 2. Schematic representation of the method used for distinguishing O_2^--induced strand breaks from total strand breaks in human granulocytes stimulated with PMA. Addition of low levels of catalase eliminates extracellular H_2O_2 and prevents about 50% of the strand breaks. High concentrations of catalase have no further effect. The remaining breaks can be prevented with SOD or other O_2^--specific reagents.

EXPERIMENTAL METHODS

The calcium-/magnesium-free salt solution used to suspend and treat cells (BSS) contained (in mM) NaCl,137; KCl,5; HEPES buffer,10; glucose,5.5; pH 7.4. $CaCl_2$ and $MgSO_4$ were added where indicated. Experimental procedures for isolation of granulocytes from human volunteers, treatment of cells with PMA and inhibitors in the presence of catalase, and measurement of DNA strand breaks and O_2^- levels have been described elsewhere.[11,14] Catalase was present in all experiments at 20 $\mu g/mL$ to eliminate the effects of extracellular H_2O_2 unless stated otherwise.

RESULTS AND DISCUSSION

O_2^- is a negatively charged molecule that enters cells through anion channels.[15–19] To investigate whether O_2^- requires entry into the cell through anion channels in the plasma membrane to cause DNA strand breaks or whether it acts at the external surface of the membrane, we tested the effect of two anion channel blockers, 4-acetamido-4'-isothiocyanatostilbene-2,2'-disulfonic acid (SITS) and 8-anilino-1-naphthalene sulfonic acid (ANSA). These compounds, at a concentration of 100–200 μM, have previously been shown to cause 80% inhibition of O_2^- release from phagocytic vacuoles of granulocytes[17] and to inhibit the passage of O_2^- through red blood cell membranes. Each was added to cells and its effects on the number of O_2^--induced strand breaks was determined after a 40-min incubation at 37°C (TABLE 1) (Birnboim, submitted for publication). Neither blocker significantly affected the induction of strand breaks by O_2^-, supporting the notion that O_2^- acts at the external cell surface and not by entering the cell.

An indication that the mechanism by which O_2^- causes DNA strand breaks in granulocytes may be novel is provided by the data presented in FIGURE 3.[12] In these experiments, we found that the rate and extent of extracellular O_2^- production by PMA-stimulated human granulocytes could be controlled by simply varying the amount of glucose added to the medium (FIG. 3a). In the presence of 5 mM glucose, cells generated approximately three-fold more O_2^- than in the absence of glucose. (Glucose is required by the hexose monophosphate shunt pathway to generate

TABLE 1. Effect of Anion Channel Blockers on Superoxide Anion-Induced DNA Strand Breaks in Human Granulocytes

	Number of DNA Strand Breaks/Cell (% ± Range)	
Concentration (μM)	SITS	ANSA
0 (PMA alone)	100	100
25	89.0 ± 6.0	102.8 ± 1.6
50	87.0 ± 6.3	117.5 ± 23.2
100	91.7 ± 4.1	110.9 ± 1.4
200	N.T.	84.6 ± 1.7

NOTE: Human granulocytes were incubated for 40 min at 37°C in BSS with 0.8 mM Mg^{2+} and 10 nM PMA either alone or in the presence of the indicated amount of anion channel blocker. Other details as in Reference 11. The average number of strand breaks and range of two experiments is shown.

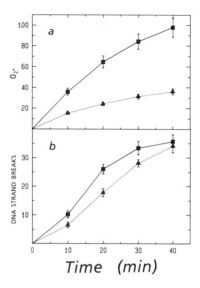

FIGURE 3. The effect of glucose on the amount of O_2^--generated (**a**) and the number of DNA strand breaks produced (**b**) in granulocytes stimulated with PMA at time 0. *Solid line:* 5 mM glucose added; *dotted line:* no glucose added.

NADPH, which is used by the membrane-bound NADPH oxidase to reduce O_2 to O_2^-; in the absence of glucose, granulocytes presumably utilize their intracellular stores of glycogen to supply glucose.) If O_2^- were to act as a source of H_2O_2 or $OH\cdot$, which in turn attacked nuclear DNA in a more direct fashion as occurs with radiation-induced breaks, then one might expect that the number of strand breaks would rise in proportion to the amount of O_2^- detected outside the cell. To determine whether this is so, the time course of DNA breakage was followed under these conditions (FIG. 3b). A large (three-fold) increase in O_2^- produced only a small increase in strand breaks, ranging from a maximum of 30% at 20 min to less than 10% at 40 min. These data allow us to reach a somewhat unexpected conclusion, that the number of strand breaks induced by O_2^- does not exceed a certain maximum, corresponding to 3000–4000 strand breaks per cell, despite a three-fold change in the amount of O_2^- generated. It should be noted that there is virtually no cell killing at the time these strand breaks are induced, as judged by the ability of the cells to exclude vital dyes.

These results suggested that O_2^- could be acting at an extracellular target, implying that a transmembrane signal might be involved to trigger strand breaks in nuclear DNA. To investigate the nature of this putative membrane target, cells were treated with PMA (and catalase) in the presence of an iron-specific chelating agent, 2 mM diethylenetriamine penta-acetate (DTPA) (FIG. 4) (unpublished experiments). DTPA had no effect or slightly stimulated the number of O_2^--induced breaks, indicating that iron was likely not involved. Desferrioxamine mesylate (Desferal) is an iron chelator with an affinity for Fe(III) which is 2–3 orders of magnitude higher than DTPA. Like DTPA, it is also a highly polar molecule which is poorly adsorbed after oral administration and likely to be taken up only slowly into cells. A low concentration of Desferal (20–40 μM) decreased the number of O_2^--induced breaks by 55% (FIG. 4). Higher concentrations of Desferal had no further effect (data not shown). Pretreating cells with 40 μM Desferal for 10 min and then removing it by washing the cells twice

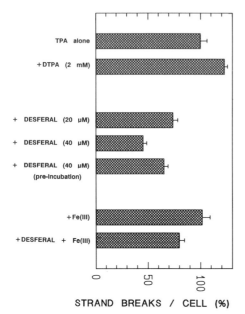

FIGURE 4. Effects of Desferal on the yield of O_2^--induced strand breaks. TPA (PMA) alone represents the number of breaks formed after a 40-min incubation at 37°C with 10 nM PMA after a 20-min preincubation and washings. +DTPA: 2 mM added together with PMA. +Desferal: added together with PMA at the indicated concentrations. +Desferal (preincubation): cells were incubated for 20 min with Desferal, then washed twice to remove Desferal and suspended in fresh medium containing PMA and incubated further for 40 min at 37°C. +Fe(III): 10 μM was added during the preincubation step. +Desferal+Fe(III): 10 μM Fe(III) was added in excess of the Desferal after 10 min of the preincubation period.

STRAND BREAKS / CELL (%)

before addition of PMA was still able to cause inhibition of the induction of strand breaks by 35%. None of the Desferal treatments affected the amount of O_2^- generated by the cells (data not shown). The final part of the figure shows that preloading cells with 10 μM Fe(III) for 10 min and then washing out Fe(III) in the presence of DTPA had no effect on O_2^--induced strand breaks. Adding back Fe(III) for 10 min after Desferal pretreatment for 10 min restored the number of O_2^--induced strand breaks only partially. Desferal is known to remove iron from ferritin and transferrin, but not hemoglobin or cytochromes. The nature of the surface iron that is affected by Desferal is unknown.

The lipophilic iron chelator, 2,2'-dipyridyl, also inhibits O_2^--induced strand breaks (unpublished experiments). Dipyridyl (100 μM) (in the presence or absence of 2 mM DTPA) inhibits strand breaks by more than 75%. However, pretreatment by the combination of chelators and subsequent removal by washing of cells does not confer continued inhibition as it does for Desferal (data not shown). This suggests that inhibition by dipyridyl is not due to chelation of intracellular iron. Furthermore, o-phenanthroline, a chelator similar to dipyridyl has the opposite effect: this agent *enhances* the yield of O_2^--induced strand breaks.[11] However, o-phenanthroline, like dipyridyl, inhibits H_2O_2-induced strand breaks. Taken together, these observations suggest strongly that Desferal acts by removing extracellular iron and not intracellular iron.

If O_2^- acts at an extracellular target to induce breaks in nuclear DNA, then a protein kinase could be involved, as is frequently the case for receptor-ligand-mediated transmembrane signals. Sodium orthovanadate is a phosphate analogue which inhibits various phosphatases including phosphoprotein phosphatases. Its effect on O_2^--induced strand breaks is shown in FIGURE 5.[12] Vanadate (100 μM) *increases* the yield of breaks by about 40%. An even more marked effect is seen in the presence of 3 μM A23187, an ionophore. A23187 alone markedly inhibits the yield of breaks (see below), but this inhibition is largely abrogated by the presence of vanadate. By contrast, vanadate *inhibits* the production of breaks by H_2O_2.

Intracellular calcium appears to be essential for the induction of breaks by O_2^-. As shown in FIGURE 5, ionophore A23187 markedly inhibits the generation of breaks in the absence of extracellular Ca^{2+}. Although it depletes intracellular Ca^{2+} under these conditions, it is not very specific for Ca^{2+} and therefore another metal cation could be involved. More specific depletion of intracellular Ca^{2+} is produced by ionophore ionomycin in combination with the Ca^{2+}-specific chelator, EGTA. Together these agents are highly specific for Ca^{2+} and they cause a 65% inhibition of strand breaks (TABLE 2) (Birnboim, submitted for publication). Interestingly, there is an increase in O_2^- production by these cells, showing once again (as in FIGURE 3) that O_2^- levels do not correlate with strand break levels. Another agent that chelates intracellular calcium and prevents strand breaks is Quin 2, an agent used as a fluorescent probe for monitoring intracellular levels (data not shown). Although intracellular Ca^{2+} depletion prevents strand break induction by O_2^-, intracellular Ca^{2+} levels do not rise significantly, as determined by FURA-2 (another fluorescence probe of intracellular Ca^{2+}), while strand breaks are being induced after exposure to PMA (data not shown).

We have also provided evidence that intracellular Mg^{2+} is required for the induction of strand breaks by O_2^- (FIG. 6) (Birnboim, submitted for publication). These experiments were carried out in the absence of extracellular Ca^{2+}. Varying the level of extracellular Mg^{2+} had little effect on the yield of O_2^--induced strand breaks (curve A). As seen also in FIGURE 5, the addition of 3 μM A23187 strongly inhibited

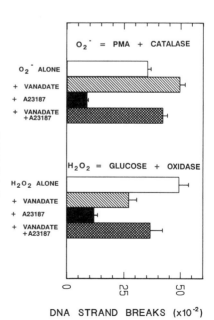

FIGURE 5. Effect of the indicated agents on the yield of O_2^-- and H_2O_2-induced DNA strand breaks in granulocytes. Where indicated, the agents were added together with PMA or glucose oxidase and strand breaks were measured after 40-min incubation. Concentrations used (μM): Vanadate, 100; A23187, 3.

TABLE 2. Effect of Ionomycin and EGTA on Superoxide Anion-Induced DNA Strand Breaks and O_2^- Production in Human Granulocytes

Treatment	Number of DNA Strand Breaks Per Cell ($\times 10^2$)	O_2^- Generated (nmol/10^6 cells/30 min)
1. PMA alone	39.7 ± 3.3 (19)	48.4 ± 9.0 (12)
2. +EGTA (4 mM)	31.9 ± 2.8 (8)	NT
3. +IONO (3 μM)	33.4 ± 2.0 (14)	58.5 ± 6.0 (8)
4. +IONO (3 μM)+EGTA (4 mM)	14.1 ± 2.9 (8)	78.3 ± 7.8 (9)

NOTE: Values shown are the average ± SEM (number of experimental observations). Granulocytes were incubated in BSS containing 0.8 mM $MgCl_2$ for 40 min at 37°C with 10 mM PMA. IONO, Ionomycin; NT, not tested. Other details as in Ref. 11.

strand break formation. However, the extent of inhibition was partially affected by the concentration of Mg^{2+} in the medium (curve B). In the complete absence of Mg^{2+}, inhibition of strand breaks was 86% and inhibition was reduced to 61% by adding 1.6 mM Mg^{2+}. A more marked effect of Mg^{2+} on the yield of strand breaks could be demonstrated in the presence of A23187 and sodium orthovanadate (curve C). As shown above (FIG. 5), the inhibition of O_2^--induced strand breaks by A23187 can be reversed by vanadate. The data of curve C indicate that the degree of reversal by vanadate is markedly influenced by the concentration of Mg^{2+}. In the absence of Mg^{2+}, there was 53% inhibition compared with the number of O_2^--induced breaks seen in the presence of 0.8 mM Mg^{2+}. The effect of Mg^{2+} is demonstrated even more clearly when EGTA is present together with vanadate and A23187 (curve D). In this case, there is greater inhibition (85%) in the absence of added Mg^{2+}, either because EGTA removes traces of Mg^{2+} or because (together with A23187) it depletes intracellular Ca^{2+} more

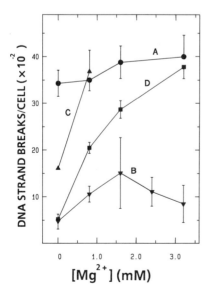

FIGURE 6. Effect of Mg^{2+} concentration on the production of O_2^--induced DNA strand breaks in PMA-stimulated human granulocytes. Cells were incubated in BSS at 37°C with 10 nM PMA for 40 min (for determination of strand breaks) in the presence of the following agents: (A) no addition; (B) A23187 (3 μM); (C) A23187 (3 μM) + vanadate (100 μM); and (D) A23187 (3 μM) + vanadate (100 μM) + EGTA (5 mM).

effectively. Higher concentrations of Mg^{2+} were needed in the presence of EGTA to restore the control number of breaks presumably because of its (relatively low) affinity for Mg^{2+} or possibly because higher concentrations of Mg^{2+} could compensate for the more complete exclusion of Ca^{2+}. In all cases where the number of strand breaks was reduced, this was accomplished by little or no effect on the amount of O_2^- produced. These experiments provide clear evidence that intracellular Mg^{2+} is required for the induction of DNA strand breaks by O_2^- in granulocytes.

Much of the interest in oxygen radicals has tended to focus upon the direct damaging effects on cells of $OH\cdot$, peroxy radicals, and reactive aldehydes that may be

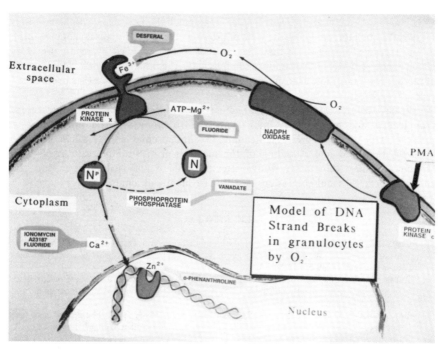

FIGURE 7. Hypothetical model of the biochemical steps involved in the induction of DNA strand breaks by O_2^- in PMA-stimulated human granulocytes.

generated by O_2^- and H_2O_2 (e.g., Reference 20 and other papers in this section), and upon other cytotoxic events[5,21] (the right limb of FIG. 1). Our work has emphasized the left limb of FIGURE 1, in particular, the mechanism by which O_2^- causes DNA strand breaks in intact granulocytes. We postulate that a similar mechanism may be operative in other cells such as early cancer cells that may be in contact with granulocyte- and macrophage-derived oxyradicals. These oxyradical species may contribute to the progression of early cancer cells to more malignant phenotypes. Our experiments strongly suggest that metabolic steps, perhaps a defined metabolic pathway, is triggered by extracellular O_2^-, which ultimately causes the observed DNA strand breaks. A speculative model is presented in FIGURE 7 to summarize our observations and to provide a framework for the design of further experiments. Addition of PMA to granulocytes activates protein kinase C, which in turn activates a membrane-bound NADPH oxidase.[22] This enzyme complex utilizes intracellular NADPH to generate extracellular O_2^-, which either dismutes to H_2O_2 and is removed by catalase (not shown) or reacts with an iron-containing moiety in the cell membrane; the reaction can be blocked by treatment with Desferal. The O_2^--iron reaction triggers a transmembrane signal, shown as activation of protein kinase x, which phosphorylates a hypothetical protein target, N. The process is inhibited by fluoride,[11] which severely depletes ATP[12] and removes Mg^{2+}. Vanadate inhibits a protein phosphatase and increases the level of the phosphorylated form, N^p. This hypothetical protein is proposed to cut DNA strands at specific locations and have a Ca^{2+} requirement. A DNA-binding protein with "Zn^{2+} fingers" may block the DNA since o-phenanthroline, a Zn^{2+} chelator, enhances the yield of breaks.[11] The DNA breaks do not constitute DNA "damage" in the sense that they are not readily rejoined by repair processes which can readily rejoin OH·-induced breaks.[23] Further studies may elucidate details of this pathway and allow us to determine whether a similar pathway exists in other cell types.

We have considered elsewhere that active oxygen species generated by intratumor granulocytes and macrophages may contribute to promotion and progression of tumors.[24] DNA strand breakage leading to chromosomal abnormalities may be involved. Improved understanding of the mechanism of strand breaks by O_2^- in granulocytes may allow us to study better the process responsible for causing the chromosome breaks and aberrations observed frequently in tumor cells.

REFERENCES

1. CROSS, C. E. 1987. Oxygen radicals and human disease. Ann. Intern. Med. **107:** 526–545.
2. HANSEN, M. F. & W. K. CAVENEE. 1988. Tumor suppressors: Recessive mutations that lead to cancer. Cell **53:** 172–173.
3. LEDER, A., P. K. PATTENGALE, A. KUO, T. A. STEWART & P. LEDER. 1986. Consequences of widespread deregulation of the c-myc gene in transgenic mice: Multiple neoplasms and normal development. Cell **45:** 485–495.
4. LANGDON, W. Y., A. W. HARRIS, S. CORY & J. M. ADAMS. 1986. The c-myc oncogene perturbs B lymphocyte development in Eμ-myc transgenic mice. Cell **47:** 11–18.
5. SCHRAUFSTATTER, I. U., P. A. HYSLOP, D. B. HINSHAW, R. G. SPRAGG, L. A. SKLAR & C. G. COCHRANE. 1986. Hydrogen peroxide-induced injury of cells and its prevention by inhibitors of poly(ADP-ribose) polymerase. Proc. Natl. Acad. Sci. USA **83:** 4908–4912.
6. HAMERS, M. N. & D. ROOS. 1985. In Oxidative Stress. H. Sies, Ed.: 351. Academic Press. Orlando, FL.
7. ROSSI, F. 1986. The O_2^--forming NADPH oxidase of the phagocytes: Nature, mechanisms of activation and function. Biochim. Biophys. Acta **853:** 65–89.
8. BRIGGS, R. T., J. M. ROBINSON, M. L. KARNOVSKY & M. J. KARNOVSKY. 1986. Superoxide production by polymorphonuclear leukocytes. A cytochemical approach. Histochemistry **84:** 371–378.
9. PICK, E., Y. BROMBERG, S. SHPUNGIN & R. GADBA. 1987. Activation of the superoxide forming NADPH oxidase in a cell-free system by sodium dodecyl sulfate. J. Biol. Chem. **262:** 16476–16483.
10. BIRNBOIM, H. C. 1982. DNA strand breakage in human leukocytes exposed to a tumor promoter, phorbol myristate acetate. Science **215:** 1247–1249.
11. BIRNBOIM, H. C. & M. KANABUS-KAMINSKA. 1985. The production of DNA strand breaks in human leukocytes by superoxide anion may involve a metabolic process. Proc. Natl. Acad. Sci. USA **82:** 6820–6824.

12. BIRNBOIM, H. C. 1988. A superoxide anion-induced DNA strand break metabolic pathway in human leukocytes: Effects of vanadate. Biochem. Cell Biol. **66:** 374–381.

13. BIRNBOIM, H. C. & W. D. BIGGAR. 1982. Failure of phorbol myristate acetate to damage DNA in leukocytes from patients with chronic granulomatous disease. Infect. Immun. **38:** 1299–1300.

14. BIRNBOIM, H. C. & J. J. JEVCAK. 1981. Fluorometric method for rapid detection of DNA strand breaks in human white blood cells produced by low doses of radiation. Cancer Res. **41:** 1889–1892.

15. LYNCH, R. E. & I. FRIDOVICH. 1978. Permeation of the erythrocyte stroma by superoxide radical. J. Biol. Chem. **253:** 4697–4699.

16. WEISS, S. J. 1982. Neutrophil-mediated methemoglobin formation in the erythrocyte. J. Biol. Chem. **257:** 2947–2953.

17. GENNARO, R. & D. ROMEO. 1979. The release of superoxide anion from granulocytes: Effect of inhibitors of anion permeability. Biochem. Biophys. Res. Commun. **88:** 44–49.

18. ROSEN, G. M. & B. A. FREEMAN. 1984. Detection of superoxide generated by endothelial cells. Proc. Natl. Acad. Sci. USA. **81:** 7269–7273.

19. TAKAHASHI, M.-A. & K. ASADA. 1983. Superoxide anion permeability of phospholipid membranes and chloroplast thylakoids. Arch. Biochem. Biophys. **226:** 558–566.

20. HALLIWELL, B. 1987. Oxygen radicals and metal ions: Potential antioxidant intervention strategies. *In* C.E. Cross, Moderator. Oxygen Radicals and Human Disease. Ann. Intern. Med. **107:** 526–545.

21. SCHRAUFSTATTER, I. U., P. A. HYSLOP, J. JACKSON & C. C. COCHRANE. 1987. Oxidant injury of cells. Int. J. Tissue React. **IX:** 317–324.

22. SHA'AG, D. & E. PICK. 1988. Macrophage-derived superoxide-generating NADPH oxidase in an amphiphile-activated, cell-free system; partial purification of the cytosolic component and evidence that it may contain the NADPH binding site. Biochim. Biophys. Acta **952:** 213–219.

23. BIRNBOIM, H. C. 1986. DNA strand breaks in human leukocytes induced by superoxide anion, hydrogen peroxide and tumor promoters are repaired slowly compared to breaks induced by ionizing radiation. Carcinogenesis **7:** 1511–1517.

24. KADHIM, S., B. F. BURNS & H. C. BIRNBOIM. 1987. In vivo induction of tumor variants by phorbol 12-myristate 13-acetate. Cancer Lett. **38:** 209–214.

DISCUSSION

M. MIRANDA (*University of L'Aquila, Italy*): Have you taken into account in your experiments the effects of divalent ion chelators on DNA repair?

H. C. BIRNBOIM (*Ottawa Regional Cancer Centre, Ottawa, Ontario, Canada*): We believe that the lack of strand break rejoining in nuclear DNA, induced by extracellular superoxide anion, is not due to faulty DNA repair mechanisms. DNA breaks due to ionizing radiation or bleomycin are rapidly repaired under conditions in which superoxide anion breaks would be very slowly "repaired." We have not tested for the effects of any of the chelators listed on radiation-induced strand breaks.

R. LOTAN (*University of Texas M.D. Anderson Hospital, Houston, Texas*): Could you please clarify why you study the effect of TPA on DNA strand breaks in granulocytes, which are nonproliferating cells and where the consequences of the DNA damage cannot easily be followed.

BIRNBOIM: It is certainly true that the consequences of DNA breaks in granulocytes are not likely to be biologically significant. We have studied this system in order to identify the metabolic pathway involved. Only in this system have we been able to monitor extracellular O_2^- and DNA strand breaks in the same cells. Generation of O_2^- in free solution (for example, xanthine-xanthine oxidase) leads almost exclusively to $H_2O_2^-$-induced breaks and no O_2^--induced breaks. We hope that what we learn in granulocytes will help us to understand the effects of neutrophil-generated oxyradicals on early tumor cells and their possible involvement in tumor progression.

G. MINOTTI (*Catholic University, Rome, Italy*): My first question deals with the differences between Desferal and phenanthroline; did you take into account that Desferal has a Fe^{2+} oxidase activity, whereas phenanthroline does not? Secondly, do you have any information on the nature of the extracellular target of Desferal? Could it be a NADH-diferric transferrin reductase?

BIRNBOIM: I am unsure about what could be expected by a putative Desferal Fe(II) oxidase activity. The inhibition of O_2^--induced strand breaks persists partially even after Desferal is washed out. *o*-Phenanthroline was mentioned primarily to indicate that it differed from dipyridyl, a similar lipophilic chelator, with respect to its effects on O_2^--induced breaks. I have no information on Desferal and NADH-diferric transferrin reductase, but it is an interesting suggestion.

P. A. RILEY (*University College and Middlesex School of Medicine, London, U.K.*): I believe that I am correct in saying that these results obtained with different iron chelators may be consequences of the redox state of the iron bound to them. Example: *o*-phenanthroline binds Fe(II) whereas DTPA binds Fe(III), so that your results with the glucose-oxidase-generated H_2O_2 could be explained by HO· generation by the Fenton reaction (in the case of *o*-phenanthroline). Since, as you pointed out, it is extremely unlikely that extracellularly generated HO· will produce DNA strand breaks, this could be seen as an inhibition in your system.

BIRNBOIM: DTPA and Desferal have similar specificities for Fe(II) compared to Fe(III), yet only Desferal inhibits O_2^--induced strands breaks. Both *o*-phenanthroline and dipyridyl inhibit H_2O_2-induced breaks, so Fe(II) could be involved in a Fenton reaction. However, *o*-phenanthroline and dipyridyl differ in that the former enhances O_2^--induced breaks whereas dipyridyl inhibits. Both lipophilic chelators have similar specificities for Fe(II) versus Fe(III).

D. COWEN (*Case Western Reserve University, Cleveland, Ohio*): First, was the effect of the indicated agonists controlled for changes in superoxide production? Otherwise changes in superoxide production could have caused the changes in DNA breakage. Second, your data indicate that DNA breakage is receptor-mediated but they do not discern between a membrane or cytosolic receptor.

BIRNBOIM: In answer to your first question: In all cases we monitor the amount of O_2^- produced so that inhibition is not due to this "trivial" reason.

Second, our data suggest that there is a saturable site for O_2^- action at the plasma membrane and that a transmembrane signal is likely needed. A cytosolic receptor seems less likely.

Oncogenes, Hormones, and Free-Radical Processes in Malignant Transformation *in Vitro*[a]

CARMIA BOREK

Radiological Research Laboratory
Department of Radiation Oncology, and
Department of Pathology/Cancer Center
College of Physicians & Surgeons
Columbia University
New York, New York 10032

Epidemiologic studies and data from experiments *in vivo* and *in vitro* support the notion that cancer is a multi-step process, in that a series of specific events is required to transform a normal cell into a malignant one.[1,2]

The multi-step model of carcinogenesis involves initiation, in which irreversible genetic alterations take place, and promotion, in which the clonal population of initiated cells is expanded and ultimately progresses to a malignancy.

Within the past decade we have witnessed changes in our understanding of the molecular origins of cancer. Much of this progress stems from the discovery of specific genes, the oncogenes, which are present in the genomes of a variety of tumor cells and are responsible for specifying many of the malignant traits of the cells.[3,4] A number of oncogenes found in tumor cells or in cells transformed *in vivo* or *in vitro* play a central role in carcinogenesis. This has been underscored by the ability of these genes to confer a malignant state on normal cells when introduced into the normal cells by means of transfection.[5-7]

While DNA is the target in carcinogenesis, the ultimate course and frequency of the neoplastic processes are determined by an interplay of endogenous and exogenous factors.[8,9] These included permissive factors which may act as co-transforming agents and potentiate carcinogenesis. The permissive factors are balanced, under normal conditions, by cellular protective factors which suppress the carcinogenic process at its various stages and antagonize the action of the permissive factors.[9]

Work in our laboratory has focused on cell transformation by radiation and chemicals using *in vitro* cell cultures.[10,11] These systems of human and rodent cells afford the opportunity to study genetic changes associated with transformation, under conditions free from host-mediated effects. They also enable us to characterize factors that potentiate or inhibit transformation at a cellular and molecular level.

[a]This investigation was supported by Grant No. CA-12536 awarded by the National Cancer Institute and by a contract from the National Foundation for Cancer Research.

ONCOGENES IN RADIATION AND CHEMICALLY
INDUCED TRANSFORMATION

Our studies on radiogenic transformation *in vitro* of mouse C3H/10T-1/2 and hamster embryo cells indiate that DNAs from the radiation-transformed cells, and from the tumors induced by injecting the cells into nude mice, contain genetic sequences with detectable transforming activity in three recipient cell lines: NIH/3T3

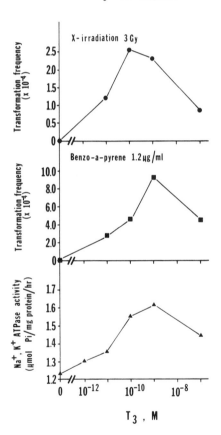

FIGURE 1. Thyroid hormone dose-dependence of radiation and chemically induced transformation and the enzyme Na/K ATPase. (From Borek *et al.*[15] Reproduced by permission.)

cells, C3H/10T-1/2 cells, and Rat-2 cells.[7] Southern blot analysis of the transfectants carrying these transforming genes (oncogenes) indicates that they are not of the *ras* gene family, which has been found in chemically transformed cells[6] and in most neoplasias including some tumors induced by radiation,[12] nor are these genes the *neu, trk, raf, abl* or *fms,* genes that have been implicated in other neoplastic states.[7]

Restriction enzyme cleavage of the DNAs indicated that an identical pattern of sensitivity and resistance to endonucleases exists among the DNAs of various independently radiation-transformed clonal lines of the same speices, but differs between the hamster and mouse cells,[7] suggesting that different oncogenes are activated by radiation in the hamster and mouse cells or that the same oncogene differs in its location on the genome of the two species.[7]

We have found that malignant transformation *in vitro* by radiation of hamster embryo cells and mouse C3H/10T-1/2 cells involves the activation of unique oncogenes, which heretofore have not been described.[7]

PERMISSIVE AND PROTECTIVE FACTORS

While DNA is the target in radiogenic and chemically induced *in vitro* transformation, a variety of endogenous and exogenous factors play a determining role in modifying the onset and expression of the neoplastic process. Physiological factors as well as cell contact modulate the expression of radiogenic transformation in genetically homogenous, cloned populations of cells.[13] Cell-cell communication between cells can be modified and may lead to an inhibition of growth in the transformed cell populations by contact with their normal counterparts.[13] Serum factors also modify cell-cell communication among transformed cells[13] and modulate their "metastasizing" phenotype *in vitro.*

HORMONES AS PERMISSIVE AND POTENTIATING FACTORS IN TRANSFORMATION

Hormones have long been known to exert an important influence on neoplastic transformation *in vivo,* yet we have limited knowledge of the underlying mechanisms at a cellular and molecular level. *In vitro,* where cells are grown and treated under defined conditions free from homeostatic events, we investigated the role of hormones in transformation of rodent cells.[14–16]

THYROID HORMONES

Our studies, using hamster embryo cells and C3H/10T-1/2 mouse cells or NRK rat cells, show that thyroid hormones play a critical role in cellular transformation by radiation, chemical carcinogens, and tumor viruses.[14–16]

Removal of thyroid hormones from the serum did not modify cell growth or survival, but made the cells refractory to transformation by X-rays, benzo[a]pyrene (BP), and N-methyl-N'-nitro-N-nitrosoguanidine (MNNG), and by the Kirsten murine sarcoma virus. When triiodothyronine (T3) was added at physiological levels (10^{12} to 10^{10}M), cell transformation was induced in a T3-dose-dependent manner (FIG. 1). The induction of transformation took place within a confined window in time. Maximum transformation was observed when the hormone was added 12 hours prior to exposure to the oncogenic agents, and it had no effect on transformation frequencies when added after exposure to radiation, chemicals, or the virus.[14–16]

The action of thyroid hormones in transformation appears to be mediated via more than one route. Our studies indicate that T3 may influence gene expression and the synthesis of a specific transformation-associated protein that may play a role in the neoplastic process.[12,13] This possibility is supported by the fact that the T3 dose-response relationship for transformation is similar to the T3 dose-dependence for a cellular protein, the enzyme Na/K^+.[14,15] In recent work using two-dimension high-resolution electrophoresis we found that T3 induced the expression of a specific cellular protein (Lambert and Borek, in preparation).

Other recent data indicate that T3 plays a role in cellular transfection by genomic DNA from radiation-transformed cells, as well as by the T24-mutated *ras* gene, indicating that the hormone regulates cellular gene expression in transformation (TABLE 1) in a similar manner to that of the tumor promoter TPA.[8]

An additional mechanism for thyroid hormone regulation of transformation resides in the ability of the hormones to regulate cellular oxygen uptake. The pro-oxidant state of the cells would then be modified by the presence of the hormones. This possibility is supported by the fact that superoxide dismutase (SOD) suppresses the effects of T3 in

TABLE 1. Thyroid Hormone Modulation of Cellular Transforming Genes

	Efficiency of Transformation		
Donor DNA	Foci/μg of DNA: Recipient	Foci/μg of DNA: Thyroid-Containing Serum	Foci/μg of DNA: Thyroid-Depleted Serum
C3H/10T-1/2	NIH/3T3	0.00	0.00
C3H/10T-1/2, X-ray-transformed	NIH/3T3	0.14	0.00
Hamster embryo, normal	NIH/3T3	0.00	0.00
Hamster embryo, X-ray: transformed	NIH/3T3	0.15	0.03
T24 mutated *ras* gene	NIH/3T3	52.4	9.5

potentiating radiogenic transformation as well as inhibiting the promoting action of TPA and teleocidin,[17] tumor-promoting agents that generate free-oxygen species.

ANTIOXIDANTS AS PROTECTIVE FACTORS IN TRANSFORMATION

The interaction of cells with radiation, both x-ray and ultraviolet (UV) light, as well as with a variety of chemicals, results in an enhanced generation of free-oxygen species and free-radical intermediates.[18] The result is a loss in the optimal cellular balance between the oxidative challenge, a source of DNA damage, and the inherent mechanisms that protect the cell from excess oxidative stress. These protectors include enzymes (SOD, catalase, peroxidases, transferases) and thiols. Also included are a variety of nutrients that directly or indirectly prevent peroxidation and autoxidation of proteins and lipids in cell membranes as well as in the nucleus. These include vitamin A, β-carotene, vitamin C, selenium, and vitamin E.[18]

In recent years, increasing evidence has implicated free-radical mechanisms in the initiation and promotion of malignant transformation *in vivo* and *in vitro*. Much of the evidence has come from the fact that the agents that scavenge free radicals directly or that interfere with the generation of free-radical-mediated events inhibit the neoplastic process. We have shown in hamster embryo cells that SOD inhibits transformation by radiation and bleomycin and suppresses the promoting action of TPA.[19] Catalase had no effect as an inhibitory agent in this cell system, perhaps beause of the inherent high level of the enzyme in the hamster cells.[19] SOD had a more dramatic inhibitory effect when maintained on the cells throughout the experiment, suggesting that later stages in the transformation process are influenced by free radicals.[19]

SELENIUM AND VITAMIN E

Other agents that qualify as important antioxidants are various examples of nutrients which are critical in controlling free-radical damage, namely, selenium, a component of glutathione peroxidase, and vitamin E, a powerful antioxidant and a component of the cell membrane.[18] We examined the single and combined effects of selenium and vitamin E on cell transformation induced in C3H/10T-1/2 cells by X-rays, benzo[a]pyrene, or tryptophan pyrolysate and on the levels of cellular scavenging systems and peroxide destruction. Incubation of C3H/10T-1/2 cells with 2.5 μM Na_2SeO_3 (selenium) or with 7 μM α-tocopheral succinate (vitamin E) 24 hours prior to exposure to X-rays or the chemical carcinogens resulted in an inhibition of transformation by each of the antioxidants with an additive-inhibitor action when the two nutrients were combined.[18] Cellular pretreatment with selenium resulted in increased levels of cellular glutathione peroxidase, catalase, and nonprotein thiols (glutathione) and in an enhanced destruction of peroxides. Cells pretreated with vitamin E did not show these biochemical effects, and the combined pretreatment with vitamin E and selenium did not augment the effect of selenium on these parameters. The results support our earlier studies showing that free-radical-mediated events play a role in radiation and chemically induced transformation. They indicate that selenium and vitamin E act alone and in additive fashion as radio-protecting and chemopreventing agents. Selenium confers protection in part by inducing or activating cellular free-radical scavenging systems and by doubling peroxide breakdown, thus enhancing the capacity of the cell to cope with oxidant stress. Vitamin E appears to confer its protection by an alternate complementary mechanism. Vitamin E acts as a chain-breaking antioxidant inhibiting the lipid peroxidation and the formation of malonaldehyde, a compound with oncogenic potential.[18]

Selenium acts as a true protector. Time-course experiments indicate that the addition of selenium at various exposures to X-rays results in a suppressive action which diminishes with time.[20]

An important determinant in the efficiency of cellular protection by inherent antioxidants lies in the interaction among various factors. The metabolic functions of vitamin E and selenium are interrelated, and selenium plays a role in the storage of vitamin E.[18] Vitamin E action is also closely related to that of vitamin C, which appears to increase its antioxidant effect.[21] We find that vitamin C acts in synergistic fashion with vitamin E to inhibit radiogenic transformation.[22]

Different organs have a different content of inherent antioxidants and these may vary from one species to another as well as from one individual to another. Thus, tissues and cells will vary in their response to oxidant stress. Adding external antioxidants may be effective in helping some cells mount a protective response while being ineffective in others.

OZONE AS A CARCINOGEN

The role of free radicals in the carcinogenic process can be inferred from the protective action of agents which scavenge free radicals at different stages of the oxidative process. However, their role can further be substantiated by the carcinogenic action of ozone, a free-radical-producing agent.[23]

Ozone, a reactive species of oxygen, is not a free radical per se. However, it interacts with a wide range of molecules on cell membranes to produce free radicals.[23] Its toxic action can be counteracted by a variety of antioxidants, thus preventing its direct oxidation of protein or polyunsaturated fatty acids.

We have evaluated whether ozone can directly transform cells *in vitro* and whether ozone as a radiomimetic agent modulates the transforming action of ionizing radiation.[23]

We found that treatment of hamster embryo and mouse C3H/10T-1/2 cells with 5 ppm O_3 for 5 min resulted in enhancing cell transformation compared to control untreated cells. When cells were first irradiated with gamma-rays and then, 2 hours later, exposed to O_3, the two agents acted synergistically in inducing cell transformation.[20]

While O_3 as an oxidant may interact with a large number of molecules to produce its effects, one of its major actions resides in its ability to peroxidize polyunsaturated fatty acids and produce malonaldehyde, which reacts with thiols, cross-links DNA and histones, and acts as an initiator in mouse skin carcinogenesis.[23]

We tested the short-term effects of 5 ppm O_3 in producing lipid peroxidation products in the hamster embryo and mouse C3H/10T-1/2 cells compared to air-treated controls. The assays were done in conjunction with the transformation experiments and used cultures parallel to the ones being exposed to O_3 for transformation assays. Lipid peroxidation as measured by the formation of thiobarbituric acid (TBA)-reactive products was assayed within 10 min after O_3 exposure.[23]

We found that malonaldehyde and malonaldehyde-like products were formed at higher levels in O_3-exposed cells as compared to controls.[23]

The finding that lipid peroxidation products are elevated in response to O_3 suggests a partial role for free-radical-mediated reactions in O_3-induced neoplastic transformation. Further support comes from our recent results indicating that the antioxidant vitamin E (α-tocopherol) inhibits O_3-induced transformation in the hamster and C3H/10T-1/2 cells.[20]

CONCLUSIONS

One of the basic conundrums in carcinogenesis evolves from our ability to unequivocally distinguish primary events associated with initiation of malignant transformation from those that function as secondary events. Thus the role of oncogenes, mutations, gene rearrangements, amplification and other DNA alterations in transformation is yet unclear. The changes that take place give rise to abnormal expression of cellular genes.

We must always be cognizant of the fact that a variety of factors may modify the neoplastic process at its various stages of development. These constitute physiological permissive or protective factors. When permissive factors prevail, such as genetic susceptibility, optimal stage in the cell cycle, optimal hormonal control, or a particular stage in differentiation, initiation of transformation will take place. By contrast, if these permissive factors do not prevail, protective factors such as free-radical scavengers will inhibit to varying degrees the onset and progression of the neoplastic process. These may be inherent cellular factors or those added externally by dietary means acting as anticarcinogens. Thus, the interplay between inherent genetic and physiological factors, and lifestyle influences which either enhance or inhibit the neoplastic process, are critical determinants in the process of multi-stage carcinogenesis and in establishing the incidence of cancer.

REFERENCES

1. FOULDS, L. 1969. Neoplastic Development, Vol. I. Academic Press. London and New York.
2. DOLE, F. R. S. & R. PETO. 1981. The Causes of Cancer. Oxford University Press.
3. BARBACID, M. 1986. Carcinogenesis **7**: 1037–1042.
4. LAND, H., L. F. PARADA & R. A. WEINBERG. 1983. Nature (London) **304**: 596–602.
5. SUKUMAR, S., V. NOTARIO, D. MARTIN-ZANCA & M. R. BARBACID. 1983. Nature (London) **306**: 658–661.
6. TABIN, C. J., S. M. BRADLEY, C. O. BARGMANN, R. A. WEINBERG, A. G. PAPAGEORGE, E. SCOLNICK, R. DHAR, D. R. LOWY & E. H. CHANG. 1982. Nature (London) **300**: 143–152.
7. BOREK, C., A. ONG & H. MASON. 1987. Proc. Natl. Acad. Sci. USA **84**: 794–798.
8. HSIAO, W. L. W., S. GATTON-CELLI & I. B. WEINSTEIN. 1984. Science **226**: 552–554.
9. BOREK, C. 1984. *In* The Biochemical Basis of Chemical Carcinogenesis. H. Greim *et al.*, Eds.: 175–188. Raven Press. New York.
10. BOREK, C. 1980. Nature **183**: 776–778.
11. BOREK, C. 1985. Pharmacol. Ther. **27**: 99–142.
12. GUERRERO, I., P. CALZADA, A. MAYER & A. PELLICER. 1984. Proc. Natl. Acad. Sci. USA. **81**: 202–205.
13. BOREK, C., S. HIGASHINO & W. R. LOEWENSTEIN. 1969. J. Membrane Biol. **1**: 274–293.
14. GUERNSEY, D. L., C. BOREK & I. S. EDELMAN. 1981. Proc. Natl. Acad. Sci. USA **78**: 5708–5711.
15. BOREK, C., D. L. GUERNSEY, A. ONG & I. S. EDELMAN. 1983. Proc. Natl. Acad. Sci. USA **80**: 5749–5752.
16. BOREK, C., A. ONG & J. S. RHIM. 1985. Cancer Res. **45**: 1702–1703.
17. BOREK, C., J. E. CLEAVER & H. FUJIKI. 1984. *In* Cellular Interaction by Environmental Tumor Promoters and Relevance to Human Cancer. H. Fujiki *et al.*, Eds.: 195. Japan Scientific Societies. Tokyo.
18. BOREK, C., A. ONG, H. MASON, L. DONAHUE & J. E. BIAGLOW. 1986. Proc. Natl. Acad. Sci. USA **83**: 1490–1494.
19. BOREK, C. & W. TROLL. 1983. Proc. Natl. Acad. Sci. USA **80**: 5749–5752.
20. BOREK, C. 1987. Br. J. Cancer **55**: 74086.
21. NIKI, E., T. SAITO, A. KAWAKAMI & Y. KAMIYA. 1984. J. Biol. Chem. **259**: 4177–4182.
22. BOREK, C. 1988. *In* Medical, Biochemical and Chemical Aspects of Free Radicals. E. Niki, Ed. Elsevier. Amsterdam. In press.

DISCUSSION

G. GUIDOTTI (*Università di Patologia Generale, Parma, Italy*): You have shown that thyroid hormones are potentiating the effect of radiation. The receptor for these hormones is likely to be by the proto-oncogene c-*erb-A*. Do you have any evidence that this gene is saturated under the conditions you have adopted? Is its expression involved in the radiation-mediated damage?

C. BOREK: We do have an effect, but the data are not yet ready to be discussed.

M. MIRANDA (*University of L'Aquila, Italy*): On transforming cells by using radiation do you expect to always find the same pattern of oncogene expression?

BOREK: We seem to have the same gene(s) activated in the C3H10T1/2-mouse-transformed cells, seen in ten individually X-ray-transformed clones and the tumors induced by the transformed cells. Similarly the hamster-embryo-transformed cells (ten clones and the tumors induced by them) show the same activated sequence(s), which differ from those in the mouse cells (the difference may be in the genes itself or in its location on the genomen). The location may differ in the two species.

H. C. BIRNBOIM (*Ottawa Regional Cancer Centre, Ottawa, Ontario, Canada*): Can you explain why concentrations of thyroid hormone (T3) above about 10^{-9}M strongly inhibit transformation?

BOREK: The thyroid receptor is saturated at physiological levels so that T3 at higher levels may produce effects other than those produced at the physiological levels of $10^{-12} - 10^{-9}$ M.

N. KRINSKY (*Tufts University School of Medicine, Boston, Mass.*): Does T3 affect vitamin E, Se, or GSH-Px levels?

BOREK: We do not know yet.

N. BERGER (*Case Western Reserve University, Cleveland, Ohio*): Thyroid hormone has a growth-stimulating effect on many cells. Can you separate a growth-promoting effect from a transformation-promoting effect for thyroid hormone?

BOREK: The first experiment we did was to study the effect of T3 on the growth rate of the cells and on cells after exposure to radiation. T3 did not affect these parameters. Its effects were specifically on transformation.

L. ERNSTER (*University of Stockholm, Stockholm, Sweden*): Could the thyroid hormone effect in enhancing cell transformation be explained simply in terms of an enhancement of oxidative metabolism—and therefore of "oxidative stress"—induced by the hormone through its known stimulation of mitochrondrial protein synthesis?

BOREK: I think that this is clearly part of the answer. T3 affects cell communication, and it induces the expression of specific proteins which may be transforming proteins and has an effect on oxidative metabolism. All these may be interrelated. We have found that SOD inhibits the T3 effects, which support the role of T3 in modifying oxidant stress, thereby acting as a potentiator of transformation by radiation and chemicals.

Phagocyte-Mediated Carcinogenesis: DNA from Phagocyte-Transformed C3H 10T1/2 Cells Can Transform NIH/3T3 Cells[a]

SIGMUND WEITZMAN,[b] CYNTHIA SCHMEICHEL,
PATRICK TURK, CRAIG STEVENS, SARA TOLSMA,
AND NOEL BOUCK

Northwestern University Cancer Center
Northwestern University Medical School
Chicago, Illinois 60611

INTRODUCTION

The association of chronic inflammation with cancer has been recognized for centuries. During the 1970s it was noted that inflammatory phagocytes produce large amounts of reactive oxygen metabolites, and that these metabolites could be toxic to a variety of cells and tissues.[1-4] More recently, many groups have observed that these oxygen species can produce genetic lesions and phenotypic changes characteristic of carcinogens, thereby leading to the hypothesis that phagocyte-generated oxidants play a central role in carcinogenesis associated with inflammation.[5-10]

We showed previously that C3H mouse-derived 10T1/2 fibroblasts (10T1/2 cells) could undergo malignant transformation when exposed to activated human phagocytes or a cell-free oxygen radical-producing system.[11] Still unknown, however, is the nature of the specific genetic targets critical for transformation by this mechanism. In this paper we describe our initial efforts to identify transforming genes in the DNA extracted from these phagocyte-transformed 10T1/2 cells.

MATERIALS AND METHODS

Two standard cell lines were used in these experiments. 10T1/2 cells (clone 8) derived originally from C3H mouse fibroblasts were used as targets in the primary transformation experiments.[12] NIH/3T3 cells were used as recipients in the DNA transfection experiments (subclone kindly provided by Thomas Fitting of Scripps Clinic and Research Foundation).

[a]This work was supported in part by Grant CA27306 from the National Institutes of Health and by the Amoco Foundation.

[b]Address for correspondence: Sigmund Weitzman, M.D., Section of Hematology/Oncology, Department of Medicine, Northwestern University Medical School, Olson 8524, 303 E. Chicago Avenue, Chicago, Illinois 60611.

The transformation of the 10T1/2 cells by oxidants generated by activated human neutrophils has been described.[11] In brief: 10T1/2 cells were grown in Eagle's basal medium with Earle's salts supplemented with 10% fetal calf serum in 100-mm plastic dishes. Human neutrophils suspended in Krebs Ringer's phosphate buffer were layered over the 10T1/2 cells for one hour at 37°C. The 10T1/2 cells were washed, removed from the dishes, and then processed according to one of two different protocols. In one method, the 10T1/2 cells were immediately injected subcutaneously into nude mice (directly after removal of the cells from the original treatment dishes). Tumors that appeared were excised and placed in tissue culture. With the second protocol, no passage through mice occurred; after removal from the treatment dishes, the cells were plated into new 100-mm dishes. After six to eight weeks transformed foci began to appear. Foci were then isolated by aspiration into plastic cloning cylinders and plated into new dishes.

Genomic DNA was prepared from individual tumors or transformed foci by the methods of Maniatis et al.[13] DNA was also prepared from native 10T1/2 and NIH/3T3 cells for use as negative controls. For positive controls, DNA from the MCA-16[14] cell line (a chemically transformed 10T1/2 cell line kindly provided by Dr. J. Landolph of the University of Southern California) and the T24 bladder cancer cell line[15] were used.

The DNA transfection was accomplished by using the calcium phosphate co-precipitation method, as described by Wigler et al.[16] NIH/3T3 cells in Dulbecco modified Eagle's medium (DMEM) supplemented with 10% donor calf serum (from Flow Laboratories, McLean, VA or from Whitaker M.A. Bioproducts, Walkerville, MD) were seeded at a density of 7×10^5 cells per 60-mm dish. Twenty-four hours after seeding, dishes were incubated with between 20 and 60 micrograms of the appropriate calcium phosphate–DNA co-precipitate for four hours. In some experiments the cells were co-transfected with the plasmid pSV2 neo (5 µg/dish), which confers resistance to the antibiotic G418, thereby providing a selectable marker for successful transfection. The cells were then exposed to 20% DMSO in HEPES buffered saline solution for 2 minutes, washed, and re-fed. Twenty-four hours later, each dish was trypsinized and seeded into four 100-mm dishes. These dishes were then fed every three days and at confluence the serum supplement was reduced from 10% to 3%. Seventeen to twenty-six days after transfection, several transformed foci were isolated by means of cloning cylinders. All dishes were then stained with 0.5% crystal violet or 2% Giemsa and scored for the presence of transformed foci. Isolated foci were expanded in culture for extraction of transformant DNA. In those experiments with pSV2 neo was used, foci were expanded in the antibiotic G418 (Geneticin, 400 µg/ml, Sigma Chemical Corp.).

DNA extracted from the transfected NIH/3T3 transformants was then subjected to Southern blot analysis. Fifteen-microgram aliquots of each DNA sample were digested with a minimum of 10 units of restriction endonuclease per microgram of DNA under conditions recommended by the manufacturer. Samples were then subjected to electrophoresis overnight at 1.5 volts/cm in 1.0% agarose gels in 89 mM Tris-borate, 89 mM boric acid, and 20 mM EDTA. DNA was transferred by capillarity to Hybond filters (Amersham Corp.) according to the manufacturer's protocol. Blots were hybridized with DNA probes labeled with ^{32}P by the random primer method.[17] After hybridization, filters were washed and subjected to autoradiography at $-70°$. In other experiments, DNA from the original transformed 10T1/2 cells was also subjected to Southern blot analysis.

Oncogene probes were obtained from Oncor, Inc., Gaithersburg, MD, and included v-Ha-*ras*, v-Ki-*ras*, and v-*raf*. Restriction endonucleases were obtained from either BRL Laboratories or Boehringer-Mannheim Laboratories. Oligonucleotide primers for the labeling reactions were purchased from Pharmacia, and Klenow polymerase from BRL. The v-Ha-*ras* probe used was a 730-bp *Sst*I–*Pst*I fragment originally from rat; the v-Ki-*ras* probe was a 618-bp *Sst*II–*Hinc*II fragment originally from rat; and the v-*raf* probe was a 290-bp *Xho*I–*Sst*I fragment originally from mouse.

The *neu* probe was a 400-bp *Bam*HI fragment from rat, kindly provided by the laboratory of Dr. R. Weinberg.

TABLE 1. Transforming Activity of DNA from Neutrophil-Treated 10T1/2 Fibroblasts

Donor DNA	Foci/μg DNA[a] ($\times 10^{-2}$)
10T1/2	4 ± 1
T24	24 ± 9
MCA-16	14 ± 6
A[b]	29 ± 15
B	8 ± 2
C	5 ± 1

[a]Mean ± SE of five individual experiments.
[b]A, B, and C represent DNA from unrelated isolates of phagocyte-transformed 10T1/2 cells, each obtained from separate transformation experiments.

RESULTS

Results of five separate transfection experiments are summarized in TABLE 1. As shown, DNA from two of three phagocyte-transformed samples tested showed transforming activity more than twice that of the control 10T1/2 DNA. One isolate had transforming potency even greater than that of the standard positive control, human T24 DNA, while others were considerably less active.

The observation that the transfected DNA samples contained transforming sequences led us to screen the samples for some of the genetic alterations of mouse cells previously described to transform NIH/3T3 cells. Several groups have described mutations in codons 12 or 61 of the Ha-*ras* proto-oncogene in mouse cells.[18–21] Codon-61 mutations appear to predominate in studies using chemical carcinogens as transforming agents, but both codon-12 and -61 mutations are described.[18] The restriction enzyme *Xba*I can be used to detect activating mutations in the 61st codon of mouse Ha-*ras*, while the restriction enzymes *Taq*I or *Pst*I can be used to detect activating mutations in codon 12.[18,20] FIGURE 1 shows autoradiographs of Southern blots of DNA samples from 10T1/2 cells transformed by phagocytes and studied for Ha-*ras* mutations. As shown, no *Xba*I or *Taq*I polymorphisms were observed. *Pst*I digests also failed to reveal any Ha-*ras* polymorphisms (not shown). To verify that Ha-*ras* activation was not involved, we performed the same analysis on DNA samples

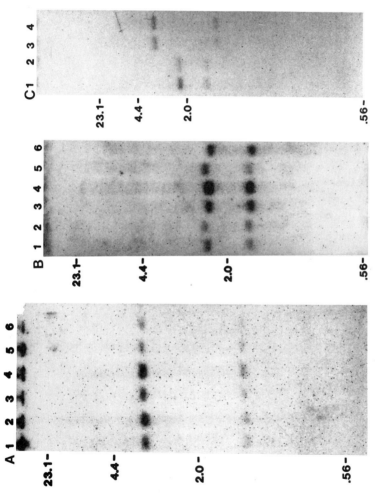

FIGURE 1. Southern blot analysis for Ha-*ras* mutation. (A) DNA samples from 10T1/2 cells were digested with both *Bam*HI and *Xba*I and hybridized with an Ha-*ras* probe. *Lane* 1: control 10T1/2 DNA; *lane* 2: MCA-16 (a chemically transformed 10T1/2 cell line) DNA; *lane* 3: DNA from phagocyte-transformed isolate "D"; *lane* 4: DNA from phagocyte-transformed isolate "C"; *lane* 5: DNA from phagocyte-transformed isolate "B"; *lane* 6: DNA from phagocyte-transformed isolate "A". (B) DNA samples from 10T1/2 cells were digested with *Taq*I and hybridized to a Ha-*ras* probe. *Lanes* are the same as in A. (C) DNA samples from NIH/3T3 cells were digested with *Taq*I (*lanes* 1 and 2) or *Bam*HI/*Xba*I (*lanes* 3 and 4) and hybridized to a Ha-*ras* probe. *Lanes* 1 and 3: control NIH/3T3 DNA; *lanes* 2 and 4: DNA from cells transformed by transfection with DNA from phagocyte-transformed 10T1/2 isolate "A". (The same results were observed using DNA from NIH/3T3 cells transformed by transfection with DNA from the other phagocyte-transformed 10T1/2 cell isolates [not shown]).

extracted from NIH/3T3 cells which had been transformed by transfection with the DNA from phagocyte-transformed 10T1/2 cells. As shown in FIGURE 1, there is no evidence of polymorphism of Ha-*ras* in the recipient transfected, transformed cells, suggesting that Ha-*ras* was not transferred in these experiments.

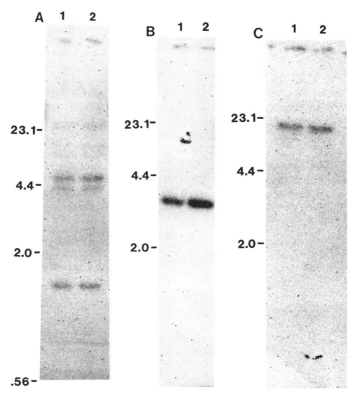

FIGURE 2. Southern blot analysis of transformed NIH/3T3 cells for evidence of transfection of Ki-*ras*, *raf*, or *neu* oncogenes. (**A**) DNA samples were digested with *Hind*III and hybridized to a Ki-*ras* probe. *Lane* 1: control NIH/3T2 DNA; *lane* 2: DNA from cells transformed by transfection with DNA from phagocyte-transformed isolate "A". (**B**) DNA samples were digested with *Eco*RI and hybridized to a v-*raf* probe. Lanes are the same as in **A**. (**C**) DNA samples were digested with *Eco*RI and hybridized to a *neu* probe. Lanes are the same as in **A**.

Similar experiments were performed with DNA samples extracted from NIH/3T3 cells transformed after transfection with DNA from the other phagocyte-transformed isolates (not shown), and the results were the same as those shown here.

Other groups, also using mouse cells, have reported mutational activation of the Ki-*ras* oncogene with either radiation or chemically induced transformation.[22,23] This appears to occur less frequently than Ha-*ras* mutation. FIGURE 2 shows results of Southern blot analysis of samples of DNA extracted from transfected, transformed NIH/3T3 cells digested with *Hind*III (as described by others[23,24]), and hybridized with a Ki-*ras* probe. Again, no polymorphisms were observed, suggesting that activation of the Ki-*ras* gene was not responsible for transformation of the NIH/3T3 cells in the transfection experiments.

We also hybridized Southern blots of the DNA from the transformed recipient NIH/3T3 cells with two other oncogene probes which have been reported to be able to transform NIH/3T3 cells, *raf* and *neu*. No differences were observed between control 3T3 DNA and the transformed cell samples (FIG. 2).

DISCUSSION

The major findings described in this paper are that: (1) DNA from 10T1/2 cells transformed by phagocyte-generated oxidants can produce transformation after transfection into NIH/3T3 cells, and (2) the oncogenes most commonly detected as mutationally activated by the NIH/3T3 system, Ha-*ras* and Ki-*ras*, do not appear to be involved in the transformation observed here. Furthermore, the transforming potency of the DNA from different isolates of phagocyte-transformed 10T1/2 cells can vary markedly. While one isolate can transform more efficiently than human T24 DNA, a standard positive control containing an activated *ras* oncogene, transforming activity of DNA from some other foci was lower, and indistinguishable from background in one of the three cases.

The finding that the 10T1/2-derived DNA samples can cause transformation is strong evidence that one of the ultimate lesions in phagocyte-induced transformation is heritable and stable in the DNA of some of the 10T1/2 target cells. The variability of transforming potency is of interest, as it held true with two different DNA preparations across five experiments. Similar results have been reported when a variety of methycholanthrene-transformed 10T1/2 cell lines were examined by this assay.[14] Possible explanations for the range of potencies observed include the following: (1) There are different genes involved in each of the respective oxidant-induced transformants; (2) the same gene is involved but at a different site or sites; or (3) the same gene is altered at the same site, but the specific nature of the alteration varies (analogous to different types of mutations at the same codon of the *ras* gene).

In the majority of the published investigations involving transformation of mouse cells *in vitro* or carcinogenesis in mice *in vivo*, evidence is cited of mutational activation of either the Ha-*ras* or Ki-*ras* protooncogene.[18–23] The most commonly reported specific lesion appears to be a mutation in codon 61 of Ha-*ras*.[18–20] In our studies, we found no evidence of involvement of Ha-*ras* or Ki-*ras* in the observed transformation. We also failed to find evidence of transfection *raf* or *neu* sequences in the NIH/3T3 transformants. Although Guerrero *et al.* have found Ki-*ras* activation in radiation-induced mouse lymphomas,[23] Borek *et al.*, investigating 10T1/2 cells transformed by radiation, published some findings similar to ours.[24] They found that DNA from radiation-transformed 10T1/2 cells could transform NIH/3T3 (and Rat-2) cells. Further, they found that the transforming sequences were not Ha-, Ki-, or N-*ras*, *neu*, *raf*, *abl*, *fms* or *trk*.[24] Borek *et al.*, also showed that transforming ability could be maintained if the transfected DNA was digested with the restriction enzymes *Bam*HI and *Xho*I, but was lost when *Eco*RI, *Kpn*I, or *Hind*III were used. We have not yet performed similar experiments, but it would be of interest to find out whether the transforming genes with which we are working are in any way related to those produced by radiation, since oxygen radical-related species may be involved in the mechanisms of transformation produced by both radiation and phagocytes.

We are now proceeding with studies that should allow further elucidation and cloning of the genes involved in transforming the NIH/3T3 cells discussed here. These investigations should provide useful information regarding the mechanisms of phagocyte-induced malignant transformation.

ACKNOWLEDGMENTS

We are grateful to N. Johnson and G. Groth for secretarial assistance.

REFERENCES

1. BABIOR, B. 1984. Blood **64:** 959–966.
2. FRIDOVICH, I. 1983. Ann. Rev. Pharmacol. Toxicol. **23:** 239–257.
3. HALLIWELL, B. & J. M. C. GUTTERIDGE. 1984. Biochem. J. **219:** 1–14.
4. HYSLOP, P. A., D. B. HINSHAW, W. A. HALSEY, I. V. SCHRAUFSTATTER, R. D. SAUERHEBER, R. G. SPRAGG, J. JACKSON & C. G. COCHRANE. 1988. J. Biol. Chem. **263:** 1665–1675.
5. WEITBERG, A. B., S. A. WEITZMAN, M. DESTREMPES, S. A. LATT & T. P. STOSSEL. 1983. N. Engl. J. Med. **308:** 26–30.
6. BIRNBOIM, H. C. 1983. *In* Radioprotectors and Anticarcinogens, O. F. Nygaard & M. G. Simic, Eds.: 539–556. Academic Press. New York.
7. CERUTTI, P. 1985. Science **227:** 375–381.
8. KENSLER, T. W. & B. G. TAFFE. 1986. Adv. Free Rad. Biol. Med. **2:** 347–387.
9. YAMASHINA, K., B. E. MILLER & G. H. HEPPNER. 1986. Cancer Res. **46:** 2396–2401.
10. GORDON, L. I. & S. A. WEITZMAN. 1988. *In* The Respiratory Burst and Its Physiological Significance. R. Strauss & A. J. Sbarra, Eds. Plenum. New York. In press.
11. WEITZMAN, S. A., A. B. WEITBERG, E. P. CLARK & T. P. STOSSEL. 1985. Science **227:** 1231–1233.
12. REZNIKOFF, C. A., J. BERTRAM, D. W. BRANKOW & C. HEIDELBERGER. 1973. Cancer Res. **33:** 3239–3249.
13. MANIATIS, T., E. F. FRITSCH & J. SAMBROOK. 1982. *In* Molecular Cloning, A Laboratory Manual. Cold Spring Harbor Laboratory. Cold Spring Harbor, NY.
14. SHIH, C., B. Z. SHILO, M. P. GOLDFARB, A. DANNENBERG & Ŗ. A. WEINBERG. 1979. Proc. Natl. Acad. Sci. USA **76:** 5714–5718.
15. REDDY, E. P., R. K. REYNOLDS, E. SANTOS & M. BARBACID. 1982. Nature **300:** 149–152.
16. WIGLER, M., A. PELLICER, S. SILVERSTEIN & R. AXEL. 1978. Cell **14:** 725–731.
17. FEINBERG, A. B. & B. VOLGELSTEIN. 1983. Anal. Biochem. **132:** 6–7.
18. QUINTANILLA, M., K. BROWN, M. RAMSDEN & A. BALMAIN. 1986. Nature **322:** 78–80.
19. WISEMAN, R. W., S. J. STOWERS, E. C. MILLER, M. W. ANDERSON & J. A. MILLER. 1986. Proc. Natl. Acad. Sci. USA **83:** 5825–5829.
20. BIZUB, D., A. W. WOOD & A. M. SKALKA. 1986. Proc. Natl. Acad. Sci. USA **83:** 6048–6052.
21. SMITH, G. J. & J. W. GRISHAM. 1987. Biochem. Biophys. Res. Commun. **147:** 1194–1199.
22. PARADA, L. F. & R. A. WEINBERG. 1983. Mol. Cell. Biol. **3:** 2298–2301.
23. GUERRERO, I., P. CALZADA, A. MAYER & A. PELLICER. 1984. Proc. Natl. Acad. Sci. USA **81:** 202–205.
24. BOREK, C., A. ONG & H. MASON. 1987. Proc. Natl. Acad. Sci. USA **84:** 794–798.

DISCUSSION

N. BERGER (*Case Western Reserve University, Cleveland, Ohio*): As you know the Philadelphia chromosome in chronic myelocytic leukemia involves a rearrangement of the *sis* and *abl* oncogenes. Do you think your findings support a peculiar sensitivity of the *abl* oncogene in granulocytes?

S. WEITZMAN: While we at first thought that the restriction polymorphism we noted in c-*abl* was due to a rearrangement, this is probably not the case. It would appear that methylation changes are responsible for the observed polymorphism.

R. LOTAN (*University of Texas M.D. Anderson Cancer Center, Houston, Texas*): Which changes in oncogenes did you expect to find in the phagocyte-transformed 10T1/2 cell?

WEITZMAN: We had expected to find an altered *ras* oncogene for two reasons: first, because most transforming genes in the NIH/3T3 system turn out to be mutationally activated *ras* genes, and second, because others had reported *ras* activation in mouse cells transformed *in vitro* and *in vivo*. This does not appear to be the case in our cells, however.

C. BOREK (*Columbia University, New York, N.Y.*): Do you find changes in transcription of the *abl* gene as we do in the radiation-transformed cells?

WEITZMAN: In our initial screening with Northern blots, total *abl* transcription does not appear to be increased in the transformants. However, the mouse c-*abl* gene appears to generate at least four distinct messages by 5′ switching, and we have not yet looked for alterations in the relative proportions of these different transcripts.

Radicals in Melanin Biochemistry[a]

PATRICK A. RILEY

Department of Chemical Pathology
University College, and
Middlesex School of Medicine
London W1P 6DB, England

INTRODUCTION

Melanins are bathochromic aromatic polymers of uncertain structure.[1,2] Melanin is widespread in both the animal and plant kingdoms. Although in some cases the function of melanin is apparent, as in such cases as protective coloring, the camouflage afforded by squid ink, and the protection of seeds and spores by a coating of melanin, the evolutionary significance of the pigment is often unclear.[3,4] Some melanins are formed by the autoxidation of polyhydric phenol precursors, as seems to be the case for neuromelanin, and these pigments represent byproducts of unrelated metabolic pathways. However, in most cases, melanin is formed by a distinct biochemical pathway involving oxidation by a specific enzyme. I shall confine my remarks to the tyrosinase-catalyzed process of melanogenesis. This enzyme-catalyzed oxidative pathway appears to be the basis for the generation of pigments ranging from black (eumelanins) to yellow (pheomelanins). There is some inverse correlation between the darkness of the pigment and the sulphur content, which is attributed to the formation of cysteine adducts during the polymerization.[5,6] Structural differences introduced by the inclusion of cysteine residues in the polymer do not readily explain the differences in color. It has been suggested that thiol adducts act as chain terminators in the polymerization process,[2] and there is evidence that pheomelanosomes are different in size and shape in comparison with eumelanosomes.[7] Thus, the color of melanized structures may be a function of the size and aggregation of the melanin granules.[8]

THE MELANOGENIC PATHWAY

The major pathway of melanogenesis involves the progressive oxidation of the amino acid, tyrosine. In vertebrates the process takes place in specialized organelles in special pigment-forming cells, embryologically derived from the neural crest, known as melanocytes. The main steps in the process of melanogenesis were analyzed by Raper sixty years ago.[9] A number of refinements to the scheme and some additional reactions have been defined since then.[10] The Raper-Mason scheme of eumelanogenesis (FIG. 1) leads to the generation of what may be termed indolic melanin. The details of the polymerization process remain to be elucidated, but indole-5,6-quinone-2-carboxylic

[a]Part of this work was supported by the Medical Research Council, the National Foundation for Cancer Research, and the Maurice Elton Memorial Fund.

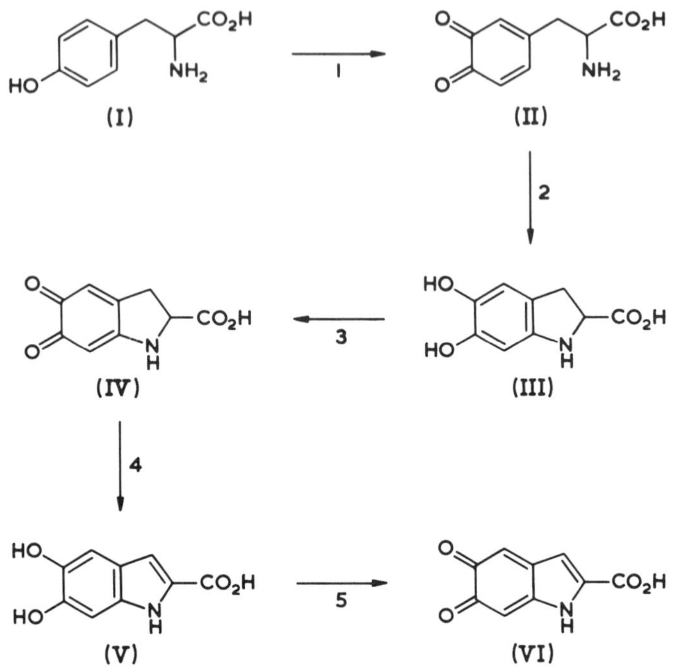

FIGURE 1. Schematic outline of the eumelanogenic pathway. Reaction 1 is the oxy-tyrosinase-catalyzed conversion of tyrosine (I) to dihydroxyphenylalanine quinone (dopaquinone) (II). Reaction 2 is the intramolecular rearrangement involving the Michael addition of the side chain amino group to the C6 of the ring to give the indolene compound, cyclodopa (III). Reaction 3 is the oxidation of cyclodopa to give dopachrome (IV). Dopachrome is often depicted in the hydrogen-shifted quinone-imine isomeric form. Reaction 4 is an intramolecular hydrogen shift to give 5,6 dihydroxyindole-2-carboxylic acid (V). In some conditions a decarboxylation occurs to give 5,6 dihydroxyindole. Reaction 5 is an oxidation step converting 5,6,dihydroxyindole-2-carboxylic acid to the corresponding orthoquinone (5,6 indolequinone-2-carboxylic acid) (VI).

acid (5,6IQ2CA) is rapidly incorporated into melanin. The incorporation of other constituents of the oxidative pathway have been proposed, including 5,6-dihydroxy-indole (and the corresponding quinone) formed by the decarboxylation of dopachrome, but the current view is that 5,6 IQ2CA is the major precursor of melanin. Although it has been shown that melanogenic intermediates other than tyrosine are oxidized by tyrosinase,[11-14] the rate constants for oxidations subsequent to the formation of dopaquinone[15,16] indicate that the major oxidation step catalyzed by tyrosinase is from tyrosine to dopaquinone, and that subsequent oxidations occur by redox exchange with dopaquinone, which generates dopa in the reaction mixture. Thus, it is possible that the entire oxidative system is driven by a cycle of alternate reduction of dopaquinone to dopa and reoxidation of dopa by tyrosinase to dopaquinone. This has important implications with regard to the activation of tyrosine oxidation by tyrosinase, which is unable to bind oxygen when the two copper ions at the active site are in the Cu (II) state.[17] However, this met-form of the enzyme can oxidize dopa by accepting two electrons which reduce the active-site copper to Cu (I), thus recruiting previously unutilized potential of the enzyme to oxidize tyrosine. Some of the important regulatory aspects of this phenomenon have been recently examined.[18]

Tyrosinase-catalyzed oxidation is the major and rate-limiting step in melanogenesis. The other reactions, consisting of reductive rearrangements (e.g., reactions 2 and 4 in FIG. 1) and polymerization, proceed spontaneously. The rearrangement of dopachrome is accelerated by reducing agents and the existence of an enzyme catalyzing this step has been proposed[19] although its significance in the melanogenic pathway has not yet been clarified. The activity of tyrosinase is diminished as the melanization of the melanosome proceeds, suggesting that the enzyme is modified by its product.

The coexistence in the reaction mixture of several quinones and hydroquinones, in addition to stimulating two-electron redox reactions of the type alluded to, also permits the formation of semiquinones by single-electron exchange.[16] The major species generated are likely to be dopa semiquinone, cyclodopa semiquinone, and indole-2-carboxylic acid semiquinone (FIG. 2), which may be formed by either homologous or heterologous redox couples. At present the relative rates and probabilities of these reactions are unknown.

POLYMERIZATION

The current view is that the polymerization leading to eumelanin formation occurs by free-radical reactions. These probably result from redox reactions which generate semiquinones. In the case of indoles this appears to involve carbon-centered radicals favoring the 2 and 4 positions.[20] The formation of dimeric adducts and their rearrangement to form resonance structures is illustrated in FIGURE 3. If this dimer formation is a model of polymerization that can be generalized with regard to melanin formation, it suggests that indolic melanins could be either regular with syndiotactic or isotactic monomer arrangments or irregular (atactic) polymers with a random mixture of bonding between the monomeric units. The type of polymer may have significance in relation to the physical properties of the melanin.

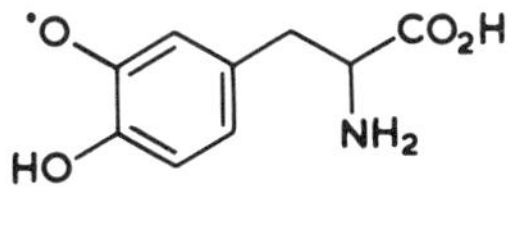

(VII)

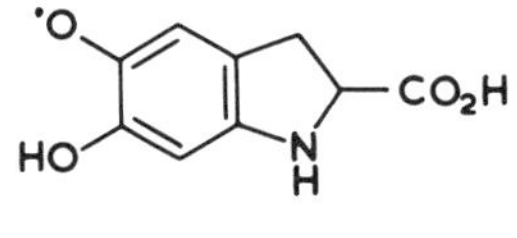

FIGURE 2. Semiquinone species that could be generated during melanogenesis. The predominant oxygen-centered radical forms of dopa semiquinone (VII), cyclodopa semiquinone (VIII), and indole-2-carboxylic acid semiquinone (IX) are shown.

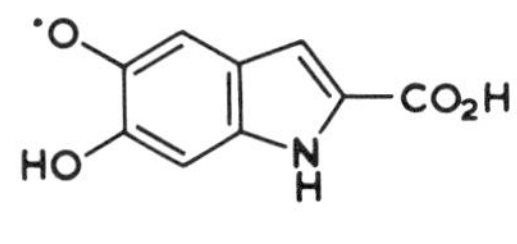

(VIII)

(IX)

SEMIQUINONE FORMATION

It will be evident that for polymerization beyond the dimer stage by a free-radical process, the oligomeric species needs to be in the form of a radical, and it is possible that this is generated by a single-electron exchange between the oligomer and a quinone such as 5,6 indolequinone-2-carboxylic acid, thus providing a mechanism for chain elongation. This could occur at either end of a linear polymer or at branch points. The polymer itself is therefore likely to be an important source of reducing equivalents in the melanogenic system. Polymerization may thus involve the incorporation of

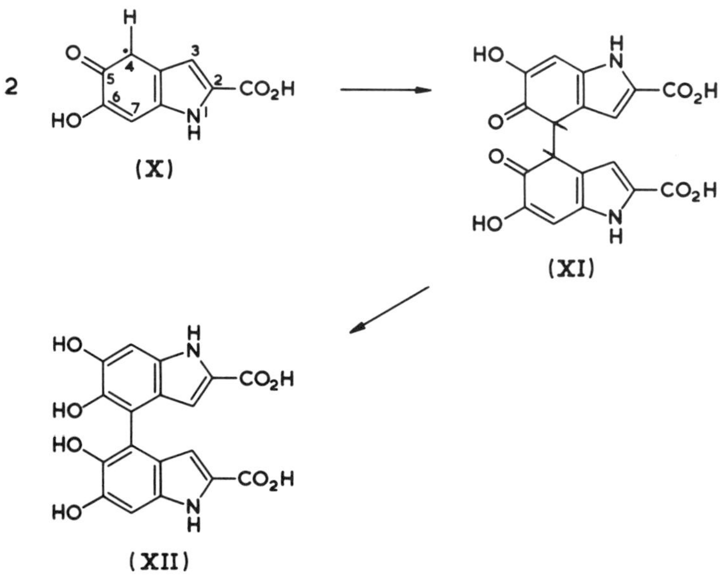

FIGURE 3. Adduct formation between carbon-centered radicals. Delocalization of the unpaired electron in the indole semiquinone (IX, FIG. 2) can give rise to carbon-centered radicals associated with the 2,3,4 and 7 positions. Adduct formation between two C4 radicals (X) is illustrated (XI) followed by rearrangement, by hydrogen shifts, to give a 4,4 dimer with a resonance structure ("melanochrome," XII).

dopaquinone and cyclodopa quinone, although restrictions on electron delocalization in these latter species will result in different bonding from that thought to be predominant in indole addition. It is possible that this feature is also significant for the structure and properties of the hypoindolic pigment formed. The melanin is deposited on and bound to a protein matrix inside the melanosome, and the fully pigmented granules so formed are transferred to surrounding structures by a process that leaves the membrane of the granule intact.[21] The pigment granules appear to have a light-protective function by virtue of the photon-absorptive capacity of the melanin.

STABLE RADICALS

The melanin polymer has long been known to exhibit stable free-radical properties because of semiquinones.[22-25] In the case of the melanin polymer, the semiquinone radicals are thought to be stabilized by resonance and the diffusional constraints of the polymer.[26] Their function appears to be to act as a sink for diffusible radical species. This can occur either by electron donation:

$$R_1OHO\cdot + R_2\cdot \rightarrow OR_1O + R_2H$$

or by electron capture:

$$R_1OHO\cdot + R_2\cdot \rightarrow R_1OHOR_2$$

The latter mechanism of adduct formation by radical annihilation is essentially similar to the proposed mechanism of co-polymerization of semiquinones.

MELANOGENESIS AND CYTOTOXICITY

If we consider the melanogenic intermediates that are present in the oxidation mixture, it is clear that several are highly reactive species. This is particularly true of the quinones. That the process of melanogenesis is potentially hazardous to cells was pointed out many years ago by Hochstein and Cohen.[27] The production of potentially toxic species in the course of melanogenesis requires the existence of certain safeguards for the cells that synthesize melanin.[28] These consist essentially of confining the process to a membrane-bound organelle (i.e., the melanosome) and the metabolic inactivation of potentially toxic species that may diffuse out of the melanosome and damage the cell. Model experiments using tyrosinase entrapped in liposomes[29] have demonstrated that negligible amounts of the normal melanogenic intermediates diffuse across a phospholipid barrier. Any melanogenic intermediates reaching the cytosol are metabolized to nontoxic derivatives that are excreted. The rapid reaction of dopaquinone with nucleophiles has long been recognized and addition products with cytosolic cysteine and glutathione are formed. The major species is 5-S-cysteinyl dopa,[30] but since this can be reoxidized it is probable that detoxification is predominantly by O-methylation.[31] O-methylation by methyl transferase is the major detoxification and excretion pathway for indolic products of the melanogenic pathway.[32] It is possible that other enzymes are also involved in this detoxification process.[33]

MELANOMA THERAPY

The possibility of utilizing the potentially cytotoxic melanogenic pathway for the development of a rational chemotherapy for malignant melanoma has received considerable attention. Most studies have been made on dihydric phenols, and a number of dopa analogues have been investigated for their toxic actions.[34] Since it appears that trihydric phenols readily undergo autoxidation[35] yielding reactive oxygen species such as superoxide and hydrogen peroxide,[36,37] it is possible that the generation of 2,4,5-trihydroxyphenylalanine by the action of tyrosinase on dihydric phenols (e.g., 2,4-dopa[38]) is the basis of their toxic action on melanogenic cells.

Our studies[39,40] have concerned the depigmenting tyrosine analogue, 4-hydroxyanisole (4-HA). This compound is oxidized by tyrosinase (FIG. 4) to the corresponding quinone.[41] The methoxy side chain excludes the possibility of cyclization, and thus the formation of hydroquinone species depends on interaction with nucleophils (which may possibly include water under alkaline conditions). The generation of the anisyl semiquinone radical probably results from such a sequence of reactions.[42] The inverse dependence of the anisyl semiquinone radical concentration on oxygen tension in the reaction system[43] is considered to be the consequence of reoxidation of the hydroquinone species by tyrosinase, since a direct interaction with oxygen has been excluded.[44] Moreover, this pathway must be considered relatively minor since there is a near 1:1 stoichiometry in oxygen utilization during 4-HA oxidation by tyrosinase.[45] The currently available data suggest that the cytotoxic damage due to oxidative stress induced by the metabolic products of 4-hydroxyanisole is due to quinone rather than semiquinone products.[46]

(XIII)   **(XIV)**

FIGURE 4. The tyrosine analogue substrate 4-hydroxyanisole (XIII) oxidized by tyrosinase to the corresponding orthoquinone (XIV). The methoxy side chain prevents endocyclization (compare dopaquinone (II) in FIGURE 1).

The inability to "endocyclize" makes the anisyl quinone more stable than the corresponding dopaquinone, and the higher lipid solubility may increase the concentration reaching the cytosol, thus producing oxidative stress to the cell by glutathione depletion or causing toxicity by interaction with thiol groups in proteins with a crucial role, such as membrane ion pumps.[47] Another factor that may be of crucial importance to the apparent specificity of the toxic action of tyrosinase oxidation products in melanoma therapy is that many malignant melanoyctes possess melanosomes with defective limiting membranes, which would permit free diffusion into the cytosol of potentially cytotoxic compounds generated by the melanogenic pathway.[48]

One of the advantages of 4-hydroxyanisole is that it is a remarkably nontoxic pro-drug. This may be partly accounted for by its rapid excretion, but from a therapeutic point of view the human pharmacokinetics are unfavorable.[49] It is important also to stress that the rationale of the therapy confines it to the treatment of pigment cell tumors with significant tyrosinase activity, which limits its potential range of usefulness in the chemotherapy of amelanotic malignant melanoma.

Although the precise mechanism of cytotoxicity of 4-HA is not yet clear, the results of preliminary clinical studies of 4-HA therapy in localized recurrent melanoma[50] have been sufficiently encouraging to warrant further investigation.

SUMMARY

Melanins are light-absorbant polymeric pigments found widely dispersed in nature. They possess many interesting physicochemical properties. One of these is the expression in the polymer of stable free radicals which appear to have a protective action in cells, probably by acting as a sink for diffusible free-radical species.

Polymer formation is thought to occur by a free-radical process in whch semiquinones are added to the chain. Semiquinones are formed by redox equilibration interactions between metabolic intermediates formed during the tyrosinase-catalyzed oxidation process.

In the continued presence of substrate, steady-state concentrations of reactive species are predicted in the reaction system, and the melanogenic pathway may be considered as potentially hazardous for pigment-generating cells. This feature has been exploited by the use of analogue substrates to generate cytotoxic species as a possible rational approach to the treatment of malignant melanoma. One such substance is 4-hydroxyanisole, the oxidation of which gives rise to semiquinone radical species. The possibility that the anisyl semiquinone initiates a mechanism leading to cell damage has not been excluded. However, the current view is that the major cytotoxicity due to the oxidation products of this compound is the result of the action of the corresponding orthoquinone.

A number of mechanisms exist for detoxifying quinones if they reach the cytosol such as O-methylation and the formation of thiol adducts with cysteine or glutathione, and these can be used as markers of melanogenesis. In general, however, only small amounts of reactive intermediates of melanogenesis escape from the confines of the melanosome, probably because of their limited lipid solubility. The selective toxic action of anisyl quinone in the treatment of melanoma may, in part, be due to membrane defects in the melanosomes of malignant melanocytes.

ACKNOWLEDGMENTS

Some of the general ideas expounded in this paper are based on themes developed in discussion with many collaborators. In particular I wish to acknowledge the contribution of Dr. E. J. Land, Dr. C. Lambert and Professor T. G. Truscott.

REFERENCES

1. NICOLAUS, R. A. 1968. Melanins. Hermann. Paris.
2. RILEY, P. A. 1980. Melanin and Melanogenesis. Pathobiol. Ann. **10**: 223–251.
3. MORRISON, W. L. 1985. What is the function of melanin? Arch. Dermatol. **121**: 1160–1163.
4. RILEY, P. A. 1977. The mechanism of melanogenesis. Symp. Zool. Soc. Lond. **39**: 77–95.
5. PROTA, G. 1980. Recent advances in the chemistry of melanogenesis in mammals. J. Invest. Dermatol **75**: 122–127.
6. DEIBEL, R. M. B. & M. R. CHEDEKEL. 1982. Biosynthetic and structural studies on pheomelanin. J. Am. Chem. Soc. **104**: 7306–7309.
7. SAKURAI, T., H. OCHIAI & T. TAKEUCHI. 1975. Ultrastructural change of melanosomes associated with agouti pattern formation in mouse hair. Devel. Biol. **47**: 466–471.

8. KURTZ, S. K., L. ALBRECHT, T. SCHULTZ & L. WOLFRAM. 1988. The physical origin of colour in Melanin Pigment dispersions (abstract). Pigment Cell Res. 1: 261.
9. RAPER, H. S. 1928. The aerobic oxidases. Physiol. Rev. 8: 245–282.
10. JOLLEY, R. L., L. M. EVANS, N. MAKINO & H. S. MASON. 1974. Oxytyrosinase. J. Biol. Chem. 249: 335–342.
11. DULIERE, W. L. & H. S. RAPER. 1930. The action of tyrosinase on certain substances related to tyrosine. Biochem. J. 24: 239–249.
12. RILEY, P. A. 1967. Histochemical demonstration of melanocytes by the use of 5,6-diacetooxyindole as substrate for tyrosinase. Nature 213: 190–191.
13. MIRANDA, M., G. URBANI, L. DiVITO & D. BOTTI. 1979. Inhibition of tyrosinase by indole compounds and reaction products. Protection by albumin. Biochim. Biophys. Acta 585: 389–404.
14. KORNER, A. & J. PAWALEK. 1982. Mammalian tyrosinase catalyses three reactions in the biosynthesis of melanin. Science 217: 1163–1165.
15. CHEDEKEL, M. R. , E. J. LAND, A. THOMPSON & T. G. TRUSCOTT. 1984. Early steps in the free radical polymerisation of 3,4 dihydroxyphenylalanine (Dopa) into melanin. J. Chem. Soc. Chem. Commun: 1170–1172.
16. LAMBERT, C., J. N. CHACON, M. R. CHEDEKEL, E. J. LAND, P. A. RILEY, A. THOMPSON & T. G. TRUSCOTT. 1988. A pulse radiolysis investigation of the oxidation of indolic eumelanin precursors and metabolites: evidence for indole quinones and subsequent intermediates. In preparation.
17. LERCH, K. 1981. Copper monooxygenase. In Metal Ions in Biological Systems, Vol. 13. H. Sigel, Ed. 143–186. Marcel Dekker. New York.
18. NAISH, S. & P. A. RILEY. 1988. Tyrosinase-catalysed oxidation of monophenols (abstract). Pigment Cell Res. 1: 288.
19. BARBER, J. I., D. TOWNSEND, D. P. OLDS & R. A. KING. 1984. Dopachrome oxidoreductase: A new enzyme in the pigment pathway. J. Invest. Dermatol 83: 145–149.
20. PROTA, G. 1988. Some New Aspects of Eumelanin Chemistry. Prog. Clin. Biol. Res. 256: 101–124.
21. RILEY, P. A. 1978. A theoretical note on the topological features of melanin transfer. Br. J. Dermatol 99: 107–110.
22. COMMONER, B., J. TOWNSEND & G. PAKE. 1954. Free radicals in biological materials. Nature 174: 689.
23. MASON, H. S., H. E. INGRAM & B. ALLEN. 1960. Free radical property of melanins. Arch. Biochem. Biophys. 86: 225.
24. LONGUET-HIGGINS, H. C. 1960. On the origin of the free radical property of melanins. Arch. Biochem. Biophys. 86: 231–236.
25. BLOIS, M. S., J. E. MALING & A. B. ZAHLAN. 1984. Electron spin resonance studies on melanin. Biophys. J. 4: 471–477.
26. SEALY, R. C., C. C. FELIX, J. S. HYDE & H. M. SWARTZ. 1980. Structure and reactivity of melanins. In Free Radicals in Biology, Vol. 4. W. A. Pryor, Ed. 209–260. Academic Press. New York.
27. HOCHSTEIN, P. & G. COHEN. 1963. The cytotoxicity of melanin precursors. Ann. N. Y. Acad. Sci. 100: 876–886.
28. SLATER, T. F. & P. A. RILEY. 1966. Photosensitization and lysosomal damage. Nature 209: 151–154.
29. MIRANDA, M., F. AMILCARELLI, A. M. RAGNELLI, A. ARCADI & A. POMA. 1988. Tyrosinase-liposome interaction; a model to study melanogenesis regulation (abstract). Pigment Cell Res. 1: 286.
30. CARSTAM, R., C. HANSSON, C. LINDBLADH, H. RORSMAN & E. ROSENGREN. 1987. Dopaquinone and addition products in cultured human melanoma cells. Acta Derm. Venereol 67: 100–105.
31. AGRUP, G., B. FALCK, C. HANSSON, H. RORSMAN, A-M. ROSENGREN & E. ROSENGREN. 1977. Metabolism of 5-S-cysteinyldopa by O-methylation. Acta Derm. Venereol 57: 309–312.
32. PAVEL, S. 1988. Eumelanin-related compounds: Their metabolism and clinical relevance. Rodopi. Amsterdam.

33. SCHALLREUTER, K. U., J. M. WOOD, E. W. BREITBART, W. KIMMIG, R. HICKS, H.
     RADLOFF & M. JÄNNER. 1988. Thioredoxin reductase in primary melanoma and
     surrounding skin (ESPCR Abst. 90). Pigment Cell. Res. 1: 290.
34. WICK, M. M. 1980. An experimental approach to the chemotherapy of melanoma. J. Invest.
     Dermatol 74: 63–65.
35. KALYANARAMAN, B., W. KORYTOWSKI, B. PILAS, T. SARNA, E. J. LAND & T. G.
     TRUSCOTT. 1988. Reaction between *ortho*-semiquinones and oxygen: Pulse radiolysis,
     electron spin resonance and oxygen uptake studies. Submitted for publication.
36. COHEN, G. & R. E. HEIKKILA. 1974. The generation of hydrogen peroxide, superoxide
     radical, and hydroxyl radical by 6-hydroxydopamine, dialuric acid and related cytotoxic
     agents. J. Biol. Chem. 249: 2447–2452.
37. GRAHAM, D. G., S. M. TIFFANY, W. R. J. BELL & W. F. GUTKNECHT. 1978. Autoxidation
     versus covalent binding of quinones as the mechanism of toxicity of dopamine, 6-
     hydroxydopamine and related compounds towards C1300 neuroblastoma cells in vitro.
     Mol. Pharmacol. 14: 644–653.
38. MORRISON, M. E. & G. COHEN. 1983. Novel substrates for tyrosinase such as precursors of
     6-hydroxydopamine and 6-hydroxydopa. Biochemistry 22: 5465–5467.
39. RILEY, P. A. 1984. Hydroxyanisole: Advances in Antimelanoma Therapy. IRL Press.
     Oxford.
40. RILEY, P. A. 1985. Radicals and Melanomas. Phil. Trans Roy. Soc. B 311: 679–689.
41. NAISH, S., J. L. HOLDEN, C. J. COOKSEY & P. A. RILEY. 1988. Major primary cytotoxic
     product of 4-hydroxyanisole oxidation by mushroom tyrosinase in 4-methoxy-*ortho*-
     benzoquinone. Pigment Cell Res. 1: 382–385.
42. NILGES, M. J., H. M. SWARTZ & P. A. RILEY. 1984. Electron spin resonance studies of
     reactive intermediates of tyrosinase oxidation of hydroxyanisole. J. Biol. Chem.
     259: 2446.
43. NILGES, M. J. & H. M. SWARTZ. 1984. Quinone and semiquinone intermediates and
     reaction of 4-hydroxyanisole and tyrosinase. *In* Hydroxyanisole: Recent Advances in
     Anti-Melanoma Therapy. P. A. Riley, Ed.: 25–34. IRL Press. Oxford.
44. COOKSEY, C. J., E. J. LAND, P. A. RILEY, T. SARNA & T. G. TRUSCOTT. 1987. On the
     interaction of anisyl-3,4-semiquinone and oxygen. Free Rad. Res. Commun. 4: 131–138.
45. DOBRUCKI, J. & P. A. RILEY. 1988. The stoichiometry of tyrosinase-catalysed oxidation of
     4-hydroxyanisole. Free Rad. Res. Commun. 4: 325–329.
46. NAISH, S., C. J. COOKSEY & P. A. RILEY. Initial oxidation product of 4-hydroxyanisole is
     4-methoxy-*ortho*-benzoquinone. Pigment Cell Res. 1: 379–381.
47. NAISH, S., J. HOLDEN & P. A. RILEY. 1988. Mechanism of toxicity of anisyl quinone.
     Pigment Cell Res. In press.
48. BOROVANSKY, J. & P. A. RILEY. 1988. The possible relationship between abnormal
     melanosome structure and cytotoxic phenomena (abstract). Pigment Cell Res. 1: 292.
49. MORGAN, B. D. G., J. HOLDEN, D. L. DEWEY & P. A. RILEY. 1984. Human pharmacokine-
     tics of 4-hydroxyanisole. *In* Hydroxyanisole: Recent Advances in Anti-Melanoma
     Therapy. P. A. Riley, Ed. IRL Press. Oxford.
50. MORGAN, B. D. G. 1984. Recent results of a clinical pilot study of intra-arterial 4-HA
     chemotherapy in malignant melanoma. *In* Hydroxyanisole: Recent Advances in Anti-
     Melanoma Therapy. P. A. Riley, Ed. IRL Press. Oxford.

DISCUSSION

C. BOREK (*Columbia University, New York*): Was the hydroxyanisole applied
locally or internally to suppress the melanomas?

P. A. RILEY (*University College, London, England*): Some initial studies (WEB-
STER, D., *et al.* 1984. *In* Hydroxyanisole: Recent Advances in Antimelanoma Therapy.

P. A. Riley, Ed. IRL Press. Oxford) reported injection into the tumor, but as far as I know 4-HA has not been applied locally, except in depigmentation experiments. Our first attempt at melanoma therapy used intravenous infusion, but we ended up giving the 4-HA intra-arterially by bolus injection.

M. MIRANDA (*University of Aquila, Italy*): Do you still believe that tyrosinase also oxidizes 5,6-dihydroxyindole?

RILEY: As you know, in 1965 we showed in *Nature* that tyrosinase can oxidize 5,6-DHI, and this has been elegantly confirmed in 1986 by the Pawalek's group at Yale in *Science*. In our view there is no doubt that DHI can be oxidized by tyrosinase, but the rate of the reaction is slow compared to the rate of the dopaquinone-driven oxidation. Therefore, provided that dopaquinone is present in the reaction system (as is the case in melanogenesis), we would expect the oxidation of DHI to be by redox exchange (LAMBERT *et al.,* in preparation).

M. MIRANDA (*University of Aquila Italy*): How have you measured the levels of the intermediates in melanin synthesis since some, like tyrosine, dopa, and dopachrome, absorb at specific wavelengths, while others do not? How have you measured the others?

RILEY: By HPLC.

First Electron Spin Resonance Evidence for the Generation of the Daunomycin Free Radical and Superoxide by Red Blood Cell Membranes[a]

J. Z. PEDERSEN,[b,c] L. MARCOCCI,[d] L. ROSSI,[c]
I. MAVELLI,[e] AND G. ROTILIO[c,f]

[c]Department of Biology
"Tor Vergata" University of Rome
Rome, Italy

[d]CNR Center for Molecular Biology
"La Sapienza" University of Rome
Rome, Italy

[e]Institute of Biology
University of Udine
Udine, Italy

INTRODUCTION

Quinone-like anthracycline antibiotics are important agents in cancer chemotherapy; however, they produce a cumulative, dose-dependent cardiotoxicity as a side effect, which severely limits their use in the treatment of neoplastic diseases. It has been postulated that free-radical mechanisms might be responsible for this toxicity. In fact, the formation of semiquinone radicals from the anthracyclines adriamycin and daunomycin has been demonstrated in cellular membrane systems[1-4] as well as in intact cells[5] by electron spin resonance (ESR) spectroscopy. The radicals have been proposed either to interact directly with cellular components (e.g., DNA[3]) or to autoxidize, giving rise to potentially toxic oxygen derivatives such as O_2^-, H_2O_2 and ultimately $\cdot OH$.[5-9]

The complexity of living cells is often a serious problem when trying to sort out the various effects caused by a drug. It is therefore useful to study a simplified system such as the red blood cell: it has a comparatively simple and well-studied metabolism and, in particular, its metabolic handling of oxygen metabolites is well described. Furthermore, the lack of DNA and RNA in mature human red blood cells precludes drug interaction with nucleic acids as a cause of any effect observed. Recently the entrapment of adriamycin in erythrocytes in order to allow a slow release of the drug to

[a]This study was partially supported by CNR Special Project "Oncologia" and by the Italian Association for Cancer Research.
[b]Supported by a grant from The Danish Natural Science Research Council.
[f]Address for correspondence: Professor G. Rotilio, Dipartimento di Biologia, II Università di Roma, Tor Vergata, Via O. Raimondo, 00173 Roma, Italy.

121

peripheral targets was proposed as a new therapeutic approach on the basis of the absence of heavy oxidative effects.[10] In view of this development it becomes important to understand the mechanisms of interaction of anthracyclines with red blood cells. A stimulation of oxidative metabolism, in the form of NADPH consumption and H_2O_2 formation,[11-13] was demonstrated after exposure of red blood cells to adriamycin or daunomycin. In addition, the generation of $\cdot OH$ by adriamycin in intact red blood cells was demonstrated by the ESR technique of spin trapping.[14] However, the direct detection of the formation of anthracycline semiquinone radicals or of O_2^- by intact red blood cells has never been reported, in contrast to the results obtained with other types of cells,[5] and the site of activation of anthracyclines by the erythrocytes has never been conclusively determined.

Recently, a stimulating report[15] pointed to the erythrocyte membrane as a site of activation of anthracyclines. The semiquinone radical of carminomycin was shown to be formed anaerobically by erythrocyte ghosts. Furthermore, daunomycin increased the susceptibility of erythrocytes to hypotonic lysis,[16,17] and the adriamycin-Fe(III) complex was found to bind to and destroy human red blood cell ghosts with the formation of O_2^- and H_2O_2.[18] Thus, the plasma membrane might behave both as an activator and as a major target of these drugs in erythrocytes.

In order to clarify the mechanism(s) of anthracycline activation and the subsequent effects on red blood cells, we have made a series of experiments using daunomycin, which has been claimed to cause damage in drug-loaded erythrocytes.[17] We here report preliminary results from this study, during which the ESR signal of the daunomycin radical generated by intact red blood cells and the related production of O_2^- were seen for the first time. The results suggest that the radical we observed is formed on the outside of the erythrocyte; no major damage to cell function and integrity was observed.

EXPERIMENTAL PROCEDURES

For the experiments shown in FIGURES 1 and 3 washed human red blood cell suspensions (5% v/v) in phosphate-buffered saline (PBS), pH 7.4, supplemented with 5 mM glucose or 1 mM NADPH were used as such or hemolyzed and incubated at 37°C under gentle stirring. Hemolysis and methemoglobin formation were measured by standard techniques[12] and expressed as percentage of total hemoglobin. H_2O_2 was measured by the extent of inhibition of catalase in the presence of 3-amino-1,2,4-triazole (AT). Catalase activity was measured spectrophotometrically.[12]

ESR experiments were done at room temperature with a Bruker ESP-300 instrument using a 40-G scan width, 1-G modulation, 25-mW microwave power, 82-ms time constant and a scan time of 84 seconds. Typically 1–2 hr of single scan accumulation was necessary to obtain a good signal-to-noise ratio. Red blood cells, hemolysates, and isolated membranes prepared according to the method of Maddy[19] were suspended in PBS at concentrations comparable to whole blood conditions. Anaerobiosis was achieved by flushing the samples with CO to remove the fraction of oxygen that otherwise might remain bound to hemoglobin. The flushing was done directly in the flat cell, via the capillary opening at its bottom.

Formation of O_2^- was demonstrated according to the method of Scarpa et al.[20] by incubating an air-saturated suspension of washed red blood cells (50% hematocrit) in the presence of 1 mM daunomycin and 0.05 mM oxidized bovine Cu,Zn-superoxide dismutase, purified as previously described.[21] The decrease in the Cu(II)-ESR signal reflects the reaction of the enzyme with O_2^- to approach the 50% reduced steady-state condition.

RESULTS AND DISCUSSION

The possibility that the red blood cell membrane might be a target for daunomycin-induced oxidative damage was investigated by incubating human erythrocytes with increasing concentrations of daunomycin (FIG. 1A). A low extent of substantially dose-

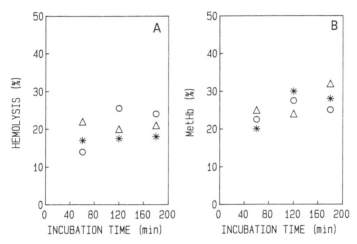

FIGURE 1. Effects of daunomycin on human erythrocyte membrane integrity (**A**) and methemoglobin formation (**B**). Human red blood cells (5% v/v) were incubated with 2.5 mM (O), 5 mM (∗) or 10 mM (△) daunomycin. Cells incubated without daunomycin did not show any hemolysis or methemoglobin formation.

and time-independent hemolysis was found. The moderate and relatively constant level of hemolysis observed does not speak in favor of a generalized damage to the membrane, but rather indicates a selective damage to only a minor fraction of the red blood cell population. To determine whether oxidative damage was occurring at the expense of hemoglobin during the exposure of erythrocytes to daunomycin, we measured the extent of methemoglobin formation. The trend of hemoglobin oxidation during incubation of human erythrocytes with various daunomycin concentrations is shown in FIG. 1B and parallels the one already depicted for the hemolysis. At most, methemoglobin formation accounted for about 30% of the total hemoglobin content. Again, no substantial dose- or time-dependency showed up. As a whole, the results in FIGURE 1 lead to the conclusion that hemolysis and methemoglobin cannot be relevant to the study of the mechanism of daunomycin action. Under our experimental conditions, erythrocyte membrane and hemoglobin do not seem to be major targets of the oxidative stress produced by daunomycin in erythrocytes and in particular of the H_2O_2 formed. In fact, not even when this oxidative injury was experimentally potentiated by AT-mediated catalase inactivation, thus causing H_2O_2 accumulation, did methemoglobin formation and hemolysis increase (results not shown). In addition, these findings question whether oxyhemoglobin may behave as an activator of daunomycin under our experimental conditions. Some reports have suggested that hemoglobin acts both as the activator of the anthracyclines and as their main target in red blood cells.[11,22] Our results are more in agreement with the data of De Flora *et al.*,[10] who reported mild oxidative damage in adriamycin-encapsulating erythrocytes, and with the observation by Bannister and Thornalley[14] that the adriamycin-stimulated · OH production in intact red blood cells was only partially decreased in the presence of carbonmonoxy hemoglobin.

To get a better insight into the mechanism(s) of activation of daunomycin in

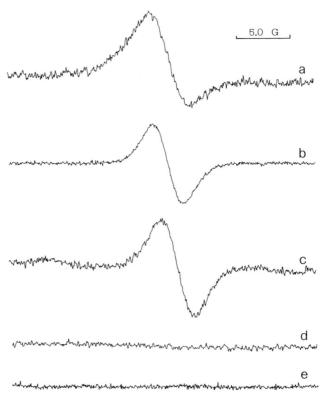

FIGURE 2. ESR spectra of daunomycin radicals formed by intact erythrocytes (**a**), xanthine oxidase (**b**), or isolated red blood cell ghosts (**c**). The electron sources for the semiquinone generation were glucose (5 mM), xanthine (1 mM) and NADPH (5 mM), respectively. Samples of erythrocyte hemolysates included 5 mM NADPH (**d**) or the xanthine-xanthine oxidase radical generating system (**e**).

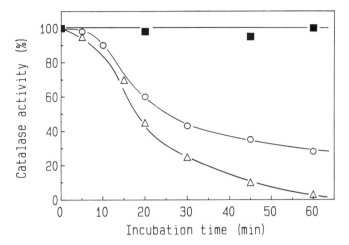

FIGURE 3. H_2O_2 formation by intact (O) or hemolyzed ($\triangle$) human red blood cells was measured in the presence of AT (50 mM) and daunomycin (2.5 mM), or in the absence of daunomycin ($\blacksquare$).

erythrocytes, we demonstrated the formation of the semiquinone by isolated human red blood cells, recording the ESR spectrum of the radical (FIG. 2a). The ESR signal was only seen in anaerobic conditions under CO atmosphere and disappeared immediately when oxygen was introduced. The identity of the radical was checked by comparison with that obtained by the xanthine-xanthine oxidase system[23] (FIG. 2b). The appearance of the semiquinone radical in samples where hemoglobin was present as the CO-derivative strongly supports the hypothesis of an alternative site of activation, although the involvement of oxyhemoglobin in anthracycline activation cannot be excluded completely. Daunomycin radicals could also be detected in samples of red blood cell ghosts (FIG. 2c), thus verifying that the finding reported for carminomycin[15] might likely be extended to the whole group of anthracycline antibiotics. In contrast, when ESR spectra of erythrocyte homogenates were recorded, the signal never appeared (FIG. 2d); moreover the presence of hemolysates prevented the accumulation of the semiquinone radical signal with the xanthine-xanthine oxidase system (FIG. 2e). Taken together, these results suggest that the radical we identified might be generated by the membrane on the outside of the red blood cell.

The formation of the radical might still occur in the hemolysates, but under anaerobic conditions it may react rapidly with intracellular components, preventing its accumulation to measurable levels. Support for this hypothesis comes from the data summarized in FIGURE 3. Daunomycin-induced redox cycling with either isolated erythrocytes or hemolysates under aerobic conditions was demonstrated. H_2O_2 was detected at very early incubation times, and its formation occurred in a time-dependent manner in whole erythrocytes and in samples of red blood cell homogenates in a comparable fashion. H_2O_2 formation seemed to be strictly dependent on daunomycin's presence, since AT alone had little effect (FIG. 3). The finding that daunomycin-dependent H_2O_2 production occurs both with hemolyzed and intact erythrocytes under aerobic conditions confirms that daunomycin-reductive activation and autoxidation occurs also in cell homogenates; thus the failure of the ESR technique in detecting the semiquinone in the absence of oxygen might be due to its quenching by intracellular components.

The disappearance of the radical upon introduction of air indicates that it reacts quickly with oxygen, yielding O_2^-. We demonstrated for the first time the daunomycin-mediated production of O_2^- by intact red blood cells. The reduction of daunomycin by red blood cells generated a flux of O_2^-, which drove externally added superoxide dismutase towards the steady-state condition. The intensity of ESR signal of the Cu(II) decreased to about 60% within 4 hours (FIG. 4). When we take into account the well-known high level of intracellular superoxide dismutase activity and the low rate of O_2^- diffusion through the membrane compared to the high rate of its dismutation, it is

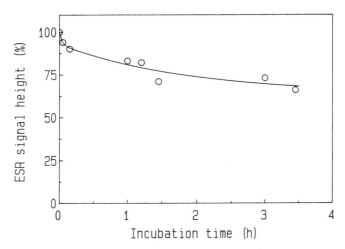

FIGURE 4. Demonstration of the generation of O_2^- by intact red blood cells upon incubation with daunomycin. The Cu(II)-site of superoxide dismutase is reduced to nonparamagnetic Cu(I) by one molecule of O_2^- and then reoxidized by another; the rates of these two reactions are almost equal and at steady state the enzyme will appear as 50% reduced.

unlikely that the flux of O_2^- observed has internal origins. In this context it is relevant that the existence of an active redox chain on red blood cell membrane has been reported, although there is very little conclusive evidence about its nature.[24] Our findings also imply that the diffusion of the radicals across the cell membrane must be slow, slow enough to allow for the accumulation of the radical outside the cell for a period of several hours, and too slow to make it possible for the radical to diffuse out in case it should be formed inside the cells without being scavenged by cellular components.

Our findings could be of more general attribution and may suggest an involvement of the plasma membrane in the activation of anthracyclines in other cell types as well. This is supported by previous reports[9,25,26] showing that adriamycin can be cytotoxic to tumor cells, acting at the plasma membrane without penetrating into the cell.

Furthermore, anthracyclines have been proved to affect many membrane activities such as diferric transferrine reductase[27] and transmembrane electron transport.[24]

ACKNOWLEDGMENTS

We are grateful to Dr. V. Malatesta (Farmitalia, Milano, Italy) for the generous supply of daunomycin hydrochloride.

REFERENCES

1. MAVELLI, I., L. ROSSI, F. AUTUORI, P. BRAQUET & G. ROTILIO. 1983. *In* Oxy Radicals and Their Scavenger System, Vol. 2. R. A. Greenweld & G. Cohen, Eds.: 326–329. Elsevier. Amsterdam.
2. DAVIES, K. J. A. & J. H. DOROSHOW. 1986. J. Biol. Chem. **261:** 3060–3067.
3. BACHUR, N. R., S. L. GORDON & M. V. GEE. 1978. Cancer Res. **38:** 1745–1750.
4. PESKIN, A. V., A. A. KOSTANTINOV & I. B. ZBARSKY. 1987. Free Rad. Res. Commun. **3:** 47–55.
5. BOZZI, A., I. MAVELLI, B. MONDOVI, R. STROM & G. ROTILIO. 1981. Biochem. J. **194:** 369–372.
6. GOODMAN, J. & P. HOCHSTEIN. 1977. Biochem. Biophys. Res. Commun. **77:** 797–803.
7. DOROSHOW, J. H. & K. J. A. DAVIES. 1986. J. Biol. Chem. **261:** 3068–3074.
8. DOROSHOW, J. H. 1986. *In* Superoxide and Superoxide Dismutase in Chemistry, Biology and Medicine. G. Rotilio, Ed.: 422–424. Elsevier. Amsterdam.
9. SINHA, B. K., A. G. KATKI, G. BATIST, K. H. COWAN & C. E. MYERS. 1987. Biochemistry **26:** 3776–3781.
10. DE FLORA, A., U. BENATTI, L. GUIDA & E. ZOCCHI. 1986. Proc. Natl. Acad. Sci. USA **83:** 7029–7033.
11. HENDERSON, C. A., E. N. METZ, S. P. BALCERZAK & A. L. SAGONE. 1978. Blood **52:** 878–885.
12. ROTILIO, G., M. R. CIRIOLO & I. MAVELLI. 1986. *In* Free Radicals and Arthritic Disease. A. J. G. Swaak & J. F. Koster, Eds.: 185–189. Eurage Rijkswijk. Amsterdam.
13. SAGONE, A. L. & G. M. BURTON. 1979. Am. J. Hematol. **7:** 97–106.
14. BANNISTER, J. V. & P. J. THORNALLEY. 1983. FEBS Lett. **157:** 170–172.
15. PESKIN, A. V. & G. BARTOSZ. 1987. FEBS Lett. **219:** 212–214.
16. SCHIOPPOCASSI, G. & H. S. SCHWARTZ. 1977. Res. Commun. Chem. Pathol. Pharmacol. **18:** 519–531.
17. SPRANDEL, U. & N. ZOLLNER. 1985. Res. Exp. Med. **185:** 77–85.
18. MYERS, C. E., L. GIANNI, C. B. SIMONE, R. KLECKER & R. GREENE. 1982. Biochemistry **21:** 1707–1713.
19. MADDY, A. H. 1966. Biochim. Biophys. Acta **117:** 193–198.
20. SCARPA, M., P. VIGLINO, D. CONTRI & A. RIGO. 1984. J. Biol. Chem. **259:** 10657–10659.
21. McCORD, J. M., & I. FRIDOVICH. 1969. J. Biol. Chem. **244:** 6049–6055.
22. BATES, D. A. & C. C. WINTERBOURN. 1982. Biochem. J. **203:** 155–160.
23. SCHREIBER, J., C. MOTTLEY, B. K. SINHA, B. KALYANARAMAN & R. P. MASON. 1987. J. Am. Chem. Soc. **109:** 348–351.
24. CRANE, F. L., I. L. SUN, M. G. CLARK, C. GREBING & H. LOW. 1985. Biochim. Biophys. Acta **811:** 233–264.
25. TRITTON, T. R. & G. YEE. 1982. Science **217:** 248–250.
26. BREDEHORST, R., M. PANNEERSELVAM & C.-W. VOGEL. 1987. J. Biol. Chem. **262:** 2034–2041.
27. SUN, I. L., P. NAVAS, F. L. CRANE, D. J. MORRÉ & H. LOW. 1987. Biochem. Int. **14:** 119–127

Oxidative Stress-Induced Plasma Membrane Blebbing and Cytoskeletal Alterations in Normal and Cancer Cells

G. BELLOMO, F. MIRABELLI, A. SALIS, M. VAIRETTI,
P. RICHELMI, G. FINARDI, H. THOR,[a]
AND S. ORRENIUS[a]

Dipartimento di Medicina Interna e Terapia Medica
Clinica Medica I
University of Pavia
27100 Pavia, Italy

[a]*Department of Toxicology*
Karolinska Institutet
S-104 01 Stockholm, Sweden

The appearance of multiple surface protrusions (also called "blebs") is one of the characteristic morphologic features of ischemic, hyperthermic, and toxic cell injury[1-3] that has been recognized in different experimental systems, including isolated and cultured cells,[2,3] isolated and perfused organs,[4] and also *in vivo*.[5] The biochemical and molecular changes responsible for plasma membrane blebbing have not been investigated in detail. Different mechanisms of bleb formation have been proposed, including alterations in intracellular thiol and Ca^{2+} homeostasis,[3] ATP depletion,[1] and disruption of the normal architecture of the lipid bilayer.[6] However, the demonstration that two well-known cytoskeletal toxins, phalloidin and cytochalasin B, induce extensive plasma membrane blebbing led to the proposal that alterations of the structure and function of the cytoskeleton could represent a common mechanism for bleb formation.[7]

In the last few years we have been actively engaged in investigating the biochemical mechanisms responsible for cell damage caused by oxidative stress generated during the metabolism of redox cycling quinones, such as 2-methyl-1,4-naphtoquinone (menadione). Irreversible injury was invariably preceded by plasma membrane blebbing, which was an early and initially reversible sign of menadione toxicity.[8] Both plasma membrane blebbing and cytotoxicity were related to menadione-induced alterations in glutathione and protein thiol homeostasis. Here we report that during oxidative stress caused by the metabolism of menadione in a variety of normal and cancer cells, marked changes in cytoskeletal protein composition and protein thiols occur which appear to be related to plasma membrane blebbing. The cytoskeletal fraction from control (untreated) and menadione-treated cells was prepared by a conventional technique using extraction of intact cells in a Trixon-X 100-containing buffer, followed by solubilization in urea/SDS, biochemical assays, and polyacrylamide gel electrophoresis (PAGE).[9]

Menadione caused plasma membrane blebbing in all cell lines investigated, although this phenomenon appeared more pronounced in cells maintained or growing in suspension, as compared to firmly adhering cells (TABLE 1). Analysis of the

TABLE 1. Plasma Membrane Blebbing and Cytoskeletal Abnormalities Resulting from Menadione-Induced Oxidative Stress in Mammalian Cells

Cell Type	Blebbing[a] (%)	Cytoskeletal Protein[b]	Cytoskeletal Protein-SH[b]	Oxidative Cross-Linking of Actin[c]
Hepatocytes in suspension	100	220	24	yes
Hepatocyte primary cultures	78	198	31	yes
Human platelets	91	142	28	yes
McCoy's (human synovial)	68	155	36	yes
HeLa (cervical carcinoma)	75	204	38	yes
GH₃ (pituitary adenoma)	64	225	40	yes
BT-20 (breast carcinoma)	52	138	42	yes
LoVo (colon carcinoma)	48	152	32	yes
THP-1 (myelomonocytic leukemia)	93	215	28	yes

NOTE: The indicated cells were incubated for 60 min with or without 200 μM menadione; then the percentage of cells exhibiting multiple surface protrusions was measured by phase contrast microscopy. The cytoskeletal fraction was extracted and processed for biochemical assays and PAGE.

[a]Evaluated as the percentage of cells exhibiting multiple surface protrusions.
[b]Expressed as the percentage of control (cytoskeleton extracted from untreated cells).
[c]Evaluated from comparative PAGE (nonreducing versus reducing conditions).

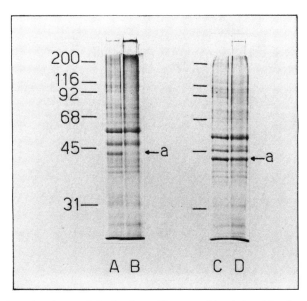

FIGURE 1. Alterations in cytoskeletal polypeptide composition induced by the metabolism of menadione in HeLa cells. HeLa cells were incubated for 60 min with (*lanes* B and D) or without (*lanes* A and C) 200 μM menadione. The cytoskeletal fraction was then prepared and analyzed by PAGE under nonreducing (*left columns*) and reducing (*right columns*) conditions. *Arrows* (**a**) indicate a protein band corresponding to actin as revealed by its molecular mass and its co-migration with an actin standard and by immunoblotting analysis with monoclonal anti-actin antibodies (not shown).

cytoskeletal fraction extracted from treated cells revealed a time- and dose-dependent increase in the amount of cytoskeletal proteins and a concomitant loss of protein thiols (TABLE 1). Addition of the thiol reductant, dithiothreitol (DTT), to menadione-treated cells largely prevented the decrease in cytoskeletal protein sulfhydryl groups, indicating that oxidation was the main mechanism responsible for this decrease (not shown). These changes were associated with the disappearance of actin and formation of high molecular weight aggregates when the cytoskeletal proteins were analyzed by PAGE under nonreducing conditions (FIG. 1 and TABLE 1). On the other hand, when the cytoskeletal proteins were treated with β-mercaptoethanol, no changes in the relative abundance of actin or formation of large molecular weight aggregates were detected in cytoskeletal preparations from treated cells. Moreover, addition of DTT to cells exposed to oxidative stress generated by the metabolism of menadione prevented both the appearance of surface blebs and the oxidative cross-linking of actin.

These findings indicate that the plasma membrane blebbing occurring in cells exposed to oxidative stress is associated with oxidation of thiol groups in actin and the formation of actin-containing aggregates. They also suggest that cytoskeleton is an important target in oxidative stress-induced cell injury.

REFERENCES

1. LEMASTER, J. J., J. DiGIUSEPPE, A. L. NIEMINEN, & B. HERMAN. 1987. Nature 325: 78–81.
2. BORRELLI, M. J., R. S. L. WONG, & W. C. DEWEY. 1986. J. Cell Physiol. 126: 181–190.
3. JEWELL, S. A., G. BELLOMO, H. THOR, S. ORRENIUS & M. T. SMITH. 1982. Science 217: 1257–1259.
4. LEMASTER, J. J., C. J. STEMKOWSKI, & S. JI. 1983. J. Cell Biol. 97: 778–786.
5. SATO, T., J. TANAKA, & Y. KONO. 1982. Lab. Invest. 47: 304–310.
6. FERREL, J. E. & W. H. HUESTIS. 1984. J. Cell Biol. 98: 1992–1998.
7. MESLAND, D. A. M., G. LOS, & H. SPIELE. 1981. Exp. Cell Res. 135: 431–435.
8. BELLOMO, G., H. THOR, L. EKLOW, P. L. NICOTERA, & S. ORRENIUS. 1987. Chem. Scr. 27A: 117–120.
9. MIRABELLI, F., A. SALIS, V. MARINONI, G. FINARDI, G. BELLOMO, H. THOR, & S. ORRENIUS. 1988. Arch. Biochem. Biophys. 264:261–269.

Membrane Modifications Induced by Superoxide Dismutase Depletion as a Model of Oxidative Stress

P. PALOZZA, E. PICCIONI, LI YAN, AND G. M. BARTOLI

Institute of General Pathology
Catholic University
Rome, Italy

Reduction of molecular oxygen to active species is believed to be the basis for several disorders associated with oxygen toxicity. In particular, free radicals are involved in cancerogenesis in relation to damaging effects on DNA, proteins, and membrane lipids. It is well established that tumor cells lack defense mechanisms against oxygen-reactive species, in particular superoxide dismutase (SOD), glutathione peroxidase, and catalase.[1] This condition probably makes tumor cells more sensitive to oxidative stress, deeply modifying cell structure and function. We have previously observed that the treatment of Wistar rats with a copper-deficient diet inhibits liver SOD and modifies membrane lipid composition. In fact, liver microsomal membranes isolated from copper-deficient animals show a fatty acid pattern more saturated with respect to control membranes and an increased resistance to peroxidation induced *in vitro*.[2] This situation is very similar to that observed in tumor cells where the level of

TABLE 1. Superoxide Dismutase Content of Erythrocytes of Control and Copper-Deficient Rats

	ng/mg Hb (mean ± SEM)	No. of Experiment	%
Control	1.20 ± 0.08	7	100
Cu-deficient	0.34 ± 0.02	8	28

SOD is related to the resistance to lipid peroxidation and to the grade of membrane phospholipid unsaturation.[3] To further investigate the relationship between SOD level and membrane composition and function, we have studied erythrocyte membranes in copper-deficient animals. As shown in TABLE 1 8-week treatment of a copper-deficient diet in Wistar rats induces an inhibition of erythrocyte SOD of 72%, which is very similar to that observed in liver. FIGURE 1 shows lipid peroxidation of erythrocyte ghosts induced *in vitro* by 1 mM *t*-BOOH and expressed as malondialdehyde production. Under these conditions membranes isolated from SOD-deficient animals are less resistant to peroxidation than are control membranes. Preliminary data show that copper deficiency does not induce changes in lipid composition in erythrocyte membranes. These data show a different behavior of red cell membranes with respect to microsomal membranes. The oxidative stress consequent to SOD depletion does not induce changes in the phospholipid fatty acid pattern observed in liver membranes, but

131

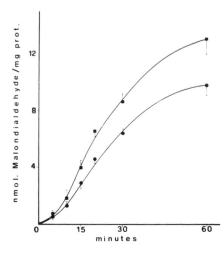

FIGURE 1. Malondialdehyde production by erythrocyte membranes from control (●) and copper-deficient (■) animals. Reaction was started by 1 mM t-BOOH. Membranes (1 mg protein/ml) were incubated at 37°C. Mean ± SEM of 3–6 experiments.

makes red cell membranes more sensitive to lipid peroxidation. This increased susceptibility to lipid peroxidation exerted by red cell membranes could be the expression of exhaustion of other antioxidant defenses. The different response to oxidative stress observed between liver microsomal membranes and erythrocyte membranes can be explained in terms of different structure and regulation between intracellular and plasma membranes.

REFERENCES

1. OBERLEY, L. W., & G. R. BUETTNER. 1979. Cancer Res. **39:** 1141–1149.
2. BARTOLI, G. M., B. GIANNATTASIO, P. PALOZZA, & A. CITTADINI. 1988. Biochim. Biophys. Acta. In press.
3. BARTOLI, G. M., S. BARTOLI, T. GALEOTTI, & E. BERTOLI. 1980. Biochim. Biophys. Acta. 620: 205–211.

Copper Complexes with Superoxide Dismutase Activity Enhance Oxygen-Mediated Toxicity in Human Erythroleukemia Cells[a]

C. STEINKÜHLER,[b] I. MAVELLI,[c] G. MELINO,[d] L. ROSSI,[e]
U. WESER,[b] AND G. ROTILIO[e,f]

[b]Physiologisch-Chemisches Institut
University of Tübingen
Tübingen, Federal Republic of Germany

[c]Institute of Biology
University of Udine
Udine, Italy

[d]Department of Experimental Medicine
[e]Department of Biology
2nd University of Rome "Tor Vergata"
Rome, Italy

INTRODUCTION

Many human tumor cells have low activity of H_2O_2-detoxifying glutathione peroxidase (GSH-Px) and catalase, though some cell lines still maintain normal levels of superoxide dismutase (SOD), which scavenges O_2^- by dismutating it to H_2O_2 and O_2. Such an alteration of the balance of the antioxygenic enzymes should result in a higher H_2O_2 steady-state concentration, thus increasing the cell sensitivity to oxidative insults.[1] This hypothesis is supported by the finding that lipophilic low molecular weight copper complexes with SOD activity display antitumor action *in vivo*,[2] probably by further increasing H_2O_2 content of the cells and producing oxidative killing of the tumor. In this context we investigated the effects of two low molecular weight active-center analogues of copper- and zinc-containing SOD, N,N'-bis(2-pyridylmethylene)-1,4- butanediamine (N,N',N'',N''')-Cu(II)-diperchlorate (PUPY), and 1,8-di(2-imidazoyl)-2,7-diazoctadiene-1,7-(N,N',N'',N''')-Cu(II)-diperchlorate (CuIm) on the human erythroleukemia cell line K562.

MATERIALS AND METHODS

PUPY and CuIm were prepared according to the method of Linss.[3] K562 human erythroleukemia cells were grown in Dulbecco's Minimal Essential Medium (Flow

[a]This study was supported by DFG Grant No. We 401/18-20 and by the CNR special project "Oncologia." C.S. was supported by a DAAD scholarship.
[f]To whom correspondence should be addressed at via Orazio Raimondo 1, 00173 Rome, Italy.

133

Lab, U.K.) supplemented with 10% heat-inactivated fetal calf serum (Flow Lab, U.K.). For experiments cells were incubated in fresh complete medium at 37°C in a humidified atmosphere containing 5% CO_2. Duroquinone was added in DMSO or hexane. Cell toxicity was measured by Trypan Blue uptake. Enzyme activities were determined and defined in units as in Ref.[4] Oxygen consumption was measured by a Clark electrode.

RESULTS AND DISCUSSION

K562 cells showed particularly high levels of Cu,ZnSOD and manganese-containing SOD activities (i.e., 4.5 ± 0.9 U/mg protein and 3.4 ± 0.1 U/mg protein respectively) when compared to other tumor cell lines,[1] while containing very low activities of catalase (42.8 ± 9.0 U/mg protein) and glutathione peroxidase (2.0 ± 0.4 mU/mg protein). Incubation of K562 cells with PUPY or CuIm caused a decrease of cell viability (FIG. 1A). $CuSO_4$ or Cu-EDTA, a copper complex without SOD activity, or the ligands of copper in the SOD-mimics alone proved to be nontoxic when added at the same concentrations as PUPY or CuIm. Addition of external catalase significantly decreased toxicity of both copper complexes, whereas the heat-inactivated enzyme had

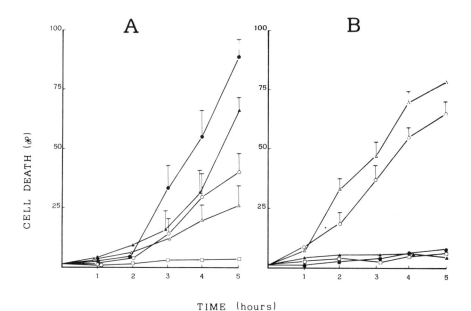

TIME (hours)

FIGURE 1. Cu complexes induced cytotoxicity in K562 erythroleukemia cells. (**A**) K562 cells (2×10^6/ml) received the following additions: □, none, or $CuSO_4$ (1 mM), or CuEDTA (1 mM); ●, PUPY (1 mM); ○, PUPY (1 mM) + catalase (1000 U/ml); ▲, CuIm (0.5 mM); △, CuIm (0.5 mM) + catalase (1000 U/ml). (**B**) K562 cells (10^6/ml) received: □, Duroquinone (0.6 mM); ●, PUPY (0.1 mM); ○, PUPY (0.1 mM) + duroquinone (0.6 mM); ▲, CuIm (0.1 mM); △, CuIm (0.1 mM) + duroquinone (0.6 mM). The values are expressed as the mean of three separate experiments.

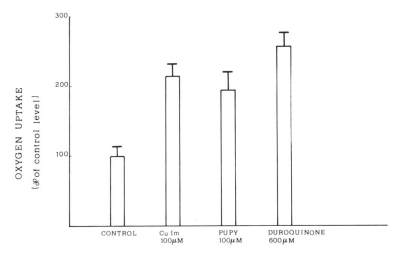

FIGURE 2. Copper complexes- and duroquinone-mediated oxygen activation with K562 cells. Oxygen consumption by K562 cells (5×10^6/ml) was monitored at 37°C in 2.5 ml of phosphate-buffered saline. Oxygen uptake by K562 cells was 12.8 ± 0.5 nmol O_2 consumed per minute per 10^6 cells. Values are the means of three separate experiments.

no effect on mortality, suggesting an involvement of H_2O_2 in cell-killing. An increased respiration of the cells in the presence of PUPY or CuIm (FIG. 2), but not of Cu-EDTA, indicates that the former compounds might be reduced by intracellular reductants (such as NADPH, GSH, or ascorbate) and generate H_2O_2 after reoxidation of the resulting Cu(I) complex by molecular oxygen.

A combination of nontoxic concentrations of either copper complexes and duroquinone, a redox-cycling agent known to produce O_2^- in cells,[5] drastically reduced cell viability (FIG. 1B). Activation of duroquinone was shown by an enhanced oxygen consumption by K562 cells upon addition of the compound (FIG. 2), which was demonstrated (data not shown) to be totally accounted for by CN^- insensitive respiration. This implies that duroquinone activation proceeds via semiquinone. The enhancement of the SOD-mimics' toxicity by duroquinone could be explained on the basis that the semiquinone radical could donate electrons to O_2, H_2O_2 or directly to the SOD-mimics, thus accelerating their redox-cycling.[6]

In conclusion the present data suggest that low molecular weight SOD-mimics increase intracellular H_2O_2 either by acting as superoxide dismutases or by undergoing a redox-cycling reaction, possibly causing depletion of intracellular reductants. H_2O_2 thus generated by the complexes seems to be involved in killing of K562 cells, which might be particularly susceptible to oxidative insult because of their low content in both catalase and GSH-Px.

REFERENCES

1. MAVELLI, I., G. ROTILIO, M. R. CIRIOLO, G. MELINO & O. SAPORA. 1987. Antioxygenic enzymes as tumor markers: A critical reassesment of the respective roles of SOD and

glutathione peroxidase. *In* Human Tumor Markers. S. Cimino, S. Birkmayer, P. Klavious, R. Pimentel & F. Salvatore, Eds. **1:** 883–888. Walter de Gruyter. Berlin.

2. OBERLEY, L. W., S. W. C. LEUTHAUSER, R. F. PASTERNACK, T. D. OBERLEY, L. SCHUTT & J. R. J. SORENSON. 1984. Agents and Actions. **15:** 535–538.
3. LINSS, M. 1986. Dissertation, University of Tübingen.
4. GALIAZZO, F., A. SCHIESSER & G. ROTILIO. 1988. Biochim. Biophys. Acta **965:** 46–51.
5. ROSSI, L., G. A. MOORE, S. ORRENIUS & P. J. O'BRIEN. 1986. Arch. Biochem. Biophys. **251:** 25–35.
6. VILE, G. F., C. C. WINTERBOURN & C. H. SUTTON. 1987. **259:** 616–626.

Antioxygenic Enzyme Activities in Differentiating Human Neuroblastoma Cells[a]

C. STEINKÜHLER,[b] I. MAVELLI,[c] G. MELINO,[d]
M. PIACENTINI,[e] L. ROSSI,[e] U. WESER,[b]
AND G. ROTILIO[e,f]

[b]Physiologisch-Chemisches Institut
University of Tübingen
Tübingen, Federal Republic of Germany

[c]Institute of Biology
University of Udine
Udine, Italy

[d]Department of Experimental Medicine, and
[e]Department of Biology
2nd University of Rome "Tor Vergata"
Rome, Italy

INTRODUCTION

The human neuroblastoma cell line SK-N-BE can be induced to differentiate by exposure to retinoic acid (RA) or α-difluoromethylornithine (DFMO) (FIG. 1). RA treatment causes neuronal-like morphologic alterations such as the formation of neurite extensions, 60% reduction of growth rate, enhanced transglutaminase activity, and an increase in putrescine, γ-aminobutyric acid, and acetylcholinesterase levels.[1] Inhibition of ornithine decarboxylase by DFMO severely affects polyamine metabolism, leading to complete growth inhibition and to an antigenic profile, suggesting a melanocytic differentiation of the cells.[1] In both models the acquisition of more differentiated characteristics, paralleled by the growth inhibition, suggests the transformation towards a less malignant phenotype.

As shown by our previous studies, tumor cells present an altered pattern of the enzymes that scavenge oxygen radicals—superoxide dismutase, catalase, and glutathione peroxidase—when compared to their nonmalignant counterparts. In particular, low levels of H_2O_2 detoxifying enzymes were detected.[2]

The present study examines whether differentiation is able to "normalize" the balance of the cell antioxidative defences.

[a]This study was supported by DFG grant We 401/18-20 and by the CNR special project "Oncologia." C.S. was supported by a DAAD scholarship.
[f]To whom correspondence should be addressed at via Orazio Raimondo 1, 00173 Rome, Italy.

137

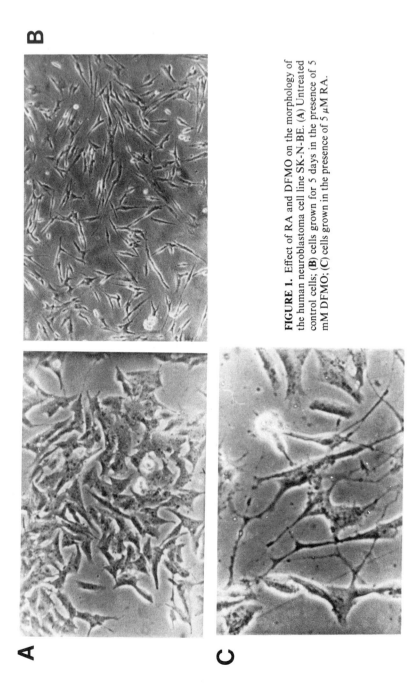

FIGURE 1. Effect of RA and DFMO on the morphology of the human neuroblastoma cell line SK-N-BE. (A) Untreated control cells; (B) cells grown for 5 days in the presence of 5 mM DFMO; (C) cells grown in the presence of 5 μM RA.

MATERIALS AND METHODS

SK-N-BE human neuroblastoma cell line, kindly provided by Dr. Lawrence Helson from the Memorial Sloan-Kettering Hospital in New York, was grown in RPMI-1640 medium (Flow Lab, U.K.) supplemented with 10% heat-inactivated fetal calf serum (Flow Lab, U.K.) in a humidified atmosphere containing 5% CO_2 at 37°C. During the induction of differentiation the growth medium, containing 5 mM DFMO or 5 μM RA, was replaced daily.

Before assaying enzyme activities, cells were mechanically removed, washed twice in phosphate-buffered saline solution and disrupted by sonication. Protein concentration was estimated by the method of Lowry. Enzyme activities were measured as reported in Ref. 3: copper- and zinc-containing superoxide dismutase (Cu,ZnSOD) and manganese-containing superoxide dismutase (MnSOD) were assayed according to the cytochrome c/xanthine oxidase/xanthine method; glutathione peroxidase (GSH-Px) activity was determined using H_2O_2 as a substrate.

TABLE 1. Activities of the Antioxygenic Enzymes During Differentiation of Human Neuroblastoma Cell Line SK-N-BE Induced by RA or DFMO

Treatment	Cu,ZnSOD (U/mg protein)[a]	MnSOD (U/mg protein)[a]	Catalase (U/mg protein)[b]	GSH-Px (mU/mg protein)[c]
Control	4.0 ± 0.8	1.9 ± 0.4	11.5 ± 1.0	2.3 ± 0.5
RA 3 days	3.2 ± 0.5	2.3 ± 0.3	10.1 ± 1.0	1.6 ± 0.1
RA 5 days	2.1 ± 0.1	3.6 ± 0.4	13.5 ± 2.0	3.4 ± 0.2
RA 7 days	1.2 ± 0.6	3.2 ± 0.5	11.3 ± 1.0	3.7 ± 1.0
DFMO 3 days	2.1 ± 0.1	2.0 ± 0.1	7.6 ± 1.8	2.5 ± 0.8
DFMO 5 days	1.9 ± 0.1	2.6 ± 0.3	8.9 ± 0.6	Not determined
DFMO 7 days	1.4 ± 0.4	3.2 ± 0.6	5.2 ± 1.0	<0.5

[a] One unit = amount of SOD required to inhibit the rate of reduction of cytochrome c by 50%.
[b] One unit = 1 μmol H_2O_2 consumed per minute.
[c] One unit = 1 μmol NADPH consumed per minute.

RESULTS AND DISCUSSION

TABLE 1 summarizes the results of the measurements of the antioxygenic enzyme activities during RA- or DFMO-induced differentitation of SK-N-BE cells.

A decrease in the activities of catalase, GSH-Px, and Cu,ZnSOD was observed during exposure to DFMO. This might be related to the inhibition of protein biosynthesis occurring during DFMO treatment[4] and to a subsequent modified protein turnover. However, MnSOD, which is mainly found in the mitochondrial matrix, increased significantly. This could be associated with an increased oxidative metabolism during differentiation. On the other hand MnSOD activity has often been reported to be modulated by nonspecific factors.[5]

During RA exposure, catalase activity remains unaffected, whereas GSH-Px showed a 60% increase. This latter finding is in line with our recent report that identifies GSH-Px as a negative tumor marker.[2] In fact, RA treatment does not result in a nonspecific inhibition of protein biosynthesis. Even though DFMO reduces the synthesis of the proteins, the decrease of Cu,ZnSOD activity is evident also during RA treatment. Thus it can be considered a true consequence of neuroblastoma cell differentiation. The acquisition of a more differentiated phenotype decreases the steady-state concentration of H_2O_2, both by decreasing Cu,ZnSOD activity and increasing GSH-Px activity. This emphasizes the role of H_2O_2 in differentiation.

REFERENCES

1. MELINO, G., M. G. FARRACE, M. P. CERU & M. PIACENTINI. 1988. Exp. Cell Res. In press.
2. MAVELLI, I., G. ROTILIO, M. R. CIRIOLO, G. MELINO & O. SAPORA. 1987. Antioxygenic enzymes as tumor markers: A critical reassesment of the respective role of SOD and glutathione peroxidase. *In* Human Tumor Markers. S. Cimino *et al.*, Eds. **1:** 883–888. Walter de Gruyter. Berlin.
3. GALIAZZO, F., A. SCHIESSER & G. ROTILIO. 1988. Biochim Biophys. Acta **965:** 46–51.
4. HÖLTTÄ, E., J. JÄNNE & T. HOVI. 1979. Biochem. J. **179:** 109–117.
5. BANNISTER, J. V., W. H. BANNISTER & G. ROTILIO. 1987. CRC Critical Reviews in Biochemistry. **22:** 111–180.

Efflux of Cytotoxic Species within Lipo-Melanosome Membrane

F. AMICARELLI, A. BONFIGLI, O. ZARIVI, A. POMA,
AND M. MIRANDA

Department of Cell Biology and Physiology
University of L'Aquila
I-67100 L'Aquila, Italy

The process of melanogenesis is potentially cytotoxic because of the generation of O_2^-, H_2O_2, quinones and semiquinones,[1] cysteinyl-L-3,4-dihydroxyphenylalanine (Cys-Dopa),[2] or 5,6-dihydroxyindole, which may also be mutagenic.[3] There is evidence that cytotoxic species are released in the cultures of melanoma cells[4] and there is evidence of Cys-Dopa in the urines of patients bearing metastatic melanoma,[5] but no direct investigation of the efflux of these substances within melanosome membrane has been carried out. The present work investigates the releases or accumulations of O_2^-, H_2O_2, Cys-Dopa, and glutathionyl-Dopa (Glu-Dopa) by lipo-melanosomes with or without cholesterol (lipo-melanosome is the liposome-entrapped tyrosinase) versus time of incubation with 5 mM Dopa in the presence of 5 mM cysteine or 5 mM glutathione as scavengers of quinones produced during melanin synthesis.[6] In our experiments lipo-melanosomes represented the minimal melanosome model.

The lipo-melanosomes, with or without cholesterol, were prepared according to the procedure described previously.[7] The lipo-melanosomes were incubated for various times with 5 mM Dopa in 0.1 M Na-phosphate buffer, pH 6.8, with or without superoxide dismutase (SOD) 10 U/ml or catalase 35 U/ml. The incubation systems and inner vesicle content were monitored for O_2^- and H_2O_2 by adding catalase, SOD, and catalase and by breaking vesicles with 0.05% TRITON X-100. The efflux and accumulation of Cys-Dopa and Glu-Dopa within lipo-melanosomes incubated for 240 min with Dopa, in the presence or absenne of cysteine or glutathione, were monitored by gel electrophoresis and concentration with concentrators (AMICON) of the incubation media and of the lipo-melanosome extracts.

FIGURE 1 shows that no evident accumulations of H_2O_2 or O_2^- occurred in the medium of lipo-melanosome incubated, in the case of lipo-melanosomes with or without cholesterol. The addition of 0.05% TRITON X-100 does not increase the O_2 signal; on the contrary, this signal decreases (FIG. 1) maybe because of a minimal remaining Dopa oxidase activity. However, the possibility exists that O_2^-, being transient, does not reach levels in our experiment detectable by means other than spin traps; but the disproportion of H_2O_2 accumulation from O_2^- should be evident unless the H_2O_2 is rapidly reduced by Dopa itself. This case suggests that melanosomes may have an autoscavenging mechanism for active oxygen. No effluxes of Cys-Dopa or Glu-Dopa from lipomelanosomes were observed either (FIG. 2), so that, unless the structure of melanosomes was deranged, pigment cells should not be at risk.

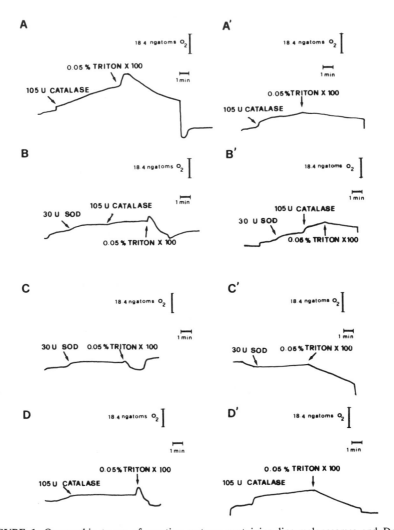

FIGURE 1. Oxygraphic traces of reaction systems containing lipo-melanosomes and Dopa. **A,B,C,D**- lipo-melanosomes; **A′,B′,C′,D′** = cholesterol-containing lipo-melanosomes. **A-A′**: incubated 120 min and then catalase is added; **B-B′**: incubated 120 min and then SOD and catalase are added; **C-C′**: incubated 120 min with catalase; **D-D′**: incubated 120 min with SOD. The same findings as at 120 min were observed at 0, 20, 40, 60, and 90 minutes. Temperature: 25°C.

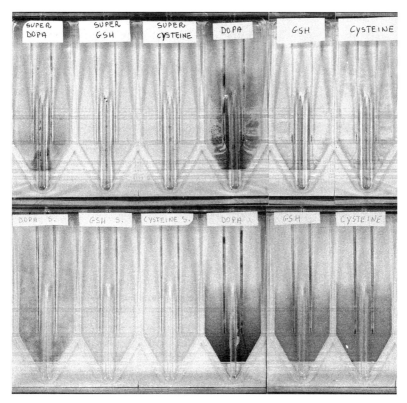

FIGURE 2. Photographs of minicon (AMICON) concentrators charged with supernatants and Triton extracts of lipo-melanosomes (*top row:* lecithin only; *bottom row:* cholesterol) incubated 240 min with 5 mM Dopa in the presence or absence of 5 mM cysteine or 5 mM glutathione. The different conditions are indicated.

REFERENCES

1. RILEY, P. A. 1980. Pathobiol. Annu. **10:** 223–251.
2. FUJITA, K. *et al.* 1980. Cancer Res. **40:** 2543–2546.
3. MIRANDA, M. *et al.* 1987. Mutagenesis **2:** 45–50.
4. HALABAN, R. & A. B. LERNER. 1977. Exp. Cell Res. **108:** 119–125.
5. AGRUP, G. *et al.* 1977. Acta Dermatol. **57:** 221–222.
6. MIRANDA, M. *et al.* 1987. Biochim. Biophys. Acta **913:** 386–394.
7. MIRANDA, M. *et al.* Colloids and Surfaces. In press.

Effect of *tert*-Butyl Hydroperoxide on Metabolism of Liver and Hepatoma[a]

S. BORRELLO, G. PALOMBINI, T. GALEOTTI,
AND G.D.V. VAN ROSSUM[b]

Institute of General Pathology
School of Medicine
Catholic University
00168 Rome, Italy

[b]*Department of Pharmacology*
Temple University School of Medicine
Philadelphia, Pennsylvania 19140

Free-radical formation catalyzed by cytochrome P-450 causes peroxidation of membrane lipids, leading to cell damage. Peroxidation is induced by *tert*-butyl hydroperoxide (t-BuOOH), as indicated by accumulation of malondialdehyde (MDA). In microsomes of liver and Morris hepatomas, MDA formation varied with the content of cytochrome P-450 (liver > 9618A > 3924A).[1] In whole cells, lipid peroxidation must secondarily alter cell functions, and those related to membranes may be especially susceptible. We have compared the sensitivity of some cell functions to t-BuOOH, using tissue slices of liver and hepatomas 9618A and 3924A.

Maintenance of the tumor lines and the experimental procedures were as described before.[1-3] Slices were preincubated at 1°C to induce a loss of K^+ and gain of water which were reversed by active transport systems of the plasma membrane upon restoration of metabolic activity at 38°C. The ultrastructure of the slices also showed an excellent recovery at 38°C.[2,4]

Production of MDA was increased when t-BuOOH in the medium exceeded 1 mM (FIG. 1), the amount varying according to microsomal cytochrome P-450.[1] Incorporation of [14C]leucine into proteins of the hepatoma slices, in the absence of t-BuOOH, was 2–4 times greater than in liver. In each case the incorporation was very sensitive to t-BuOOH, with half-maximal inhibition at 1.5 mM t-BuOOH for liver and 2–3 mM for the hepatomas. The inhibition was poorly correlated with the amount of MDA accumulating (for example, in liver and 9618A, leucine incorporation was much inhibited by 1 mM t-BuOOH, without significant increase of MDA, and inhibition of leucine incorporation ran closely parallel in the two hepatomas, despite the much smaller MDA formation by 3924A).

In each tissue, extrusion of excess water (TABLE 1) and reaccumulation of K^+ (not shown) were completely inhibited by t-BuOOH, with half-maximal inhibition at 3–5

[a]This work was supported by Consiglio Nazionale delle Ricerche, special project "Oncology" No. 86.00423.44.

mM *t*-BuOOH. Thus, these systems were rather less sensitive to *t*-BuOOH than leucine incorporation was. Slice ATP was even less sensitive, for 5 mM *t*-BuOOH was needed for a significant reduction at ATP in liver and 3924A and even 10 mM had no effect in 9618A (TABLE 1). The incubation medium contained no glucose and the ATP

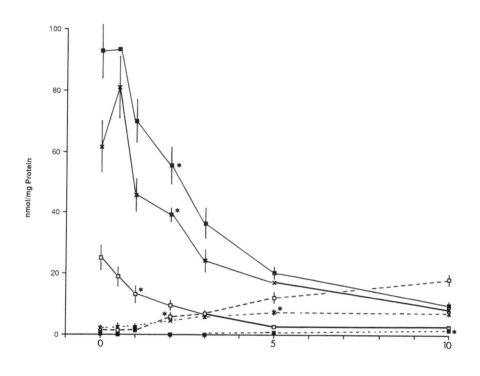

t-BuOOH (mM)

FIGURE 1. Effects of varying concentrations of *tert*-butyl hydroperoxide on protein synthesis and malondialdehyde accumulation in slices of liver and hepatomas. Tissue slices were preincubated for 90 min at 1°C followed by 60 min at 38°C in a Ringer's medium buffered with Tris[hydroxymethyl]amino methane and phosphate and containing 5 mM L-leucine labelled with ^{14}C; it was gassed with O_2.[2,3] After incubation, total-slice proteins were counted for radioactivity. Malondialdehyde was the sum of that determined in slices and medium at the conclusion of incubation. Points represent the mean ± SEM of ten samples of slices for the liver and six for each of the hepatomas; for clarity, some error bars are not shown. □, liver; ×, hepatoma 9618A; ■, hepatoma 3924A. *Solid lines* represent [^{14}C]leucine incorporated into proteins; *broken lines* represent malondialdehyde formed. *Asterisk* shows lowest concentration of *t*-BuOOH giving significant reduction below control level ($p < 0.05$).

was therefore predominantly derived from mitochondrial oxidative phosphorylation.[2,4]

We conclude: (1) The levels of MDA accumulating were a poor indication of the damage to cell functions. Perhaps only a small fraction of the MDA is directly related to the crucial peroxidative events. (2) Lower cell contents of cytochrome P-450 did not

TABLE 1. Effects of Varying Concentrations of *tert*-Butyl Hydroperoxide on Water Extrusion and ATP Contents of Liver and Hepatoma Slices

t-BuOOH (mM)	Water Extrusion[a] (kg water/kg dry wt)			ATP Content[b] (mmol/kg protein)		
	Liver	9618A	3924A	Liver	9618A	3924A
0.0	1.5 ± 0.2	1.6 ± 0.4	1.4 ± 0.2	22 ± 2	17 ± 2	54 ± 3
0.5	1.3 ± 0.2	0.8 ± 0.3	1.2 ± 0.3	21 ± 2	22 ± 3	54 ± 4
1.0	1.2 ± 0.2	1.1 ± 0.4	1.6 ± 0.1	27 ± 3	28 ± 3	45 ± 4
2.0	0.7 ± 0.3^{c}	0.5 ± 0.3^{c}	0.9 ± 0.3^{c}	25 ± 4	25 ± 2	47 ± 3
3.0	0.6 ± 0.2	0.5 ± 0.3	0.9 ± 0.4	24 ± 3	22 ± 2	43 ± 4
5.0	0.1 ± 0.2	0.4 ± 0.4	-0.2 ± 0.3	13 ± 2^{c}	20 ± 3	41 ± 4^{c}
10.0	-0.8 ± 0.7	0.4 ± 0.3	0.2 ± 0.3	8 ± 1	12 ± 3	30 ± 4
(n)	(10)	(6)	(6)	(10)	(6)	(6)

NOTE: Values represent mean $\pm$ SEM (*n* sets of slices).

[a]Difference between contents after preincubation at 1°C and after subsequent incubation at 38°C.

[b]Final content after incubation.

[c]Lowest concentration of *t*-BuOOH giving significant difference from controls ($p < 0.05$).

reduce the deleterious effects of *t*-BuOOH. (3) Functions of the endoplasmic reticulum and plasma membrane may be more sensitive than mitochondrial activity to *t*-BuOOH. The order of sensitivity of the cellular functions to *t*-BuOOH is similar to that caused by CCl_4 or 1,2-dichloroethane, both of which stimulate MDA formation.[3]

REFERENCES

1. MINOTTI, G., S. BORRELLO, G. PALOMBINI & T. GALEOTTI. 1986. Cytochrome P-450 deficiency and resistance to t-butyl hydroperoxide of hepatoma microsomal lipid peroxidation. Biochim. Biophys. Acta **876:** 220–225.
2. RUSSO, M. A., T. GALEOTTI & G. D. V. VAN ROSSUM. 1976. Metabolism-dependent maintenance of cell volume and ultrastructure in slices of Morris hepatoma 3924A. Cancer Res. **36:** 4160–4174.
3. THOMAS, L., B. DEFEO, G. D. V. VAN ROSSUM & M. F. MARIANI. 1989. Comparison of metabolic effects of carbon tetrachloride and 1,2 dichloroethane added *in vitro* to slices of rat liver. Toxicol. in Vitro. **3:** in press.
4. RUSSO, M. A., G. D. V. VAN ROSSUM & T. GALEOTTI. 1977. Observations on the regulation of cell volume and metabolic control *in vitro*. Changes in the composition and ultrastructure of liver slices under conditions of varying metabolic and transport activity. J. Membrane Biol. **31:** 267–299.

The Intracellular Homeostasis of Calcium: An Overview

ERNESTO CARAFOLI

Laboratory of Biochemistry
Swiss Federal Institute of Technology (ETH)
8092 Zürich, Switzerland

INTRODUCTION

Ca has now been accepted as a signaling agent of universal significance which controls a large number of cellular functions, among which are the synthesis and release of hormones, muscle and nonmuscle motility, and a number of membrane-linked processes (see Carafoli[1] for a recent review). The signaling function of Ca obviously requires maintenance within cells at a very low ionic concentration and mechanisms to modulate it in the vicinity of the targets of the signaling function. Signaling agents normally are regulated by biosynthesis and breakdown. For the case of Ca, evolution has selected a different control mechanism, that is, the reversible complexation by specific proteins, which are either soluble, organized in nonmembranous structures, or intrinsic to membranes. These proteins "buffer" intracellular Ca at a concentration that is at least 10,000-fold lower than in the external spaces. The cells require both high- and low-affinity regulation of Ca. The rapid and precise modulation is performed by intracellular Ca-binding proteins soluble in the cytosol, but also (and, in fact, essentially, which will be treated of below) by high-Ca-affinity proteins intrinsic to membranes, particularly those of endo(sarco)plasmic reticulum. The Ca-filtering function of the plasma membrane, which also depends on the operation of membrane-intrinsic Ca-binding proteins, is responsible for the long-term maintenance of the Ca gradient between cells and medium. The gradient of Ca across the plasma membrane is convenient to the signaling function, since the large inwardly directed Ca pressure ensures that even minor increases in the Ca permeability of the membrane produces significant swings in its intracellular concentration.

The High-Affinity Intracellular Ca-Binding Proteins

The solution of the crystal structure of the Ca-binding protein parvalbumin by Kretsinger and his associates[2] has led to the proposal of structural guidelines for the complexing of Ca by high-affinity Ca-binding proteins (see Kretsinger and Nelson[3] for a review). The parvalbumin principles demand that Ca-binding proteins contain repeat units made of two perpendicular α-helices, flanking a loop of 10–12 amino acids where Ca coordinates to 6–8 oxygen atoms of carboxylic side chains and, less frequently, to carbonyl oxygens of the peptide backbone. This Ca-binding site has been optimized in the course of evolution: only minimal variability is now tolerated in it. The parvalbumin model has become known as the EF-hand model, and has been shown to be compatible with the amino acid sequences of a number of other Ca-binding proteins;

recently, it has been directly validated by the crystal structures of other Ca-binding proteins.[4-6]

Although the soluble high-affinity Ca-binding proteins complex Ca and thus contribute to its buffering, they have an important additional function: hydrophobic domains appear on their surface upon binding Ca, and the conformational change permits the interaction with (enzyme) targets. Therefore, these proteins are essentially processors of the Ca signal, rather than intracellular Ca buffers. Even if their Ca buffering role is accessory and in any case quantitatively limited by their amount in the cell, one can nevertheless attempt to evaluate the total theoretical Ca buffering capacity of the two most important Ca-binding proteins of this type, calmodulin and troponin C. The concentration of the former is lower in muscles than in other tissues, where its total Ca-buffering capacity corresponds to approximately 8 μM Ca.[1]

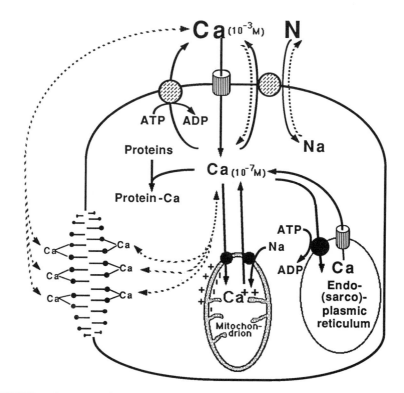

FIGURE 1. Ca-transporting systems in eukaryotic cells. This figure shows the transporting systems that have been characterized so far. The systems in the Golgi membranes and in the lysosomes are not shown because they are still very poorly known. Three are in the plasma membrane (a Ca-ATPase, a Na/Ca exchanger, which normally imports Ca but sometimes also exports it, and a Ca channel); two in endo(sarco)plasmic reticulum (a Ca-ATPase and a still poorly characterized release channel); and two are in the inner membrane of mitochondria (an electrophoretic uptake uniporter and a Ca-releasing Na/Ca exchanger). The soluble Ca-binding proteins of the cytosol and the acidic phospholipids of the plasma membrane have been included in this figure as participants in the Ca-buffering function (see text).

TABLE 1. Ca-Transporting Systems in Cell Membranes

Transporting Mode	Membrane System	Ca Affinity
ATPases	Plasma membrane Endo(sarco)plasmic reticulum (Golgi? lysosomes?)	High
Exchangers (Na/Ca)	Plasma membrane Mitochondria	Low
Channels	Plasma membrane Endo(sarco)plasmic reticulum	Low
Electrophoretic uniporters	Mitochondrial inner membrane	Low

Troponin-C, on the other hand, is more concentrated in muscles than in nonmuscle tissues; on the basis of a measured content of 2 μM in heart, and 20 μM in skeletal muscles,[7] it can be calculated that troponin-C could handle up to 6 μM and 80 μM Ca in heart and skeletal muscles, respectively.[1]

Membrane Transport of Calcium

The limitations in the Ca-buffering function described above do not apply to the Ca-binding proteins intrinsic to membranes. They complex Ca at one membrane side, transfer it across, and repeat the operation continuously, thus eliminating the problems arising from the possibly insufficient amount of protein available. Membrane-intrinsic Ca-binding (and transport) proteins thus play the dominant role in the buffering of intracellular Ca. Eukaryotic cells possess several Ca-transporting systems in the plasma membrane and in the membranes of the organelles (FIG. 1). They operate with different kinetic properties, responding to the varying demands of the functional cycle of cells: sometimes rapid, precise, and high-affinity regulation is required, at other times lower affinity and less rapid regulation (i.e., transport) may suffice. As FIGURE 1 indicates, eukaryotic cell membranes contain several Ca-transporting membrane systems, but only four basic transporting mechanisms are known: ATPases, exchangers, channels, and electrophoretic uniporters (TABLE 1). High-affinity Ca regulation depends obligatorily on ATPases, since only this transporting mode confers to the system high affinity for Ca. Lower affinity regulation may, on the other hand, "choose" from among the other modes listed in TABLE 1.

Ca Transport in the Plasma Membrane

FIGURE 2 shows the three Ca-transporting systems known to exist in the plasma membrane: the channel, the Na/Ca exchanger, and the pump. Although known for a long time, the channel, which is responsible for the penetration of Ca into cells, has been characterized at a molecular level only recently: the patch-clamp technique has permitted the study of single-channel activity and has led to the definition of its precise kinetic parameters (see Reuter[8] for a review). The study of single-channel currents has revealed a conductance of 15–25 pS, corresponding to the passage of about 3×10^6 Ca ions per second.[9] The density of the Ca channel is particularly high in the T-tubular membranes of skeletal muscles, which are thus the tissue of choice for isolation studies. The channels are controlled by the electrical potential across the plasma membrane, and begin to open as the transmembrane potential increases from the resting level of -70 mV to about -40 mV (maximal Ca currents are seen at about 0 mV). The channels are blocked by the "Ca antagonists,"[10] which have greatly advanced the understanding of the Ca channel and have recently permitted the isolation of the dihydropyridine receptor and the determination of its primary structure.[11]

Another interesting aspect of the Ca channel is its stimulation by adrenergic neurotransmitters: patch-clamp experiments on heart plasma membranes have shown that cAMP enhances the opening probability of the channel.[9] The properties described

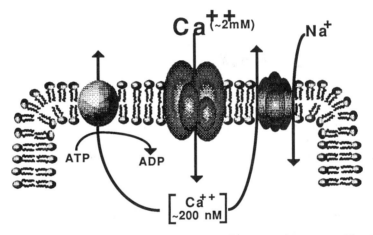

FIGURE 2. A scheme of the three plasma membrane Ca-transporting systems. The subunit composition of the Ca channel and of the Na/Ca exchanger is based on the isolation work of several workers for the former (see for example Curtis and Catterall[36] and Borsotto et al.[37]), and on a proposal by Longoni and Carafoli[38] for the latter.

so far refer to one of the types now known to exist. However, several other types different in properties have been recently recognized. Two have been identified by patch-clamp experiments in mammalian hearts,[12] including the most common, which is the L-type described above. It is dihydropyridine-sensitive and its openings produce currents of long duration. The T-type opens at more negative transmembrane potentials, is insensitive to dihydropyridines, and produces currents of shorter duration.

The Na/Ca exchanger is one of the two Ca-exporting systems of the plasma membrane (under some conditions it can also mediate the influx of Ca). It is a large-capacity, low-affinity system, particularly active in excitable tissues. The system has been studied essentially on heart and the giant axon of the squid (see Philipson[13] for a review). Early electrophysiological work has established that the system operates electrogenically, exchanging 3 Na for 1 Ca; thus, it does not only respond to the Na and Ca gradients, but also to the transmembrane electrical potential. Recent work on heart sarcolemmal vesicles[14] has shown that the system has low affinity for Ca (K_m, 1–20 μM) but high maximal rate of transport, corresponding in heart to 20 nmol per mg of sarcolemmal protein per second. The low affinity of the system for Ca is puzzling, since the free Ca concentration in the cytosol presumably never increases to 10 μM. Possibly, however, the kinetic parameters of the exchanger in isolated vesicles are different from those in the intact tissue. The activation of the exchanger by a kinase-linked phosphorylation process,[15] which decreases the K_m(Ca) of the system to about 1 μM, may be important in this context.

The Ca-ATPase of the plasma membrane interacts with Ca with high affinity (K_m about 0.5 μM), but has low total transport capacity: (about 0.5 nmol per mg of membrane protein per second in heart). The high Ca affinity of the enzyme suggests that it exports Ca from cells even when its concentration in the cytosol is at the resting submicromolar level. Thus, the Ca-ATPase probably plays the most important role in maintaining the gradient of Ca between cells and medium. The enzyme is an ATPase of the P-class, that is, it forms an aspartyl phosphate during the reaction mechanism and is inhibited by vanadate (see Schatzmann[16] for a review). In addition, it is a target of calmodulin stimulation: the interaction with calmodulin has permitted the purification of the ATPase on calmodulin columns as a polypeptide of about 140 kDa, which can be reconstituted in liposomes with optimal transport efficiency.[17] The liposomal system has permitted us to establish that the ATPase transports Ca with a 1:1 stoichiometry to ATP hydrolysis, (compare the stoichiometry of 2 for the analogous enzyme of sarcoplasmic reticulum; see below). Caroni and Carafoli[18] have found that a cAMP-linked phosphorylation process stimulates the ATPase and the associated pumping of Ca. Since cAMP also stimulates the Ca channel activity (see above), the role of cAMP is thus the overall increase of the plasma membrane Ca flux, rather than the undirectional stimulation of Ca transport.

Work on the purified Ca ATPase has established that the enzyme can be stimulated by several treatments alternative to calmodulin, among them the exposure to acidic phospholipids and polyunsaturated fatty acids, and the controlled proteolysis by a number of proteases. The proteolysis work has been instrumental in a recent comprehensive attempt in this laboratory, based on chemical sequencing and DNA recombination technology, to establish the primary structure of the enzyme and to assign functional domains in it.[19] Given the novelty of the information, details on the structure are given here.

The pump contains 1220 amino acids, corresponding to an M_r of 134,683. Asp 475 forms the acyl phosphate during the reaction cycle, and Lys 601 binds the ATP antagonist fluoroscein isothiocyanate. The calmodulin (CaM)-binding domain has been identified next to the C-terminus (residues 1100–1127) using a bifunctional, cleavable, photoactivatable cross-linker: the domain resembles the CaM-binding regions of CaM-modulated enzymes and of CaM-binding peptide venoms. The translated sequence of the ATPase contains at the N and C sides of the CaM-binding domain sequences that are very rich in Asp and Glu and that show some homology to CaM: the former may play a role in the binding of Ca and in regulating the interaction of CaM with the pump. Of interest, the pump also contains, next to the N-terminus, two 11-amino acid stretches which resemble EF-hands (residues 22–33 and 310–321), and could thus form Ca-binding sites. Ten hydrophobic domains, presumably spanning the membrane, have been identified: four are located in the N-terminal portion of the pump, and six in the C-terminal portion. The mid-portion of the pump (about 500 residues) contains no hydrophobic domains. Ser 1178, located on the C-terminal side of the CaM-binding domain, is phosphorylated by the cAMP-dependent kinase, increasing the Ca-affinity of the pump. A scheme of the architecture of the protein in the plasma membrane is shown in FIGURE 3.

Ca Transport across Intracellular Membranes

The long-term maintenance of the Ca gradient between cells and medium is thus the result of the action of the plasma membrane transport systems. However, it is

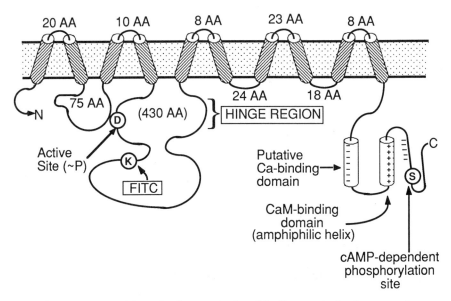

FIGURE 3. A scheme of the molecular organization of the Ca pump in the plasma membrane.

important to recognize that the plasma membrane fluxes are quantitatively minor when compared to the total amount of Ca used during the functional cycle of cells. The Ca crossing the plasma membrane triggers important intracellular events and is thus vital to cell function, but most of the Ca needed for cell activity derives from intracellular stores.

The Endo(sarco)plasmic Reticulum

Most of the early work on this membrane system has been performed cn the easily available sarcoplasmic reticulum and on its key enzyme, the Ca-ATPase. More recently, endoplasmic reticulum has seen considerable activity, due to the discovery that its Ca pool is sensitive to inositol-tris-phosphate.[20] The role of this messenger in the release of Ca from sarcoplasmic reticulum is still unclear in heart and skeletal muscle, but appears to be probable in smooth muscles.[21]

The endo(sarco)plasmic reticulum is responsible for the fine regulation of Ca in the cytosol. It contains an ATPase which has high affinity for Ca (K_m below 0.5 μM). The enzyme is very abundant in the membranes of sarcoplasmic reticulum, which thus has a large total Ca-transporting capacity. This is particularly so in fast skeletal muscles, where the rate may reach 70 nmol per mg of membrane protein per second. The sarcoplasmic reticulum ATPase has been purified by MacLennan[22] in 1970 as a single polypeptide of about 100 kDa, and has been characterized extensively in several laboratories. It belongs to the same ATPase class as that of the plasma membrane Ca pump: it forms an aspartyl-phosphate, is inhibited by vanadate, and can be reconstituted into liposomes, where it transports Ca with a 2:1 stoichiometry to the hydrolyzed ATP. Recently, its primary structure has been determined[23] on both heart and fast skeletal muscle reticulum, which contains two isoforms of the enzyme. The ATPase in heart, smooth, and slow (but not fast) skeletal muscles is modulated by the acidic proteolipid phospholamban,[24] which is a pentamer of five identical subunits of about 6 kDa, and is phosphorylated by both the cAMP and the calmodulin-dependent protein kinase. Phosphorylated phospholamban activates the ATPase-linked transport of Ca, thus transmitting to sarcoplasmic reticulum hormonal messages from the plasma membrane.

The release of Ca from heart sarcoplasmic reticulum is now the subject of very intensive investigations, also as the result of the great interest in the function of inositol-tris-phosphate in endoplasmic reticulum and in sarcoplasmic reticulum of smooth muscles. The favored mechanism for the Ca-release phenomenon in heart is the Ca-induced Ca release,[25] in which the liberation of massive amounts of Ca from the vesicles of sarcoplasmic reticulum is triggered by Ca added to the medium at concentrations below 0.1 μM. The channel that mediates the release phenomenon is now also being intensively studied. Recent work[26] has indicated that the release is activated by Ca, adenine nucleotides, and caffeine, and is inhibited by ruthenium red, Mg, protons, and calmodulin.

Much less is known on the mechanism of Ca release induced by inositol-tris-phosphate. One interesting recent development is the proposal by Volpe et al.[27] that the organelle that responds to inositol-tris-phosphate in nonmuscle tissues is not the endoplasmic reticulum proper, but a separate vesicular system possibly linked to endoplasmic reticulum, which has been termed calciosome.

Mitochondria

Mitochondria have been traditionally considered to be important in the regulation of both cytosolic and intramitochondrial Ca (see Carafoli[28] for a review). The demonstration that mitochondria only handle Ca with low affinity[30] and the finding that they contain much less Ca in situ than was generally concluded from experiments on isolated mitochondria[31] have forced a re-evaluation of their importance as cytosolic Ca regulators, a function that is now considered of minor importance. Their main Ca-related task now appears to be the regulation of matrix Ca. This is an important function because of the existence of matrix dehydrogenases which are precisely controlled by Ca (FIG. 4). Mitochondria accumulate Ca by means of an electrophoretic route which is energized by the electrical component of the proton-motive force across the inner membrane and is inhibited by ruthenium red. The route has low Ca affinity (K_m, 1–10 μM) but could nevertheless accumulate Ca at a potential optimal rate of up to 10 nmol per milligram of protein per second. *In vivo*, however, the suboptimal concentration of Ca in the cytosol and the presence of Mg, which also inhibits the route, reduce the uptake rate to a fraction of the optimum. Interestingly, mitochondria also accumulate inorganic phosphate to precipitate Ca as an insoluble salt in the matrix. This results in the damping of the changes in ionic Ca in the latter compartment, limiting the disturbance to the Ca-modulated dehydrogenases by newly accumulated Ca.

Since the mitochondrial Ca uptake process is electrophoretic, and since the electrical potential across the inner membrane is unlikely to fluctuate, the release of Ca does not occur through the reversed uptake route. It occurs through an electroneutral Na/Ca exchanger,[32] which is particularly active in heart and other excitable tissues[33] and which releases Ca at a slow rate (about 0.2–0.3 nmol per milligram of mitochondrial protein per second). The Na-promoted release route is insensitive to ruthenium red, but is blocked by some of the inhibitors of the plasma membrane Ca channel (e.g., the benzothiazepine diltiazem[34]). The energy-driven uptake route and the Na/Ca

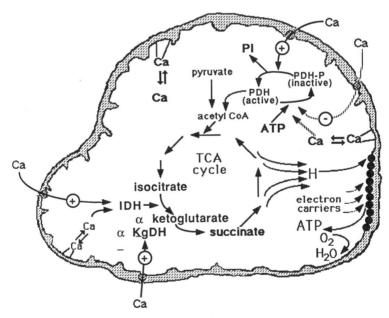

FIGURE 4. The regulation of matrix dehydrogenases by Ca. The information in the figure derives from the work of Denton and his coworkers.[39,40] PDH = pyruvic acid dehydrogenase; IDH = NAD-linked isocitric acid dehydrogenase; α-Kg-DH = α-ketoglutaric acid dehydrogenase.

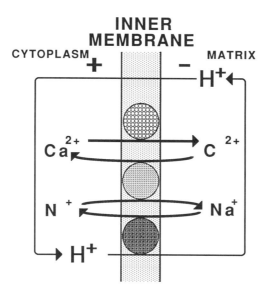

FIGURE 5. The mitochondrial Ca cycle.[35]

exchanger can be integrated into an energy-dissipating Ca cycle[35] (FIG. 5). The low overall activity of the mitochondrial Ca transport system *in vivo* limits the level of energy dissipation by the cycle.

The minor role of mitochondria in the regulation of cytosolic Ca under physiological conditions is thus essentially determined by the insufficient concentration of Ca in cells. However, cell-injuring conditions may alter the permeability properties of the plasma membrane, permitting the influx of excess Ca into the cytosol. Should this occur, the mitochondrial uptake system may become activated, leading to the precipitation of large amounts of Ca phosphate in the matrix (deposits of Ca phosphate in mitochondria of injured cells, including heart cells, are indeed routinely documented [see Carafoli[29] for a review]). If the injuring condition is removed, mitochondria slowly release the stored Ca at a rate compatible with the exporting ability of the plasma membrane systems and thus play a decisive protective role against cytosolic Ca overload.

CONCLUSIONS

The control of cell Ca is essentially performed by the reversible complexation to specific proteins. Soluble proteins contribute to Ca buffering, but are more important in the transmission of Ca information to (enzyme) targets. Membrane-intrinsic proteins play the main role in the buffering of cell Ca. They control Ca very precisely and with high affinity (ATPases) or with lower affinity (channels, exchangers, the electrophoretic uniporter of mitochondria). The endo(sarco)plasmic reticulum is responsible for the fine regulation of Ca, whereas the mitochondrial transporting systems essentially control intramitochondrial Ca. The mitochondrial system, however, also protects the cytosol against pathologic Ca increases.

REFERENCES

1. CARAFOLI, E. 1987. Intracellular calcium homeostasis. Annu. Rev. Biochem. **56:** 395.
2. KRETSINGER, R. H. & C. E. NOCKOLDS. 1973. Carp muscle calcium-binding protein. J. Biol. Chem. **248:** 3313.
3. KRETSINGER, R. H. & D. J. NELSON. 1977. Calcium in biological systems. Coord. Chem. Rev. **18:** 29.
4. SZEBENYI, D. M. E., OBENDORF, S. K. & MOFFAT, K. 1981. Structure of vitamin D-dependent calcium binding protein from bovine intestine. Nature **294:** 327.
5. HERZBERG, O. & M. N. G. JAMES. 1985. Structure of the calcium regulatory muscle protein troponin-C at 2.8. A resolution. Nature **313:** 665.
6. BABU, Y. S., J. S. SACK, T. J. GREENHOUGH, C. E. BUGG, A. R. MEANS & W. J. COOK. 1985. Three-dimensional structure of calmodulin, Nature **315:** 37.
7. GRAND, R. J. A., S. V. PERRY & R. A. WEEKS. 1979. Troponin-C like proteins (calmodulin) from mammalian smooth muscle and other tissues. Biochem. J. **177:** 521.
8. REUTER, H. 1984. Ion channels in cardiac cell membranes. Annu. Rev. Physiol. **46:** 473.
9. REUTER, H., C. F. STEVENS, R. W. TSIEN & G. YELLEN. 1982. Properties of single calcium channels in cardiac cell culture. Nature **297:** 501.
10. FLECKENSTEIN, A. 1973. Calcium Antagonism in Heart and Smooth Muscle. Wiley, New York.
11. TANABE, T., H. TAKESHIMA, A. MIKAMI, V. FLOCKERZI, H. TAKAHASHI, K. KANGAURA, M. KOJIMA, H. MATSUO, T. HIROSE, & S. NUMA. 1987. Primary structure of the receptor for calcium channel blockers from skeletal muscle. Nature **328:** 313.
12. NILIUS, B., P. HESS, J. B. LANSMANN & R. W. TSIEN. 1985. A novel type of cardiac calcium channel in ventricular cells. Nature **316:** 443.
13. PHILIPSON, K. D. 1985. Sodium-calcium exchange in plasma membrane vesicles. Annu. Rev. Physiol. **47:** 561.
14. REEVES, J. P. & J. L. SUTKO. 1979. Sodium-calcium exchange in cardiac membrane vesicles. Proc. Natl. Acad. Sci. USA **76:** 590.
15. CARONI, P. & E. CARAFOLI. 1983. The regulation of the Na^+/Ca^{2+} exchanger of heart sarcolemma. Eur. J. Biochem. **132:** 451.
16. SCHATZMANN, H. 1982. The calcium pump of erythrocytes and other animal cells. In Membrane Transport of Calcium. E. Carafoli, Ed.: 41. Academic Press. London.
17. NIGGLI, V., E. S. ADUNYAH, J. T. PENNISTON, & E. CARAFOLI. 1981. Purified (Ca^{2+}-Mg^{2+})-ATPase of the erythrocyte membrane; reconstitution and effect of calmodulin and phospholipids. J. Biol. Chem. **256:** 395.
18. CARONI, P. & E. CARAFOLI. 1981. Regulation of the Ca^{2+}-pumping ATPase of heart sarcolemma by a phosphorylation-dephosphorylation process. J. Biol. Chem. **256:** 9371.
19. VERMA, A. K., A. G. FILOTEO, D. R. STANFORD, E. D. WIEBEN, J. T. PENNISTON, E. S. STREHLER, R. FISCHER, R. HEIM, S. VOGEL, S. MATHEWS, M. A. STREHLER-PAGE, P. JAMES, T. VORHERR, J. KREBS & E. CARAFOLI. 1988. Complete primary structure of a human plasma membrane calcium pump. J. Biol. Chem. **263:** 14152.
20. STREB, H., R. F. IRVINE, M. J. BERRIDGE & I. SCHULZ. 1983. Release of Ca^{2+} from a non-mitochondrial intracellular store in pancreatic acinar cells by inositol-1,4,5-triphosphate. Nature **306:** 66.
21. SOMLYO, A. V., M. BOND, A. P. SOMLYO & A. SCARPA. 1985. Inositol tris phosphate-induced calcium release and contraction in vascular smooth muscle. Proc. Natl. Acad. Sci. USA **82:** 5231.
22. MACLENNAN, D. H. 1970. Purification and properties of an adenosine triphosphatase from sarcoplasmic reticulum. J. Biol. Chem. **245:** 4508.
23. MACLENNAN, D. H., C. J. BRANDL, B. KORCZAK & N. M. GREEN. 1985. Amino-acid sequence of a $Ca^{2+} + Mg^{2+}$-dependent ATPase from rabbit muscle sarcoplasmic reticulum, deduced from its complementary DNA sequence. Nature **316:** 696.

24. TADA, M., M. A. KIRCHBERGER & A. M. KATZ. 1975. Phosphorylation of 22.000-Dalton component of the cardiac sarcoplasmic reticulum by adenosine 3′,5′-monophosphate-dependent protein kinase. J. Biol. Chem. **250:** 2640.

25. FABIATO, A. & F. FABIATO. 1975. Contractions induced by a calcium triggered release of calcium from the sarcoplasmic reticulum of single skinned cardiac cells. J. Physiol. **249:** 457.

26. MEISSNER, G., and HENDERSON, J. S. 1987. Rapid calcium release from sarcoplasmic reticulum vesicles is dependent on calcium and is modulated by Mg^{2+}, adenine nucleotide, and calmodulin. J. Biol. Chem. **262:** 3065.

27. VOLPE, P., K. H. KRAUSE, G. HASHIMOTO, F. ZORZATO, T. POZZAN, J. MELDOLESI & D. P. LEW. 1988. "Calcisome," a cytoplasmic organelle: The inositol 1,4,5-trisphosphate-sensitive Ca^{2+} store of non-muscle cells? Proc. Natl. Acad. Sci. USA **85:** 1091.

28. CARAFOLI, E. 1982. The transport of calcium across the inner membrane of mitochondria. In Membrane Transport of Calcium. E. Carafoli, Ed.: 109. Academic Press. London.

29. CARAFOLI, E. 1982. Membrane transport and the regulation of the cell calcium levels. In Pathophysiology of Shock, Anoxia, and Ischemia. R. A. Cowley, B. F. Trump, Eds.: 95. Williams & Wilkins. Baltimore, MD.

30. CROMPTON, M., E. SIGEL, M. SALZMANN, & E. CARAFOLI. 1976. A kinetic study of the energy-linked influx of Ca^{2+} into heart mitochondria. Eur. J. Biochem. **69:** 429.

31. SOMLYO, A. P., A. V. SOMLYO & H. SHUMAN. 1979. Electron probe analysis of vascular smooth muscle, composition of mitochondria, nuclei, and cytoplasm. J. Cell Biol. **81:** 316.

32. CARAFOLI, E., G. TIOZZO, F. LUGLI, F. CROVETTI & C. KRATZING. 1974. The release of calcium from heart mitochondria by sodium. J. Mol. Cell. Cardiol. **6:** 361.

33. CROMPTON, M., R. MOSER, H. LÜDI & E. CARAFOLI. 1978. The interrelations between the transport of sodium and calcium in mitochondria of various mammalian tissues. Eur. J. Biochem. **82:** 25.

34. VAGHY, P. L., J. D. JOHNSON, M. A. MATLIB, T. WANG & A. SCHWARZ. 1982. Selective inhibition of Na^+-induced Ca^{2+} release from heart mitochondria by diltiazem and certain other Ca^{2+} antagonist drugs, J. Biol. Chem. **257:** 6000.

35. CARAFOLI, E. 1979. The calcium cycle of mitochondria. FEBS Lett. **104:** 1.

36. CURTIS, B. M. & W. A. CATTERALL. 1985. Phosphorylation of the calcium antagonist receptor of the voltage-sensitive calcium channel by cAMP-dependent protein kinase. Proc. Natl. Acad. Sci. USA, **82:** 2528.

37. BORSOTTO, M., R. I. NORMAN, M. FOSSET & M. LAZDUNSKI. 1984. Eur. J. Biochem. **14:** 449.

38. LONGONI, S. & E. CARAFOLI. 1987. Identification of the Na^+/Ca^{2+} exchanger of calf heart sarcolemma with the help of specific antibodies. Biochem. Biophys. Res. Commun. **145:** 1059.

39. DENTON, R. M., P. J. RANDLE & B. R. MARTIN. 1972. Stimulation by calcium ions of pyruvate dehydrogenase phosphate phosphatase. Biochem. J. **128:** 161.

40. DENTON, R. M., D. A. RICHARDS & J. G. CHIN. 1978. Calcium ions and the regulation of NAD^+-linked isocitrate dehydrogenase from the mitochondria of rat heart and other tissues. Biochem. J. **176:** 899.

DISCUSSION

R. S. SOHAL (*Southern Methodist University, Dallas, Texas*): How does oxidative stress affect intracellular calcium-regulating systems? Which component is most susceptible to oxidative stress?

E. CARAFOLI (*Swiss Federal Institute of Technology, Zurich, Switzerland*): Very little has been done on possible direct effects of oxygen radicals on any of the membrane-linked Ca^{2+}-transporting systems. Mitochondrial uncoupling would, of course, result in a decreased ability to accumulate and retain Ca^{2+}.

L. CANTLEY (*Tufts University, Boston, Massachusetts*): Is the mitochondrial Na^+/Ca^{2+} antiporter electrogenic and does an Na^+ gradient exist across the mitochondrial inner membrane?

CARAFOLI: There is a 2–3-fold gradient of Na^+ across the inner mitochondrial membrane, Na^+ being higher in the cytoplasm. As for the exchanger, it is normally assumed to be electroneutral, which makes good sense, considering the large negative electrical potential inside mitochondria.

H. WOHLRAB (*Boston Medical Research Institute, Boston, Massachusetts*): Do you believe that mitochondria act as an effective calcium buffer in cardiac ischemia?

CARAFOLI: Yes, I believe so—heart ischemia is but one of the conditions in which mitochondria would experience increased Ca^{2+} concentrations in their environment. They would react to it by accumulating (and precipitating) Ca^{2+} and phosphate.

P. A. RILEY (*University College of London, London, England*): I was very interested in your proposed structure for the plasma membrane calcium pump. It can be shown that large increases in intracellular calcium are produced by exposure of cells to external thiol reagents. Are these exposed thiols in your protein important for the activity of the pump?

CARAFOLI: The plasma membrane Ca^{2+} pump is sensitive, like most enzymes, to thiol reagents. However, no S-containing amino acids critically involved in the active site have been identified so far.

L. ERNSTER (*University of Stockholm, Stockholm, Sweden*): You did not include in your cartoon of the "mitochondrial Ca^{2+} cycle" the $NAD(P)^+$-dependent efflux of Ca^{2+} first described by Vercesi and Lehninger and subsequently shown by Richter and his colleagues to proceed via an ADP-ribose activated protein. Would you care to comment on this?

CARAFOLI: I did not refer to the $NAD(P)^+$-dependent release phenomenon because according to many specialists it is linked to nonspecific damage of the inner membrane. It is, however, a very interesting system, which may turn out to be of primary importance in nonexcitable tissues.

Intracellular Source(s) of $[Ca^{2+}]_i$ Transients in Nonmuscle Cells

A. MALGAROLI,[a,b,c] S. HASHIMOTO,[a,b] F. GROHOVAZ,[a]
G. FUMAGALLI,[a] T. POZZAN,[d] AND J. MELDOLESI[a,b]

[a]Department of Pharmacology
CNR Center of Cytopharmacology
University of Milan
Milan, Italy

[b]Scientific Institute S. Raffaele
20132 Milan, Italy

[d]Institute of General Pathology
University of Ferrara
Ferrara, Italy

INTRODUCTION

A host of cellular functions are known to be regulated by the elevation of the cytosolic Ca^{2+} concentration $[Ca^{2+}]_i$, from the resting level around 10^{-7} M to higher values, usually in the micromolar range.[1,2] For many years, the Ca^{2+} responsible for these changes has been known to originate from at least two different sources: the extracellular medium, where $[Ca^{2+}]$ is millimolar; and intracellular storage organelle(s), whose calcium content can approach 10^{-1} M. Traditionally, some specific functions (for example, the release of neurotransmitters at nerve terminals) have been attributed primarily to the stimulation of Ca^{2+} influx via the activation of voltage-gated Ca^{2+} channels; others (for example, striated muscle contraction) have been ascribed to release from intracellular store(s). In virtually all nonmuscle cells this latter phenomen has been recognized to be triggered by inositol-1,4,5-trisphosphate (Ins-P_3), a second messgener generated after the activation of a variety of surface receptors via the hydrolysis of a plasma membrane phospholipid, phosphatidylinositol-4,5-biosphosphate.[3] Activation of these same receptors causes increased Ca^{2+} influx through channels different from voltage-operated Ca^{2+} channels (VOCS), which have not yet been characterized in detail.[3,4] Other problems that at the moment appear still open are whether Ins-P_3 is the only intracellular messenger capable of inducing the release of Ca^{2+} from intracellular stores in nonmuscle cells; whether the intracellular-evoked release of Ca^{2+} originates from one or multiple types of storage organelles; the cytological identification of these organelles; the effects the two processes (Ca^{2+} influx and release from intracellular stores) exert on the subcellular distribution of free Ca^{2+} in the various regions of the cytosol ($[Ca^{2+}]_1$ topology). In this short review we will summarize the recent findings from our laboratories on these various problems.

---

[a]Address for correspondence: Dr. J. Meldolesi, Department of Pharmacology, Scientific Institute S. Raffaele, Via Olgettina 60, 20132 Milan Italy.

MECHANISMS OF INTRACELLULAR Ca^{2+} RELEASE

As already mentioned, no doubt exists at present about the ability of Ins-P_3 to induce Ca^{2+} release from an intracellular store (the nature of which is discussed in the following section), thus causing the transient increase of $[Ca^{2+}]_i$.[3] A number of other agents (such as GTP and arachidonic acid) have also been suggested to induce intracellular Ca^{2+} release, but the physiological relevance of these effects in intact cells has not been demonstrated yet. Up to very recently, the intracellular Ca^{2+} release process was believed to be quite simple; that is, activation by Ins-P_3 of a channel located in the membrane of an intracellular organelle, followed by the discharge—up to the complete emptying—of the stored calcium. The introduction of single cell studies has revealed, however, that in many cells $[Ca^{2+}]_i$ is not stable, but rather oscillates. In some cases, these oscillations could be unambiguously demonstrated to

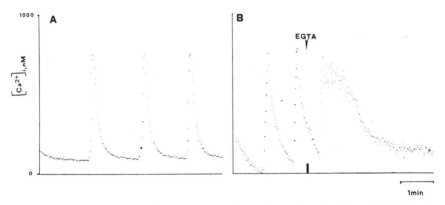

FIGURE 1. Fluctuations of $[Ca^{2+}]_i$ in unstimulated rat chromaffin cells. (**A**) A cell loaded with fura-2 has been studied while incubated in a complete, Ca^{2+}-containing medium. (**B**) Another cell was first incubated in the same medium, which was then supplemented with excess EGTA (*arrow*) to chelate $[Ca^{2+}]_0$. Notice that the $[Ca^{2+}]_i$ fluctuations did not subside immediately after EGTA addition, but persisted for a few minutes, indicating an intracellular origin.

depend on Ca^{2+} influx, in particular on the generation of spontaneous action potentials, with activation of VOCS.[5–7] In others cases, however, oscillations induced by moderate activation of receptors coupled to PIP_2 hydrolysis persisted for some time after withdrawal of Ca^{2+} from the medium, thus demonstrating their intracellular origin.[7,8] Finally, in rat chromaffin primary cultures (FIG. 1), fluctuations from intracellular stores were observed in a large proportion (70%) of cells even without any stimulation; they became almost general (over 90%) after a moderate depolarization of the plasma membrane (incubation with 10–15 mM KCl); and they disappeared temporarily after more pronounced depolarization.[9] It appears therefore that previously unsuspected "oscillators" exist in the membrane of intracellular Ca^{2+} storage organelle(s) (or elsewhere in the cell), which may trigger the generation of $[Ca^{2+}]_i$ transients. The treatments that have been reported to induce (or increase the probability of) these transients—slight depolarization; moderate receptor activation—might be lumped together by the ability to moderately increase $[Ca^{2+}]_i$. Thus, the oscillations could be due to a Ca^{2+}-induced Ca^{2+} release, a process different in its trigger mechanisms (and possibly even in the organelles involved, [see the following section] from the Ins-P_3-induced Ca^{2+} release.

INTRACELLULAR Ca^{2+} STORES: THE CALCIOSOME

Many of the considerations discussed so far relate to the nature of the intracellular store responsible for the rapid uptake and release of Ca^{2+} in response to Ins-P_3 generation (and possibly to other signals). Up to very recently, the Ins-P_3-sensitive store was generally believed to reside in the endoplasmic reticulum (ER),[3,10] a complex endomembrane system present in almost all eukaryotic cells which is responsible for a variety of other functions, such as protein and lipid synthesis and drug metabolism. Such an identification was based primarily on negative criteria, that is, on the demonstration that other structures (mitochondria, plasma membrane) are not involved in the Ins-P_3-triggered Ca^{2+} release process.[3,10] A variety of reasons that, in our opinion, strongly argue against the straight identification of the entire ER as the Ins-P_3-sensitive stores have already been discussed elsewhere by us and others, although with some divergent conclusions.[11–13] In fact, others believe the store to reside not in the entire ER, but in a discrete portion of that system, which, however, has been identified neither on structural nor on biochemical grounds.[11,12] In contrast, we have proposed that the store might be accounted for by a distinct organelle that we have discovered in a variety of nonmuscle cells and that we have named the calciosome.[13] In view of the present uncertainties about the ER subcompartment hypothesized by others, and the still incomplete characterization of the calciosome, these two hypotheses might not be as drastically alternative as they immediately appear. Indeed, calciosomes have been identified because of their specific composition, in particular, of their expression of components similar to those of the sarcoplasmic reticulum (SR) of muscle fibers (see Ref. 13 and below). Without the assistance of these markers, calciosomes would have been impossible to distinguish from other smooth-surfaced vacuoles, that is, elements that are often lumped together with tubules and cisternae under the name of smooth ER. We believe, on the contrary, that the true ER has a precise molecular identity, and that calciosomes (and possibly other structures not yet identified because of the lack of appropriate markers) should be regarded as separate cytological entities.

Herein we will summarize the present information about calciosomes. Expression of protein(s) that are recognized by affinity-purified antibodies raised against calsequestrin (CS, the low-affinity, high-capacity Ca^{2+}-binding protein contained within the terminal cisternae of the striated muscle SR) has been observed in a variety of tissues (liver, pancreas, brain, adrenals, kidney), isolated cells (platelets, granulocytes), and cell lines (HL60, PC12).[13–16] In addition to the immunologic similarity, the protein(s) from at least some of these sources resemble CS also in molecular weight, in their ability to bind Ca^{2+}, to be metachromatically stained by carbocyanine dyes, and to migrate electrophoretically in a pH-dependent fashion.[13] In all cells so far investigated, the CS-like protein(s) have been found to be concentrated within a population of apparently discrete, small vacuoles, 70–250 nm in diameter, distributed

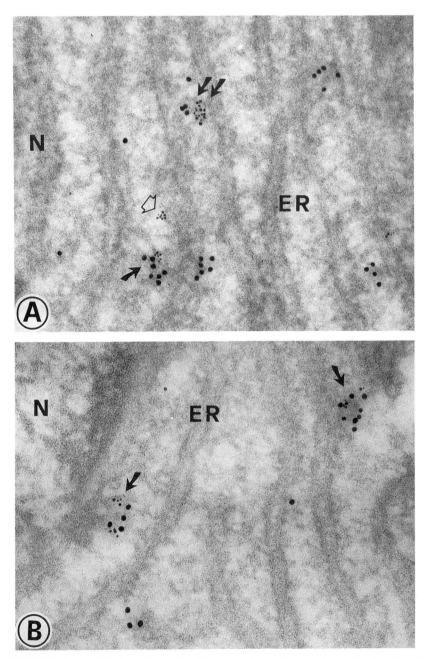

FIGURE 2. Immunolabeling of liver ultra-thin cryosections with anticalsequestrin (small gold) and anti-Ca^{2+}-ATPase (large gold) antibodies. Labeling with either marker is localized over small structures (membrane-bound vacuoles, calciosomes), shown here in between endoplasmic reticulum (ER) cisternae, but present also at other cell sites. *Solid arrows* point to dually labelled calciosomes (more than 50% in an analyzed population of ~500 organelles). The calciosome marked by *two arrows* shows a partial dissociation of the two labels. Other calciosomes are labelled for either the Ca^{2+}-ATPase or calsequestrin (*open arrow*) only. Magnification ×92,000.

throughout the entire cytoplasm: in between ER cisternae, in the Golgi area, and beneath the plasma membrane (FIG. 2). In different cell types the size of the entire calciosome pool accounts for 0.5–1.5% of the total cytoplasmic volume.[13–15] Double-label high-resolution immunocytochemistry showed calciosomes to be endowed with a membrane protein recognized by antibodies against the Ca^{2+}-ATPase of the muscle SR; this protein was not found in the ER cisternae.[14] Calciosomes, on the other hand, failed to express markers of the ER (cytochrome P450 and NADH cytocrome b_5 reductase), secretory granules (secretogranin II) and endo-lysosomes (luminal acidity)[13,14] (FIG. 2). Cell fractionation experiments, carried out in HL60 cells and brain tissue,[13,16] demonstrated a tendency to codistribution of the CS-like protein(s) with the biochemical markers of the Ins-P_3-sensitive Ca^{2+} store: ATP-dependent, high-affinity Ca^{2+} uptake, and Ins-P_3-triggered Ca^{2+} release. Markers for other subcellular structures (ER, mitochondria, Golgi complex, lysosomes, secretory granules, and plasma membrane) were at least partially dissociated from the markers of the Ca^{2+} store.[13,16] Taken together, the evidence so far available appears consistent with the view of the calciosome as a "simplified" or "primitive" version of the same specific calcium storage organelle which is fully developed in striated muscle fibers, the SR. Because of their endowment with the high affinity Ca^{2+}-ATPase and CS, and the predicted expression of both an Ins-P_3 receptor and an Ins-P_3-operated Ca^{2+} channel, calciosomes might be exquisitely equipped for the generation of rapid, receptor-initiated $[Ca^{2+}]_i$ transients. It should be acknowledged, however, that up to now calciosomes have not been isolated as a pure subcellular fraction. Therefore their characterization is still incomplete, and their identification with the Ins-P_3-sensitive Ca^{2+} store only tentative.

Calciosomes might not be the only type of rapidly exchanging intracellular Ca^{2+} pool. Indeed (as already mentioned in the preceding section), evidence for the existence of yet another process, a Ca^{2+}-induced Ca^{2+} release sensitive to caffeine, has been recently reported in sympathetic and sensory neurons.[15–18] Initial evidence suggests that the processes induced by Ins-P_3 and $[Ca^{2+}]_i$ may be differently located within the cells, and might therefore be due to different organelles.[18] The search for another, completely new intracellular Ca^{2+} storage organelle is therefore expected to begin in the near future.

TOPOLOGY OF $[Ca^{2+}]_i$ INCREASE

In view of the various mechanisms by which Ca^{2+} can be increased in the cytosol (influx from the extracellular medium; release from internal stores), the possibility of different $[Ca^{2+}]_i$ distribution patterns after the application of different stimuli had been repeatedly considered in the past few years. In particular, the increases brought about by surface channel opening were supposed to generate gradients with maxima initially concentrated beneath the plasma membrane.[19,20] Recent results obtained by digital image analysis of cells loaded with the $[Ca^{2+}]_i$ indicator fura-2[21] have borne out these early predictions. In our laboratory, PC12 pheochromocytoma cells were initially treated with bradykinin in a Ca^{2+}-free medium containing EGTA, and switched to a medium containing 2 mM Ca^{2+} 2 min later. From previous experiments in cell suspensions[22] we knew that the initial application of the peptide causes a transient

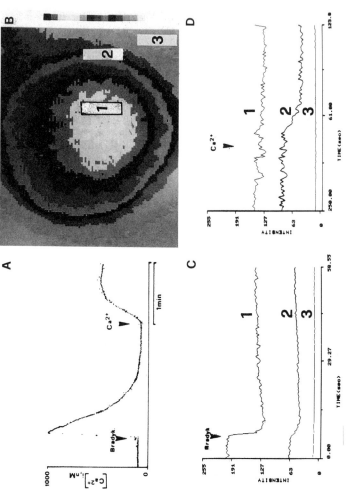

FIGURE 3. $[Ca^{2+}]_i$ changes in single PC12 cells treated with bradykinin with and without Ca_o^{2+}. (A) Time-course of $[Ca^{2+}]_i$ in a cell first exposed to bradykinin (100 mM) in a Ca^{2+}-free EGTA-containing medium, and then switched to a Ca^{2+}-containing medium (2 mM free $[Ca^{2+}]_o$) in the continuous presence of the peptide. Under these conditions the first $[Ca^{2+}]_i$ peak originates from intracellular stores, the second from influx across the plasma membrane. **B, C** and **D** refer to imaging experiments. (**B**) PC12 cell loaded with fura-2 (excitation wavelength: 380 nm) with recording windows positioned over the center (1), the periphery (2), and over the extracellular space (3). Bradykinin stimulation of intracellular Ca^{2+} release (i.e., the $[Ca^{2+}]_i$ increase induced by the peptide applied in the Ca^{2+}-free medium) occurs simultaneously in the center as well as at the periphery of the cell (downward deflections in the fluorescence traces 1 and 2, respectively). In contrast, the $[Ca^{2+}]_i$ increase due to bradykinin stimulation of Ca^{2+} influx (**D**) is limited to the periphery of the cell (*trace 2*, peripheral window) and does not reach the center (*trace 1*, window over the cell center). The *lower trace* (3) refers to the extracellular window.

$[Ca^{2+}]_i$ increase due to Ins-P$_3$-induced release from internal stores, which subsides in 1.5 min. The reintroduction of Ca^{2+} in the medium, on the other hand, causes a further, persistent $[Ca^{2+}]_i$ increase, which is also due to bradykinin since it is blocked by B$_2$ receptor antagonists,[22] but is entirely dependent on Ca^{2+} influx (FIG 3A). The imaging data reported in FIGURES 3 B and C demonstrate that these two phases of $[Ca^{2+}]_i$ increase are not only different in their mechanisms and time-course, but also differently located within the cells. In fact, the initial Ins-P$_3$-induced redistribution from the stores appears to occur almost simultaneously in the entire cell, as revealed by the parallelism of the traces monitoring fura-2 fluorescence from windows positioned over different cell areas (FIG 3C). In contrast, the $[Ca^{2+}]_i$ increase due to bradykinin-induced Ca^{2+} influx was clearly restricted to the periphery of the cell, with minimal involvement of the central cytoplasm (FIG 3C).

The imaging results obtained by us and others provide a number of suggestions about problems still open in the field. In particular, the rapid and generalized increase of $[Ca^{2+}]_i$ observed after Ins-P$_3$-induced release of Ca^{2+} from intracellular stores by bradykinin strongly suggests that these stores are localized throughout the cell, and not concentrated beneath the plasma membrane, as previously suggested by others.[11] In this respect it is interesting that calciosomes have indeed a widespread distribution as described in detail in the preceding section and Refs. 13–15.

Another interesting point concerns the physiology of $[Ca^{2+}]_i$ increases. Clearly, an increase concentrated beneath the plasma membrane is expected to affect local events (for example, exocytosis) to an extent much greater than could have been predicted from total cell measurements, where $[Ca^{2+}]_i$ is averaged out over the entire cell volume.

FUTURE PROSPECTS

The direct measurement of $[Ca^{2+}]_i$ at the single cell and the subcellular (digital imaging) level has already yielded a large body of information and many exciting ideas. The newly recognized aspects of the intracellular Ca^{2+} homeostasis—in particular, the existence of cellular entities that appear specifically involved in rapid Ca^{2+} storage and release (calciosomes and possibly others)—can foster a number of important developments in the near future. Specifically, two areas seem to us particularly promising: The first is in classical cell biology. Specific markers for the various Ca^{2+} storage organelles are being, and presumbly will continue to be, identified. The localization of these markers within the cells by immunocytochemistry and the concomitant application of other techniques of cell biology are expected to provide direct information about the nature of the various Ca^{2+} pools that still remain elusive in cytological terms. An example along this line is the calciosome, whose detailed composition, function, and general distribution in various cells and tissues are now under intense investigation. The second area is physiological. The role of Ca^{2+} as an intracellular messenger needs to be reinterpreted. $[Ca^{2+}]_i$ fluctuations and their underlying mechanisms, the topology of $[Ca^{2+}]_i$ changes, and the processes of cross-talk between receptors, channels, and intracellular Ca^{2+} stores are all examples of phenomena, unsuspected until recently, in which substantial progress is now possible. The present, rapid expansion of knowledge and understanding in the field might ultimately lead to rational explanations of the Ca^{2+} paradox, that is, the involvement of one and the same messenger in the regulation of an almost endless list of diversified and variously timed events, localized at virtually all subcellular sites, in all kinds of cells.

ACKNOWLEDGMENTS

The fundamental contribution of our colleagues P. Volpe, D.P. Lew, K.H. Krause, and B. Bruno in the work on calciosomes is gratefully acknowledged.

REFERENCES

1. CARAFOLI, E. !987. Annu. Rev. Biochem. **56:** 395–433.
2. MILLER, R. J. 1988. Trends Neurosci. In press.
3. BERRIDGE, M. J. 1987. Annu. Rev. Biochem. **56:** 159–193.
4. MELDOLESI, J. & T. POZZAN. 1987. Exp. Cell Res. **171:** 271–283.
5. SCHLEGEL, W., B. P. WINIGER, P. MOLLARD, P. VACHER, F. WUARIN, G. R. ZAHND, C. B. WOLLHEIM & B. DUFY. 1987. Nature **329:** 719–721.
6. MALGAROLI, A., L. VALLAR, F. REZA ELAHI, T. POZZAN, A. SPADA & J. MELDOLESI. 1987. J. Biol. Chem. **262:** 13920–13927.
7. JACOB, R., J. E. MERRIT, T. J. HALLAM & T. J. RINK. 1988. Nature **335:** 40–45.
8. WOODS, N. M., K. S. R. CUTHBERSON & P. H. COBBOLD. 1987. Cell Calcium **8:** 79–100.
9. MALGAROLI, A. & J. MELDOLESI. 1989. In preparation.
10. STREB, H., E. BAYERDORFFER, W. HAASE, R. F. IRVINE & I. SCHULZ. 1984. J. Membrane Biol. **81:** 241–253.
11. PUTNEY, J. W. 1987. Trends Pharmacol. Sci. **8:** 481–486.
12. WILLIAMSON, J. R., S. K. JOSEPH, K. E. COLL, A. P. THOMAS, A. VERHOEVEN & M. PRENTKI. 1986. *In* New Insights into Cell and Membrane Transport Processes. G. Poste & S.T. Crooke, Eds.: 217–247, Plenum Press. New York.
13. VOLPE, P., K. H. KRAUSE, S. HASHIMOTO, F. ZORZATO, T. POZZAN, J. MELDOLESI & D. P. LEW. 1988. Proc. Natl. Acad. Sci. USA **85:** 1091–1095.
14. HASHIMOTO, S., B. BRUNO, D. P. LEW, T. POZZAN, P. VOLPE & J. MELDOLESI. 1988. J. Cell Biol. In press.
15. MELDOLESI, J., P. VOLPE & T. POZZAN. 1988. Trends Neurosci. **11:** 449–452.
16. VOLPE, P., B. H. ALDERSON, C. A. DETTBARN, P. PALADE, B. BRUNO & J. MELDOLESI. 1988. 18th Annual Meeting of the Society of Neurosciences, Toronto, Ontario, Canada. Abstract 372.12.
17. LIPSCOMBE, D., D. V. MADISON, M. POENIE, H. REUTER & R. W. TSIEN. 1988. Proc. Natl. Acad. Sci. USA **85:** 2398–2402.
18. THAYER, S. A., D. A. EWALD, T. M. PERNEY & R. J. MILLER. 1988. J. Neurosci. In press.
19. SIMON, S. M. & R. R. LLINAS. 1985. Biophys. J. **48:** 485–498.
20. DI VIRGILIO, F., D. MILANI, A. LEON, J. MELDOLESI & T. POZZAN. 1987. J. Biol. Chem. **262:** 9189–9195.
21. GRYNKIEWICZ, G., M. POENIE & R. Y. TSIEN. 1985. J. Biol. Chem. **260:** 3440–3445.
22. FASOLATO, C., A. PANDIELLA, J. MELDOLESI & T. POZZAN. 1988. J. Biol. Chem. In press.

DISCUSSION

A. BENEDETTI (*Istituto Patologia Generale, Siena, Italy*): As far as the liver goes, we have recently shown that endoplasmic reticulum vesicles do release Ca^{2+} after Ins-P$_3$ addition. We are pretty sure that they do indeed contain glucose 6-phosphatase, which is considered a strict marker for endoplasmic reticulum. This is not consistent—at least in the liver—with your data. Can you comment?

J. MELDOLESI (*Istituto S. Raffaele, Milan, Italy*): Any evidence in subcellular fractions is open to the question of fraction purity. Just as an example I will remind you that the bulk of the Ins-P$_3$ binding has been recently reported to be recovered in a liver plasma membrane fraction (Guillemette *et al.* 1988. J. Biol. Chem. **263:** 4541–4548). At this stage I think it is well demonstrated that most of the Ins-P$_3$-sensitive calcium pool is recovered by many centrifugation procedures in the microsomal fraction. However, only a fraction of the microsomal Ca^{2+} is releasable by Ins-P$_3$. This leaves open the possibility that the Ins-P$_3$-sensitive pool resides in the minor component. Indeed, in HL60 cells we have recently demonstrated that markers of the ER (and other subcellular structures: Golgi, plasma membrane, lysosomes, secretory granules) can be dissociated in large part by centrifugation and free-flow electrophoresis from the high-affinity ATP-dependent Ca^{2+} uptake and Ins-P$_3$-dependent Ca^{2+} release, whereas calsequestrin is not. This has led us to propose the possibility that calciosomes (and not the ER) might be the target organelle of Ins-P$_3$ activity; (Volpe *et al.* Proc. Natl. Acad. Sci. USA **85:** 1091–1095). Because of the complexity and multiplicity of Ca^{2+} storage, I feel that a final conclusion cannot be made at the present time. Thus the calciosome hypothesis, although consistent with our experimental evidence, still remains hypothetical.

R. S. SOHAL (*Southern Methodist University, Dallas, Texas*): Have you examined the relationship between the distribution of calciosomes and the physiological variations in different cellular types pertaining to demands for calcium handling? Does the number of calciosomes vary under different experimental conditions related to calcium load of cells?

MELDOLESI: In all cells so far investigated, calciosomes have been found to be distributed rather evenly in the cytoplasm. The size of the calciosome pool (percentage of cytoplasmic volume) does not change much from cell to cell, as is the case with the size of InS-P$_3$ sensitive Ca^{2+} pool. As far as changes in the distribution of calciosomes, preliminary data in collaboration with O. Stendahl and Daniel P. Law suggest a possible gathering of the organelles towards the cell surface after various kinds of stimulations in neutrophils.

B. SZWERGOLD (*Fox Chase Cancer Center, Philadelphia, Pennsylvania*): A related question: can you estimate the amount of Ca^{2+} sequestered in these organelles?

MELDOLESI: Given our estimate of the fractional volume of calciosomes at 0.5% to 1%, and assuming that this calsequestrin has the same Ca^{2+}-binding capacity as the muscle protein, we can make a preliminary estimate of the Ca^{2+} that can be sequestered within calciosomes as 0.4 mmol Ca^{2+} per liter of tissue.

The H$^+$-ATP Synthase of Mitochondria in Tissue Regeneration and Neoplasia[a]

SERGIO PAPA AND FERDINANDO CAPUANO

Institute of Medical Biochemistry and Chemistry
University of Bari
Bari, Italy

INTRODUCTION

The mitochondrial H$^+$-ATPase complex, H$^+$-ATP synthase (EC 3.6.1.34), is a membrane-associated enzyme that utilizes the electrochemical proton gradient generated by the respiratory chain to produce most of the ATP necessary to the cell.

The ATP synthase is structurally and functionally organized in three parts: the catalytic sector, or F$_1$,[1] which catalyzes ATP synthesis or ATP hydrolysis; the membrane sector, or F$_o$,[2] which functions as a transmembrane proton translocator during the catalytic cycle of the enzyme; and a stalk, which connects F$_o$ and F$_1$ and is involved in the coupling between chemical catalysis and proton translocation.

The subunit composition of the ATP synthase is complex, the *Escherichia coli* enzyme being the simplest defined so far. This prokaryotic complex appears to be composed of eight polypeptide subunits.[3] Five of them, α, β, γ, δ and ϵ subunits, in the F$_1$ sector and three, a, b and c subunits, in the F$_o$ sector. On the other hand, the subunit composition of mammalian ATP synthase is not yet established. It has been reported, in fact, that the bovine heart enzyme contains between 14 and 15 different polypeptides.[4,5] The F$_1$ moiety of mammalian ATP synthase has an overall composition and organization similar to that of prokaryotic F$_1$. The subunit composition of F$_o$ remains unclear, however. The mammalian F$_o$ contains subunits that have no counterparts in *Escherichia coli*. Among these, there are the inhibitor protein that binds to β-F$_1$[6] and Factor 6, which is involved in binding of F$_1$ to F$_o$.[7]

The genes for the *Escherichia coli* enzyme are located in the *unc* or *atp* operon,[8] whereas the subunits of the mammalian enzyme are encoded partly by nuclear and partly by mitochondrial genes. Thus, normal biosynthesis of ATP synthase in eukaryotes requires a concerted expression of genes located in two different genomes. The F$_1$ subunits,[9] OSCP, Factor 6, the inhibitor protein,[10] subunit c, and a subunit of 25-kDa of F$_o$ (F$_o$I)[11] (see also Ref. 12) are encoded by nuclear DNA, whereas two subunits of F$_o$, ATPase 6 and A6L, are the products of mitochondrial genes.[13] There appear to be two genes for subunit c,[14,15] and more pseudogenes for OSCP,[10] which may be differently expressed in various mammalian tissues.

This study concerns the alteration of structure and function of the mitochondrial H$^+$-ATPase complex that can be observed in liver regeneration and in Morris hepatoma 3924A, a rapidly growing tumor, poorly differentiated and characterized by a high rate of aerobic glycolysis.

---

[a]This study was supported by a grant from the Italian National Research Council, Special Project "Oncology," under Contracts 86.000516.44 and 87.01387.44.

Regeneration of rat liver involves reorganization of gene expression.[16] It has also been shown that in certain tumors there is a deficiency in the ATPase activity[17-22] which may result from a lower content of the mitochondrial ATP synthase,[17,21] an enhanced content of the ATPase inhibitor protein,[22] or from defects in the coupling function of the enzyme.

Our group has observed that in early rat liver regeneration there occurs a marked decrease of ATPase activity and an increase of oligomycin-sensitive proton conductivity, which can be correlated with a decrease of the F_1 content in mitochondria during regeneration.[23,24]

Kinetic analysis of ATP hydrolase activity in submitochondrial particles isolated from Morris hepatoma 3924A, and the differential effect of oligomycin on the hydrolase activity and on the passive proton conduction through the F_o seem to indicate a defect in the functional connection between F_1 and F_o.

METHODS

Partial Hepatectomy

Male Wistar rats (300 g) were anesthetized and the median and left lateral lobes of the liver excised (approximately 65–75% of the total liver mass). The rats were maintained on a normal diet for 24 hours (12-hour light/12-hour dark cycle) and were then sacrificed by decapitation. The mitochondria were isolated according to the method of Pedersen.[25] Sham-operated rats were subjected to the same treatment without excision of the liver.

Preparation of Mitochondria from Morris Hepatoma 3924A

Morris hepatoma 3924A tumors were transplanted into both hind legs of inbred ACI/T rats and harvested 3 to 4 weeks after transplantation. Mitochondria were isolated according to a previously described method.[26]

Preparation of Submitochondrial Particles

"Inside out" submitochondrial particles (ESMP) were obtained by exposure of mitochondria to ultrasonic energy in the presence of EDTA at pH 8.5 as in Lee and Ernster.[27]

Determination of ATPase Activity

ATPase activity was determined in the presence of added pyruvate kinase, phosphoenolpyruvate, and lactate dehydrogenase by following NADH oxidation spectrophotometrically at 340 nm in a thermostatically controlled reaction cell at 30°C. The reaction mixture consisted of 250 mM sucrose, 50 mM KCl, 5 mM MgCl$_2$, 20 mM Tris/HCl, pH 7.5, 0.025 mM NADH, 0.5 μg rotenone, 1 mM phosphoenolpyruvate, 2.5 units lactate dehydrogenase, and 2 units pyruvate kinase in a final volume of 1 ml. The reaction was started by the addition of ATP at the concentrations reported in the legends to the figures.

Measurement of Proton Translocation

ESMP (3 mg protein/ml) were incubated in 1.5 ml of reaction mixture containing 250 mM sucrose, 30 mM KCl, 0.5 μg valinomycin/mg protein, 0.2 mg/ml purified catalase, and 20 mM succinate as respiratory substrate, pH 7.5, 25°C. Respiration-driven proton translocation was activated by repetitive pulse of 1–3% H_2O_2 (5 μl/ml) and the pH of the suspension was monitored potentiometrically with a Beckman combination electrode.

Gel Electrophoresis and Immunodecoration

Twelve to twenty percent polyacrylamide gradient gels were prepared essentially as described in Laemmli.[28] Electrotransfer to nitrocellulose sheets and subsequent immunoblotting were carried out as in Buckle et al.[23]

RESULTS

Regenerating Rat Liver

Mitochondria from regenerating liver show an increase of state IV respiratory rate and a decrease of the respiratory control ratio as compared to control mitochondria

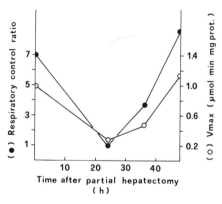

FIGURE 1. Respiratory control ratios of mitochondria and ATPase activity of ESMP prepared during rat liver regeneration. Respiration was measured by incubating 3 mg of mitochondrial protein in a 1-ml solution containing 75 mM sucrose, 50 mM KCl, 0.5 mM K-EDTA, 1 mM $MgCl_2$, 1 mM KH_2PO_4, rotenone (1 μg/mg mitochondrial protein) and 30 mM Tris HCl, pH 7.4. Respiration, measured polarographically, was induced by the addition of 20 mM succinate, and 50 μM ADP was added to stimulate respiration. ATPase activity was measured spectrophotometrically by following NADH oxidation at 340 nm as described in the METHODS section. ●, respiratory control ratios; O, ATPase activity (expressed as V_{max}, μmol · $\min^{-1}$ · mg $prot^{-1}$). The K_m, obtained from Lineweaver and Burk plots, amounted to 0.17 mM ± 0.02 both in control and regenerating liver.[24] (From Buckle et al.[24])

FIGURE 2. Densitometric traces of immunoblots of antibody-treated submitochondrial particles from control rat liver and rat liver after 24 hours of regeneration. Preparation of SDS-PAGE 12–20% gradient polyacrylamide gels containing 3.5 M urea and subsequent electrotransfer (immunoblotting) onto nitrocellulose sheets were carried out as described in the METHODS section. Photographs of immunoblots were developed onto translucent sheets which were then scanned using a Gelman DCD-16 densitometer. The relative absorbance was estimated by calculating the area within the band of the photograph of the nitrocellulose sheet developed onto translucent sheets. ●, ESMP from control rat liver; ○, ESMP from regenerating rat liver (24 hours). (From Buckle et al.[24])

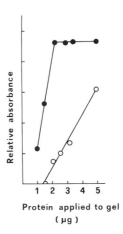

isolated from rat liver of sham-operated rats. As shown in FIGURE 1, the respiratory control ratio reached the minimum value in mitochondria isolated 24 hours after partial hepatectomy. This depression, however, disappeared with time and after 48 hours of liver regeneration, the respiratory control ratio returned to control levels. The decrease of the respiratory control was essentially due to enhancement of state IV respiration, which was brought to control values by the addition of oligomycin (see Ref. 24). Submitochondrial particles isolated from regenerating rat liver showed a decrease of the ATPase activity which exhibited a time-dependent profile similar to that observed for the decrease of the respiratory control (FIG. 1). V_{max} values of ATP hydrolysis dropped to 25% of control after 24 hours of regeneration and then returned to control values 48 hours after partial hepatectomy.

Through the use of antibodies against bovine heart F_1 which reacted with α- and β-subunits of rat liver ATPase, it has been possible to estimate the F_1 content of submitochrondrial particles from regenerating rat liver. When compared with control submitochondrial particles, the amount of material reacting with the antibody against F_1 was markedly lower in submitochondrial particles isolated after 24 hours of regeneration (FIG. 2).

In submitochondrial particles the aerobic relaxation of transmembrane proton gradient set up by respiration occurs through the membrane F_o sector of the H⁺-ATPase.[29] The polypeptide composition of the F_o sector varies from species to species,[4,30] although the N,N'-dicyclohexylcarbodiimide [cHxN)$_2$C]-binding protein or subunit c,[31] which appears to be directly involved in proton translocation,[2,4,5] is a constant component. FIGURE 3 shows the inhibitory effect of (cHxN)$_2$C on the overall rate of anaerobic relaxation of respiratory $\Delta\mu$H⁺. It can be noted that the overall rate of anaerobic proton release from submitochondrial particles, expressed as $1/t_{1/2}$, was higher in regenerating liver than in control, being more than twice in the absence of the inhibitor. In the presence of an excess of (cHxN)$_2$C, the proton conductivity was, however, reduced to the same low level in submitochondrial particles from both control and regenerating (24-hr) rat liver.

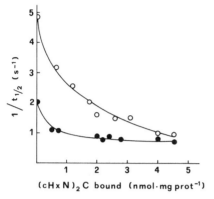

FIGURE 3. Effect of $(cHxN)_2C$ on the overall rate of anaerobic relaxation of respiratory $\Delta\mu H^+$ in ESMP. ESMP (3 mg prot/ml) were incubated in 1.5 ml of the reaction mixture described in the METHODS section. $^{14}C(cHxN)_2C$ was added at the concentrations indicated, and at given time intervals aliquots were removed into ice-cold trichloroacetic acid (5% w/v) and the precipitated protein was washed and counted in a Packard Tricarb liquid scintillation spectrometer. ●, ESMP from control rat liver; O, ESMP from regenerating (24 hours) rat liver. (From Buckle et al.[24])

Morris Hepatoma 3924A

It has been reported that mitochondria isolated from transplantable hepatomas reveal a number of biochemical differences with regard to normal and host liver. Among these, one concerns the respiratory characteristics.[17] We have used the Morris hepatoma 3924A mitochondria to assess the properties of the H^+-ATPase complex.

FIGURE 4 shows the saturation kinetics, plotted according to Lineweaver and Burk, of ATP hydrolase activity in submitochondrial particles from rat liver and Morris hepatoma. Statistical evaluation of V_{max} and K_m values determined in different submitochondrial particle preparations is summarized in TABLE 1. It can be seen that the V_{max} value for ATP hydrolysis in hepatoma particles did not differ significantly from that in control particles. On the other hand, the K_m value for ATP was considerably higher in hepatoma than in liver particles. The same situation was also found for the ATPase activities of particles uncoupled as a consequence of a freeze-thaw treatment[36] (TABLE 2).

FIGURE 5 presents the Eadie-Hofstee plots of the kinetics of ATP hydrolysis in submitochondrial particles from control liver mitochondria and hepatoma mitochon-

FIGURE 4. Double reciprocal plot of ATP hydrolase activities of EDTA submitochondrial particles from control rat liver and Morris hepatoma 3924A. ESMP (50 μg of protein per ml) were preincubated in the sucrose medium described in METHODS. Final volume was 1 ml, pH 7.5, and temperature 30°C. ●, rat liver; O, Morris hepatoma 3924A.

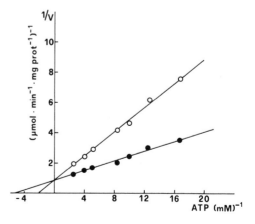

TABLE 1. ATPase Activity of Submitochondrial Particles (ESMP) from Rat Liver and Morris Hepatoma 3924A

ESMP	V_{max} (μmol $\cdot$ min^{-1} $\cdot$ mg prot^{-1})	K_m (mM)
Rat liver	1.13 ± 0.12	0.17 ± 0.01
Morris hepatoma 3924A	1.08 ± 0.10	0.49 ± 0.02

NOTE: ATPase activity was measured as μmol phosphoenolpyruvate hydrolized as described in the METHODS section, in which experimental details can be found. V_{max} and K_m were calculated from Lineweaver-Burk plots. Means ± SE from six experiments.

dria. The plots show the reported cooperative behavior of the ATP hydrolase activity.[37] This cooperative pattern is explained on the basis of two (or three) catalytic alternating sites.[38] It can be noted that the degree of the cooperativity was diminished for the hepatoma ATP hydrolase activity as compared to the control enzyme, as indicated by the enhancement of the apparent high-affinity constant(s) while the low-affinity constant remained practically unchanged.

Oligomycin depresses the catalytic activity of F_1 as a consequence of direct inhibition of proton conduction through the F_o sector of the H⁺-ATPase.[39] Titration curves for the inhibitory effect of oligomycin on the ATP hydrolase activity in control and hepatoma submitochondrial particles are shown in FIGURE 6. It can be noted that ten times more oligomycin was required to cause half-maximal inhibition in hepatoma as compared to control liver. In control particles 50% inhibition of hydrolytic activity occurred at 0.21 μg of oligomycin per milligram of particle protein (FIG. 6A), whereas in hepatoma the same extent of inhibition was reached at 2.7 μg of oligomycin per milligram of protein (FIG. 6B).

FIGURE 7 shows the effect of increasing oligomycin concentration on the overall rate of passive proton backflow through the F_o. It may be seen that there was no difference in the rate of proton conduction between control and hepatoma particles and that the oligomycin titer for half-maximal inhibition was, in both types of particles, close to 0.1 μg per milligram of protein.

DISCUSSION

The results presented show that in the course of liver regeneration there are parallel changes in the respiratory control of mitochondria and in the ATPase activity. Thus, at

TABLE 2. ATPase Activity of Submitochondrial Particles (ESMP) from Rat Liver and Morris Hepatoma 3924A after Freeze-Thaw Treatment

ESMP	V_{max} (μmol $\cdot$ min^{-1} $\cdot$ mg prot^{-1})	K_m (mM)
Rat liver	4.00	0.15
Morris hepatoma 3924A	3.73	0.69

NOTE: Submitochondrial particles were frozen for 24 hr at -70°C then thawed and assayed for ATPase activity. For experimental conditions see the legend to FIGURE 4 and the METHODS section.

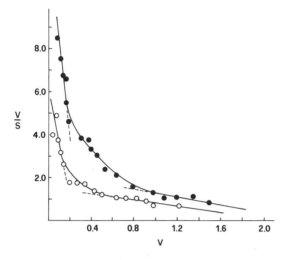

FIGURE 5. Eadie-Hofstee plots of the kinetics of ATP hydrolysis by ESMP from control rat liver and Morris hepatoma 3924A. ESMP (100 μg of protein per ml) were incubated as decribed in METHODS. The ATP concentration ranged from 0.01 to 1.8 mM. ●, control rat liver; O, Morris hepatoma 3924A.

24 hours of regeneration, when mitochondrial respiration was completely uncoupled (state IV respiration equal to state III), the ATPase activity reached a minimum value.

The direct determination by immunoblotting of the F_1 content demonstrates that in submitochondrial particles isolated 24 hours after partial hepatectomy there was a reduced amount of the catalytic moiety of the H^+-ATPase with regard to control particles. Thus the depression of ATPase activity can be directly correlated to diminution in the F_1 content.

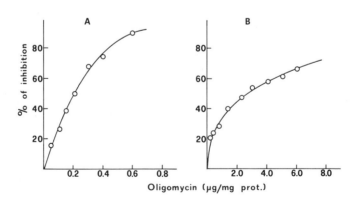

FIGURE 6. Titration of the inhibitory effect of oligomycin on ATPase activities in ESMP from rat liver (A) and Morris hepatoma 3924A (B). ESMP were preincubated for 5 min with oligomycin at the concentration given on the abscissa. For experimental conditions see METHODS. Half-maximal inhibition occurred at 0.21 μg of oligomycin per mg of protein for rat liver and 2.7 μg per mg of protein from Morris hepatoma.

Other data[23] showed that in intact mitochondria from regenerating rat liver there was a decrease in β-subunit content as compared to controls. Loss of F_1 during the preparation of submitochondrial particles from regenerating rat liver was therefore excluded.

Our data suggest that in regenerating rat liver there exist F_0 sectors lacking their F_1 counterparts.

The effect of $(cHxN)_2C$ on passive proton conductivity indicates that there is no significant difference in the content of $(cHxN)_2C$-binding protein in the F_0 sector between regenerating and control liver.

The transient decrease of F_1 content, occurring at the initial phase of regeneration, suggests that in this phase there can be changes either in the synthesis of F_1 or in the uptake into mitochondria of F_1 subunits. This results in ATPase complexes lacking F_1 moieties that constitute a proton pore through the F_0 sector of the mitochondrial membrane; this in turn dissipates the electrochemical proton gradient set up by respiration. This effect is transitory, and 48 to 72 hours of regeneration restores the full complement of F_1.

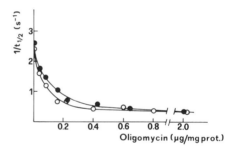

FIGURE 7. Effect of oligomycin on the overall rate of proton conduction in EDTA submitochondrial particles from rat liver and Morris hepatoma 3924A. For experimental conditions see the legend to FIGURE 6 and METHODS. Half-maximal inhibition occurred at 0.1 μg of oligomycin per mg of protein from rat liver and 0.08 μg per mg of protein for Morris hepatoma. ●, rat liver; ○, Morris hepatoma 3924A.

It has been reported that analogies exist in the metabolic behavior of regenerating liver, embryofetal tissues, and neoplasia.[16,40,41] Our investigations on the hydrolase activity in submitochondrial particles isolated from Morris hepatoma 3924A show a significant decrease in the affinity for ATP as compared with that of control rat liver. This difference was also observed in submitochondrial particles uncoupled by freeze-thaw treatment. The change in affinity for ATP observed in hepatoma when compared with control particles should not depend on difference in F_1 content since the V_{max} for ATP hydrolysis has the same value in both hepatoma and liver submitochondrial particles. The lower affinity for ATP may result from an alteration of the catalytic domain of F_1 and appears to reflect a decrease in the cooperative behavior of the enzyme.

The catalytic cycle in F_1 is compulsorily coupled to proton conduction in F_0,[32] but the overall rate of proton backflow through the F_0 had the same value in both hepatoma and rat liver submitochondrial particles.

Oligomycin is a specific inhibitor of the proton-conducting pathway in the F_0 sector of ATPase, and the extension of the inhibitory effect to ATP hydrolysis (or synthesis) appears to involve one $(OSCP)^{42}$ or more proteins at the F_1–F_o junction.[43] The oligomycin concentration required in hepatoma particles for half-maximal inhibition of the ATP hydrolase activity was one order higher than that required in control particles. On the other hand, 50% inhibition of passive proton conduction in F_o occurred in hepatoma particles at the same concentration, causing 50% inhibition of proton conduction in control particles. Thus, the change observed in hepatoma particles in the inhibitory titer of oligomycin for the ATP hydrolase activity cannot be ascribed to alterations of F_o subunits directly involved in oligomycin-sensitive proton conduction.[12]

These observations would therefore suggest that in H^+-ATPase complex from hepatoma, whereas subunits involved in the proton-conducting pathway are not affected, there is an alteration of one or more of the subunits (OSCP, F6, γ, δ and ϵ), which are thought to be involved in the functional interaction between the two sectors of the complex.

Further work is in progress to identify the subunits involved in the observed change of the ATPase activity and the role it may play in the shift of energy metabolism to a fermentative pattern, which is characteristic of many rapidly growing tumors.

ACKNOWLEDGMENT

We wish to thank Prof. T. Galeotti for providing the Morris hepatoma 3924A.

REFERENCES

1. PULLMAN, M. E., H. S. PENEFSKY, A. DATTA & E. RACKER. 1960. J. Biol. Chem. **235:** 3322–3329.
2. KAGAWA, Y. & E. RACKER. 1966. J. Biol. Chem. **241:** 2461–2466.
3. FOSTER, D. L. & R. H. FILLINGAME. 1979. J. Biol. Chem. **254:** 8230–8236.
4. PAPA, S., K. ALTENDORF, L. ERNSTER & L. PACKER, Eds. 1984. H^+-ATPase (ATP Synthase): Structure, Function, Biogenesis. The F_oF_1 Complex of Coupling Membranes. ICSU Press. Miami/Adriatica Editrice. Bari, Italy.
5. GODINOT, C. & A. DIPIETRO. 1986. Biochemie **68:** 367–374.
6. PULLMAN, M. E. & G. C. MONROY. 1963. J. Biol. Chem. **238:** 3762–3769.
7. KANNER, B. I., R. SERRANO, M. A. KANDRACH & E. RACKER. 1976. Biochem. Biophys. Res. Commun. **69:** 1050–1056.
8. WALKER, J. E., M. SARASTE & N. J. GAY. 1984. Biochim. Biophys. Acta **758:** 164–200.
9. SENIOR, A. E. 1979. In Membrane Protein Energy Transduction. A. R. Capaldi & M. Decker, Eds.: 223–278. Marcel Dekker. New York.
10. WALKER, J. E., N. J. GAY, S. J. POWELL, M. KOSTINA & M. R. DYER. 1987. Biochemistry **26:** 8613–8619.
11. WALKER, J. E. & M. J. RUNSWICK. 1987. J. Mol. Biol. **197:** 89–100.
12. HOUSTEK, J., J. KOPECKY, F. ZANOTTI, F. GUERRIERI, E. JIRILLO, G. CAPOZZA & S. PAPA. 1988. Eur. J. Biochem. **173:** 1–8.
13. FEARNLEY, I. M. & J. E. WALKER. 1986. EMBO J. **5:** 2003–2008.
14. GAY, N. J. & J. E. WALKER. 1985. EMBO J. **4:** 3519–3524.
15. WALKER, J. E., A. L. COZENS, M. R. DYER, J. M. FEARNLEY, S. J. POWELL & M. J. RUNSWICK. 1987. In Bioenergetics: Structure and Function of Energy Transducing Systems. T. Ozawa and S. Papa, Eds.: 167–178. Japan Sci. Soc. Press. Tokyo. Springer Verlag. Berlin.
16. URIEL, J. 1979. Adv. Cancer Res. **29:** 127–174.
17. PEDERSEN, P. L. 1978. Progr. Expr. Tumor Res. **22:** 190–274.

18. BARBOUR, R. L. & M. P. CHAN. 1978. J. Biol. Chem. **253:** 367–376.
19. KNOWLES, A. F. & N. O. KAPLAN. 1980. Biochim. Biophys. Acta **590:** 170–181.
20. KNOWLES, A. F. 1982. Biochim. Biophys. Acta **681:** 62–71.
21. NELSON, B. D., F. KABIR, J. KOLAROV, K. LUCIAKOVA, S. KUZELA, N. LATRUFFE & M. LINDEN. 1984. Arch. Biochem. Biophys. **234:** 24–30.
22. LUCIAKOVA, K. & S. KUZELA. 1984. FEBS Lett. **177:** 85–88.
23. BUCKLE, M., F. GUERRIERI & S. PAPA. 1985. FEBS Lett. **188:** 345–351.
24. BUCKLE, M., F. GUERRIERI, A. PAZIENZA & S. PAPA. 1986. Eur. J. Biochem. **155:** 439–455.
25. PEDERSEN, P. L. 1977. Anal. Biochem. **80:** 401–408.
26. SCHREIBER, J. R., W. X. BALCAVAGE, M. P. MORRIS & P. L. PEDERSEN. 1970. Cancer Res. **30:** 2497–2501.
27. LEE, C. P. & L. ERNSTER. 1968. Eur. J. Biochem. **3:** 391–400.
28. LAEMMLI, U. K. 1970. Nature **227:** 680–685.
29. PANSINI, A., F. GUERRIERI & S. PAPA. 1978. Eur. J. Biochem. **92:** 545–551.
30. SENIOR, A. E. & J. C. WISE. 1983. J. Membr. Biol. **73:** 105–124.
31. WALKER, J. E., V. L. J. TYBULEWICZ, G. FALK, N. J. GAY & A. HAMPE. 1984. In H$^+$-ATPase (ATP Synthase): Structure, Function, Biogenesis. The F$_o$F$_1$ Complex of Coupling Membranes. S. Papa, K. Altendorf, L. Ernster & L. Packer, Eds.: 1–14. ICSU Press. Miami/Adriatica Editrice. Bari, Italy.
32. PAPA, S. & F. GUERRIERI. 1981. *In* Chemiosmotic Proton Circuits in Biological Membranes. V. P. Skulacev and P. C. Hinkle, Eds.: 459–470. Addison Wesley. Reading, MA.
33. PAPA, S., F. GUERRIERI, F. ZANOTTI & R. SCARFÒ. 1984. *In* Biological Membranes: Information and Energy Transduction in Biological Membranes. C. L. Bolis, E. J. M. Helmreich, and H. Passow, Eds.: 187–197. Alan R. Liss. New York.
34. CELIS, H. 1980. Biochem. Biophys. Res. Commun. **92:** 26–31.
35. SCHINLER, H. & N. NELSON. 1982. Biochemistry **21:** 5787–5794.
36. KASCHNITZ, R. M., Y. HATEFI & H. P. MORRIS. 1976. Biochim. Biophys. Acta **449:** 224–235.
37. CROSS, R. L., C. GRUBMEYER & H. S. PENEFSKY. 1982. J. Biol. Chem. **257:** 12101–12105.
38. SENIOR, A. E. 1988. Physiol. Rev. **68:** 177–231.
39. FILLINGAME, R. H. 1980. Annu. Rev. Biochem. **49:** 1079–1113.
40. BAGGETTO, L., D. C. GAUTHERON & C. GODINOT. 1984. Arch. Biochim. Biophys. **232:** 670–678.
41. ENRICH, C. & C. G. GAHMBERG. 1985. FEBS Lett. **181:** 12–16.
42. MacLENNAN, D. H. & A. TZAGALOFF. 1968. Biochemistry **7:** 1603–1610.
43. PENEFSKY, H. S. 1985. Proc. Natl. Acad. Sci. USA **82:** 1589–1593.

DISCUSSION

L. ERNSTER (*University of Stockholm, Sweden*): Have you analyzed the hepatoma F$_o$F$_1$-ATPase with respect to factor F$_6$? In reconstitution studies with beef heart F$_o$F$_1$-ATPase we have found (SANDRI, WOJTCRAK & ERNSTER. 1985. Arch. Biochem. Biophys. **239:** 595) that lack of F$_6$ results in a partial loss of oligomycin sensitivity—a situation similar to the one that you find with the hepatoma ATPase.

S. PAPA (*University of Bari, Italy*): We are exploring which subunit of the stalk is involved in the change of oligomycin sensitivity of the catalytic process in F$_1$ in hepatomas, and F$_6$ is, indeed, a possible candidate.

G. DELICOSTANTINOS (*University of Athens Medical School, Greece*): The hepatoma cell's mitochondria are characterized by a high cholesterol-to-phospholipid ratio, which means a decrease in membrane fluidity. How can you correlate the allosteric changes of ATPase activity with the changes of the membrane fluidity?

PAPA: There is some evidence that changes in the lipid composition of the mitochondrial membrane may affect the catalytic activities of the H^+-ATP synthase. This is indeed another factor to be taken into account when changes in the ATPase activity are observed.

G. FISKUM (*George Washington University School of Medicine, Washington, DC*): Although alterations in the kinetics of hepatoma mitochondrial ATPase are apparent in assays performed with submitochondrial particles, the rate of phosphorylating respiration and the respiratory control ratio of mitochondria from hepatomas and other tumors are not significantly different from those of mitochondria isolated from normal tissues.

PAPA: The standard assay conditions for the phosphorylating activity of isolated mitochondria are substantially different from those met under steady-state *in vivo* conditions. The system is forced to make ATP so that certain more subtle alterations, which may be critical *in vivo*, are missed in the *in vitro* tests. Our studies are, indeed, directed to characterize specific alterations in the catalytic and coupling activity and to correlate this with possible alterations in the biosynthesis and proper assembly of the ATP-synthase complex in tumor cells.

G. GUIDOTTI (*University of Parma, Italy*): In the inner mitochondrial membrane, an anion channel seems to protect the cell from hyperoxia and perhaps anoxia by delaying the permeability to hydroxyl ions. Does the activation of this channel change during the early periods of liver regeneration, thus contributing to the diminuation of oxidative phosphorylation you have mentioned?

PAPA: The enhanced proton conductivity we observe in the early phase of liver regeneration and the consequent decrease of respiratory control are fully suppressed by oligomycin. Thus, they are apparently due only to enhancement of proton conductivity by the F_o sector of the H^+-ATP synthase.

Ionic Signals in the Action of Growth Factors

WOUTER H. MOOLENAAR

The Netherlands Cancer Institute
1066 CX Amsterdam, the Netherlands

INTRODUCTION

Growth factors regulate the proliferation of cells *in vivo* and in culture by binding to specific high-affinity receptors on the cell surface. Receptor occupancy triggers a cascade of biochemical and physiological changes in the target cell, which ultimately (after 10–20 hours) leads to stimulation of DNA synthesis and cell division. Among the immediate consequences of growth factor–receptor interaction are (*a*) the activation of a protein tyrosine kinase activity that is intrinsic to the receptors for epidermal growth factor (EGF), platelet-derived growth factor (PDGF) and insulin-like growth factor 1(IGF-1), and so on; (*b*) the phospholipase-C-mediated breakdown of inositol phospholipids generating several signal molecules; and (*c*) the activation of various ion transport systems in the plasma membrane and changes in intracellular ionic composition.[1] This chapter will focus on the generation of ionic signals in the action of growth factors. Mitogen-induced alterations in the levels of cytoplasmic free $Ca^{2+}([Ca^{2+}]_i)$ and H^+ are of special interest because these ions are thought to serve as second messengers in activated cells. During the past few years our understanding of the ionic changes in stimulated cells has increased dramatically. The molecular mechanisms underlying ionic signaling have been largely unraveled and much effort is being devoted to determining how rapid increases in $[Ca^{2+}]_i$ and pH_i may contribute to a proliferative response.

Activation of Na^+/H^+ Exchange

Most, if not all, mitogens activate an otherwise quiescent Na^+/H^+ exchanger in the plasma membrane of their target cells (reviewed in Ref. 2). This Na^+/H^+ exchanger, which is sensitive to the diuretic amiloride and its analogues, is normally involved in the close regulation of steady-state pH_i by virtue of its high sensitivity to cytoplasmic H^+. Most cells maintain their pH_i at 7.0–7.2, which is well above the electrochemical equilibrium value of 6.4 predicted by the Nernst equation from a transmembrane potential of -60 mV (at a pH of 7.4).

The functioning of the plasma membrane Na^+/H^+ exchanger and its normal housekeeping role in pH_i regulation are usually assessed by acutely acidifying the cytoplasm and monitoring the consequent recovery of pH_i back to its steady-state level. While basal Na^+/H^+ exchange activity is normally very low and exactly balances the acidifying effects of $H+$ influx and metabolic acid production, Na^+/H^+ exchange is dramatically accelerated as soon as the cytoplasm is acidified. Cytoplasmic H^+ is then rapidly extruded from the cell, causing pH_i to return to its basal value within a few minutes. Although Na^+/H^+ exchange is driven, in thermodynamic terms, by the steep transmembrane Na^+ gradient, the major determinant of the Na^+/H^+ exchange rate is cytoplasmic $[H^+]$. It is generally assumed that cytoplasmic H^+ acts as an allosteric activator of the Na^+/H^+ exchanger by binding to a regulatory site. The protonation of this site sets the exchanger in motion, while a functionally separate H^+ transport site mediates H^+ efflux once the exchanger is activated. It appears that growth factors and activators of protein kinase C (phorbol esters) exploit this control mechanism to shift steady-state pH_i in an alkaline direction. Growth factors act by increasing the sensitivity of the Na^+/H^+ exchanger for cytoplasmic H^+, thereby raising steady-state pH_i by 0.1–0.3 pH unit. This mitogen-induced shift in pH_i sensitivity presumably reflects an increase in the apparent affinity of the regulatory site for H^+. It is easy to imagine that some ionizable group at the regulatory site acquires a greater pK_a because its immediate environment becomes more negatively charged, for example, by receptor-mediated phosphorylation either via protein kinase C or through the intrinsic receptor tyrosine kinase.

A growth factor-induced shift in pH_i of 0.2–0.3 units would be expected to affect the role of numerous pH-sensitive processes in the cell. Perhaps the most convincing demonstration of a signaling role of pH_i in growth control has been made with mutant fibroblasts that lack a functional Na^+/H^+ exchanger.[3] Pouysségur et al. have elegantly shown that below a certain threshold value (around 7.0–7.2) pH_i becomes limiting for cell proliferation and, furthermore, that one of the critical pH_i-dependent steps in activated fibroblasts appears to be the initiation of protein synthesis.[3]

The involvement of the Na^+/H^+ exchanger in such a carefully regulated process as cell growth raises the possibility that the exchanger in autonomously growing tumor cells may be subject to uncontrolled activation. Indeed, recent work by Bierman et al.[5] on pH_i regulation in embryonal carcinoma (EC) cells shows that the Na^+/H^+ exchanger in mouse P19 EC cells is constitutively activated and fails to respond to any extracellular stimulus. These pluripotent cells have no requirement for exogenous growth factors and exhibit a highly transformed phenotype. Upon differentiation, however, the cells lose their transformed phenotype and they become dependent on the presence of external growth factors. This differentiation process, induced by either retinoic acid or DMSO, is accompanied by "de-activation" of the Na^+/H^+ exchanger and a resultant drop in steady-state pH_i of up to 0.5 unit (Bierman et al., manuscript in preparation). Kinetic analysis indicates that the high resting pH_i of undifferentiated EC cells is attributable to an alkaline shift in the pH_i sensitivity of the Na^+/H^+ exchange rate as compared with that in differentiated cells. It thus seems as if signal pathways normally utilized by growth factors are constitutively operative in autonomously growing, undifferentiated EC cells. The biochemical nature of this pathway remains to be elucidated.

Ca^{2+} Signaling and Phosphoinositide Hydrolysis

There is now general agreement that many, but not all, growth factors evoke the phospholipase-C-mediated breakdown of inositol phospholipids, resulting in the formation of diacylglycerol, which activates protein kinase C and inositol-1,4,5-trisphosphate (IP_3 (1,4,5)), which triggers the release of Ca^{2+} from internal stores, as well as various other inositol polyphosphates.

We have analyzed and compared the actions of EGF, on the one hand, and of Ca^{2+}-mobilizing hormones (bradykinin, histamine), on the other, on both ionositol phosphate formation and calcium signaling in responsive human carcinoma cells.[6,7] EGF is a rather weak inducer of phospholipase C activity, evoking a rather small and variable release of intracellularly stored Ca^{2+}, while accumulation of IP_3 (1,4,5) is barely detectable. In contrast, bradykinin and histamine are among the most potent activators of phospholipase C, resulting in larger Ca^{2+} signals and substantial increases in inositol phosphate levels. Almost immediately after addition of these hormones to A431 or HeLa cells, the levels of IP_3 (1,4,5) start to increase, reaching peak values of ~10 times the basal level within 10–15 sec, which coincides with the $[Ca^{2+}]$ peak. The formation of $IP_3(1,4,5)$ is a very transient phenomenon, and our results support the view that the (1,4,5) isomer is rapidly metabolized either to IP_2 through phosphatase

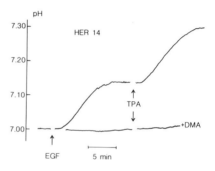

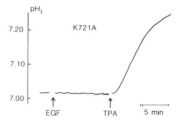

FIGURE 1. Changes in pH_i after addition of EGF (50 ng/ml) or TPA (50 ng/ml) in the presence or absence of dimethylamiloride (DMA, 20 μM) in 3T3 cells expressing either wild-type (HER 14) or kinase-defective (K721A) human EGF receptor.

action or to $IP_3(1,3,4)$ via phosphorylation/dephosphorylation with IP_4 as intermediate. The latter inositol polyphosphate may be involved in regulating Ca^{2+} influx, but its precise role has not been fully clarified to date.[8] Relatively large amounts of IP_5 and IP_6 are found in A431, HeLa and neuroblastoma cells,[6] but IP_5/IP_6 levels remain virtually unaltered after receptor stimulation, suggesting that these inositol phosphates are not directly involved in cellular signaling.

The accumulation of inositol phosphates in response to bradykinin levels off in less than 2 minutes. This apparent desensitization is not observed with EGF as a stimulus; in general, the bradykinin-dependent pattern of stimulation is significantly different from that observed in EGF-stimulated A431 cells, suggesting that separate mechanisms of inositol lipid hydrolysis are involved.

Regardless of the distinct mechanisms by which growth factors raise $[Ca^{2+}]_i$, it is widely accepted that Ca^{2+} has an important role as second messenger which regulates or modulates numerous cellular activities. The transient rise in $[Ca^{2+}]_i$ is likely to trigger a sequence of early cellular changes occurring within minutes of receptor stimulation. Of special interest is the finding that artificial elevation of $[Ca^{2+}]_i$ by means of ionophores mimics growth factors in rapidly inducing the expression of the c-*fos* and c-*myc* proto-oncogenes,[9] while suppression of the growth-factor-induced Ca^{2+} signal can inhibit c-*fos* expression, at least in the action of bradykinin and histamine.[10]

Mutational Analysis of EGF Receptor-Mediated Ionic Signaling

To explore the importance of the receptor tyrosine kinase in the action of EGF, we have used several different EGF receptor mutants expressed in NIH-3T3 cells that contain undetectable amounts of endogenous EGF receptors. The wild-type human EGF receptor expressed in these cells behaves like the native EGF receptor and is able to stimulate all the known responses to EGF including inositol phosphate formation and ionic signaling (FIGS. 1–3).[11,12]

Of special interest is a point mutant receptor in which Lys-721, a key residue in the presumed ATP-binding region, is replaced by an alanine residue. This K721A mutant receptor expressed in 3T3 cells lacks protein tyrosine kinase activity, but is normally processed and expressed on the cell surface. However, the kinase-deficient receptor is unable to stimulate both early responses such as activation of Na^+/H^+ exchange, Ca^{2+}

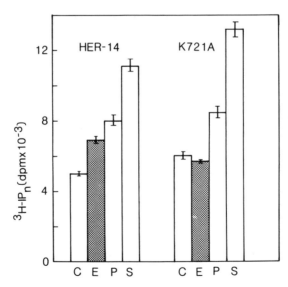

FIGURE 2. Inositol phosphate formation induced by 50 ng/ml EGF (E), 10 ng/ml PDGF (P) and 10% fetal calf serum (FCS) in wild-type (HER-14) and kinase-defective (K721A) EGF receptor transfectants. For experimental details see Ref. 11.

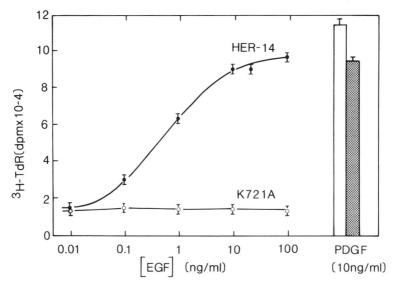

FIGURE 3. EGF- and PDGF-induced DNA synthesis as measured by [³H]thymidine incorporation in HER-14 cells (*solid circles, shaded bar*) and K721A mutant cells (*open circles, open bar*). Details are in Ref. 11).

signaling, inositol phosphate formation, and late responses such as DNA synthesis and focus formation.[11] Hence, tyrosine kinase activity is essential, although not necessarily sufficient, for EGF receptor signal transduction. In other words, EGF-induced phosphorylation of specific cellular substrates on tyrosine residues is a critical step in the pleiotropic response to EGF. The inability of the kinase-defective receptor to stimulate phosphoinositide hydrolysis is particularly intriguing, since it suggests that regulatory proteins in this pathway, such as G-proteins or phospholipase C, may serve as substrates and thus be regulated by protein tyrosine kinases. Such a mechanism contrasts with the accepted mode of action of "classical" Ca^{2+}-mobilizing hormones, such as bradykinin and histamine, whose receptors are linked to G-protein intermediates without any evidence for tyrosine phosphorylation's being required for the activation process.

In addition to testing the K721A EGF receptor mutant, we have also tested a truncated receptor that lacks 63 amino acid residues from the C-terminal end, including two major autophosphorylation sites.[13] This mutant receptor appeared to be indistinguishable from the wild-type receptor in terms of EGF-dependent ionic signaling, inositol phosphate formation, and DNA synthesis (unpublished results and Ref. 13). These results imply that receptor autophosphorylation is not essential for a normal signal-transducing response to EGF.

It is obvious that the use of site-directed mutagenesis combined with transfections into cultured cells provides a promising novel approach to the analysis of the various domains of the EGF receptor and their role in signal transduction.

REFERENCES

1. MOOLENAAR, W. H., L. H. K. DEFIZE & S. W. DE LAAT. 1986. J. Exp. Biol. **124:** 359–373.
2. MOOLENAAR, W. H. 1986. Annu. Rev. Physiol. **48:** 363–376.
3. POUYSSÉGUR, F., C. SARDET, A. FRANCHI, G. L'ALLEMAIN & S. PARIS. 1984. Proc. Natl. Acad. Sci. USA **81:** 4833–4837.
4. CHAMBARD, J. C. & J. POUYSSÉGUR. 1986. Exp. Cell Res. **164:** 282–294.
5. BIERMAN, A. J., L. G. J. TERTOOLEN, S. W. DE LAAT & W. H. MOOLENAAR. 1987. J. Biol. Chem. **262:** 9621–9628.
6. TILLY, B. C., P. VAN PARIDON, I. VERLAAN, K. WIRTZ, S. W. DE LAAT & W. H. MOOLENAAR. 1987. Biochem. J. **244:** 129–135.
7. TILLY, B. C., P. VAN PARIDON, I. VERLAAN, S. W. DE LAAT & W. H. MOOLENAAR. 1988. Biochem. J. **252:** 857–863.
8. IRVINE, R. F. & M. MOOR. 1986. Biochem. J. **240:** 917–920.
9. BRAVO, R., J. BURCKHARDT, T. CURRAN & R. MULLER. 1985. EMBO J. **4:** 1193–1198.
10. KRUIJER, W., S. VAN GENESSEN & W. H. MOOLENAAR. 1988. Submitted for publication.
11. MOOLENAAR, W. H., A. J. BIERMAN, B. C. TILLY, I. VERLAAN, A. M. HONEGGER, A. ULLRICH & J. SCHLESSINGER. 1988. EMBO J **7:** 707–710.
12. SCHLESSINGER, J. 1988. Biochemistry **27:** 3119–3123.
13. SCHLESSINGER, J. 1986. J. Cell Biol. **103:** 2067–2072.

DISCUSSION

A. SCARPA (*Case Western Reserve University, Cleveland, Ohio*): What is the origin of the large changes in single cell potential recordings? Did you correlate this with changes in the free calcium of single cells?

W. H. MOOLENAAR (*The Netherlands Cancer Institute, Amsterdam, the Netherlands*): The hyperpolarizations are due to opening of Ca^{2+}-activated K^+ channels. We have used electrophysiological techniques to monitor oscillatory behavior rather than fura-2 digital imaging.

S. PAPA (*University of Bari, Italy*): Do you agree with the proposal of Pouysségur that protein kinase C activates the Na^+/H^+ antiport by modifying its allosteric response to cytosolic pH?

MOOLENAAR: Yes; in fact Dr. Pouysségur and others have nicely confirmed our finding that growth factors activate Na^+/H^+ exchange by increasing its sensitivity to cytoplasmic H^+.

C. KLEE (*National Cancer Institute, Bethesda, Maryland*): What is the difference in the mechanism of activation of phospholipase C by histamine and EGF which leads to and explains differences in levels of IP_3 and DAG?

MOOLENAAR: Receptor tyrosine kinase activity is essential for activation of phospholipase C in the action of EGF. "Classical" Ca^{2+}-mobilizing receptors do not seem to have intrinsic tyrosine kinase activity and communicate to phospholipase C through interactions with a specific G-protein.

B. SZWERGOLD (*Fox Chase Cancer Center, Philadelphia, Pennsylvania*): As you know, we have been arguing for a number of years that in physiological bicarbonate-buffered media an increase in pH_i is not a signal or even a correlate of cell activation. I am therefore gratified that this point has been acknowledged. I am nevertheless disturbed that you continue to refer to an increase in pH_i as a consequence of mitogenic activation. I think we need to clear up the confusion surrounding this tissue by stating unequivocally that intracellular alkalinization does not occur in physiological media and is therefore not a signal or even a correlate.

MOOLENAAR: A rise in pH_i brought about either by growth factors or by HCO_3-dependent H^+ pumping is essential but not sufficient for an optimal mitogenic response. So cytoplasmic alkalinization does not serve as a trigger but is a permissive event in growth factor action.

L. CANTLEY (*Tufts University, Boston, Massachusetts*): I was surprised at your data showing that Quin 2 buffering of cytosolic Ca^{2+} completely blocks histamine induction of c-*fos* message. One might expect that DG produced from P_i turnover would still induce c-*fos* via activation of protein kinase C. Does Quin 2 buffering block TPA induction of c-*fos* message?

MOOLENAAR: That is of course a critical experiment that has to be done yet. The point I wish to make, however, is that EGF does not use its Ca^{2+} signal to induce c-*fos*, whereas Ca^{2+}-mobilizing hormones like histamine and bradykinin do.

G. DUBYAK (*Case Western Reserve University, Cleveland, Ohio*): Is there potentiation of bradykinin- and/or histamine-induced phospholipase C activation in cells pretreated with EGF?

MOOLENAAR: No, there is not. In fact, the inositol phosphate responses to EGF and bradykinin/histamine are perfectly additive, at least in A431 cells.

W. R. LOEWENSTEIN (*University of Miami, Florida*): How long do your Ca^{2+} oscillations last when agonist is continuously present, and precisely what agonists will elicit it.

MOOLENAAR: Histamine, but not EGF, elicits long-lasting Ca_i^{2+} oscillations in single HeLa cells as evidenced by direct intracellular recording. These oscillations last for up to 30 minutes and can be mimicked by microinjection of GTPγS.

S. ADAMO (*University of L'Aquila, Italy*): Do you know of any evidence suggesting that the autophosphorylative activity of the EGF-R may be separated from the external substrate tyrosinekinase activity?

MOOLENAAR: No, there is no evidence for that. There is only one cytoplasmic tyrosine kinase domain that is responsible for both autophosphorylation and substrate phosphorylation.

Early Cellular Responses to the Activation of a Mitogenic/Oncogenic Viral K-RAS Protein

JON P. DURKIN, BALU CHAKRAVARTHY,[a]
DOUG FRANKS,[a] GEOFFREY MEALING,
JEAN-LOUIS SCHWARTZ, ROGER TREMBLAY,
AND JAMES F. WHITFIELD

Cellular Oncology Group
Division of Biological Sciences
National Research Council of Canada
Ottawa, Canada K1A 0R6

[a]Department of Pathology
Faculty of Health Sciences
University of Ottawa
Ottawa, Canada K1H 8M5

GTP-binding RAS proteins, ranging in size from 21 kDa to 40 kDa, have been found in all eukaryotes from budding yeast and slime molds to mammalian cells.[1–5] Normal cellular RAS proteins participate in driving normal mammalian cells through the cell cycle by operating specifically during the later stages of the G_1 phase[6–9] and perhaps during the G_2 phase.[10] Unlike cellular RAS proteins, which promote but cannot initiate cell cycle transit, "activated" oncogenic forms of the RAS proteins are able to initiate and then promote transit through the cell cycle. Thus, microinjecting the oncogenic T24H-RAS protein from the EJ human bladder carcinoma into quiescent NIH 3T3, NRK or REF-2 cells stimulates them to transit the G_1 phase and initiate DNA replication.[11] Also, as we shall see, the oncogenic K-RAS protein of Kirsten sarcoma virus stimulates quiescent *ts*KSV-NRK cells to proliferate in the absence of exogenous growth factors.[12] Point-mutated *ras* genes and their oncogenic RAS protein products have been found in a substantial proportion of human tumors and therefore have been implicated in the development of human cancer.[13] Clearly finding out how "activated" RAS proteins start quiescent cells cycling should contribute substantially to our understanding of the origins of some human neoplasms.

Membrane-associated RAS proteins appear to influence the transduction of receptor signals by modulating the activation of membrane-associated signaling enzymes such as adenylate cyclase and phosphoinositidase C.[14–18] The signal given by the binding of a growth factor, such as PDGF, to its receptor displayed on the cell surface is transduced by a sequence of events which is only now beginning to be understood (SCHEME 1). In many cases, the binding of a growth factor to the extracellular domain of the receptor causes the rapid phosphorylation of the receptor's cytoplasmic domain at a tyrosine residue.[19] It is generally believed that receptor tyrosine phosphorylation triggers the events leading to DNA replication by a mechanism which involves increased hydrolysis of phosphatidylinositol 4,5-bisphosphate

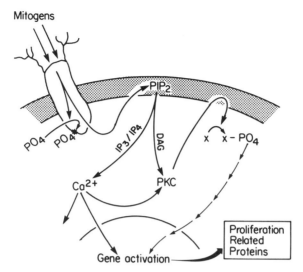

SCHEME 1. A simplified view of the transduction of a mitogenic signal from receptors displayed on the cell surface to early-acting proliferogenic genes. DAG, diacylglycerols; IP_3, inositol 1,4,5-trisphosphate; IP_4, inositol 1,3,4,5-tetrakisphosphate.

(PIP_2).[20,21] The primary PIP_2 degradation products, inositol 1,4,5-trisphosphate (IP_3) and diacylglycerols (DAG), profoundly affect the cell. IP_3 triggers the release of Ca^{2+} from the endoplasmic reticulum and its phosphorylated derivative inositol 1,3,4,5-tetrakisphosphate (IP_4) opens membrane Ca^{2+} channels.[22–24] The intracellular Ca^{2+} surge, which is a characteristic feature of most, if not all, mitogenic signals, is capable of stimulating many different cellular activities. It is the established view that the DAG's produced by PIP_2 hydrolysis collaborate with the Ca^{2+} surge to translocate protein kinase C (PKC) from the cytosol to the plasma membrane, where it is activated. Once associated with membrane, PKC phosphorylates as yet undefined substrates which are important for a number of cellular processes including proliferation.[24,25]

A clue to why "activated" RAS proteins are oncogenic may rest in how they stimulate quiescent cells to proliferate, even in the total absence of exogenous growth factors. Do they directly or indirectly generate Ca^{2+} and membrane-associated PKC signals like mitogenic growth factors[22–24]? Indeed, there is some indirect evidence that they do,[26–28] but here we present the first direct demonstration that one of these proteins, a temperature-sensitive variant of the viral K-RAS protein, does trigger a cytosolic Ca^{2+} transient and a brief surge of membrane-associated PKC activity.

MATERIALS AND METHODS

Cell Quiescence and Stimulation

Uninfected NRK cells and *ts*KSV-NRK cells were the gift of M. Scolnick (Merck Sharpe and Dohme, West Point, PA). The *ts*KSV-NRK cell line was produced by infecting NRK cells with the temperature-sensitive, transformation-defective ts 371 variant of Kirsten sarcoma virus.[12] Stock cultures, maintained at 36°C in a complete medium consisting of 90% DMEM–10% bovine calf serum (BCS), were replated in the same medium at a density of 2.75 × 10³ cells per cm² and incubated at 40°C for 48 hr. The cells were then put into a quiescent G_0 state by incubation in serum-free medium (DMEM/F12 [1:1] containing 10 mM HEPES, pH 7.2) for 48 hr at 41°C. These quiescent tsKSV-NRK cells were stimulated to resume proliferation by either (1) adding serum to the medium at a final concentration of 10% at 41°C; or (2) lowering the temperature from 41°C to a permissive 36°C without adding any exogenous mitogens.[10,12]

Measurement of DNA Synthetic Activity

DNA synthetic activities were estimated from the proportion of cells whose nuclei were labeled during an exposure to [³H]thymidine, as described previously.[12]

Free-Intracellular Ca^{2+} Measurement

Uninfected NRK and tsKSV-NRK cells, grown on No. 1 circular glass coverslips, were put into a quiescent G_0 state by serum deprivation for 48 hr at 41°C. At 0 hr, cultures were removed from the 41°C incubator, rinsed in HEPES-buffered simplified medium at room temperature (NaCl 140 mM, KCl 5 mM, CaCl₂ 2.5 mM, MgCl₂ 1.1 mM, glucose 5.6 mM, HEPES 10 mM, pH 7.3) and immediately placed on the spectrofluorimeter (CM-3 cation measurement, Spex Inc., Newark, NJ) microscope stage (IM35, Zeiss Instruments) at room temperature. Five to ten cells were then pressure-microinjected (Eppendorf 5242 microinjector, Hamburg, West Germany) with a 0.5–1.0-µm tip glass pipette filled with 1 mM Fura 2 pentapotassium salt in high-potassium saline solution (KCl 140 mM, MgCl₂ 2 mM, HEPES 10 mM, pH 7.2) under 200× phase-contrast magnification. The chamber was rinsed three times with fresh simplified medium and the optical magnification was increased to 630× using a Zeiss 63×/1.25 N.A. oil immersion Plan-Neofluar objective. An appropriate pin-hole diameter was selected to limit the emission field to a single cell. The whole procedure took about 10 minutes.

Fluorescence measurements were performed at room temperature on single cells for up to 1 hr using 350- and 390-nm excitation wavelengths alternating every second. The emitted light was passed through a 505-nm interference filter (10-nm bandwidth) and recorded in the photon-counting mode by the detector mounted on the microscope. Background fluorescence and cell autofluorescence counts were subtracted from the raw data corrected for lamp-intensity fluctuations. The ratios of the fluorescence intensities of the two wavelengths were then translated into free intracellular calcium concentrations.[29] The intracellular calcium level was evaluated by use of the following formula: $[Ca^{2+}]_i = K_d \times (F_{min}/F_{max}) \times (R - R_{min})/(R_{max} - R)$, where R, R_{min} and R_{max} are the fluorescence ratios recorded during the experiment (R) and during calibration tests on unlysed cells using 4 µM ionomycin (R_{max}) and 10 mM EGTA at pH 8.2 (R_{min}).[30] F_{max} and F_{min} are the corresponding fluorescence intensities for the 390-nm excitation and K_d is the Fura 2 dissociation constant at room temperature (135 nM).

Fura 2 was purchased from Molecular Probes Inc. or Calbiochem, ionomycin from Calbiochem, EGTA from Baker, and PDGF from Collaborative Research Inc.

Isolation of Phosphotyrosine-Containing Proteins

The media from cultures of quiescent *ts*KSV-NRK cells at 41°C were removed and replaced with phosphate-free DMEM for 15 min. The cells were then incubated with 1 mCi of [^{32}P]orthophosphate in the above medium for 3 hr at 41°C. At the indicated times during this incubation some cultures were shifted to 36°C to activate the mitogenic/oncogenic viral RAS protein. After the 3-hr incubation some cultures at 41°C were stimulated for 7 min with receptor-grade PDGF (1 nM) or EGF (3 nM) purchased from Collaborative Research Inc. The medium was removed and the cells were lysed in the dish at 0°C with 1 ml of lysis buffer (10 mM Tris, pH 7.6, 1% triton X-100, 5 mM EDTA, 50 mM NaCl, 30 mM sodium orthovanadate, 50 mM sodium fluoride, 100 μM sodium orthovanadate, 100 μM PMSF and 0.1% bovine serum albumin) per 10^6 cells. The cell extract was clarified by centrifugation, and phosphotyrosine-containing proteins were immunopurified using the anti-phosphotyrosine antibody method described by Frackelton *et al.*[31] The immunopurified proteins were separated on 7% polyacrylamide-SDS gels followed by autoradiography of the dried gels.

The Measurement of Membrane-Associated Protein Kinase C (PKC) Activity

Membrane-associated PKC activity was directly measured in NRK and *ts*KSV-NRK cells, without the prior extraction, purification, and reconstitution of the enzyme with phospholipid required for the current standard method,[25,32,33] using a novel procedure recently developed in our laboratory.[34] This assay involves the PKC-catalyzed phosphorylation of a heat-stable, 85-kDa cytosolic protein from S49 lymphoma cells shown to be a specific and physiological PKC substrate.

Substrate Preparation. S49 lymphoma cells, cultured in medium containing 90% RPMI and 10% horse serum, were hypotonically lysed by vortexing for 1 min in solution containing 1 mM NaHCO$_3$ and 5 mM MgCl$_2$, pH 7.5. The lysate was clarified by centrifugation at 280,000 $\times$ *g* for 10 min. The cytosol fraction was incubated at 95°C for 5 min and the precipitated material was removed by centrifugation. This heat-treated cytosol was used directly as the substrate source in the PKC assay.

Membrane Preparation. Monolayers of NRK and *ts*KSV-NRK cells which had been washed twice in PBS and incubated in lysis buffer (1 mM NaHCO$_3$, 5 mM MgCl, pH 7.5) for 5 min, were scraped from the dish with a rubber policeman and vortexed for 1 min. After nuclei were removed by centrifugation at 500 $\times$ *g* for 5 min, the membranes were sedimented at 280,000 $\times$ *g* for 10 min and used directly in the PKC assay.

PKC Measurement. Membranes (30 μg) were washed with lysis medium and resuspended in 50 μl of reaction buffer (1 mM NaHCO$_3$, 5 mM MgCl$_2$, 200 μM Na orthovanadate, 200 μM Na pyrophosphate, 2 mM Na fluoride, 2 μM CaCl$_2$), 100 μM Tris-HCl, pH 7.5. The reaction was started by adding 50 μl of S49-cytosol containing 20 μM ATP and 2 μCi ^{32}P-ATP. The tubes were incubated at 37°C for 10 min and the reaction was then terminated by adding EGTA to 1 mM. Membranes were removed by centrifugation at 280,000 $\times$ *g* and the level of phosphorylation of the 85-kDa substrate in the supernatant was determined by separating the cytosolic proteins on 10% polyacrylamide-SDS gels followed by autoradiography.

RESULTS AND DISCUSSION

At the permissive 36°C, *ts*KSV-NRK cells are transformed because they produce active viral K-RAS protein and consequently they can proliferate in serum-free medium without exogenous growth factors.[12] When incubated at 41°C these cells become phenotypically "normal" because the thermolabile viral K-RAS protein is inactivated.[12] At this higher, nonpermissive temperature the cells acquire a need for exogenous growth factors and arrest in a quiescent G_0 state when incubated in serum-free medium.[12] Reactivating the viral K-RAS protein in these quiescent cells by lowering the temperature from the nonpermissive 41°C to a permissive 36°C stimulates them to transit the G_1 phase, replicate DNA, and finally divide despite the lack of exogenous mitogens (FIG. 1). The G_0 to S-phase transit triggered by the viral K-RAS protein seemed to be similar to that triggered by adding serum to the serum-starved quiescent cultures at 41°C (FIG. 1). However, the cellular response to the viral K-RAS signal differs from the response to serum factors in three ways. Firstly, while the reactivated viral protein is mitogenic, it does not stimulate overall RNA and protein syntheses as do serum mitogens.[12] Nevertheless, the RAS-stimulated cells do produce a

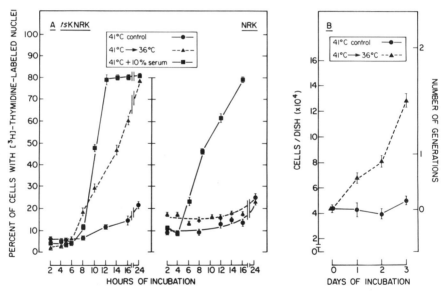

FIGURE 1. The ability of quiescent, serum-deprived *ts*KSV-NRK cells, but not uninfected NRK cells, to transit G_1 phase and initiate DNA synthesis (**A**) and divide (**B**) after a 41°C to 36°C shift. Cultures were rendered quiescent in G_0 phase by incubation in serum-free medium for the previous 48 hr at 41°C as described in the MATERIALS AND METHODS section. At 0 hr, cultures were either (1) left unstimulated at 41°C, (2) stimulated to transit G_1 phase by adding serum to 10% at 41°C, or (3) shifted from 41°C to 36°C without the addition of serum. Then: (**A**) cultures were labeled continuously with [³H]thymidine from 0 hr and at the indicated times their DNA-synthetic activities were determined autoradiographically as described in MATERIALS AND METHODS; or (**B**) cell numbers were measured at various times with a Coulter electronic cell counter. The points are means ± SEM of at least three determinations.

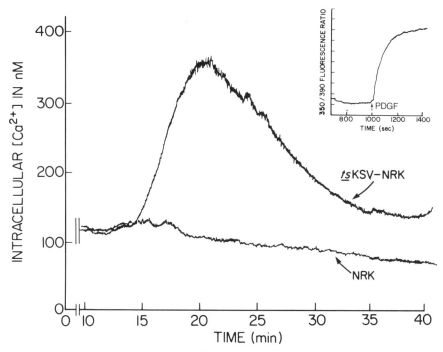

FIGURE 2. The activation of the viral RAS protein in quiescent *ts*KSV-NRK cells causes a transient three-fold increase in free-intracellular Ca^{2+} levels. Uninfected NRK and *ts*KSV-NRK cells grown on coverslips were rendered quiescent at G_0 by serum-deprivation after 48 hr at 41°C. At 0 hr, small groups of cells were quickly microinjected at room temperature with 1 mM Fura 2 and change in the free cytosolic Ca^{2+} of individual cells was measured at room temperature with a dual-wavelength excitation microspectrofluorimeter (Spex Inc.) as described in MATERIALS AND METHODS. The data shown are representative results from seven of ten separate experiments on *ts*KSV-NRK cells and seven of seven experiments on uninfected NRK cells. *Inset:* The rapid change in free intracellular Ca^{2+} levels upon the addition of PDGF (1 nM) to the culture medium of quiescent NRK cells microinjected with Fura 2.

select group of RNA transcripts and proteins which are required for G_1 transit and initiation of DNA replication.[12] Secondly, while serum mitogens must act in tandem to drive the cell through the G_1 phase,[35] the reactivated viral RAS protein is a complete mitogen in that it can by itself initiate and then promote cell proliferation.[12] Finally, viral K-RAS-stimulated cells can transit the G_1 phase and initiate DNA replication despite an extreme Ca^{2+}-deficiency, which prevents serum-stimulated cells from doing so.[12]

A rapid and transient increase in cytosolic free Ca^{2+} is part of the first response of cells to growth factors such as EGF and PDGF.[22-24] Indeed, adding PDGF to cultures of quiescent NRK cells in serum-free medium triggers an almost instantaneous increase in the cytosolic free Ca^{2+} concentration (FIG. 2). Reactivating the viral K-RAS protein in quiescent *ts*KSV-NRK cells in serum-free medium by dropping the temperature from 41°C to 36°C causes a transient, though delayed, three-fold increase in the cytosolic Ca^{2+} concentration which does not occur in temperature-shifted, uninfected NRK cells (FIG. 2). Typically, this RAS-induced Ca^{2+} surge occurs 15 to 20 min after the RAS-reactivating temperature shift (FIG. 2). This delay reflects either the time needed to build up the viral K-RAS protein to a critical level, or to accumulate and secrete enough RAS-induced autocrine factors (such as TGFα or PDGF-like factors) to activate surface receptors and trigger the signal transduction pathway. However, 15 minutes seems to be too little time for the induction and production of significant amounts of an autocrine factor(s).

EGF and PDGF are the two principal mitogens for NRK cells and homologs of both of these are secreted by KSV-infected cells.[36–39] If the delayed Ca^{2+} surge is due to such autocrine factors, reactivating the viral K-RAS protein should stimulate the autophosphorylation of tyrosine residues of EGF or PDGF receptors. This was tested with an anti-phosphotyrosine antibody[31,40,41] which binds to both activated EGF (data not shown) and activated PDGF receptors in cell membrane fractions (FIG. 3). It is apparent from FIGURE 3 that reactivating the viral K-RAS protein is not accompanied by the tyrosine phosphorylation of the 170-kDa EGF receptor or the 180-kDa PDGF receptor or, for that matter, any other receptor. Thus, as might be expected from the small amount of time it takes to generate the Ca^{2+} signals, it seems that the reactivated K-RAS protein directly triggers the Ca^{2+} and PKC surges from inside the cell.

However, autocrine factors may play a later role in the mitogenic response, because the response is blocked by protamine sulfate (FIG. 4), which is known to inhibit the mitogenic actions of PDGF and FGF, but not EGF, by preventing ligand-receptor binding.[42–45] This inhibitory action of protamine sulfate is not the result of nonspecific

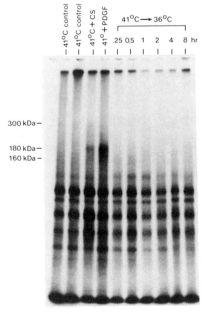

tsKSV−NRK

FIGURE 3. The reactivation of p21 in quiescent tsKSV-NRK cells does not cause the tyrosine phosphorylation of surface receptors. All cultures were metabolically labeled with [^{32}P]orthophosphate for 3 hr before the cells were harvested. Quiescent cultures of tsKSV-NRK cells in serum-free medium were either (1) left untreated at 41°C, (2) treated with serum or PDGF (to a final concentration of 10% and 1 nM, respectively) for 7 min at 41°C, or (3) shifted from 41°C to a viral RAS protein-activating 36°C for the indicated times before being lysed. Phosphotyrosine-containing proteins were isolated using the anti-phosphotyrosine antibody procedure described in MATERIALS AND METHODS. The immunopurified proteins were separated on 7% polyacrylamide-SDS gels followed by autoradiography of the dried gels.

FIGURE 4. The inhibition of the viral RAS-stimulated, but not serum-stimulated, G_1 transit by protamine sulfate. Cultures of *ts*KSV-NRK cells were rendered quiescent at G_0 by serum deprivation at 41°C. The cultures were then treated with the indicated concentrations of protamine sulfate and the cells stimulated to transit G_1 phase by adding serum to 10% at 41°C (■), or by activating the viral RAS protein by lowering the temperature to 36°C (▲). The data are presented as the total proportion of cells that entered S-phase during the first 24 hr after stimulation, relative to the proportion of cells that entered S-phase during the same period in the absence of protamine sulfate. The points are the means ± SEM of 4 separate determinations.

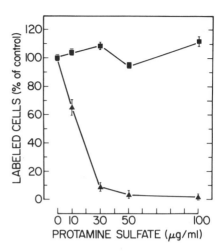

toxicity because it does not inhibit the mitogenic actions of EGF or serum, nor does it block the mitogenic response of *ts*ASV-NRK cells to reactivation of their mitogenic/oncogenic viral pp60[v-src] protein kinase.[46] However, protamine sulfate probably does not specifically affect the first signal from the reactivated K-RAS protein, because the compound can inhibit the initiation of DNA replication even when added several hours later in the G_1 phase (data not shown).

In the past few years PKC has emerged as an enzyme which supposedly plays a key role in a myriad of cellular activities including proliferation.[22,24] To a large extent, the importance of PKC in these activities has been inferred mainly from the responses of cells to the phorbol ester TPA (12-*O*-tetradecanoylphorbol-13-acetate), which is believed to be a specific PKC activator.[47] While it is certain that TPA activates PKC by causing the enzyme's translocation from the cytosol to membranes,[47] it is far from certain that all of the cellular responses to TPA action are mediated by PKC. Thus, PKC might not do all that it is believed to do. The reliance on TPA as an indicator of PKC movement and activation has arisen mainly from the lack of a reliable and physiologically relevant *in vitro* assay for the enzyme and the lack of specific inhibitors. The only currently available *in vitro* PKC assay method is a complicated, laborious procedure that requires that the physiologically important membrane-associated PKC be solubilized by detergent, partially purified by column chromatography, reconstituted in an artificial lipid (phosphatidyl serine) environment necessary for enzymatic activity, and then assayed using a nonspecific and nonphysiological histone preparation.[23,32,33] Understandably, investigators are reluctant to use this procedure, which measures the enzyme's activity under conditions that at best only poorly represent the state of PKC in its native, membrane-associated environment.

We have remedied these problems by developing a simple *in vitro* assay that measures PKC in its native, membrane-associated state using a specific, physiological substrate present in high concentrations in a heat-denatured cytosolic fraction of S49T-lymphoma cells. This new method was used to measure PKC activity in membranes from *ts*KSV-NRK cells at various times after the viral RAS-reactivating 41°C to 36°C temperature shift (FIG. 5). As can be seen in FIGURE 5, membrane associated PKC activity transiently increases within 1 hr after the temperature shift. By contrast, no increase in membrane-associated PKC activity occurs in uninfected NRK cells after they are shifted from 41°C to 36°C (FIG. 5). In contrast to TPA, which causes PKC to redistribute to an EGTA-nonextractable membrane fraction,[47] the viral RAS protein-induced increase in membrane PKC is substantially reduced by EGTA, which suggests that most of the enzyme is not integrated into the membrane (data not shown). The translocation of PKC to membranes is a Ca^{2+}-dependent process,[47] and it is entirely possible that the increase in membrane-associated PKC in tsKSV-NRK cells stimulated by the viral K-RAS protein is at least partly the result of translocation triggered by the Ca^{2+} surge.

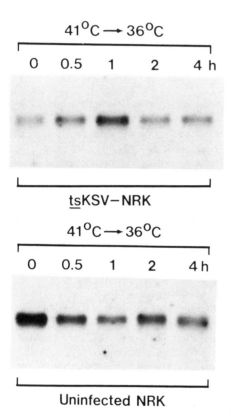

FIGURE 5. The reactivation of the viral RAS protein in quiescent tsKSV-NRK cells rapidly stimulated membrane-associated PKC activity. Uninfected NRK and tsKSV-NRK cells were rendered quiescent by incubation in serum-free medium for 48 hr at 41°C as described in MATERIALS AND METHODS. Membrane-associated PKC was measured in these quiescent cells, and in cells shifted from 41°C to 36°C for the indicated times, by the novel *in vitro* assay described in MATERIALS AND METHODS.

A rapid increase in both total RNA and total protein synthesis is a characteristic response of quiescent cells to serum growth factors.[12,48] It has been reported that PDGF stimulates transcription of as much as 0.1 to 0.3% of the 3T3 fibroblast genome.[48] By contrast, the viral K-RAS protein initiates G_1 transit of tsKSV-NRK cells without increasing total RNA or protein synthesis, but it does direct the synthesis of new RNA transcripts and proteins which are needed for G_1 transit and initiation of DNA synthesis.[12] Thus, in a sense, the viral RAS protein induces cell proliferation in a more

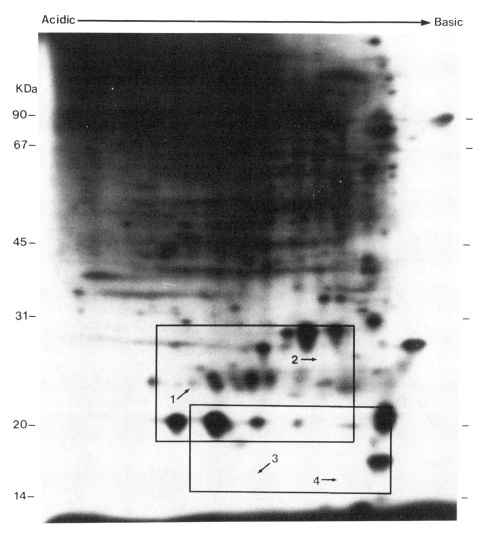

FIGURE 6. Two-dimensional gel electrophoresis of ^{32}P-labeled proteins from lysates of quiescent *ts*KSV-NRK cells at 41°C. Cultures of quiescent, serum-deprived *ts*KSV-NRK cells at 41°C were: left unstimulated at 41°C; stimulated to transit G_1 at 41°C by adding serum to 10%; or stimulated to transit G_1 by dropping the temperature to a p21-activating 36°C. Quiescent cultures and cultures at various times of the stimulation were exposed to [^{32}P]orthophosphate (50 mCi/dish) in phosphate-free medium for 45 min and in each case 200,000 cpm of acid-precipitable counts from cell lysates were subjected to two-dimensional gel electrophoresis.[49] This figure is an autoradiograph of proteins from quiescent *ts*KSV-NRK cells, and the *arrows* inside the two boxed regions indicate those changes in protein phosphorylation resulting from viral RAS activation. These regions are displayed in more detail in FIGURE 7.

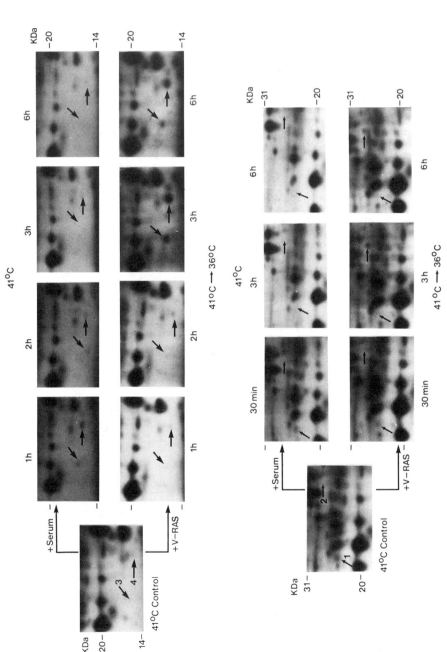

FIGURE 7. The activation of the viral RAS protein induced only four discernible changes in protein phosphorylation during G$_1$ transit. Quiescent, serum-deprived *ts*KSV-NRK cells at 41°C were stimulated to start transiting G$_1$ either by adding serum at 41°C or by lowering the temperature to a viral RAS-activating 36°C. Cellular proteins were labeled by a 45-min exposure to [^{32}P]orthophosphate immediately before the cells were lysed at the indicated times. In each case 200,000 cpm of ^{32}P-labeled proteins in the cell lysates were subjected to two-dimensional gel electrophoresis as described

selective manner than does serum.[12] This difference is also illustrated by the patterns of protein phosphorylation in cells stimulated by serum and the viral RAS protein. When quiescent *ts*KSV-NRK cells in ^{32}P-containing medium are stimulated by serum at 41°C, between 25 and 30 changes in protein phosphorylation can be resolved by two-dimensional gel electrophoresis of total cell extracts (data not shown). Probably only a few of these serum-induced phosphorylations are specifically related to proliferation. By contrast, only four changes in protein phosphorylation can be resolved in *ts*KSV-NRK cells after the RAS-reactivating 41°C to 36°C temperature shift (FIG. 6). None of these changes occurs in uninfected NRK cells after the same temperature shift (data not shown). Three of the changes induced by K-RAS reactivation, two phosphorylations (proteins 3 and 4), and one dephosphorylation (protein 1) also occur in quiescent *ts*KSV-NRK cells after stimulation by serum (FIG. 7), which suggests they may be related to proliferation. While proteins 3 and 4 are promptly but transiently phosphorylated in the serum-stimulated *ts*KSV-NRK cells, they are strongly and persistently phosphorylated in the *ts*KSV-NRK cells stimulated by the viral K-RAS protein (FIG. 7). If the transient phosphorylation of these two proteins is an important event for serum-stimulated proliferation, then the intense and persistent phosphorylation of these proteins in RAS-stimulated cells might contribute to the protein's oncogenic activity. Finally, protein 2 was phosphorylated only in the RAS-stimulated cells, which might indicate a relation to some nonproliferative viral or transforming function.

In conclusion, we have seen that an oncogenic viral K-RAS protein switches quiescent *ts*KSV-NRK cells into the proliferative mode where they make and phosphorylate a small set of proteins, replicate DNA, and ultimately divide. The viral protein seems to trigger these events with a signal which, like signals from various mitogenic factors, consists of a cytosolic Ca^{2+} transient and a brief surge of membrane-associated PKC activity. Since normal RAS proteins are known to couple a variety of receptors to the PIP_2-hydrolyzing phosphoinositidase C, and since activated RAS proteins by themselves increase basal levels of PIP_2 breakdown products, it would be reasonable to suppose that the reactivated K-RAS protein in our study by itself triggers the Ca^{2+}- and membrane-associated PKC signal surges by activating PIP_2 hydrolysis, and that these signals start the events leading to DNA replication and division.

ACKNOWLEDGMENTS

We wish to thank Gauri Muradia and Philip MacDonald for excellent technical assistance and Christine Gobey for preparing the manuscript.

REFERENCES

1. GIBBS, J. B., I. S. SIGAL & E. M. SCOLNICK. 1985. Biochemical properties of normal and oncogenic *ras* p21. Trends Biochem. Sci. **10:** 350–353.
2. NEWELL, P. C., G. N. EUROPE-FINNER, N. V. SMALL & G. LIU. 1988. Inositol phosphates, G-proteins and *ras* genes involved in chemotactic signal transduction of *Dictyostelium*. J. Cell Sci. **89:** 123–127.
3. SHIH, T. Y. & M. O. WEEKS. 1984. Oncogenes and cancer: The p21 *ras* gene. Cancer Invest. **2:** 109–123.

4. TOUCHOT, N., P. CHARDIN & A. TAVITIAN. 1987. Four additional members of the *ras* gene superfamily isolated by an oligonucleotide strategy: Molecular cloning of YPT-related cDNAs from a rat brain library. Proc. Natl. Acad. Sci. USA **84:** 8210–8214.

5. UNO, I., K. MATSUMOTO & T. ISHIKAWA. 1986. Functions of the *CYR*1 and *RAS* genes in yeast. *In* Yeast Cell Biology. J. Hicks, Ed.: 113–123. Alan R. Liss Inc. New York.

6. CAMPISI, J., H. E. GRAY, A. B. PARDEE, M. DEAN & G. E. SONNENSHEIN. 1984. Cell cycle control of c-*myc* but not c-*ras* expression is lost following chemical transformation. Cell **36:** 241–247.

7. GOYETTE, M., C. J. PETROPOULOS, P. R. SHANK & N. FAUSTO. 1984. Regulated transcription of c-Ki-*ras* and c-*myc* during compensatory growth of rat liver. Mol. Cell. Biol. **4:** 1493–1498.

8. MULCAHY, L. S., M. C. SMITH & D. W. STACEY. 1985. Requirement for *ras* proto-oncogene function during serum-stimulated growth. Nature **313:** 241–243.

9. THOMPSON, N. L., J. E. MEAD, L. BRAUN, M. GOYETTE, P. R. SHANK & N. FAUSTO. 1986. Sequential protooncogene expression during rat liver regeneration. Cancer Res. **46:** 3111–3117.

10. DURKIN, J. P. & J. F. WHITFIELD. 1987. The viral Ki-*ras* gene must be expressed in the G_2 phase if *ts* Kirsten sarcoma virus-infected NRK cells are to proliferate in serum-free medium. Mol. Cell Biol. **7:** 444–449.

11. FERAMISCO, J. R., M. GROSS, T. KAMATA, M. ROSENBURG & R. W. SWEET. 1984. Microinjection of the oncogene form of the human H-*ras* (T-24) protein results in rapid proliferation of quiescent cells. Cell **38:** 109–117.

12. DURKIN, J. P. & J. F. WHITFIELD. 1986. Characterization of G_1 transit induced by the mitogenic-oncogenic viral Ki-*ras* gene product. Mol. Cell. Biol. **6:** 1386–1392.

13. NISHIMURA, S. & T. SEKIYA. 1987. Human cancer and cellular oncogenes. Biochem. J. **243:** 313–327.

14. CHIARUGI, V. P., F. PASQUALI, S. VANUCCHI & M. RUGGIERO. 1987. Point-mutated p21^{ras} couples a muscarinic receptor to calcium channels and polyphosphoinositide hydrolysis. Biochem. Biophys. Res. Commun. **141:** 591–599.

15. FRANKS, D. J., J. F. WHITFIELD & J. P. DURKIN. 1987. A viral K-*ras* protein increases the stimulability of adenylate cyclase by cholera toxin in NRK cells. Biochem. Biophys. Res. Commun. **147:** 596–601.

16. KONISHI-IMAMURA, L., M. NODA & Y. NOMURA. 1987. Alteration by v-Ki-*ras* in NaF, cholera toxin and forskolin-induced adenylate cyclase activation in NIH/3T3 fibroblast cells. Biochem. Biophys. Res. Commun. **146:** 47–52.

17. MATSUMOTO, K., I. UNO & T. ISHIKAWA. 1985. Genetic analysis of the role of cAMP in yeast. Yeast **1:** 15–24.

18. WAKELHAM, M. J. O., S. A. DAVIES, M. D. HOUSLAY, I. MCKAY, C. J. MARSHALL & A. HALL. 1986. Normal p21^{N-ras} couples bombesin and other growth factor receptors to inositol phosphate production. Nature **323:** 173–176.

19. SIBLEY, D. R., J. F. BENOVIC, M. G. CARON & R. J. LEFKOWITZ. 1988. Phosphorylation of cell surface receptors: A mechanism for regulating signal transduction pathways. Endocrine Rev. **9:** 38–56.

20. BERRIDGE, M. J., J. P. HESLOP, R. F. IRVINE & K. D. BROWN. 1984. Inositol trisphosphate formation and calcium mobilization in Swiss 3T3 cells in response to platelet derived growth factor. Biochem. J. **222:** 195–201.

21. CHU, S.-H., C. J. HOBAN, A. J. OWEN & R. P. GEYER. 1985. Platelet-derived growth factor stimulates rapid polyphosphoinositide breakdown in fetal human fibroblasts. J. Cell. Physiol. **124:** 391–396.

22. BERRIDGE, M. J. 1987. Inositol lipids and cell proliferation. Biochim. Biophys. Acta **907:** 33–45.

23. MELDOLESI, J. & T. POZZAN. 1987. Pathways of Ca^{2+} influx at the plasma membrane: Voltage-, receptor-, and second messenger-operated channels. Exp. Cell Res. **171:** 271–283.

24. WHITFIELD, J. F., J. P. DURKIN, D. J. FRANKS, L. P. KLEINE, L. RAPTIS, R. H. RIXON, M.

SIKORSKA & P. R. WALKER. 1987. Calcium, cyclic AMP and protein kinase C-partners in mitogenesis. Cancer Metastasis Rev. **5**: 205–250.

25. HALSEY, D. L., P. R. GIRARD, J. F. KUO & P. J. BLACKSHEAR. 1987. Protein kinase C in fibroblasts. Characteristics of its intracellular location during growth and after exposure to phorbol esters and other mitogens. J. Biol. Chem. **262**: 2234–2243.

26. BAR-SAGI, D. & J. R. FERAMISCO. 1986. Induction of membrane ruffling and fluid-phase pinocytosis in quiescent fibroblasts by *ras* proteins. Science **233**: 1061–1068.

27. CHUA, C. C. & R. L. LADDA. 1986. Protein kinase C and non-functioning EGF receptor in K-*ras* transformed cells. Biochem. Biophys. Res. Commun. **135**: 435–444.

28. FLEISCHMAN, L. F., S. B. CHAHWALA & L. CANTLEY. 1986. Ras-transformed cells: altered levels of phosphatidylinositol-4,5-bisphosphate and catabolites. Science **231**: 309–312.

29. GRYNKIEWICZ, G., M. POENIE & R. Y. TSIEN. 1985. A new generation of Ca^{2+} indicators with greatly improved fluorescence properties. J. Biol. Chem. **260**: 3440–3450.

30. COBBOLD, P. H. & T. J. RINK. 1987. Fluorescence and bioluminescence measurement of cytoplasmic free calcium. Biochem. J. **248**: 313–328.

31. FRACKELTON, A. R., P. M. TREMBLE & L. T. WILLIAMS. 1984. Evidence for the platelet-derived growth factor-stimulated tyrosine phosphorylation of the platelet-derived growth factor receptor *in vivo*. J. Biol. Chem. **259**: 7909–7915.

32. BOYNTON, A. L., J. F. WHITFIELD & L. P. KLEINE. 1983. Ca^{2+}/phospholipid-dependent protein kinase activity correlates to the ability of transformed liver cells to proliferate in Ca^{2+}-deficient medium. Biochem. Biophys. Res. Commun. **115**: 383–390.

33. TAKAI, Y., A. KISHIMOTO, M. INOUE & Y. NISHIZUKA. 1977. Studies on a cyclic nucleotide-independent protein kinase and its proenzyme in mammalian tissues. J. Biol. Chem. **252**: 7603–7609.

34. CHAKRAVARTHY, B. R., D. J. FRANKS, J. F. WHITFIELD & J. P. DURKIN. 1988. Novel method of assaying protein kinase C in isolated cell membranes. FASEB J. **2**: A351.

35. O'KEEFE, E. J. & W. J. PLEDGER. 1983. A model of cell cycle control: Sequential events regulated by growth factors. Mol. Cell Endocrinol. **31**: 167–186.

36. BUICK, R. N., J. FILMUS & A. QUARONI. 1987. Activated H-*ras* transforms rat intestinal epithelial cells with expression of α-TGF. Exp. Cell Res. **170**: 300–309.

37. DERYNCK, R. 1986. Transforming growth factor-α: Structure and biological activities. J. Cell. Biochem. **32**: 293–304.

38. OWEN, R. D. & M. C. OSTROWSKI. 1987. Rapid and selective alterations in the expression of cellular genes accompany conditional transcription of Ha-v-*ras* in NIH 3T3 cells. Mol. Cell. Biol. **7**: 2512–2520.

39. OZANNE, B. R., R. J. FELTON & P. L. KAPLAN. 1980. Kirsten murine sarcoma virus transformed cell lines and a spontaneously transformed rat cell-line produce transforming factor. J. Cell. Physiol. **105**: 163–180.

40. DANIEL, T. O., P. M. TREMBLE, A. R. FRACKELTON & L. T. WILLIAMS. 1985. Purification of the platelet-derived growth factor receptor by using an anti-phosphotyrosine antibody. Proc. Natl. Acad. Sci. USA **82**: 2684–2687.

41. BISHAYEE, S., A. H. ROSS, R. WOMER & C. D. SCHER. 1986. Purified human platelet-derived growth factor receptor has ligand-stimulated tyrosine kinase activity. Proc. Natl. Acad. USA **83**: 6756–6760.

42. HUANG, J. S., S. S. HUANG, B. B. KENNEDY & J. F. DEVEL. 1982. Platelet-derived growth factor. Specific binding to target cells. J. Biol. Chem. **257**: 8130–8136.

43. HUANG, J. S., J. NISHIMURA, S. S. HUANG & J. F. DEUEL. 1984. Protamine inhibits platelet derived growth factor receptor activity but not epidermal growth factor activity. J. Cell. Biochem. **26**: 205–220.

44. NEUFELD, G. & D. GOSPODAROWICZ. 1987. Protamine sulfate inhibits mitogenic activities of the extracellular matrix and fibroblast growth factor, but potentiates that of epidermal growth factor. J. Cell. Physiol. **132**: 287–294.

45. GOSPODAROWICZ, D., N. FERRARA, L. SCHWEIGERER & G. NEUFELD. 1987. Structural characterization and biological functions of fibroblast growth factor. Endocrine Rev. **8**: 95–114.

46. DURKIN, J. P. & J. F. WHITFIELD. 1987. Evidence that the viral Ki-RAS protein, but not the
     pp60$^{v\text{-}src}$ protein of ASV, stimulates proliferation through the PDGF receptor. Biochem.
     Biophys. Res. Commun. **148:** 376–383.
47. ASHENDEL, C. L. 1985. The phorbol ester receptor: A phospholipid-regulated protein
     kinase. Biochim. Biophys. Acta **822:** 219–242.
48. COCHRAN, B. H., A. C. REFFEL, M. A. CALLAHAN, J. N. ZULLO & C. D. STILES. 1984. Cell
     cycle genes regulated by platelet-derived growth factor. *In* Cancer Cells. G. F. Vande
     Wande, A. J. Levine, W. C. Topp & J. W. Watson, Eds.: Vol. **1:** 51–56. Cold Spring
     Harbor Laboratory. Cold Spring Harbor, NY.
49. DURKIN, J. P. & J. F. WHITFIELD. 1985. The selective induction of a small number of
     proteins during G_1 transit results from the mitogenic action of pp60$^{v\text{-}src}$ in *ts* ASV-infected
     rat cells. J. Cell. Physiol. **125:** 51–60.

DISCUSSION

G. GUIDOTTI (*University of Parma, Italy*): Is there any evidence that AP-1 is
phosphorylated after the activation of RAS in your system?

DR. DURKIN: No, I don't think so. All of the phosphorylated proteins we have
identified are low molecular weight; much smaller than AP-1.

L. CANTLEY (*Tufts University, Boston, Massachusetts*): When the *ts* mutant
KSV-NRK cells are shifted to the permissive temperature, do you see an increase in
bradykinin binding? Do protein synthesis inhibitors prevent the Ca^{2+} elevation and
PKC translocation that occur in response to the shift to the permissive temperature?

DURKIN: Clearly, if the increases in Ca^{2+} and PKC translocation upon p21
reactivation are not affected by protein synthesis inhibitors, then the possibility that
autocrine factors are responsible for these events is ruled out. Experiments designed to
answer these questions are currently being carried out. We have not looked at the
effect of p21 reactivation on bradykinin binding.

C. KLEE (*National Institutes of Health, Bethesda, Maryland*): Have you checked
that the rise in Ca^{2+} concentration after temperature shift is required for the effect of
RAS on DNA synthesis?

DURKIN: This is the burning question that needs to be answered, not only in our
system, but for the Ca^{2+} effects induced by growth factors and other mitogenic stimuli.
Now that the universality of the Ca^{2+} surge upon mitogenic stimuli has been agreed
upon, attention will necessarily be turned to determining whether it really means
anything to proliferation.

R. LOTAN (*University of Texas M.D. Anderson Cancer Center, Houston, Texas*):
Do the four proteins that are phosphorylated after activation of p21 by temperature
shift exist in the cell or are they newly synthesized?

DURKIN: We have evidence that the phosphorylations/dephosphorylations that we
see upon p21 reactivation are unaffected by the presence of cycloheximide.

Magnesium and Cell Proliferation[a]

MICHAEL E. MAGUIRE

Department of Pharmacology
School of Medicine
Case Western Reserve University
Cleveland, Ohio 44106

INTRODUCTION

Although much less is known about the intracellular roles of Mg^{2+} than of Ca^{2+}, it is increasingly clear that free magnesium ion (Mg_i^{2+}) has specific and important roles to play in regulation of cell growth, homeostasis and response to exogenous stimulatory agents.

For example, with regard to Mg^{2+} regulation of cell growth, cultured cell lines derived from lymphomas are weakly if at all tumorigenic, but become extremely tumorigenic after growth in Mg^{2+}-deficient medium.[1,2] Mg^{2+}-deficiency in rats results in the induction of a specific tumor, a solid lymphoma, in 25% of the animals.[3,4] No other type of tumor appears to be increased. As another example, human lung fibroblasts in culture exhibit specific and independent Mg^{2+} and Ca^{2+} dependencies for proliferation and cell survival with the cation concentration required for proliferation being at least an order of magnitude greater than that required for cell survival. Upon transformation with SV-40, Ca^{2+} requirements for proliferation and survival are unchanged. However, the transformed cells are now able to proliferate at the Mg^{2+} concentration that previously allowed only survival, suggesting a specific role for Mg^{2+} in normal proliferation.[5]

With regard to the role(s) of Mg^{2+} in regulation of cell homeostasis and response to exogenous stimulatory agents, there are now abundant examples of important cellular processes and enzymatic systems that can be activated or inhibited specifically by free Mg^{2+}. Such interaction with Mg^{2+} occurs in these systems at specific cationic sites for Mg^{2+} independently of any interactions with Ca^{2+}. Examples include acetyl CoA synthetase,[6] 5'-nucleotidase,[7] phosphorylase kinase,[8] phosphoribosyl pyrophosphate transferase,[9] phosphoglucomutase,[10] NaK-ATPase,[11,12] Ca^{2+}-transport ATPase of sarcoplasmic reticulum,[13-15] H^+-ATPases of the mitochondrial membrane,[16-19] and adenylate cyclase.[20-24] In addition, computer simulation studies by Garfinkel and colleagues have shown that metabolic flux through cardiac muscle glycolytic pathways is highly sensitive to free Mg^{2+}.[25,26]

Studies from this laboratory have shown that Mg^{2+} influx but not efflux is regulated by a variety of hormones without any concomitant effects on Ca^{2+} flux, transport of monovalent cations, or changes in intracellular pH or Ca^{2+}.[27-33] In

[a]This work was supported by U.S. Public Health Service Grant GM 26340 and National Science Foundation Grant DCB8706562. M.E.M. was an Established Investigator of the American Heart Association with partial support from the Northeast Ohio Chapter during portions of this work.

201

addition, intracellular Mg^{2+} exchanges much more slowly in most cell types than does Ca^{2+}. Indeed, during growth arrest of lymphocytes, intracellular Mg^{2+} essentially cannot be exchanged with extracellular cation, whereas Ca^{2+} remains freely exchangeable.[30]

It is of especial note that not only is Mg^{2+} exchange between the cell and the medium quite slow, but also hormonal regulation of Mg^{2+} transport occurs over a completely different time scale than that observed for Ca^{2+}, cyclic AMP, and inositol polyphosphates. Changes in the intracellular concentration of these latter agents occur rapidly and transiently, are terminated within seconds to a few minutes, and desensitize almost as rapidly. In sharp contrast, alterations in Mg^{2+} influx may last for 1–2 hours or more after addition of hormone and do not begin to desensitize for many minutes to hours after addition of hormone.[27,29] Thus changes in Mg^{2+} flux are prolonged and could provide a different type of signal to the cell.[32,33]

These data have led us to postulate a role for Mg^{2+} as a regulatory ligand in the same sense that Ca^{2+}, cyclic AMP, and inositol phosphates are regarded. However, we suggest that, unlike these latter agents, which regulate cellular processes acutely, Mg^{2+} is a "chronic" regulatory ligand whose intracellular free concentration determines the magnitude or range over which other agents are able to act, much as the gain mechanism acts on an amplifier.[32,33]

To investigate this hypothesis further, we have undertaken three major experimental projects, the latter two of which will be discussed in this paper. First, we are studying the mechanism of Mg^{2+} flux across the cell membrane through the cloning, expression, and reconstitution of Mg^{2+} transport systems. To date we have cloned, expressed, partially sequenced, and characterized three distinct Mg^{2+} transport systems from the gram-negative bacterium *Salmonella typhimurium*.[34] In addition, we have identified and partially characterized three and possibly more genes that appear to mediate a direct Mg^{2+} regulation of gene expression. Our preliminary data suggest that Mg^{2+}, like oxygen or carbon source, is a fundamental regulator of bacterial metabolism and gene expression.[34,35]

Second, there is a great need for a method to determine intracellular free Mg^{2+} concentration on a physiologically relevant time scale. Thus we have been developing cell-permeable Mg^{2+}-selective indicators. Some preliminary data are presented below.

Finally, we have developed a model of Mg^{2+}-deficiency using our standard cell line, the murine S49 lymphoma, so that we can investigate the mechanistic basis for Mg^{2+} regulation of hormone response and possibly cell growth. Data to date on these studies are presented in this paper.

MATERIALS AND METHODS

Materials

Dulbecco's modified essential medium (DMEM) and Mg^{2+}-deficient DMEM were purchased from Flow Laboratories. Horse serum was inactivated at 54°C for 30 min. "Mg^{2+}- and Ca^{2+}-free" serum was prepared by dialyzing horse serum for 24 hours against 100 volumes of Puck's saline G solution (PSG) containing 1.0 mM EDTA, repeated once. The serum was then dialyzed twice more against 100 volumes of

PSG. Unless noted, Ca^{2+} was then re-added to its normal 1.8 mM. Wild-type S49 cells grow normally in medium containing this dialyzed serum if Mg^{2+} is added at the usual 0.8 mM. Further, such cells respond normally to hormonal stimulation. $^{28}Mg^{2+}$ was obtained from Brookhaven National Laboratory (Upton, NY).

S49 Lymphoma

The murine S49 lymphoma line is grown in DMEM with 8–10% heat-inactivated horse serum in 10% CO_2. This medium normally contains 0.8 mM Mg^{2+}. Lowering the Mg^{2+} concentration to 0.2 mM causes extensive cell death and a prolonged growth arrest. To isolate Mg^{2+}-deficient S49 clones, wild-type cells were grown in DMEM with horse serum containing 0.4, 0.3 and finally 0.2 mM total Mg^{2+} for sequential periods of 2 weeks. After 2 weeks in 0.2 mM Mg^{2+}, the remaining cells were cloned in soft agar in medium containing 30 μM Mg^{2+}. Colonies arising after several days were grown for 2 weeks in DMEM/horse serum with 50 μM Mg^{2+} and then subcloned in soft agar with 50 μM Mg^{2+}. These clones were subsequently maintained in DMEM/horse serum containing 50 μM Mg^{2+}. Clones isolated in this manner were able to grow with a doubling time of 50–60 hours with >90% viability and their growth rate was not altered after 2 weeks growth in DMEM/horse serum containing the normal 0.8 mM Mg^{2+}.

All intact cell experiments with either wild-type or Mg^{2+}-deficient cells were done in DMEM without horse serum or $NaHCO_3$ and supplemented with 20 mM NaCl, 20 mM Na-HEPES, pH 7.4 (at 37°C) and 0.1 mM Mg^{2+}. Control experiments done at 0.05 or 0.8 mM extracellular Mg^{2+} showed no significant differences.

Other Assays

Mg^{2+} transport, cyclic AMP determinations, and compartmentation analysis were performed as previously described.[27-31] Additional methodological details are given in the figure legends. Dye synthesis will be reported elsewhere (manuscript in preparation).

RESULTS

Intracellular Mg^{2+} Indicators

Our current cell-permeable Mg^{2+} indicators are derivatives of tropolone (2-hydroxycycloheptatrienone). Our current proptotype TROP-5 (FIG. 1A) has a K_a of 0.7 mM for Mg^{2+} and 2.5 mM for Ca^{2+} (FIG. 1B). In addition the maximal absorbance shift induced by Mg^{2+} is about twice that induced by Ca^{2+}. H^+ and monovalent cations have little influence on absorption. Several other indicators related to TROP-5 show similar characteristics. Using vesicles isolated from rabbit sarcoplasmic reticulum, we can easily detect Mg^{2+} release due to MgATP hydrolysis (FIG. 2). Ca^{2+} added at 200 μM has no effect on absorbance but addition of MgATP to provide substrate for the reticular Ca-ATPase leads to activation of Ca^{2+}-pumping and concomitant hydrolysis

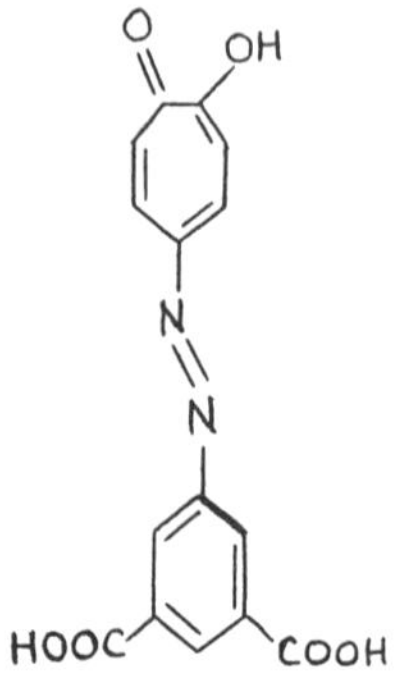

FIGURE 1A. Structure of TROP-5.

of MgATP to Mg^{2+} and ADP as shown by the change in absorbance. Addition of ionomycin, preventing intravesicular buildup of Ca^{2+}, effectively stimulates the Ca-ATPase, giving an increased rate of hydrolysis and thus release of free Mg^{2+}. From these data, we can calculate that in the presence of 0.2 mM Ca^{2+} and 0.2 mM MgATP, we can detect a change in Mg_i^{2+} concentration of 5–10 μM.

Both the acetoxymethyl and propyloxymethyl esters of TROP-5 have been synthesized and shown to be taken up into both cultured lymphoma cells and bacteria (*S. typhimurium*). The propyloxymethyl ester is hydrolyzed significantly better than the acetoxymethyl ester and remains trapped with the cell for at least an hour with negligible leakage. We are currently pursuing use of these dyes as dynamic indicators of Mg_i^{2+}.

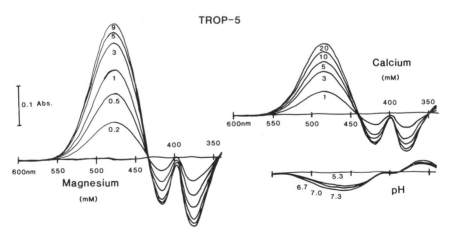

FIGURE 1B. Difference spectrum of TROP-5 in presence of differing Mg^{2+}, Ca^{2+} and H^+ concentrations. Spectra were recorded on a Cary 14 spectrophotometer at a dye concentration of 10 μM at room temperature in Buffer A (120 mM KCl, 20 mM NaCl, 10 mM K-HEPES, pH 7.1 at 20°C.)

Development of Mg^{2+}-Deficient Cell Models

Mg^{2+}-deficiency has been studied in rodents for decades.[36-38] However, model cell systems, either derived from rodent tissue or from eukaryotic cultured cell systems, have not been as thoroughly developed.[1,2] We decided to compare the effects of Mg^{2+}-deficiency on our model cultured cell system, the murine S49 T-lymphoma cell, with effects on a comparable cell type, T-lymphocytes, from Mg^{2+}-deficient rats. Derivation of Mg^{2+}-deficient S49 clones is outlined in the MATERIALS AND METHODS section. T-Lymphocytes from male Sprague-Dawley rats were obtained by maintaining rats weighing 125 gm on a Mg^{2+}-deficient diet for 8 weeks. In the rat model major signs of Mg^{2+}-deficiency appear within 3 days and are stable after 3–4 weeks. Animals were not pair-fed, although control experiments were performed with siblings. The

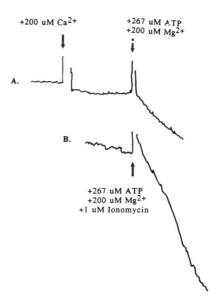

FIGURE 2. Response of TROP-5 to activation of the Ca^{2+}-ATPase pump in sarcoplasmic reticulum vesicles. Rabbit sarcoplasmic reticulum vesicles were incubated at room temperature in Buffer A in the presence of 10 μM TROP-5. Absorbance was monitored at 476 nm at a chart speed of 3 cm/min. Addition of reagents caused less than a 1% volume change. (A) Effect of sequentially adding 200 μM $CaCl_2$ followed by a mixture of MgATP (200 μM $MgCl_2$ plus 267 μM ATP). (B) Effect of adding 200 μM $CaCl_2$ (shown in A) followed by addition of 1 μM ionomycin plus Mg ATP as in (A). Addition of ionomycin alone had no effect.

major goal of the studies with the rat T-lymphocytes was to determine whether the effects of Mg^{2+}-deficiency were comparable with the effects on Mg^{2+}-deficient S49 cells. Since, as will be discussed below, the effects of Mg^{2+}-deficiency were essentially identical in the two classes of deficient cells, we have adopted the Mg^{2+}-deficient S49 cell as our model.

Characteristics of Mg^{2+}-Deficient Cells

The characteristics of Mg^{2+} influx in T-lymphocytes and S49 cells were determined using $^{28}Mg^{2+}$ (TABLE 1). The maximal velocity of the influx system was not altered by Mg^{2+}-deficiency in the case of the S49 lymphoma cell, but was slightly decreased in the Mg^{2+}-deficient T-lymphocytes. However, in both cell types, the

TABLE 1. Lymphocyte Mg^{2+} Transport

Cell Type	K_m (mM)	V_{max} (pmol/min/10^7 cells)
S49 lymphoma cells ($n = 3$)		
Wild-type	0.25	200
Mg^{2+}-deficient	0.07	190
Splenic T-lymphocytes ($n = 1$)		
Control	0.30	300
Mg^{2+}-deficient	0.10	150–200

apparent K_m of the transporter for Mg^{2+} was decreased approximately three-fold, implying that the cells were transporting Mg^{2+} more efficiently. The mechanism of this effect is unclear, but could involve either a change in physical properties of the transporter, due perhaps to membrane fluidity alterations that might accompany deficiency or of a different basal regulatory status of the transporter. In general, however, the Mg^{2+} transport system of the Mg^{2+}-deficient cells is sufficiently active to provide Mg^{2+} when the cation is present.

In normal S49 lymphoma cells we have previously compared the intracellular distribution of Mg^{2+} and Ca^{2+} using digitonin permeabilization.[30] These data may be summarized as follows. In actively growing or growth-arrested cells, Ca^{2+} is freely exchangeable, in that 100% of intracellular Ca^{2+} can exchange with medium Ca^{2+} within about 4 hours. In contrast, in actively growing cells, Mg^{2+} exchange is limited, with only about 65% of intracellular Mg^{2+} exchanging even after 44 hours. This degree of exchange is, within experimental error, equivalent to the amount of Mg^{2+} that would be expected to be taken up simply because of the increase in cell volume caused by

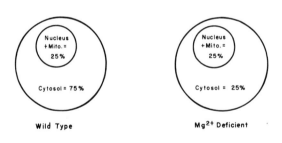

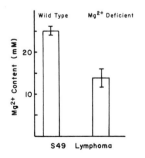

FIGURE 3. Schematic diagram of compartmentation analysis of normal and Mg^{2+}-deficient S49 cells and T-lymphocytes as percentage of wild-type Mg^{2+} content. Total Mg^{2+} was measured by atomic absorption. Additional details are presented in the text.

growth and division. Compartmentation of Mg^{2+} is most clearly shown in growth-arrested cells, in which only about 3% of total cell Mg^{2+} is exchangeable. This exchange reaches equilibrium within the first 2 hours of incubation and remains unchanged for at least 44 hours. These data imply that intracellular Mg^{2+} and Ca^{2+} are regulated quite differently.[30] The total cell Mg^{2+} in the S49 cell is equivalent to about 25 mM, of which 75% is cytosolic and 25% is nuclear plus mitochondrial as assessed by digitonin permeabilization. In the Mg^{2+}-deficient S49 cells, total intracellular Mg^{2+} is decreased about 50%, but all of this decrease occurs in the cytosolic fraction since the nuclear/mitochondrial fraction has approximately the same absolute Mg^{2+} content as Mg^{2+}-replete cells as assessed by digitonin permeabilization data (schematically presented in FIGURE 3). Ca^{2+} content of Mg^{2+}-deficient cells is approximately normal. Exchange of Mg^{2+} assessed with $^{28}Mg^{2+}$ indicates that less

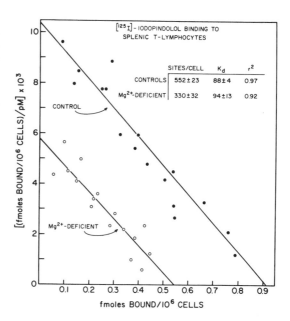

FIGURE 4. Beta-adrenergic receptor density in T-lymphocytes from control and Mg^{2+}-deficient rats. Receptor density was measured in intact cells using [^{125}I]iodohydroxybenzylpindolol ([^{125}I]-IHYP). IHYP is a relative hydrophobic receptor probe which measures total cell receptor whether used with intact or broken cells. Receptor assay and synthesis of the ligand were as previously described.[20,23] The data are a composite of experiments with lymphocytes from five pairs of control and Mg^{2+}-deficient rats. In all cases the control lymphocytes had approximately twice as many beta-receptors as did Mg^{2+}-deficient lymphocytes, although the receptor density varied slightly between lymphocyte preparations.

than 5% of total intracellular Mg^{2+} is exchangeable with extracellular Mg^{2+}, a value similar to that of control cells. Thus Mg^{2+}-deficient cells still exhibit the extreme Mg^{2+} compartmentation of normal cells.

A major difference in the Mg^{2+}-deficient cells is a deficit in their ability to respond to hormonal stimulation. Mg^{2+}-deficient cells, whether T-lymphocytes (FIG. 4) or S49 lymphoma cells (data not shown), have about 50% of the wild-type level of beta-adrenergic receptors, measured either on a per cell or protein basis. Antagonist affinity is unchanged, confirming a true density decrease. Such a decrease in receptor density might be expected to be accompanied by a decrease in effect of agonist stimulation. However, the decrease in cyclic AMP accumulation seen upon isoproterenol addition is striking (FIG. 5), much greater than the 50% decrease in receptor density. In wild-type S49 cells, cyclic AMP is stimulated almost 100-fold upon isoproterenol addition

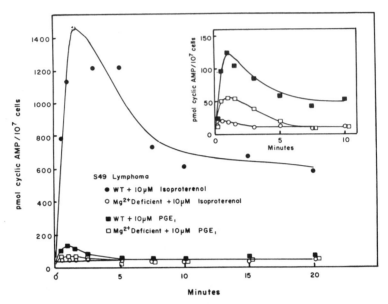

FIGURE 5. Effect of isoproterenol and PGE_1 on cyclic AMP accumulation in control and Mg^{2+}-deficient intact S49 lymphoma cells. Agonist was added at a concentration 10-fold higher than maximal for control cells (see also FIGURE 6). Cyclic AMP levels shown are total amounts of cyclic AMP (intracellular plus extracellular) to obviate possible effects of Mg^{2+}-deficiency on cyclic AMP transport. The assay medium also contained 0.1 mM RO20-1724 to inhibit phosphodiesterase.

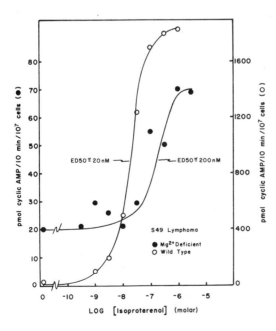

FIGURE 6. Isoproterenol dose-response curve in control and Mg^{2+}-deficient S49 lymphoma cells. Cyclic AMP was assayed after a 3-min incubation (see also FIGURE 5).

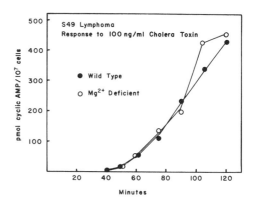

FIGURE 7. Effect of cholera toxin on cyclic AMP accumulation in control and Mg^{2+}-deficient S49 lymphoma cells. Cholera toxin was added at a supramaximal concentration of 100 ng/ml, and cell aliquots were taken at the indicated time for assay of cyclic AMP.

whereas in the Mg^{2+}-deficient cells only a three-fold stimulation is seen. Mg^{2+}-deficient cells also show additional defects in beta-receptor-mediated responses. Intracellular cyclic AMP levels return to basal levels much faster in Mg^{2+}-deficient cells. Resting levels of cyclic AMP are decreased from about 30 pmol/10^7 cells in wild-type versus 10–15 pmol/10^7 cells in the Mg^{2+}-deficient clone. In addition, there is a right shift of approximately 10-fold in ED50 for isoproterenol (FIG. 6). Finally, beta-receptor-mediated inhibition of Mg^{2+} influx is absent in Mg^{2+}-deficient cells (data not shown).

Since the decrease in beta-receptor-mediated responses is quantitatively much greater than the decrease in receptor number, the most obvious hypothesis is that coupling between a G-protein and its effector system might also be defective and contributing to the quantitative discrepancy. Two pieces of data argue against this conclusion, however. First, although the wild-type cells respond less vigorously to PGE_1 than to isoproterenol, PGE_1 response in Mg^{2+}-deficient cells is reduced only about 50% compared to that in wild-type cells, whereas the much more robust beta-receptor

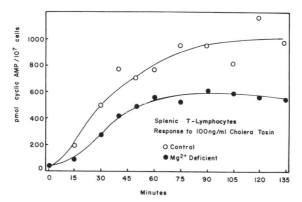

FIGURE 8. Effect of cholera toxin on cyclic AMP accumulation in T-lymphocytes from control and Mg^{2+}-deficient rats. Cholera toxin was added at a supramaximal concentration of 100 ng/ml and cell aliquots were taken at the indicated time for assay of cyclic AMP.

response is almost abolished (FIG. 5). Second, the cholera toxin response in Mg^{2+}-deficient cells of either S49 lymphoma (FIG. 7) or T-lymphocytes (FIG. 8) is not significantly diminished compared to that of control cells. This argues that G-protein–adenylate cyclase interaction is essentially intact and quantitatively similar in normal versus Mg^{2+}-deficient cells. Thus the deficit appears to lie in the interaction between the beta-receptor and in intracellular effector systems, presumably one or more G-proteins.

The lack of hormonal responsiveness in Mg^{2+}-deficiency is not limited to lymphocytes. In preliminary experiments with isolated ventricular myocytes, we have shown that phosphorylation of any of several proteins is less sensitive to beta-adrenergic agonist from Mg^{2+}-deficient animals than from controls (FIG. 9). This suggests that defective hormonal responsiveness may be widespread in Mg^{2+} deficiency.

Finally, in several experiments with Mg^{2+}-deficient S49 cells, we have been unable to reverse any of the effects of Mg^{2+}-deficiency on hormone response by (1) performing the experiments in high extracellular Mg^{2+} concentrations, (2) acutely growing the Mg^{2+}-deficient cells in normal Mg^{2+} medium for 2 hours before assay, or (3) growing

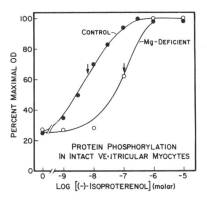

FIGURE 9. Isoproterenol dose-response curve in ventricular myocytes from control and Mg^{2+}-deficient rats. Myocytes were isolated and the phosphorylation assay was performed as described by Onorato and Rudolph.[47] The data shown are for total incorporation of ^{32}P into protein, but similar data were obtained using phosphorylation of myosin light chain kinase as determined from densitometry of SDS-polyacrylamide gels.

the Mg^{2+}-deficient cells in normal Mg^{2+} medium for 2 weeks before assay (data not shown). Thus the altered phenotype evidenced by the isolated Mg^{2+}-deficient S49 clones is quite stable.

DISCUSSION

As a primary disease, Mg^{2+}-deficiency is rare; however, it is increasingly recognized as an accompanying factor in a wide variety of disease states or as the result of drug-induced wasting of Mg^{2+}.[39] In addition, Mg^{2+}-deficiency is recognized as a contributing factor to tumorigenicity in a number of systems. For example, Hass and colleagues have shown that Mg^{2+}-deficiency leads to a 20–25% incidence of lymphoma in rodents. No other tumor type is increased in frequency and Mg^{2+}-deficiency further protects against tumors induced by various chemical carcinogens.[2,4] Lichtman and colleagues[1,2] have shown that Mg^{2+}-deficient lymphocytes in culture become far more tumorigenic than the original Mg^{2+}-replete lymphoma they were derived from. Finally, postulated roles for Mg^{2+} in the cell cycle and cell division lend further support to the concept that Mg^{2+} may play an important role in cell proliferation.[40–42]

In this laboratory it was discovered several years ago[27] that Mg^{2+} influx but not Mg^{2+} efflux is hormonally regulated and that this regulation is independent of hormonal regulation of other cation fluxes. Indeed, the beta-adrenergic-receptor-modulated inhibition of Mg^{2+} influx is not mediated through cyclic AMP, but appears to be due to interaction of the beta-receptor and a G-protein independent of adenylate cyclase[27,29] We have since shown that a wide variety of cell types and hormone receptors are involved in regulation of Mg^{2+} flux and that intracellular compartmentation of Mg^{2+} is radically different than that of Ca^{2+}.[30] These data led us to hypothesize that Mg^{2+} is a chronic regulatory agent that determines the degree to which other "acute" regulatory agents like Ca^{2+} and cyclic AMP are able to act.[32,33]

In order to investigate any potential regulatory role for Mg^{2+} in cell proliferation, it became necessary to develop new tools with which to study Mg^{2+} "metabolism." These have included isolation of Mg^{2+} transport genes as a prelude to study of the mechanism of Mg^{2+} transport.[34,35] Reported in this paper are our preliminary results in two other areas: development of cell-permeable Mg^{2+} dyes and the initial characterization of a model cell system for study of Mg^{2+}-deficiency.

Tropolone binds Mg^{2+} better than Ca^{2+} and with an affinity of Mg^{2+} close to the expected intracellular free Mg^{2+} concentration. We have synthesized more than 20 derivatives of tropolone similar to the structure shown as TROP-5 (FIG. 1A). These dyes have the desired properties of a Mg^{2+} dye in that they are insensitive to Ca^{2+} and relatively insensitive to H^+ in the intracellular physiological range. Moreover, above approximately 50 mM monovalent cation, the dyes are totally insensitive to changes in Na^+ and/or K^+ concentrations. We estimate from the data shown in FIGURE 2 that we can detect changes of 5–10 μM Mg^{2+} against a background of 200 μM Mg^{2+} and 200 μM Ca^{2+} in a buffer containing 120 mM K^+ and 20 mM Na^+. Additional dyes recently synthesized should increase this sensitivity several-fold and increase the absolute absorbance and fluorescence by up to an order of magnitude. Preliminary data with the acetoxymethyl ester of TROP-5 indicate that, unlike the Ca^{2+} dyes Quin II and Fura 2, this ester is not rapidly hydrolyzed. However, experiments with the propyloxymethyl and higher ester derivatives show rapid hydrolysis. Thus, our preliminary data indicate that we have several excellent candidates for a functional intracellular Mg^{2+} indicator.

The S49 lymphoma cell line was chosen for development of a model of Mg^{2+}-deficiency because of extensive experience in this and many other laboratories with this cell line and because of the Mg^{2+}-sensitivity of several of its cellular processes, such as activation of adenylate cyclase. In addition, Mg^{2+} "metabolism" is better understood in the S49 line than in any other eukaryotic cell.[27–33]

Mg^{2+}-deficiency initially affects growth rate, but this can be overcome in some cells in some as yet unknown manner. Initially isolated clones grew with a doubling time approximately three times that of the wild-type cell, but subclones of the initial isolates grow with doubling times of 22 hours, where is only slightly longer than the 17-hour doubling time of the wild-type cell. We have not yet detected any other differences in these two types of Mg^{2+}-deficient clones; they have the same Mg^{2+} content, transport Mg^{2+} at the same rate, and lack hormonal responsiveness, particularly to beta-adrenergic stimulation. Presumably, the cell has compensated at some point distal to the immediate actions of Mg^{2+} at the transport and hormonal response level. The growth and hormonal response characteristics of the Mg^{2+}-deficient cells appear to be quite stable characteristics. None of the results reported herein can be reversed by acute (2 hours) or chronic (up to 14 days) Mg^{2+}-repletion in the growth medium. It is not known whether the altered characteristics of the Mg^{2+}-deficient clones are due to altered normal gene expression of Mg^{2+}-deficient clones or due to one or more mutations. The stability of the phenotype and previous isolation of other response mutants of the S49 cell line by similar phenotypic stress[32] argues, however, that the Mg^{2+}-deficient clones are able to survive Mg^{2+}-deficiency due to mutation rather than altered normal gene expression.

Of most initial interest is the extreme deficiency observed in hormone response in the Mg^{2+}-deficient clones. Beta-adrenergic receptor density is decreased approximately 50%. Previous work by Bourne and colleagues[44] has shown that primary beta-receptor-deficiency in the S49 cell leads to an approximately linear decrease in response with the decrease in receptor density. Thus, a 50% receptor-deficiency should lead to a corresponding 50% decrease in cyclic AMP synthesis and accumulation. The data indicate, however, that the wild-type cell response is decreased by more than 95% in the Mg^{2+}-deficient cell. This might be explained by a deficiency in G-protein and/or adenylate cyclase; however, cyclic AMP accumulation in response to cholera toxin is at best only slightly decreased, a result clearly insufficient to explain the loss of the beta-receptor-mediated response. Perhaps even more puzzling is the comparatively modest 50% decrease in response to PGE^1, suggesting some degree of specificity for the loss of beta-receptor-mediated response. These results cannot be explained by any simple pattern of deficient coupling between receptor, G-protein, and adenylate cyclase. We are currently investigating whether specific subclasses of G-proteins, known to exist in the S49 cell,[45] may be altered by Mg^{2+}-deficiency.

We have fewer data on the effect of Mg^{2+}-deficiency on Mg^{2+} "metabolism." Mg^{2+} transport is not remarkably changed, the only major effect being a modest increase in affinity of the transport system for Mg^{2+}. However, beta-adrenergic inhibition of Mg^{2+} influx is no longer seen. The more robust effect of phorbol ester to stimulate Mg^{2+} influx[46] has not yet been tested in these cells. Mg^{2+} content is decreased about 50% in the Mg^{2+}-deficient clones, but Ca^{2+} content is not demonstrably changed. Moreover, the compartmentation pattern observed in the wild-type S49 cell is not changed for either Mg^{2+} or Ca^{2+}. Given that the wild-type cell exhibits extremely slow exchange of its Mg^{2+} content, this is not particularly surprising for Mg^{2+} since the deficient cell would be expected to hold on as tightly as possible to its remaining Mg^{2+}. It is of interest that the entire decrease in intracellular Mg^{2+} appears to be cytosolic with the presumed mitochondrial and nuclear content of Mg^{2+} remaining approximately constant. Such a severe deficiency in the cytosol suggests that the Mg^{2+}-deficient cell may have a severe metabolic derangement since so many critical metabolic enzymes require not only MgATP but also free Mg^{2+}. These data further suggest that even mild Mg^{2+} deficiency may effect cell metabolism out of proportion to the degree of deficiency if the deficiency occurs within a single intracellular compartment.

Finally, our data indicate that Mg^{2+}-deficiency in a cultured cell line mimics the results of Mg^{2+}-deficiency in cells isolated from Mg^{2+}-deficient rats. The hormonal response of T-lymphocytes from Mg^{2+}-deficient rats exhibits the same decrease as that exhibited by the Mg^{2+}-deficient S49 clones. Moreover, protein phosphorylation in ventricular myocytes from Mg^{2+}-deficient rats also exhibits a deficiency in hormone response as manifested in a right shift in the dose-response curve. Thus Mg^{2+}-deficiency has widespread effects in an organism.

While the data presented in this paper do not establish a definitive role for Mg^{2+} in cell regulation or proliferation, they nevertheless provide provocative evidence that Mg^{2+} has numerous roles within the cell and that further study of intracellular free Mg^{2+} and of the cell's Mg^{2+} "metabolism" will be worthwhile. It will be of particular interest to examine the effects of growth and other hormonal stimulation on free Mg^{2+} in these and other cell types and to determine whether the rather drawn out and chronic nature of alterations in Mg^{2+} metabolism elicited by various hormones supports our hypothesis of a regulatory role for intracellular free Mg^{2+}.

SUMMARY

Although studies in mammalian cells and yeast suggest that Mg^{2+} plays an important role in cell growth and hormone response, intracellular roles of Mg^{2+} are poorly understood. Thus, we are developing methods to study Mg^{2+} regulation of growth and hormonal response. Preliminary data using cell-permeable Mg^{2+} indicators based on tropolone suggest the feasibility of the dynamic and selective determination of intracellular free Mg^{2+} concentration. "Mg^{2+}-deficient" cell lines have also been developed. Murine S49 lymphoma cells in normal 0.8 mM Mg^{2+} medium double in 17 hours, but die when placed in 0.2 mM Mg^{2+} medium. Two classes of S49 clones have been isolated which grow in 30 μM Mg^{2+} with doubling times of 22 and 60 hours. Although total cell Mg^{2+} is decreased by 50%, the decrease is selective since cytoplasmic Mg^{2+} is decreased 75% while particulate Mg^{2+} is unchanged. Hormonal response in the Mg^{2+}-deficient cells is defective. Cyclic AMP accumulation in response to beta-adrenergic receptor activation is decreased more than 95%. In contrast, the Mg^{2+}-deficient cells lose only about 50% of their response to PGE_1 receptor activation, retain 50% of their beta-receptors, and accumulate cyclic AMP in response to cholera toxin at the wild-type rate. Mg^{2+} transport also occurs at the wild-type rate, but with a slightly higher affinity and is no longer hormone-sensitive. Ca^{2+} content is normal or slightly high. T-lymphocytes isolated from rats made Mg^{2+}-deficient for 8 weeks give similar results, indicating that the Mg^{2+}-deficient S49 lymphoma cell clones are a good model for Mg^{2+}-deficiency. The data suggest that lack of Mg^{2+} causes growth abnormalities and leads to markedly altered receptor-G-protein coupling, but may have less effect on G-protein-adenylate cylase interaction.

ACKNOWLEDGMENTS

I wish to thank Margaret Paskevich and Paula Jacobs for expert technical assistance and Dr. Robert Grubbs for advice and assistance with the Mg^{2+} transport experiments. Nina Klein was of invaluable help with the Mg^{2+}-deficient rats. Dr. Stephen Rudolph provided assistance on the myocyte phosphorylation experiment.

REFERENCES

1. BRENNAN, J. K., M. MANSKY, G. ROBERTS & M. A. LICHTMAN. 1975. In Vitro 11: 354–360.
2. BRENNAN, J. K. & M. A. LICHTMAN. 1973. J. Cell. Physiol. 82: 101–112.
3. MCCREARY, P. A., H. BATTIFORA, B. M. HAHNEMAN, G. H. LAING & G. M. HASS. 1967. Blood 29: 683–690.
4. HASS, G. M., G. H. LAING, R. M. GALT & P. A. MCCREARY. 1981. Magnesium Bull. 3: 5–10.
5. MCKEEHAN, W. L. & R. G. HAM. 1978. Nature 275: 756–758.
6. GUYNN, R. W., L. T. WEBSTER, JR. and R. L. VEECH. 1974. J. Biol. Chem. 249: 3248–3254.
7. SULLIVAN, J. M. & J. B. ALPERS. 1971. J. Biol. Chem. 246: 3057–3063.
8. KING, M. M. & G. M. CARLSON. 1981. J. Biol. Chem. 256: 11058–11064.
9. SWITZER, R. L., 1971. J. Biol. Chem. 246: 2447–2458.
10. PECK, E. J., JR. & W. L. RAY, JR. 1971. J. Biol. Chem. 246: 1160–1167.
11. FLATMAN, P. W. & V. L. LEW. 1980. J. Physiol. 307: 1–8.
12. KURIKI, Y., J. HALSEY, R. L. BILTONEN & E. RACKER. 1976. Biochemistry 15: 4956–4961.
13. EPSTEIN, M., Y. KURIKI, R. L. BILTONEN & E. RACKER. 1976. Biochemistry 19: 5564–5568.
14. JONES, L. R. 1979. Biochem. Biophys. Acta 557: 230–242.
15. MARTONOSI, A. 1969. J. Biol. Chem. 244: 613–620.
16. ADOLFSEN, R. & E. N. MOUDRIANAKIS. 1978. J. Biol. Chem. 253: 4380–4388.
17. ASKARI, A., W. H. HUANG & P. W. MCCORMICK. 1983. J. Biol. Chem. 258: 3453–3460.
18. SENIOR, A. E. 1979. J. Biol. Chem. 254: 11319–11322.
19. SKULACHEV, V. P. 1981. In Chemiosmotic Proton Circuits in Biological Membranes. V. P. Skulachev & P. C. Hinkle, Eds.: 3–48. Addison-Wesley. Reading, MA.
20. BIRD, S. J. & M. E. MAGUIRE. 1978. J. Biol. Chem. 253: 8826–8834.
21. BOCKAERT, J., B. CANTAU & M. SEBBEN-PEREZ. 1984. Mol. Pharmacol. 26: 180–186.
22. CECH, S. Y., W. C. BROADDUS & M. E. MAGUIRE. 1980. Mol. Cell Biochem. 33: 67–92.
23. CECH, S. Y. & M. E. MAGUIRE. 1982. Mol. Pharmacol. 22: 267–273.
24. IYENGAR, R. & L. BIRNBAUMER. 1982. Proc. Natl. Acad. Sci. USA 79: 5179–5183.
25. GARFINKEL, L. & D. GARFINKEL. 1985. Magnesium 4: 60–72.
26. ACHS, M. J., D. GARFINKEL & L. P. OPIE. 1982. Am. J. Physiol. 243: R389–R399.
27. MAGUIRE, M. E. & J. J. ERDOS. 1980. J. Biol. Chem. 255: 1030–1035.
28. ERDOS, J. J. & M. E. MAGUIRE. 1980. Mol. Pharmacol. 18: 379–383.
29. ERDOS, J. J. & M. E. MAGUIRE. 1983. J. Physiol. 337: 351–371.
30. GRUBBS, R. D., S. D. COLLINS & M. E. MAGUIRE. 1985. J. Biol. Chem. 259: 12184–12192.
31. GRUBBS, R. D., C. A. WETHERILL, K. KUTSCHKE & M. E. MAGUIRE. 1984. Am. J. Physiol. 248: C51–C57.
32. MAGUIRE, M. E. 1984. Trends Pharmacol. Sci. 5: 73–77.
33. GRUBBS, R. D. & M. E. MAGUIRE. 1987. Magnesium 6: 113–127.
34. HMIEL, S. P., M. D. SNAVELY, C. G. MILLER & M. E. MAGUIRE. 1986. J. Bacteriol. 168: 1444–1450.
35. SNAVELY, M. D., C. G. MILLER & M. E. MAGUIRE. Submitted for publication.
36. TUFTS, E. V. & D. M. GREENBERG. 1938. J. Biol. Chem. 122: 693–714.
37. CHUTKOW, J. G. 1965. J. Lab. Clin. Med. 75: 912–926.
38. LANDIN, W. E., F. M. KENDALL & M. F. TANSY. 1979. J. Pharmaceut. Sci. 68: 978–983.
39. ALTURA, B. M., H. J. DURLACH & M. SEELIG. 1987. Magnesium in Cellular Processes and Medicine. Karger. Basel.
40. FLATMAN, P. W. 1984. J. Membrane Biol. 80: 1–14.
41. WALKER, G. M. 1986. Magnesium 5: 9–23.
42. WALKER, G. M. & J. H. DUFFUS. 1983. Magnesium 2: 1–16.
43. BOURNE, H. R., G. F. CASPERSON, C. VAN DOP, M. E. ABOOD, B. B. NIEDERMAN, F. STEINBERG & N. WALKER. 1984. Adv. Cyclic Nucl. Protein Phosph. Res. 17: 199–205.

44. JOHNSON, G. L., H. R. BOURNE, M. K. GLEASON, P. COFFINO, P. A. INSEL & K. L. MELMON, 1979. Mol. Pharmacol. **15:** 16–27.
45. WOOLKALIS, M. J. & D. R. MANNING. 1987. Mol. Pharmacol. **32:** 1–6.
46. GRUBBS, R. D. & M. E. MAGUIRE. 1986. J. Biol. Chem. **261:** 12550–12554.
47. ONORATO, J. J. & S. A. RUDOLPH. 1981. J. Biol. Chem. **256:** 10697–10703.
48. ROGERS, T. A. 1961. J. Cell. Comp. Physiol. **57:** 119–121.
49. ROGERS, T. A. & P. E. MAHAN. 1959. Proc. Soc. Exp. Biol. Med. **100:** 235–239.

DISCUSSION

TONI SCARPA (*Case Western Reserve University, Cleveland, Ohio*): What are the amounts (or concentrations) of ATP in wild-type and Mg^{2+}-deficient S49 cells?

MICHAEL MAGUIRE (*Case Western Reserve University, Cleveland, Ohio*): The wild-type S49 cell has about 6 mM ATP, approximately 90% of which is bound with Mg^{2+}, according to ^{31}P-NMR measurements. The Mg^{2+}-deficient cells have slightly less total ATP, about 4.5–5 mM. Given the severe drop in cytosolic Mg^{2+} in the Mg^{2+}-deficient cells, I would expect the actual amount of Mg-ATP formed to be significantly decreased.

CLAUDE KLEE (*National Cancer Institute, Bethesda, Maryland*): The effect of Mg^{2+}-deficiency on protein phosphorylation and specifically on myosin light chain kinase is consistent with the high Mg^{2+} requirement of myosin light chain kinase *in vivo*. What is the total Mg^{2+} content in cells of animals fed Mg^{2+}-deficient diets?

MAGUIRE: It depends on the specific tissue examined. This is something rather unusual about Mg^{2+}-deficiency. On chronic Mg^{2+}-deficient diets, some tissues, such as brain and some types of muscle, show little if any decrease in Mg^{2+} content. Other tissues, such as liver and blood cells, show rapid and large decreases in Mg^{2+} content on such diets. This pattern is not altered even after 20–30 weeks on such diets. The studies by Greenberg,[36] Chutkow,[37] and Tansy[38] contain tables showing specific tissue contents, while the papers of Rogers and colleagues[48,49] contain rate data for tissue exchange of Mg^{2+} obtained using $^{28}Mg^{2+}$. This lack of exchange in some tissues emphasizes our compartmentation data, where actively growing lymphocytes are able to exchange intracellular Mg^{2+} but growth-arrested lymphocytes seem completely incapable of significant Mg^{2+} exchange with the environment.[30]

JON DURKIN (*National Research Council of Canada, Ottawa, Ontario, Canada*): First, you mentioned that S49 lymphoma cells have a high extracellular Mg^{2+}-dependency for proliferation. Have you looked at the Mg^{2+}-dependency of the adenylate cyclase minus (cyc^-) S49 variant to determine whether the Mg^{2+}-dependency of S49 cells is due to its effects on adenylate cyclase? Second, most cells have a high dependency on extracellular Mg^{2+} for proliferation which is attenuated or completely lost upon neoplastic transformation. On this basis I don't understand how a greatly reduced requirement of S49 lymphoma cells for extracellular Ca^{2+} relative to Mg^{2+} bears any relevance to the role of extracellular Ca^{2+} for normal cell growth.

MAGUIRE: We have not looked specifically at cyc^-; however, in wild-type S49 cells and in the Mg^{2+}-deficient clones, acute changes in extracellular Mg^{2+} concentration between 0.05 and 20 mM for up to 2 hours have no effect on basal cyclic AMP levels or hormonal stimulation of adenylate cyclase as measured by cyclic AMP accumulation. In the Mg^{2+}-deficient clones, repletion of Mg^{2+} in the growth medium for periods of up to 14 days also has no effect on cyclic AMP metabolism. Further, I wouldn't expect such a dependence in S49 cells precisely because the Mg^{2+}-dependence (dose-response curve) of the adenylate cyclase is shifted so dramatically by hormone in this cell. In the basal state, the half-maximal concentration of free Mg^{2+} required for activation is 2–3 mM. In the hormonally activated state, that concentration is shifted by at least 2 orders of magnitude, into the low micromolar range.[20-23] Surely, even the Mg^{2+}-deficient cells have this level of free Mg^{2+}.

With regard to Mg^{2+}- versus Ca^{2+}-dependency, first, I'm not arguing that Ca^{2+} has nothing to due with cell proliferation, only that Mg^{2+} is also important. Second, I do *not* know that extracellular Ca^{2+} is not required for cell proliferation in the S49 cell. What we have done is only to show that by eliminating *added* Ca^{2+}, S49 cell growth is unaltered; by not adding Ca^{2+}, we obviously still have not eliminated the inevitable contaminant Ca^{2+} present in the reagents. This is normally about 20 μM (measured by atomic absorption) in our media. Thus we haven't made any determination of the actual Ca^{2+}-dependence for growth and proliferation. Regardless, the data do show a remarkably different effect of Mg^{2+} versus Ca^{2+}. In essence, reducing extracellular Mg^{2+} concentration four-fold to 0.2 mM halts proliferation, while reducing extracellular Ca^{2+} 90-fold (1.8 mM to 20 μM) has no effect. I also draw your attention to the paper by McKeehan and Ham[5] which shows a specific loss of a Mg^{2+} but not a Ca^{2+} requirement upon SV-40 transformation of human lung fibroblasts.

PATRICK RILEY (*University College, London, England*): In view of the data that you cited for *Schizosaccharomyces pombe* by Walker and Duffus,[41,42] have you considered the possibility that Mg^{2+} is required only in relation to a specific process in the cell cycle, such as cytokinesis?

MAGUIRE We've certainly considered such possibilities from the point of view of potential routes for cloning Mg^{2+} transport genes. For example, given the data of Walker and Duffus,[41,42] one would certainly expect a Mg^{2+} transport gene to show up as a *cdc* mutant in *Saccharomyces, Schizosaccharomyces,* and other yeast species. The approach we're actually taking is somewhat simpler, however, and that is to use a strain of *S. typhimurium* that we've constructed in which all three Mg^{2+} transport genes have been deleted. The strain requires 100 mM Mg^{2+} for optimal growth, whereas the wild-type strain grows on less than 10 μM Mg^{2+}. I think we have a good chance of picking up a yeast Mg^{2+} transport gene by seeing whether something from a yeast genomic library enables growth of the Mg^{2+}-dependent *S. typhimurium* strain on low Mg^{2+}. Alternatively, we could screen selected available yeast *cdc* mutants to see whether high concentrations of extracellular Mg^{2+} enable cell cycle progression. If either approach works, then we have a way to approach your question.

BEN SZWERGOLD (*Fox Chase Cancer Center, Philadelphia, Pennsylvania*): You mentioned that some of the Mg^{2+}-deficient cells are quite fragile. How do you account for this fragility? Are there any differences in lipid profiles?

MAGUIRE: We haven't looked at lipid profiles yet, although we hope to. The cells are not any more fragile than some of the other S49 variants such as the cyc^- mutant. I expect that Mg^{2+}-deficiency causes quite pleiotropic effects, including some degree of weakening of the membrane structure, simply because there's not enough Mg^{2+} present to form the appropriate intermolecular ionic bonds. It is fairly common in my experience for any cell to become fragile when either extracellular Mg^{2+} or Ca^{2+} is chelated, regardless of whether the other cation is present.

G. NERI (*Università D'Annunzio, Chieti, Italy*): Do you think that, by lowering Mg^{2+} concentration in the S49 cells, you are forcing a few cells to adapt to such new conditions or rather that you are selecting cells that are already adapted to growth in low Mg^{2+}?

MAGUIRE: Essentially, you're asking whether the Mg^{2+}-deficient cells are real mutants or whether they have simply phenotypically adapted to be able to grow in low Mg^{2+}. The answer is that we haven't asked, by formal fluctuation analysis, whether they're mutants or not. By analogy with the many other S49 mutants, most of which were isolated by similar phenotypic stress selection procedures.[23,43,44] I would expect our clones to be actual mutants. This is supported, but hardly proved, by the observation that growing the Mg^{2+}-deficient clones for 14 days in medium with normal Mg^{2+} concentration does not alter their phenotype.

Activation of the Inositol Phospholipid Signaling System by Receptors for Extracellular ATP in Human Neutrophils, Monocytes, and Neutrophil/Monocyte Progenitor Cells

GEORGE R. DUBYAK,[a] DANIEL S. COWEN,[b]
AND HILLARD M. LAZARUS[c]

Departments of [a]Physiology/Biophysics,
[b]Pharmacology, and [c]Medicine, and
The Ireland Cancer Center
Case Western Reserve University
School of Medicine
Cleveland, Ohio 44106

BACKGROUND

Extracellular ATP, at micromolar/nanomolar concentrations, has been shown to induce significant functional changes in a wide variety of normal and transformed cell types (reviewed in Ref. 1). In most cases, these actions of ATP can be functionally distinguished from those elicited in response to occupation of the well-characterized A_1 and A_2 adenosine receptors. Thus, a growing body of data suggests the existence of specific cell surface receptors for extracellular ATP. While ATP is present in millimolar concentrations in the cytosol of all eukaryotic cell types, extracellular levels of the nucleotide will normally be maintained at extremely low levels by several mechanisms. First, there is minimal permeation of either ATP^{4-} or $MgATP^{2-}$ (the predominant cytosolic form) across lipid bilayers. Second, ubiquitous ecto-ATPases and ecto-phosphatases rapidly and efficiently hydrolyze extracellular nucleotides.[1] Thus, as is the case for any putative signaling agent, appreciable levels of extracellular ATP should occur only transiently and in response to specific physiological and/or pathologic conditions. Three major sources of extracellular ATP may be considered. First, ATP can be co-packaged in both adrenergic and cholinergic neurotransmitter granules and thus released during neurotransmission into synaptic spaces.[3] Second, cytosolic ATP stores can be released by sudden breakage of intact cells, as might occur

ABBREVIATIONS: fMet-Leu-Phe or FMLP: formyl methionyl leucyl phenylalanine; PAF: platelet activating factor; InsP: inositol monophosphate; $InsP_2$: inositol biphosphate; $InsP_3$: inositol triphosphate; $InsP_4$: inositol tetrakisphosphate; PMA: 4- -phorbol 12 -myristate 13-acetate; PI-PLC: phosphatidyl inositol—phospholipase C; GTP S: guanosine 5'-O-(3-thiotriphosphate); HEPES: 4-(2-hydroxyethyl)-1-piperazineethanesulfonic acid; ETGA: [ethyl-enebis)oxyethylenenitrilo)]tetraacetic acid; G-protein: guanine nucleotide regulatory protein; LTB_4: leukotriene B_4; FAB: French, American, British Classification of human leukemias.

during rupture of blood vessels and other tissue injury. Finally, ATP, which is also copackaged with serotonin in platelet granules, can be locally released in significant amounts during platelet activation.[4] These latter two sources suggest that significant amounts of extracellular ATP may locally accumulate at sites of thrombus formation and infection/inflammation. If so, the possibility may be considered that extracellular ATP can modify the function of the phagocytic cell types (neutrophils, monocytes, macrophages) characteristic of inflammatory sites. Putative ATP receptors that trigger Ca^{2+} mobilization and prostacyclin release have also been reported in certain types of endothelial cells,[5] the vessel elements with which phagocytes interact at vascular lesions.

In several cell types,[6-8] extracellular ATP can induce rapid activation of the inositol phospholipid/protein kinase C signaling cascade and thus produce significant changes in cytosolic $[Ca^{2+}]$, $[H^+]$, and inositol phosphate/diacylglycerol levels. This transmembrane signaling system has been implicated in the regulation or modulation of a broad spectrum of integrated biological responses. In phagocytic leukocytes, activation of this signaling pathway is among the earliest events triggered by chemotactic agonists and phagocytic stimuli (reviewed in Refs. 9 and 10). Considerable progress has been made in the biochemical characterization of this signaling cascade in phagocytic cells, particularly with regard to: (1) the role of a pertussis-toxin-sensitive guanine nucleotide-binding protein in receptor-phospholipase C (PLC) coupling; (2) the inhibitory effects of protein kinase C activation on receptor-PLC coupling; and (3) the roles of inositol phospholipid-derived messenger molecules (diacylglycerol, inositol phosphates, and Ca^{2+}) in triggering or modulating chemotaxis, secretion, phagocytosis, and superoxide release.

The function of the phospholipase C/ protein kinase C signaling system in phagocytic white blood cells is also of interest with respect to its possible role in myelomonocytic differentiation. In this regard, numerous studies have employed the HL60 human promyelocytic leukemia cell line (reviewed in Refs. 11 and 12). Exposure of these cells to a variety of pharmacologic and physiological agents, including phorbol esters, exogenous diacylglycerol analogs, and exogenous cyclic nucleotides, can trigger differentiation along either the neutrophil or monocyte/macrophage pathways. Significantly, this list includes agents that interact with specific elements of receptor-mediated signal transduction pathways. The results of the differentiation studies have prompted in-depth investigation of the adenylate cyclase and the protein kinase C systems in undifferentiated HL60 cells. Conversely, the effects of agents that activate the phospholipase C signaling cascade in these cells have not been reported to date. This contrasts with the numerous studies demonstrating that differentiation of HL60 cells is accompanied by enhanced expression of several classes of receptors that coupled to the inositol phospholipid signaling system; these include receptors for formylated chemotactic peptides.[11]

We report here that extracellular ATP, at nanomolar/micromolar levels, can activate the inositol phospholipid signaling cascade in both undifferentiated and differentiated HL60 cells. Extracellular ATP was also found to trigger mobilization of intracellular Ca^{2+} stores in freshly isolated human neutrophils, human monocytes, and in normal and leukemic myelomonocytic progenitor cells. These observations suggest that expression of ATP (also known as P_2-purinergic) receptors coupled to the inositol phospholipid signaling system may be a common feature of both precursor and mature human phagocytic leukocytes. In this paper, we review three major aspects of the transmembrane signaling actions elicited by these ATP receptors in human phagocytic leukocytes: (1) the basic characteristics of ATP-induced changes in cytosolic $[Ca^{2+}]$ as observed in undifferentiated and differentiated HL60 cells; (2) the expression, or lack thereof, of ATP-induced signaling functions in a broad range of human leukocytes and leukocyte precursor cells; and (3) the mechanisms underlying the activation of inositol phospholipid breakdown by these ATP receptors, with special emphasis on the regulatory roles of pertussis-toxin-sensitive and pertussis-toxin-insensitive G-proteins.

RESULTS AND DISCUSSION

Effects of Extracellular ATP on Cytosolic [Ca²⁺] in Undifferentiated and Differentiated HL60 Cells

FIGURE 1A illustrates the typical changes in cytosolic $[Ca^{2+}]$ observed in fura-2-loaded, undifferentiated HL60 cells when exposed to 3 μM extracellular ATP. In the absence of significant free extracellular $[Ca^{2+}]$, ATP induced a very rapid (complete within 5 sec) increase in cytosolic $[Ca^{2+}]$ from the basal level of approximately 100

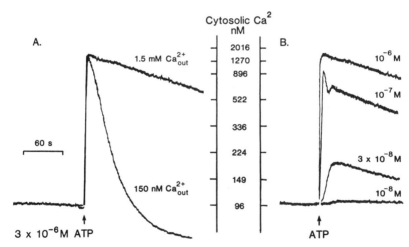

FIGURE 1. ATP-induced changes in the cytosolic $[Ca^{2+}]$ of undifferentiated HL60 human promyelocytic leukemia cells: Effects of extracellular $[Ca^{2+}]$ and [ATP]. Cytosolic $[Ca^{2+}]$ was measured in fura-2-loaded, undifferentiated HL60 cells as described in Ref. 6. Each transient was recorded from a separate aliquot of cells (3×10^5/ml) preincubated (with continuous stirring) at 37°C for 5 minutes prior to the addition of the indicated concentrations of extracellular ATP. **(A)** Ca^{2+} transients elicited by 3 μM ATP when the cells were incubated in either medium containing 1.5 mM extracellular $[CaCl_2]$ or in medium wherein the extracellular $[Ca^{2+}]$ was reduced to approximately 150 nM (by the addition of 3 mM EGTA) 5 seconds prior to ATP addition. **(B)** Ca^{2+} transients elicited by submicromolar [ATP]. In both **(A)** and **(B)** recorded transients are graphically superimposed to facilitate comparison. These ATP-induced transients were recorded from the same preparation of cells but are representative of the results obtained with more than 100 separate preparations of fura-2-loaded cells.

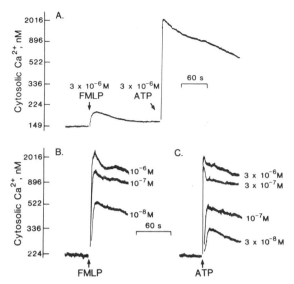

FIGURE 2. Comparative effects of ATP and chemotactic peptide on the cytosolic [Ca^{2+}] of undifferentiated and differentiated HL60 cells. A culture of undifferentiated HL60 cells was split into two 25-ml fractions (5×10^5/ml); one fraction was supplemented with 0.5 mM dibutyryl cyclic AMP, while the other was left untreated. Both fractions were then incubated under tissue culture conditions for an additional 48 hours, and the cells from each fraction were then washed and loaded with fura 2. Ca^2-dependent fura 2 fluorescence was recorded as described during incubation of cell aliquots at 37°C. These results are representative of five similar experiments using separate preparations of dibutyryl cAMP-treated cells. (**A**) An aliquot of the untreated HL60 cells was sequentially exposed to 3 μM of the chemotactic peptide, fMet-Leu-Phe (FMLP), followed by 3 μM ATP. (**B**) Separate aliquots of the treated HL60 cells were exposed to the indicated concentrations of ATP. Three recorded transients are graphically superimposed for comparison. (**C**) Additional aliquots of the treated cells were exposed to the indicated concentrations of FMLP. Five recorded transients are graphically superimposed for comparison.

nM to a peak level exceeding 1 μM; this increase was followed by a rapid (complete within 60 sec) decay to the prestimulus level. Conversely, in the presence of 1.5 mM extracellular [Ca^{2+}], the rapid, initial increase was followed by a sustained phase of elevated cytosolic [Ca^{2+}], which only gradually decreased over the next several minutes. This sustained phase, but not the initial, rapid increase, was completely blocked by inclusion of 20 μM LaCl$_3$ in the normal Ca^{2+}-containing medium (data not shown). This inhibitory effect of La^{3+} was not mimicked by specific antagonists (nifedipine, D600) of voltage-sensitive Ca^{2+} channels (data not shown). The apparent absence of voltage-sensitive Ca^{2+} channels was also suggested by the inability of elevated extracellular KCl (to 50 mM) to induce significant changes in cytosolic [Ca^{2+}] or to modify the ATP-induced transients. Thus, in HL60 cells, extracellular ATP can activate both mobilization of intracellular Ca^{2+} stores and increased influx of Ca^{2+} through voltage-insensitive channels in the plasma membrane. HL60 cells appear to be unusually sensitive to extracellular [ATP] in that the threshold concentration varies between 3 and 10 nM (FIG. 1B). Five- to seven-fold increases in peak [Ca^{2+}]

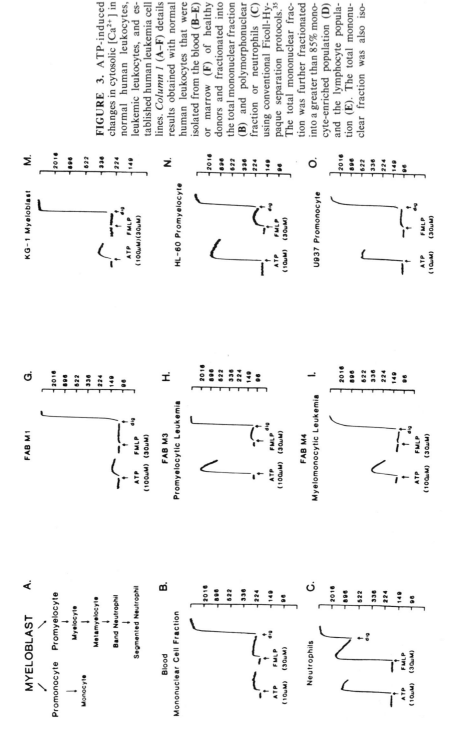

FIGURE 3. ATP-induced changes in cytosolic $[Ca^{2+}]$ in normal human leukocytes, leukemic leukocytes, and established human leukemia cell lines. *Column 1* (**A–F**) details results obtained with normal human leukocytes that were isolated from the blood (**B–E**) or marrow (**F**) of healthy donors and fractionated into the total mononuclear fraction (**B**) and polymorphonuclear fraction or neutrophils (**C**) using conventional Ficoll-Hypaque separation protocols.[35] The total mononuclear fraction was further fractionated into a greater than 85% monocyte-enriched population (**D**) and the lymphocyte population (**E**). The total mononuclear fraction was also iso-

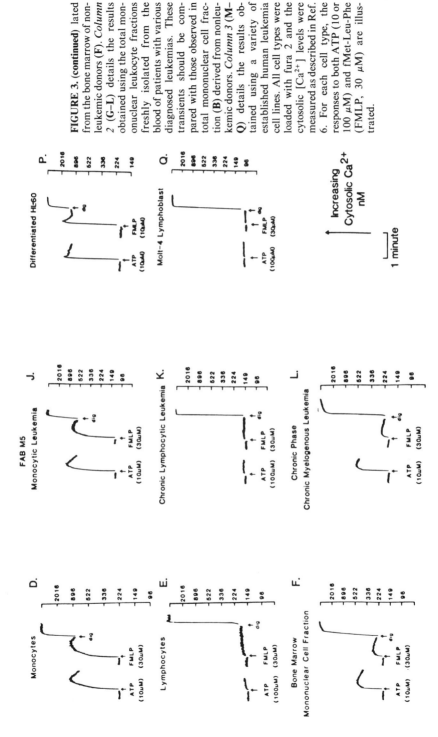

FIGURE 3. (continued) lated from the bone marrow of non-leukemic donors (**F**). *Column 2* (**G–L**) details the results obtained using the total mononuclear leukocyte fractions freshly isolated from the blood of patients with various diagnosed leukemias. These transients should be compared with those observed in total mononuclear cell fraction (**B**) derived from nonleukemic donors. *Column 3* (**M–Q**) details the results obtained using a variety of established human leukemia cell lines. All cell types were loaded with fura 2 and the cytosolic [Ca²⁺] levels were measured as described in Ref. 6. For each cell type, the responses to both ATP (10 or 100 µM) and fMet-Leu-Phe (FMLP, 30 µM) are illustrated.

were invariably triggered by 50–100 nM ATP; [ATP] > 1 μM produced maximal effects.

Previous studies have demonstrated that in uninduced HL60 cultures, the majority of cells can be morphologically classified as myeloblasts or promyelocytes; a small ($<10\%$) fraction appears as more mature myeloid cell types. In HL60 cell cultures, the expression of large numbers of receptors for formylated chemotactic peptides is greatly increased under culture conditions that induce differentiation of the cells along either the neutrophil or monocyte pathway. FIGURE 2 compares the changes in cytosolic [Ca^{2+}] triggered by either ATP or the chemotactic peptide formyl-methionyl-leucyl-phenylalanine (fMet-Leu-Phe) in undifferentiated HL60 cells and in cells differentiated by a 48-hour exposure to 0.5 mM dibutyryl cyclic AMP, a condition known to induce expression of large numbers of surface fMet-Leu-Phe receptors in $>90\%$ of the cultured cell population. In the undifferentiated cells (FIG. 2A), 3 μM fmet-Leu-Phe produced only a small (0.5-fold) increase in cytosolic [Ca^{2+}]. Conversely, 3 μM ATP triggered a 13-fold increase. While the small increase triggered by fMet-Leu-Phe could be explained by assuming that 100% of cells respond to the peptide by producing only a modest increase in [Ca^{2+}], another explanation is that a low percentage of cells respond with a large (>1 μM) peak increase in [Ca^{2+}]. Flow cytometry studies using indo-loaded HL-60 cells have confirmed the former mechanisms (see the article by Cowen *et al.* in this volume). In the dibutyryl cyclic AMP-differentiated cells, [fMet-Leu-Phe] > 100 nM triggered Ca^{2+} transients with peak changes in 1–2 μM range (FIG. 5B). The EC_{50} for fMet-Leu-Phe-induced Ca^{2+} mobilization was approximately 10 nM. Significantly, these differentiated HL60 cells also responded to extracellular ATP with Ca^{2+} transients that were indistinguishable, with regard to time courses and magnitudes, from those triggered by fMet-Leu-Phe. The dose-response relationship characterizing ATP-induced Ca^{2+} mobilization in these differentiated cells ($EC_{50} = 150$ nM) was virtually identical to that observed in the undifferentiated cells. Thus, the changes in cytosolic [Ca^{2+}] elicited by extracellular ATP are virtually identical to those triggered by a very well-characterized Ca^{2+}-mobilizing agonist (i.e., fMet-Leu-Phe) in mature phagocytic leukocytes.

Effects of Extracellular ATP on Cytosolic [Ca²⁺] in Normal and Leukemic Leukocytes

FIGURE 3 and TABLE 1 summarize our observations regarding the ability of extracellular ATP to trigger changes in cytosolic [Ca^{2+}] in a broad range of normal and leukemic human leukocytes, leukocyte progenitor cells, and established human white blood cell lines. As shown in FIGURES 3C and D, micromolar concentrations of extracellular ATP triggered significant increases in the cytosolic [Ca^{2+}] of neutrophils and monocytes. In both cell types, Ca^{2+} transients induced by ATP were equal in magnitude to those elicited by supramaximal concentrations of fMet-Leu-Phe (30 μM), platelet activating factor (90 nM) (not shown), and LTB_4 (300 nM) (not shown). In contrast, normal human lymphocytes (FIG. 3E) were totally unresponsive to either extracellular ATP or fMet-Leu-Phe. In total blood mononuclear cell fractions (FIG. 3B) composed of approximately 90% lymphocytes and 10% monocytes, ATP elicited small, 1.4-fold mean increases in cytosolic [Ca^{2+}] equal in magnitude to those elicited by fMet-Leu-Phe. Removal of monocytes (by adherence) resulted in near total

TABLE 1. Ca^{2+} Transients in Leukocytes and Leukocyte Precursors

Cell Type	Basal [Ca2+]	Peak Change in Cytosolic [Ca²⁺] Agonist			n
		100 µM ATP	10 µM ATP	30 µM FMLP	
Neutrophils	176 ± 17 nM	1729 ± 226 nM	1662 ± 74 nM	1797 ± 258 nM	(3)
Monocytes	197 ± 23 nM	1388 ± 397 nM	1136 ± 344 nM	1447 ± 254 nM	(3)
Lymphocytes	150 ± 12 nM	156 ± 9 nM	N.D.	167 ± 3 nM	(3)
Normal blood (mononuclear)	205 ± 16 nM	282 ± 17 nM	270 ± 20 nM	280 ± 14 nM	(7)
Normal mononuclear (bone marrow)	170 ± 28 nM	455 ± 75 nM	357 ± 39 nM	212 ± 22 nM	(6)
Chronic myelogenous leukemia (chronic phase)	180 ± 20 nM	1173 ± 375 nM	1062 ± 403 nM	365 ± 93 nM	(4)
KG-1 (myeloblasts)	142 ± 19 nM	265 ± 40 nM	168 ± 18 nM	142 ± 19 nM	(7)
FAB M1 (myeloblasts)	136 ± 17 nm	263 ± 97 nM	179 ± 30 nM	138 ± 19 nM	(3)
HL-60 (promyelocytes)	200 ± 20 nM	>200 nM	>2000 nM	303 ± 32 nM	(3)
Promyelocytic leukemia (FAB M3)	67 nm	1269 nM	672 nM	96 nM	(1)
U937 (promonocytes)	135 ± 5 nM	872 ± 67 nM	724 ± 98 nM	159 ± 10 nM	(4)
Myelomonocytic leukemia (FAB M4)	143 ± 8 nM	629 ± 301 nM	390 ± 222 nM	185 ± 68 nM	(2)
Monocytic leukemia (FAB M5)	123 ± 38 nM	856 ± 580 nM	817 ± 639 nM	656 ± 339 nM	(2)
B-cell chronic lymphocytic leukemia	157 ± 9 nM	166 ± 10 nM	165 ± 10 nM	164 ± 10 nM	(7)
Acute lymphocytic leukemia (FAB L2)	156 nm	176 nm	156 nM	156 nm	(1)

NOTE: Cells were loaded with fura-2 and cytosolic [Ca^{2+}] was measured as described in FIGURE 1. Each data point represents the mean ± standard error of n experiments.

elimination of both ATP- and fMet-Leu-Phe-responsive cells. This further suggested that the ATP responsiveness in mononuclear cell fractions of normal peripheral blood was restricted to the monocyte subpopulation.

Similar studies were performed with total mononuclear cell fractions isolated from the bone marrow of normal donors (FIG. 3F) and from the marrow of nonleukemic patients undergoing autologous bone marrow transplantation. Such marrow mononuclear cell preparations comprise lymphocytes, monocytes, erythroid precursors, lymphocyte precursors, monocyte precursors, and those neutrophil precursors that are less mature than band neutrophils. Extracellular ATP (10–100 μM) stimulated mean three-fold rises in cytosolic [Ca^{2+}] in contrast to the 1.2-fold mean increase induced by 3 μM fMet-Leu-Phe. Furthermore flow cytometric measurements of the cytosolic [Ca^{2+}] in single cells revealed that the percentage of ATP-responsive cells in these marrow-derived populations was greater than the percentage of fMet-Leu-Phe responsive cells (not shown). These results suggested the presence in bone marrow of leukocyte precursor cells that could be stimulated by ATP but not by fMet-Leu-Phe.

In order to determine which types of hemopoietic progenitor cells expressed this Ca^{2+}-mobilizing response to ATP, a number of established human leukemic cell lines were examined. HL-60 promyelocytes and U937 promonocytes exhibited large ATP-induced Ca^{2+} transients in contrast to the much more modest responses of fMet-Leu-Phe (FIGS. 3N and O). Flow cytometric studies (see Cowen et al. in this volume) revealed that 10 μM ATP stimulated mean 6.5-fold increases in cytosolic [Ca^{2+}] in 91 $\pm$ 1% ($n = 3$) of U937 cells, whereas only 21 $\pm$ 4% ($n = 3$) of the cells responded to 30 μM fMet-Leu-Phe, producing a 1.2-fold mean increase in the average cytosolic [Ca^{2+}]. Similarly, 94 $\pm$ 1% ($n = 3$) of HL-60 cells responded to 10 μM ATP producing greater than 10-fold mean increases in cytosolic [Ca^{2+}]. Conversely, only 69 $\pm$ 7% ($n = 3$) of these HL-60 cells responded to 30 μM fMet-Leu-Phe so as to produce a mean 1.5-fold increase in average cytosolic [Ca^{2+}]. As previously discussed, differentiated HL-60 cells responded to ATP and fMet-Leu-Phe, with equivalent large increases in [Ca^{2+}] (FIG. 3P). In contrast to these effects of ATP in HL-60 and U937 cells, cells from the less mature KG-1 myeloblastic line (FIG. 3M) exhibited only a small, 2-fold mean increase in cytosolic [Ca^{2+}] upon stimulation with 100 μM ATP; these cells were completely unresponsive to 30 μM fMet-Leu-Phe. Since KG-1 cells did contain large intracellular Ca^{2+} stores (which could be released by the Ca^{2+} ionophore ionomycin) the modest Ca^{2+} mobilization triggered by ATP suggested either: (1) that most cells responded to ATP by mobilizing only a small fraction of their ionomycin-released Ca^{2+} stores and/or (2) that only a minor subpopulation of the cells were responsive to ATP. In fact, flow cytometric measurements revealed that only 20 $\pm$ 1% ($n = 3$) of KG-1 cells responded to 100 MM ATP, and that the increases in cytosolic [Ca^{2+}] induced in those cells were small. The less mature KG-1A cell line (a variant of the KG-1 cell line)[12] was almost completely unresponsive to 100 μM ATP (as was observed with KG-1 cells, these KG-1A cells were also unresponsive to 30 μM fMet-Leu-Phe). This suggests perhaps that only a subpopulation of KG-1 cells, slightly more mature than the majority of cells in culture, had acquired responsiveness to ATP. Significantly, mobilization in response to stimulation with ATP was restricted to human leukemic cell lines exhibiting the phenotype of neutrophil/monocyte precursors. Cells from the K562 erythroblastic leukemia line (not shown), the Molt-4 T-cell lymphoblastic leukemia line (FIG. 3Q), and the HuT-78 T-cell lymphoma line (not shown) were completely unresponsive to ATP (and fMet-Leu-Phe).

To further define the particular stage of differentiation at which neutrophil and monocyte precursors acquire ATP responsiveness, we examined cells isolated from the peripheral blood of patients with various types of leukemia. The total mononuclear cell fraction of such blood samples was predominated by hemopoietic progenitor cells at specific stages of differentiation (the cell composition depending on the type of leukemia). Therefore, as with the established cell lines, large numbers of immature cells, normally found only in the bone marrow, were obtainable for study. Similar to the KG-1 myeloblastic cells, myeloblasts isolated from the blood of patients with leukemia of the French-American-British[13] classification M1 (FAB M1) responded to 100 μM ATP with small, 2-fold mean increases in cytosolic [Ca^{2+}], and were unresponsive to 30 μM fMet-Leu-Phe (FIG. 3G). Furthermore, intracellular Ca^{2+} stores mobilized by ATP represented only a small portion of total ionomycin-releasable intracellular stores. However, myeloid blasts (morphologically undifferentiated cells) isolated from the blood or bone marrow of three patients in the blast crisis[14] of chronic myelogenous leukemia exhibited differing responses to ATP. One hundred (100) μM ATP stimulated small increases in cytosolic [Ca^{2+}] of 1.1-fold and 2.2-fold, respectively, in blasts from two patients, but a large 5.8-fold rise in blasts from a third patient (not shown). This variation may have been the result of differences in the degree of maturity of cells isolated from individual patients. In fact, it has been demonstrated that meyloid blasts isolated from patients in the blast crisis of chronic myelogenous leukemia vary widely in their expression of other markers of myeloid differentiation.[15]

Normal differentiation towards neutrophils proceeds from myeloblast to promyelocyte to myelocyte to metamyelocyte to band to segmented neutrophil (FIG. 3A). At all post-myeloblast stages of differentiation, freshly isolated neutrophil progenitor cells exhibited large ATP-induced increases in [Ca^{2+}], characterized by complete mobilization of ionomycin-releasable intracellular Ca^{2+} stores. Blood mononuclear cells (containing 43% promyelocytes and 40% myeloblasts) isolated from a patient with promyelocytic leukemia (FAB-M3) responded to 100 μM ATP with a 19-fold rise in cytosolic [Ca^{2+}], but with only a 1.4-fold rise in response to fMet-Leu-Phe (FIG. 3H). Blood mononuclear cell populations comprising myeloblasts, promyelocytes, myelocytes, and metamyelocytes isolated from patients in the chronic phase[14] of chronic myelogenous leukemia responded to ATP with mean 5.5-fold increases in cytosolic [Ca^{2+}], whereas fMet-Leu-Phe induced only 2-fold mean rises (FIG. 3L). Significantly, this pattern of large ATP-induced increases in cytosolic [Ca^{2+}] and much more moderate fMet-Leu-Phe-induced [Ca^{2+}] rises is similar to that seen in mononuclear cells isolated from normal bone marrow. This was anticipated since both cell populations contain neutrophil precursors at the same stages of differentiation. Conversely, a more mature cell population composed of primarily bands and segmented neutrophils, isolated from the blood of one of these patients with chronic-phase chronic meylogenous leukemia (after removal of mononuclear cells), responded to ATP and fMet-Leu-Phe with large transients of equal magnitude (data not shown).

Alternative to granulocytic maturation, myeloblasts are capable of differentiating towards promonocytes and then to mature monocytes (FIG. 3A). Flow cytometric studies showed that 76% of the mononuclear blood cells isolated from a patient with leukemia classified as FAB-M4, containing 90% blasts (myeloblasts and promonocytes) and 10% lymphocytes, responded to ATP whereas fewer than 1% responded to fMet-Leu-Phe (see Cowen *et al.* in this volume). The magnitude of the Ca^{2+} transients elicited by ATP (FIG. 3I) in these cells (a 3-fold rise in cytosolic $[Ca^{2+}]$), and also in the U937 cells, was less than that elicited in neutrophils or HL-60 cells. This was the result of much smaller intracellular Ca^{2+} stores in the monocytic type cells (not shown). Apparently, as monocytic progenitor cells mature, the cells begin responding to ATP with large influxes of extracellular Ca^{2+} that contribute significantly to the peak heights of transients. Consequently, mononuclear blood cells, such as mature monocytes, isolated from patients with the more differentiated monocytic leukemia (FAB-M5), responded to both ATP and fMet-Leu-Phe with large Ca^{2+} transients (FIG. 3J) despite relatively modest intracellular stores of releasable Ca^{2+}.

Significantly, the ability of leukemic blood cells to mobilize Ca^{2+} in response to stimulation with ATP was restricted to cells isolated from patients with myelogenous leukemias. Blood cells isolated from patients with B-cell chronic lymphocytic leukemia (FIG. 3K and TABLE 1) and cells isolated from a patient with acute lymphocytic leukemia (TABLE 1) were unresponsive to both 100 μM ATP and 30 μM fMet-Leu-Phe. Therefore, as shown with the cell lines, responsiveness to ATP is not a general characteristic of transformed cells.

Typical of agonists whose receptors are the inositol phospholipid signaling system, ATP evoked Ca^{2+} transients resulting from both release of intracellular Ca^{2+} stores and influx of extracellular cation; this was observed in all ATP-responsive cells. Significantly, ATP stimulated larger Ca^{2+} transients in HL-60 cells, U937 cells, KG-1 cells, and KG-1a cells than were evoked in response to fMet-Leu-Phe or LTB_4 (data not shown). Similarly, HL-60 cells and U937 cells responded to ATP with Ca^{2+} transients larger in magnitude than those elicited by 90 nM PAF (data not shown).

Thus, in contrast to other activators of inositol phospholipid breakdown in mature, circulating neutrophils and monocytes, ATP was found to activate Ca^{2+} mobilization in neutrophil and monocyte precursor cells as undifferentiated as myeloblasts. The ability of such immature cells to respond to ATP raises the possibility that activation of phospholipase c by ATP may have a role in differentiation. In fact, we have observed that ATP is capable of modulating the differentiation of HL-60 and U937 cells under cell culture conditions.[37] Significantly, several types of secretory granules contain high concentrations of ATP which are released during stimulation,[1,3,4] and considerable evidence suggests that vascular endothelial cells may themselves release ATP.[16] It is tempting to speculate that marrow stromal cells may release ATP in the microenvironments of the bone marrow at concentrations sufficient to modulate normal neutrophil/monocyte differentiation.

In contrast to this highly speculative role of the ATP receptors in neutrophil and monocyte progenitor cells, a possible functional role(s) of these receptors in mature circulating phagocytes is supported by several observations. The peak magnitudes of Ca^{2+} transients elicited by maximally activating concentrations of ATP in neutrophils, monocytes and differentiated HL-60 cells were equivalent to those observed when the same cell types were stimulated with fMet-Leu-Phe, LTB_4, or PAF. Receptors for these latter three agonists have been shown to be coupled to the inositol phospholipid signaling system. These agonists are also well characterized with regard to their ability to either fully activate phagocytes (in the case of fMet-Leu-Phe) or to prime and potentiate the response of phagocytes to other inflammatory stimuli, such as immune complexes, opsonized particles, and formylated peptides. It appears likely that activation of the inositol phospholipid signaling cascade by extracellular ATP may have a similar, physiological role in modulating the integrated inflammatory responses of neutrophils and monocytes. Both Ward *et al.*[17] and Kuhns *et al.*[18] have reported that micromolar ATP primes neutrophils for enhanced superoxide release in response to stimulation with immune complexes and fMet-Leu-Phe. Moreover, *in vitro* studies by Ward *et al.*[19] have demonstrated that activated platelets (suspended at cytocrits present in blood) can release the quantities of ATP necessary to elicit this priming action. Finally, our own preliminary studies have indicated that ATP is a potent activator of azurophilic granule release in differentiated HL-60 cells.

Mechanisms Underlying the Activation of Inositol Phospholipid Breakdown by Extracellular ATP: Roles of G-Proteins and Protein Kinase C

The above findings raised the possibility that similar molecular mechanisms might mediate transmembrane signaling by both fMet-Leu-Phe receptors and the putative P_2-purinergic receptors for extracellular ATP. The general molecular mechanisms underlying the coupling between receptors and the polyphosphoinositide-specific phospholipase C (PI-PLC) have been, as yet, only partially defined. In virtually all cell types tested (reviewed in Ref. 20), incubation with tumor-promoting phorbol esters or certain exogenous diacylglycerols produces significant inhibition of both agonist-induced Ca^{2+} mobilization and inositol phospholipid breakdown. These observations have led to speculation that protein kinase C activation modulates one or more of the reactions mediating signal transduction by Ca^{2+}-mobilizing agonists. The assay of phosphoinositide hydrolysis in a wide variety of broken cell and isolated plasma membrane preparations has also indicated an obligatory role for a regulatory GTP-binding protein(s) (G-protein) in the action of most Ca^{2+}-mobilizing agonists (reviewed in Refs. 20 and 21). However, identification and characterization of the putative regulatory protein(s) has been complicated by what appears to be the heterogeneous nature of the G-proteins involved in receptor-PLC coupling.[20] In particular, measurement of agonist-induced PLC activation and/or Ca^{2+} mobilization in a wide variety of cells treated with cholera toxin or pertussis toxin has produced a spectrum of cell- and receptor-type-specific phenomena. In the majority of cell types tested, neither toxin produces significant modification of receptor/PI-PLC coupling. Conversely, pertussis toxin treatment of certain cell types, including many with hemopoietic origins,[20] has been shown to produce significant uncoupling of certain receptor classes from PI-PLC activation. Cells exhibiting this pertussis-toxin-sensitive regulation of inositol phospholipid breakdown include human neutrophils.[22]

The HL60 cell line provides a unique cell system to further characterize the role of pertussis-toxin G-proteins in mediating transmembrane signaling by P_2-purinergic receptors for extracellular ATP. The activation of phospholipase C by chemotactic receptors in human phagocytic cells is probably the best characterized of the pertussis-toxin-sensitive systems. The inhibitory effects of protein kinase C on this response in HL60 cells[23,24] are similar to those observed in human neutrophils.[25] Similarly, the role of a pertussis-toxin-sensitive GTP-binding protein activated by these receptors has been extensively studied on both differentiated HL60 cells[23,26] and in plasma membranes isolated from these cells[24,27,28]; the functional characteristics of this G-protein appear identical to those described in studies utilizing neutrophils or neutrophil membranes.[22,25,29] Furthermore, a novel, heterotrimeric GTP-binding protein, which includes a 40-kDa substrate for ribosylation by pertussis toxin, has recently been purified from both undifferentiated[30] and differentiated HL60 cells.[31] The alpha-subunit of this G-protein, which has been named both G_{HL} and G_C, can be immunologically distinguished from the alpha-subunits of either G_i or G_o.[32] Finally, on the basis of its ability to act as substrate for both pertussis toxin and cholera toxin, this HL60-derived protein appears identical to a G-protein expressed by neutrophils and monocytes.[33] It appears likely that G_{HL}/G_C functionally couples chemotactic receptors to the PI-PLC effector system. Because previous investigations have used the signaling cascade triggered by such receptors as a hallmark for comparison with the biochemical responses triggered by receptors for other chemoattractants, such as platelet-activating factor, leukotriene B_4, and complement factor C5a, we have applied a similar approach to the characterization of transmembrane signaling by extracellular ATP in both undifferentiated and differentiated HL60 cells. Particular emphasis was directed towards elucidating the roles of protein kinase-C-sensitive and pertussis-toxin-sensitive reactions on the inositol polyphosphate accumulation and Ca^{2+} mobilization triggered by extracellular ATP.

FIGURE 4 compares the Ca^{2+} transients triggered in differentiated HL-60 cells by

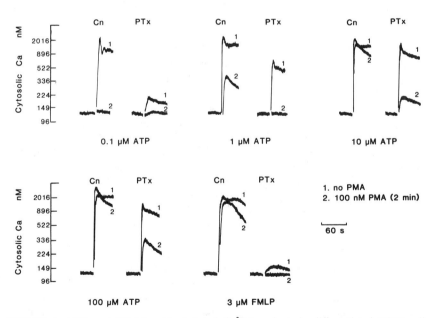

FIGURE 4. ATP- and fMet-Leu-Phe-induced Ca^{2+} transients in differentiated HL60 cells treated with pertussis toxin. Differentiation of HL60 cells was induced by incubation with 0.5 mM dibutyryl cyclic AMP as previously described. Parallel samples from a single culture of differentiated HL60 cells were preincubated for 3 hours in the absence or presence of 100 ng/ml pertussis toxin (PTx). Each cell sample was then identically processed in parallel for measurement of cytosolic $[Ca^{2+}]$. ATP-induced Ca^{2+} transients in the control (Cn) and toxin-treated cells (PTx) were recorded after no additional treatment (1) or after an additional 2-minute exposure to 100 nM PMA. (2) These data are representative of three similar experiments.

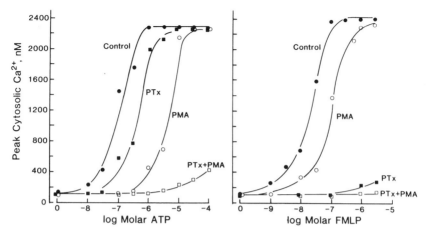

FIGURE 5. Dose-response relationships comparing the inhibitory effects of pertussis toxin treatment or phorbol ester treatment on ATP- and chemotactic peptide-induced Ca^{2+} transients in differentiated HL60 cells. Peak changes in cytosolic $[Ca^{2+}]$ were calculated from the transients illustrated in FIGURE 4. The dose-response relationships characterizing the response to ATP or fMet-Leu-Phe (FMLP) are shown for untreated cells (●), pertussis toxin (PTx)-treated cells (■), PMA-treated cells (○), and cells treated with both pertussis toxin and PMA (□). These data are representative of those obtained in three separate experiments.

several concentrations of ATP with those produced by a supramaximal concentration of fMet-Leu-Phe (3 μM). These transients were monitored in control cells, cells treated with 100 nM PMA for 3 minutes, cells treated with 100 ng/ml pertussis for 3 hours, and cells treated with both toxin and phorbol ester; the dose-response relationships describing these agonist-induced changes in cytosolic $[Ca^{2+}]$ are detailed in FIGURE 5A. As was also observed in undifferentiated cells (not shown), treatment of the differentiated HL-60 cells with either pertussis toxin alone or PMA alone significantly inhibited the Ca^{2+} mobilization triggered by submaximal amounts of ATP; in both cases, the inhibition could be overcome upon addition of higher concentrations of Ca^{2+} (EC_{50} = 0.1, 0.8 and 8 μM for control, toxin-treated, and PMA-treated cells, respectively). Treatment with both pertussis toxin and PMA resulted in >85% inhibition of the Ca^{2+} mobilization elicited by supramaximal concentration of ATP (100 μM). A different pattern of inhibitions characterized the effects of the toxin and PMA on fMet-Leu-Phe-induced change in cytosolic $[Ca^{2+}]$ (FIGS. 4 and 5B). In the PMA-treated cells, the fMet-Leu-Phe dose-response relationship was characterized by less than a single log unit shift in EC_{50} (100 nM versus the 200-nM control value). Conversely, Ca^{2+} mobilization in response to supramaximal concentrations (up to 3 μM) of fMet-Leu-Phe was inhibited by >95% in the cells treated only with pertussis toxin; no response could be elicited in cells treated with both the toxin and the phorbol ester.

Treatment of the differentiated HL60 cells with pertussis toxin also resulted in complete blockage of fMet-Leu-Phe-induced InsP$_3$ and InsP$_2$ accumulation (FIG. 6B), while producing a substantial (>70%), but still incomplete, inhibition of the InsP$_3$/InsP$_2$ accumulation triggered by maximally activating concentrations of ATP (FIG.

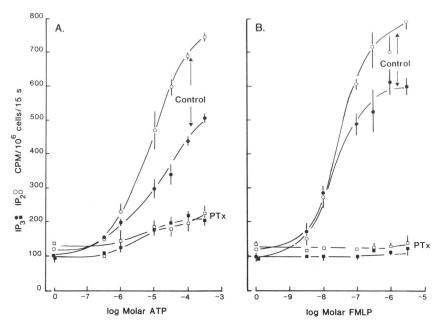

FIGURE 6. Dose-response relationships comparing ATP- and fMet-Leu-Phe-induced changes in inositol polyphosphate accumulation in control and pertussis toxin-treated HL60 cells (differentiated). HL60 cells were simultaneously labeled with [³H]inositol and differentiated by exposure to 0.5 mM dibutyryl cyclic AMP for 60 hours. Parallel samples of the [³H]inositol-labeled preparations of the differentiated HL60 cells were then incubated in the absence or presence of 100 ng/ml pertussis toxin for 3 hours. Each fraction was then washed, distributed in 0.5-ml aliquots containing 10^6 cells, and incubated at 37°C. Individual aliquots were then treated with the indicated [ATP] (**A**) or [fMet-Leu-Phe] (**B**) for 15 seconds prior to the quenching with perchloric acid. Radioactivity associated with the InsP₃ (IP₃: *solid symbols*) and InsP₂ (IP₂: *open symbols*) fractions was measured as described in Ref. 6. Data show the accumulation measured in control cells (O, ●) and the pertussis toxin (PTx)-treated cells (□, ■). Each data point represents the mean ± standard error of duplicate determinations from a single experiment. Similar data were obtained in two additional experiments.

6A). In control cells, a similar efficacy (5- to 6-fold increase in 15 sec) characterized InsP₃ accumulation in response to either ATP or fMet-Leu-Phe. Conversely, ATP (>10 μM), but not fMet-Leu-Phe (up to 3 μM), could still elicit a 2-fold increase in InsP₃ levels in the toxin-treated cells. When the time-course of ATP-induced inositol phosphate accumulation was monitored over several minutes (FIG. 7), pertussis toxin treatment produced a very significant, but invariably partial inhibition of ATP-mediated increases in InsP₃, InsP₂, and InsP. This did not appear to be due to inadequate ADP-ribosylation of target proteins by the toxin since, over the same time course, fMet-Leu-Phe-induced increases in inositol phosphate-triggered signaling were completely abolished (not shown). Moreover, the inhibition of both ATP- and fMet-Leu-Phe-induced InsP₃ accumulation by the toxin was produced over the same range of toxin concentrations (FIG. 8). The response of the cells to either agonist was half-maximally inhibited after a 3-hour exposure to 5 ng/ml of the toxin; maximal inhibition was observed in cells treated with >10 ng/ml of the toxin. Significantly, incubation with higher concentrations of the toxin did not eliminate the residual 2-fold increase in $InsP_3$ accumulation elicited by extracellular ATP (FIG. 8A).

Several possible mechanisms may underlie the residual ATP-induced activation of inositol phospholipid breakdown in pertussis-toxin-treated HL60 cells. The first possibility is based on the assumption that HL60 cells express only one G-protein, which is coupled to the PI-PLC. That is, both chemotactic peptide receptors, and the P_2-purinergic receptors would induce activation of PI-PLC only via interactions with the pertussis-toxin-sensitive G_{HL}/G_C protein.[30,31] In this case, however, ribosylation of the 40-kDa alpha-subunit would be insufficient to completely block coupling between P_2-purinergic receptors and PI-PLC. Thus, it may be possible that while ribosylation results in *de facto* prevention of the interaction of G_{HL}/G_C with chemotactic peptide

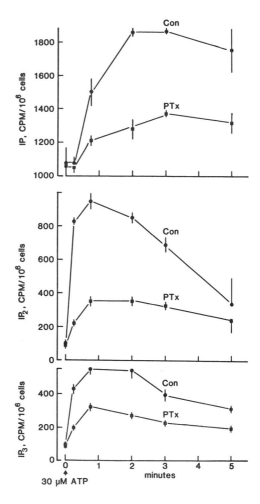

FIGURE 7. Inhibition of ATP-induced accumulation of inositol phosphates in undifferentiated HL60 cells treated with pertussis toxin. Parallel samples of a [^{3}H]inositol labeled preparation of undifferentiated HL60 cells were incubated in the absence or presence of 100 ng/ml pertussis toxin for 3 hours. Each fraction was then washed, distributed to 0.5-ml aliquots containing $1-2 \times 10^6$ cells, and incubated at 37°C. Individual aliquots of control (con, ●) or pertussis-toxin-treated cells (PTx, ■) were then treated with 30 μM ATP for the indicated times prior to the quenching with perchloric acid. Radioactivity associated with the monophosphate- (IP), biphosphate- (IP_2), and triphosphate- (IP_3) esters of inositol was measured as described in Ref. 6. Each point represents the mean ± standard error of triplicate determinations from a single experiment.

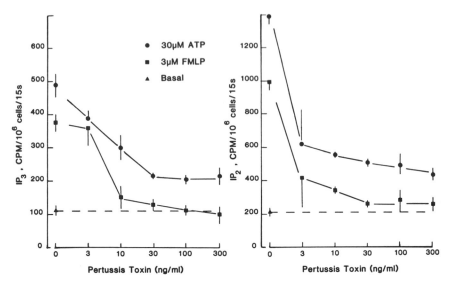

FIGURE 8. Dose-response relationship characterizing the inhibitory action of pertussis toxin on ATP- and fMet-Leu-Phe-induced inositol polyphosphate accumulation in differentiated HL60 cells. HL60 cells were differentiated and [^{3}H]inositol-labeled as described in FIGURE 6. Six parallel aliquots of the cell preparation were incubated with the indicated concentrations of pertussis toxin for 3 hours at 37°C. Each aliquot was then further subdivided into 0.5-ml aliquots containing 10^6 cells. Each of these was then treated with either 30 μM ATP ($\bullet$), 3 μM fMet-Leu-Phe ($\blacksquare$), or no agonist ($\blacktriangle$) for 15 seconds prior to perchloric acid quenching; radioactivity associated with the InsP$_3$ (IP$_3$: *left panel*) and InsP$_2$ (IP$_2$: *right panel*) fractions was determined. Data points represent the mean ± standard error of triplicate determinations from a single experiment.

receptors, it may induce only a destabilization of the interaction with P$_2$-purinergic receptors.

A second possible mechanism is based on the assumption that HL60 cells can express two distinct G-proteins, which are functionally coupled to the PI-PLC but that only one (G$_{HL}$/G$_C$) is a pertussis toxin substrate. Given this model, the existing data suggest that the chemotactic peptide receptor can only interact with G$_{HL}$/G$_C$. Conversely, the P$_2$-purinergic receptors can interact with both G$_{HL}$/G$_C$ and the other pertussis-toxin-insensitive G-protein. Results from a number of studies provide at least circumstantial support for this possibility. In hepatocytes, P$_2$-purinergic receptors have been shown to activate inositol phospholipid breakdown via a pertussis-toxin-insensitive mechanism.[7,8] Thus, there is ample evidence that P$_2$-purinergic receptors similar (on the basis of nucleotide specificity) to those expressed in HL60 cells can activate phospholipase C via pertussis-toxin-insensitive mechanisms. The question remains whether HL60 cells do indeed express pertussis-toxin-insensitive G-proteins which are coupled to phospholipase C. Uhing *et al.*[30] have recently isolated a number of GTP-binding proteins from undifferentiated HL60 cells. Although these include G$_s$ and two pertussis toxin substrates, there was a major GTP S-binding protein with a molecular weight of about 23 kDa.

Finally, the possibility should be considered that the pertussis-toxin-insensitive mobilization of intracellular Ca^{2+} stores by ATP may be mediated by some mechanism alternative to the $InsP_3$-mediated release process. The ability of GTP per se to induce Ca^{2+} release from subcellular stores has been well documented,[34] although the physiological significance of this action in intact cells remains unknown.

The combined inhibitory effect of treatment with pertussis toxin and protein kinase C activators was synergistic when Ca^{2+} mobilization was the measured variable. In either undifferentiated (not shown) or differentiated (FIG. 5A) HL60 cells treated with both agents, ATP-induced Ca^{2+} mobilization was drastically inhibited, even at the highest agonist concentrations employed. It is important to note that activation of protein kinase C itself appears to produce only a partial inhibition of both ATP- and fMet-Leu-Phe-induced signaling events (FIGS. 4 and 5). Given the similar actions of protein kinase C in partially inhibiting both ATP- and fMet-Leu-Phe-induced Ca^{2+} mobilization and $InsP_3$ accumulation (not shown) in HL60 cells, it is reasonable to consider whether the kinase is affecting (1) a signal transduction element(s) unique (e.g., the receptor moiety itself) to the respective P_2-purinergic and chemotactic peptide signaling systems or (2) a signal transduction element(s) common (e.g., the phospholipase effector, or a coupling G-protein) to both receptor-mediated signaling systems. Recent studies by Kikuchi *et al.*[23] on differentiated HL60 cells and by Smith *et al.*[25] on human neutrophils have indicated that protein kinase C activation has no significant effects on the ligand-binding properties of chemotactic peptide receptors. In contrast to the apparent lack of a protein kinase C effect on chemotactic peptide receptor–G-protein interaction in HL60 cells[23] or neutrophils,[25] there was a very significant alteration of G-protein–phospholipase C interaction in both cell types. These results indicate that the inhibitory effects of protein kinase C on the coupling of receptors to the phospholipase C effector system may be primarily mediated by a kinase-induced modification of the G-protein–phospholipase C interactions (FIG. 9

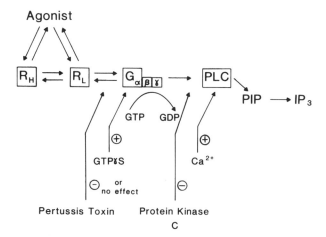

FIGURE 9. Simplified schematic view of receptor-mediated activation of phospholipase C in human phagocytic leukocytes: probable inhibitory sites of pertussis toxin and protein kinase C.

and Refs. 21 and 25). Thus, the synergistic inhibitory effects of pertussis toxin and protein kinase activation on ATP-induced signaling may involve multiple structural modifications (ATP-ribosylation and phosphorylation) of the coupling G-protein(s); these modifications may act in concert to prevent interaction of the G-protein(s) with both the ATP receptor moiety and the phospholipase C effector.

SUMMARY

We have presented evidence indicating that P_2-purinergic receptors may activate the polyphosphoinositide-phospholipase C in HL60 cells via the mediation of a pertussis-toxin-sensitive GTP-binding protein, which also mediates the actions of chemotactic peptide receptors in these and other phagocytic white blood cells. However, our data also suggest that these same receptors can be coupled to the phospholipase via an additional pertussis-toxin-insensitive mechanism. This latter finding raises the possibility that undifferentiated HL60 cells express two distinct GTP-binding proteins coupled to phospholipase C; one of these is very likely to be the G_{HL}/G_C protein recently isolated from this cell line. Significantly, the data of Oinuma et al.[31] and Falloon et al.[32] indicate that expression of the 40-kDa alpha-subunit/toxin substrate increases upon differentiation of HL60 cells along the granulocyte pathway. It would be interesting to determine whether expression of the putative pertussis-toxin-insensitive G-protein decreases with differentiation of these and other myelomonocytic progenitor cells. Such studies, which are now in progress, should be facilitated by the fact that the P_2-purinergic receptors appear to be expressed in myelopoietic cells from the promyelocytic/promonocytic stages through the terminally differentiated stages represented by circulating neutrophils and monocytes.

REFERENCES

1. GORDON, J. L. 1986. Biochem J. 233: 309–319.
2. GORDON, E. L., J. D. PEARSIN, & L. L. SLAKEY. 1986. J Biol. Chem. 261: 15496–15504.
3. HELLE, K. B., H. LAGERCRANTZ & L. STJARNE. 1971. Acta Physiol. Scand. 81: 565–567.
4. FURAMI, M. H., J. S. BAUER, G. J. STEWART, & L. SALGONICOFF 1978. J. Cell. Biol. 77: 389–399.
5. PIROTTON, S., RASPE, D. DEMOLLE, C. ERNAUX, & J-M. BOEYNAEMS. 1987. J Biol. Chem. 262: 17461–17466.
6. DUBYA, G. R., 1986. Arch Biochem. Biophys. 245: 84–95.
7. CHAREST, R., P. F. BLACKMORE & J. H. EXTON. 1985. J. Biol. Chem. 260: 15789–15794.
8. OKAJIMA, F., Y. TOKUMITSU, Y. KONDO, & M. UI. 1987. J. Biol. Chem. 262: 13483–13490.
9. VERGHESE, M. W., L. CHARLES, L. JAKOI, S. B. DILLON & R. SNYDERMAN. 1987. J. Immunol 138: 4374–4380.
10. SMITH, C. D., C. C. COX & R. SNYDERMAN. 1986. Science. 232: 97–100.
11. COLLINS, S. J. 1987. Blood 70: 1233–1244.
12. KOEFFLER, H. P. 1983. Blood 62: 709–721.
13. BENNET, J. M., D. CATOVSKY, M. T. DANIEL & G. FLANDRIN. 1986. Br. J. Haematol. 33: 451–458.
14. SOKAL, J. E., M. BACCARANI, D. RUSSO & S. TURA. 1976. Sem. Hematol. 25: 49–61.
15. GRIFFIN, J. D., R. F. TODD, III, J. RITZ, L. M. NADLER, G. P. CANNELLOS, D. ROSENTHAL,

M. GALLIVAN, R. P. BEVERIDGE, H. WEINSTEIN, D. KARP & S. F. SCHLOSSMAN. 1983. Blood **61:** 85–91.
16. BORN, G. V. R. & M. A. A. KRATZER. 1984. J. Physiol. **354:** 419–429.
17. WARD, P. A., T. W. CUNNINGHAM, K. K. MCCULLOCH & K. J. JOHNSON. 1988. Lab. Invest **58:** 438–477.
18. KUHNS, D. B., D. G. WRIGHT, J. NATH, S. S. KAPLAN & R. E. BASFORD. 1988. Lab. Invest **58:** 448–453.
19. WARD, P. A., K. K. MCCULLOCH, S. H. PHAN, J. POWELL & K. J. JOHNSON. 1988. Lab. Invest. **58:** 37–47.
20. ABDEL-LATIF, A. A. 1986. Pharmacol. Rev **38:** 228–272.
21. GILMAN, A. G. 1987. Annu. Rev. Biochem. **56:** 615–650.
22. COCKCROFT, S. 1987. Trends Biochem. Sci. **12:** 75–79.
23. KIBUCHI, A., O. KOZAWA, Y. HAMAMORI, K. KAIBUCHI & Y. TAKAI. 1986. Cancer Res. **46:** 3401–3406.
24. KIKUCHI, A., K. IKEDA, O. KOZAWA & Y. TAKAI. 1987. J. Biol. Chem. **262:** 6766–6770.
25. SMITH, C. D., R. J. UHING & R. SNYDERMAN. 1987. J. Biol. Chem. **262:** 6121–6127.
27. KIKUCHI, A., O. KOZAWA, K. KAIBUCHI, T. KATADA, M. UI & Y. TAKAI. 1986. J. Biol. Chem. **261:** 11558–11562.
28. BRANDT, S. J., R. W. DOUGHERTY, E. G. LAPETINA & J. E. NIEDEL. 1985. Proc. Natl. Acad. Sci. USA **82:** 3277–3280.
29. SMITH, C. D., R. J. UHING & R. SNYDERMAN. 1987. J. Biol. Chem. **262:** 6121–6127.
30. UHING, R. J., P. G. POLAKIS & R. SNYDERMAN. 1987. J. Biol. Chem. **262:** 15575–15579.
31. OINUMA, M., T. KATADA & M. UI. 1987. J. Biol. Chem. **262:** 8347–8353.
32. FALLOON, J., H. MALECH, G. MILLIGAN, C. UNSON, R. KAHN, P. GOLDSMITH & A. SPIEGEL. 1986. FEBS Lett. **209:** 352–356.
33. VERGHESE, M., R. J. UHING & R. SNYDERMAN. 1986. Biochem. Biophys. Res. Cummun. **138:** 887–894.
34. UEDA, T., S-H. CHUEH, M. W. NOEL & D. L. GILL. 1986. J. Biol. Chem. **261:** 3184–3192.
35. BOYUM, A. 1968. Scand J. Clin Invest. **21** (Suppl. 97):77–89.
36. COWEN, D. S., M. BERGER & G. R. DUBYAK. Submitted for publication.

DISCUSSION

C. FRANCHESCHI (*University of Modena, Italy*): Do you have an idea of the possible physiological role of the putative receptor for ATP that you described?

G. DUBYAK (*Case Western Reserve University, Cleveland, Ohio*): In neutrophils and differentiated HL60 cells, extracellular ATP (in the micromolar range) potentiates chemotactic peptide-induced superoxide release. By itself, ATP treatment does not induce superoxide release. Thus, in mature phagocytes ATP appears to act as a priming rather than an activating inflammatory agonist; in this sense, the actions of ATP are similar to those produced by leukotriene B_4 (LTB_4) and platelet-activating factor (PAF). This putative priming action of ATP at inflammatory/thrombus sites may be significant given the large amounts of ATP secreted by the activated platelets that characterize such sites.

G. DELICONSTANINOS (*University of Athens, Greece*): The leukemia lymphocytes are characterized by a high activity of ecto-ATPase as compared to normal lymphocytes. How can you correlate the extracellular levels of ATP you used with the increment of intracellular Ca^{2+}? Can you attribute a physiological role of ecto-ATPase in lymphocytes?

DUBYAK: Ecto-ATPases have at least two roles: first, the rapid degradation of released ATP so as to deactivate ATP-induced signaling actions, and second, the degradation of ATP/ADP as an initial step in the production of extracellular *adenosine,* which also can interact with cell surface receptors present in certain cell types. The observation that extracellular ATP did not elicit Ca^{2+} transients in CLL-lymphocytes, which do express high levels of ecto-ATPase, suggests that the putative ATP receptor is distinct from that of the ecto-ATPase.

Receptor-Activated Ion Channels in Neuroblastoma Cells

E. MANCINELLI, C. GOTTI, E. SHER,
A. FERRONI, AND E. WANKE

Department of General Physiology and Biochemistry, and
Department of Pharmacology
University of Milan
Milan, Italy

INTRODUCTION

A human neuroblastoma line of replicating cancer cells was used as the biological substrate for studying the mechanism of differentiation induced by chemical drug treatment.[1-3] During prolonged application of two inductors, retinoic acid (RET) and cAMP,[4] we observed the changes in the open-close kinetics of the acetylcholine receptor (AChR) channel.[5]

The main effect observed after drug treatment was a consistent decrease in the number of spontaneous openings relative to those observed after cholinergic activation. This leads us to conclude that there is a more physiological activity of this excitatory receptor channel otherwise strongly depolarizing the proliferating cells.

RESULTS AND DISCUSSION

Recordings of single-channel inward currents from openings of the AChR cationic channel are shown in FIGURE 1. An extensive statistical analysis of the currents in 13 cells is summarized in TABLE 1. The single-channel conductances and lifetimes were calculated together with the relative frequency of appearance. Different values were found for spontaneous openings (SPONT) or for those during ACh application (ACh). In the former case a test (for the reversal potential) was always performed in order to be sure that the single inward current belonged to the AChR cationic channel family. A large (51.6 pS) ACh-activated channel was present only in normal proliferating cells at day 1, when spontaneous openings were absent. In the other three conditions tested at 11 days (normal, RET, and cAMP) we could observe two levels of conductance. In the drug-treated cells the conductance is well confined in the ranges 18.6–21.3 and 32.6–34 pS, but in proliferating cells, these levels differ with (17 and 29 pS) or without (26.7 and 37.9 pS) transmitter activation. In general the lifetime of the channels of the nontreated cells is longer than that of the differentiated cells and the smallest level of conductance is shorter. In the treated cells, the ACh-activated channel stays open for a longer time than the spontaneous channel does; this is not true for the proliferating cells. In the cAMP-differentiated cells the frequency of appearance of the two conductances is reversed when ACh is applied.

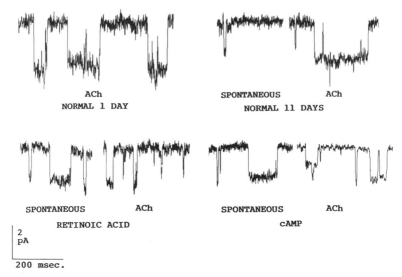

FIGURE 1. Single-channel inward current in the seven indicated different conditions of treatment (normal: 1 day; normal: 11 days; retinoic-acid- and cAMP-treated: 11 days) and of activation (with or without ACh, 100 μM). The cells of the human neuroblastoma line SK-N-BE were cultured (in a 35-mm petri dish) in Eagle's Minimum Essential Medium (MEM) containing 10% fetal calf serum (FCS), 100 IU/ml penicillin, and 0.1 mg/ml streptomycin. The culture medium was substituted, at day 2, after addition of differentiating drugs (1μM retinoic acid [RET]; 1 mM N^6-O^2-dibutyryl cyclic adenosine 3'-5' monophosphate [Bt$_2$cAMP]) and changed every 3 days.[6] For our experiments we used normal cells (1 day in culture), and cells after 11 days of culture in the modified (added inducers) medium. The cells were patch-clamped, at 37°C, in the whole-cell configuration at a membrane potential of -70 mV.[7] The compositions of external bath and patch pipette solutions, respectively, were the following (in mM): 140 NaCl, 2.8 KCl, 2 MgCl$_2$, 1 CaCl$_2$, 5 glucose, 10 HEPES/NaOH at pH 7.2; 140 K-Aspartate, 5 MgCl$_2$, 4 CaCl$_2$, 10 EGTA (calculated pCa was 7).

TABLE 1. Single-Channel Lifetimes (τ_0) and Conductances at Membrane Potential of -70 mV (Means ± SE) and Relative Frequencies of Appearance (%) in Four Experimental Conditions

		τ_0 (msec)	Conductance (pS)	Percent
Normal: 1 day	ACh	2.46 ± 0.30	28.6 ± 2.43	5
($n = 2$)		34.82 ± 3.29	51.6 ± 0.43	95
Normal: 11 days	SPONT	12.99 ± 6.87	26.7 ± 1.86	41
($n = 3$)		61.72 ± 14.63	37.9 ± 1.43	59
	ACh	4.08 ± 1.39	17.0 ± 2.57	32
		29.70 ± 15.97	29.0 ± 2.00	68
Retinoic acid	SPONT	7.38 ± 0.81	18.6 ± 0.29	43
($n = 4$)		13.63 ± 1.72	33.6 ± 0.43	57
	ACh	12.26 ± 2.32	21.3 ± 0.43	45
		13.16 ± 1.51	34.0 ± 0.43	55
cAMP	SPONT	4.05 ± 1.00	19.3 ± 0.71	11
($n = 4$)		7.68 ± 0.86	32.7 ± 0.29	89
	ACh	12.79 ± 0.05	21.1 ± 0.14	75
		15.93 ± 1.73	32.6 ± 0.29	25

In summary, the excitatory action of the nicotinic receptor channel is strongly depressed after drug treatment because (1) the spontaneous openings are reduced and (2) the half-life of the openings is selectively prolonged by the transmitter.

REFERENCES

1. DE LAAT, S. W. & P. T. VAN DER SAAG. 1982. The plasma membrane as a regulatory site in growth and differentiation of neuroblastoma cells. Int. Rev. Cytol. **74:** 1–54.
2. GUPTA, M., M. D. NOTTER, S. FELTEN & D. M. GASH. 1985. Differentiation characteristics of human neuroblastoma cells in the presence of growth modulators and antimitotic drugs. Dev. Brain Res. **19:** 21–29.
3. GOTTI, C., E. SHER, D. CABRINI, G. BONDIOLOTTI, E. WANKE, E. MANCINELLI & F. CLEMENTI. 1987. Cholinergic receptors, ion channels, neurotransmitter synthesis, and neurite outgrowth are independently regulated during the in vitro differentiation of human neuroblastoma cell line. Differentiation **34:** 144–155.
4. RUPNIAK, M. T., G. REIN, J. F. B. POWELL, T. A. RYDER, S. CARSON, S. POVERY & B. T. HIL. 1984. Characteristics of a new human neuroblastoma cell line which differentiates in response to cyclic adenosine 3'-5' monophosphate. Cancer Res. **44:** 2600–2607.
5. CHANGEUX, J. P., A. DEVILLERS-THIÉRY & D. CHEMOUILLI. 1984. Acetylcholine receptor: An allosteric protein. Science **255:** 1335–1345.
6. GOTTI, C., E. WANKE, E. SHER, D. FORNASARI, D. CABRINI & F. CLEMENTI. 1986. Acetylcholine-operated ion channel and bungarotoxin binding site in a human neuroblastoma cell line reside on different molecules. Biochem. Biophys. Res. Commun. **37:** 1141–1147.
7. HAMILL, O. P., A. MARTY, E. NEHER, B. SAKMANN & F. J. SIGWORTH. 1981. Improved patch-clamp techniques for high-resolution current recording from cells and cell free membrane patches. Pflugers Arch. **391:** 85–100.

Fibronectin Hyperpolarizes the Plasma Membrane Potential of Murine Erythroleukemia Cells

A. ARCANGELI, M. R. DEL BENE, L. RICUPERO,
L. BALLERINI, AND M. OLIVOTTO

Università degli Studi di Firenze
Istituto di Patologia Generale
50134 Florence, Italy

INTRODUCTION

The differentiation of murine erythroleukemia (MEL) cells involves a modulation of their plasma membrane potential (V_m), which is essentially regulated by Ca^{2+}-dependent K^+ channels (K_{Ca}).[1-3] The latter are the only hyperpolarizing channels in MEL cells and are apparently activated by the cell contact with the culture surface, an activation antagonized by short-range factors shed by the cells.[1,2]

Evidence is provided here that pretreatment of the culture surface with fibronectin (FN) produces a marked V_m hyperpolarization in MEL cells seeded at low cell density onto this surface (CD).

MATERIALS AND METHODS

Cell culture and V_m measurements, based on [^{3}H]tetraphenylphosphonium (^{3}H-TPP$^+$) accumulation ratio, were carried out as previously described,[1,2] except that fetal calf serum was substituted by FN-free bovin serum albumin (BSA). FN was freshly purified from human plasma according to the method of Rouslahti *et al.*[4]

RESULTS AND DISCUSSION

FIGURE 1 shows that pretreatment of the culture surface with FN produces a dose-dependent hyperpolarization of V_m in MEL cells seeded at low CD (3×10^4 cells/cm^2). In view of the recent finding that an FN-enriched substratum potentiates MEL cell differentiation, this effect emphasizes the linkage between this process and V_m modulation in these cells.

As illustrated in FIGURE 2, the hyperpolarizing effect of FN disappears when the cells are seeded at high CD (3×10^5/cm^2) namely, when the factors shed by the cells onto the culture surface are sufficient to abolish the surface-dependent activation of K_{Ca}.[1] An important implication of this fact is that FN-induced hyperpolarization is apparently mediated by K_{Ca}. However, when the hyperpolarization produced by FN in FIGURES 1 and 2 is compared, it is evident that this process is quantitatively similar

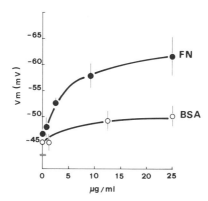

FIGURE 1. The hyperpolarizing effect of FN. Cells, harvested from a preparatory culture at 2×10^6/ml, were seeded into 25-cm² Sterilin flasks. These flasks had been previously incubated at 37°C for 1 hour with 3 ml of RPMI medium containing FN at the concentrations indicated in abscissa. The control flasks were treated with the same amount of RPMI containing BSA at the concentrations indicated in abscissa. At the end of this treatment, the flasks were washed twice with PBS and left to dry at room temperature. Cell inoculum was, for each flask, 8 ml of a cell suspension in RPMI plus BSA (250 μg/ml), containing 10^5 cells/ml (CD = 3×10^4/cm²). Flasks were then incubated at 37°C for 1 hour. Experimental points refer to V_m measurements obtained by adding ^{3}H-TPP$^+$ directly into the flasks and are means ± SEM of triplicate samples.

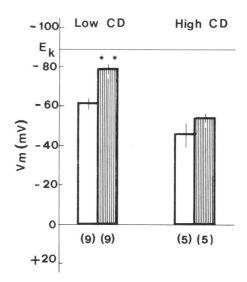

FIGURE 2. The hyperpolarization produced by FN is abolished at high CD by factors shed by MEL cells onto the culture surface. Experimental conditions were the same as those reported in FIGURE 1, except that the cells were seeded at low CD (3×10^4/cm²) or high CD (3×10^5/cm²). FN concentration used for the pretreatment of the flasks was in any case 20 μg/ml. *Bars* are means ± SEM of a number of experiments listed in parentheses. *Open bars* = control; *shaded bars* =FN-treated flasks; E_k = electrochemical equilibrium potential of potassium. ** = $p <$ 0.001 (Student's *t* test).

(about 20 mV) whatever the starting V_m of the cells. This suggests that the FN hyperpolarizes this potential through a "saturable" mechanism, probably mediated by the FN binding to a limited number of receptors on the plasma membrane.[6]

REFERENCES

1. ARCANGELI, A. & M. OLIVOTTO. 1986. Plasma membrane potential of murine erythroleu-kemia cells: Approach to measurement and evidence for cell-density dependence. J. Cell. Physiol. **127**: 17–27.
2. ARCANGELI, A., L. RICUPERO & M. OLIVOTTO. 1987. Commitment to differentiation of murine erythroleukemia cells involves a modulated plasma membrane depolarization through Ca^{2+}-activated K^+ channels. J. Cell. Physiol. **132**: 387–400.
3. ARCANGELI, A., E. WANKE, M. OLIVOTTO, S. CAMAGNI & A. FERRONI. 1987. Three types of ion channels are present on the plasma membrane of Friend erythroleukemia cells. Biochem. Biophys. Res. Commun. **146**: 1450–1457.
4. ROUSLAHTI, E., E. G. HAYMAN, M. D. PIERSHBACHER & E. ENGVALL. 1982. Fibronectin: Purification, immunochemical properties and biological activities. Methods Enzymol. **82**: 803–831.
5. PATEL, V. P. & H. F. LODISH. 1987. A fibronectin matrix is requested for differentiation of murine erythroleukemia cells into reticulocytes. J. Cell Biol. **105**: 3105–3118.
6. PATEL, V. P. & H. F. LODISH. 1986. The fibronectin receptor on mammalian erythroid precursor cells: Characterization and developmental regulation. J. Cell Biol. **102**: 449–456.

Evoked Effects of PGE$_2$ and PGA$_2$ on Lipid Fluidity and Ca^{2+}-Stimulated ATPase of Walker-256 Tumor Microsomal Membranes

G. DELICONSTANTINOS,[a] L. KOPEIKINA,[a] AND
G. RAMANTANIS[b]

[a]Department of Experimental Physiology
University of Athens Medical School
Athens, GR-115 27 Greece

[b]Department of Surgical Oncology
"Saint Savvas" Hospital
Athens, Greece

INTRODUCTION

Prostaglandins (PGs) of the E and A series have been shown to cause growth inhibition of cultured cells and to induce differentiation in several cases.[1,2] It has been suggested that PG-induced growth inhibition is mediated by intracellular cyclic AMP.[3] Additionally, Ca^{2+}-dependent adenylate cyclase has been found to be stimulated by calmodulin (CaM), thus increasing the intracellular level of cAMP in cancer cells.[4]

In our previous studies on the effect of PGE$_2$ and PGF$_{2\alpha}$ on the activity of some membrane-bound enzymes, we indicated that these compounds can alter the membrane fluidity, causing functional changes in the allosteric properties of integral enzymes.[5,6] This report focuses on the modulations of lipid dynamics and lipid–protein interactions in microsomal membranes, isolated from Walker-256 tumor, induced by PGE$_2$ and PGA$_2$ using as functional parameter the activity of the integral enzyme Ca^{2+}-stimulated ATPase.

MATERIALS AND METHODS

Walker-256 tumors were implanted subcutaneously in male Wistar rats with cell suspensions (3×10^6 cells) as previously described.[7] Twelve days after implantation the tumor was completely excised and weighed. Tumor microsomal membranes were prepared as previously described.[8]

Ca^{2+}-Stimulated ATPase Activity. Walker-256 tumor microsomal membranes were preincubated with various amounts of PGE$_2$ and PGA$_2$ (0.25–10 μM) at 25° for 30 min. Ca^{2+}-stimulated ATPase activity of microsomal membranes was estimated in a reaction mixture containing: Tris-HCl 40 mM pH 7.4; KCl 100 mM; MgCl$_2$ 2 mM; K$_2$ · ATP 2 mM; NaN$_3$ 5 mM; CaCl$_2$ 100 μM and EGTA 100 μM with traces of [γ-^{32}P]ATP; indomethacin 50 μM; and 0.25 mg of microsomal protein in a final volume of 1.0 ml. Incubations were carried out at 37°C for 45 min. The reaction was started by the addition of ATP and stopped by adding to the mixture an acidified suspension of activated charcoal (British Drug Houses).

Steady-State Fluorescence Polarization Studies. Lipid fluidity was assessed by the steady-state fluorescence polarization of 1,6-diphenyl-1,3,5-hexatriene (DPH) according to the method described by Shinitzky and Barenholz.[9] Measurements were made in an AMINCO SPF-500 spectrofluorometer with polarization accessory for fluorescence polarization measurements at excitation wavelength 360 nm and emission wavelength 430 nm. The polarization of fluorescence was expressed as the fluorescence anisotropy, r, and the anisotropy parameter $[(r_0/r) - 1]^{-1}$ was calculated using the value of $r_0 = 0.365$ for DPH as previously described.[10]

RESULTS AND DISCUSSION

The evidence presented here demonstrates that PGs decreased the Ca^{2+}-stimulated ATPase of the endoplasmic reticulum of Walker-256 tumor (FIG. 1). At low to

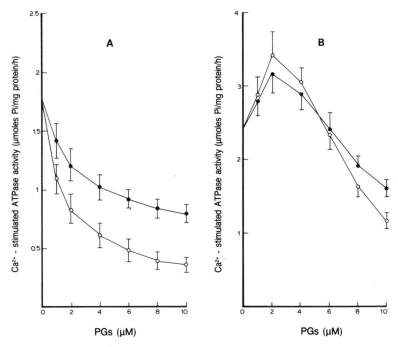

FIGURE 1. Effect of PGE_2 (●) and PGA_2 (○) on Walker-256 tumor microsomal Ca^{2+}-stimulated ATPase activity in the absence (**A**) and presence (**B**) of 20 nM calmodulin. Aqueous solutions of PGE_2 and PGA_2 were preincubated with microsomes in a medium consisting of 5 mM Tris · HCl/1.15% KCl, pH 7.4, and 50 μM indomethacin for 30 min at 25°C. Enzyme activity was estimated as described in the MATERIALS AND METHODS section. Values are means ± SD of three different experiments.

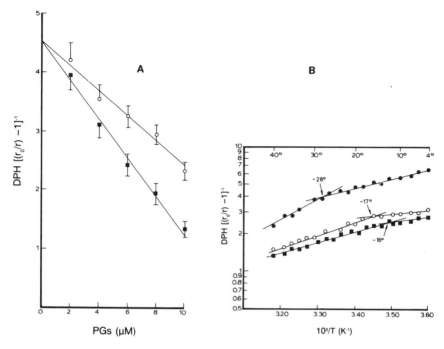

FIGURE 2. (A) DPH mobility expressed as fluorescence anisotropy $[(r_o/r) - 1]^{-1}$ in Walker-256 microsomes as a function of PGE_2 (O) and PGA_2 (■) concentrations at 25°C. Measurements were performed in duplicate for each sample. Decrease in fluorescence anisotropy reflects an increase in membrane fluidity. Each point represents the means ± SD of three different experiments. (B) Temperature-dependence of the fluorescence anisotropy of DPH in control-treated (●), PGE_2-treated (O), and PGA_2-treated (■) (10 μM) Walker-256 microsomes. The ordinate is the fluorescence anisotropy and the abscissa is the reciprocal of the absolute temperature. This experiment is representative of three that were performed. The straight lines were fitted by the method of least-squares.

moderate PGE_2 and PGA_2 concentrations an activation of the enzyme predominates in the presence of CaM, which clearly indicates that there is a competitive interaction of PGs and CaM for the CaM-binding sites of the enzyme. It seems possible that PGE_2 and PGA_2, besides inactivating the Ca^{2+}-stimulated ATPase, also decrease the CaM affinity to the enzyme. The decrease of Ca^{2+}-stimulated ATPase activity was observed with micromolar concentrations of PGE_2 and PGA_2 comparable to those that produce antiproliferative action in cultures of tumor cells.[2] The decrease in Ca^{2+}-stimulated ATPase activity amounted to approximately 55% and 75% for the PGE_2 and PGA_2, respectively, with a free calcium concentration of ~10 μM. The calcium-dependence of the effect indicates, however, that the decrement is greater at lower concentrations of calcium. Because intracellular concentrations of free calcium are between 1 and 0.01 μM, it is likely that the effects of these PGs are even greater under those conditions. The membrane functional effects of the PGs appear to be mediated through changes in the membrane fluidity. Evidence for the changes in the fluidity of the microsomal membranes induced by PGE_2 and PGA_2 was obtained by the decrease in the fluorescence anisotropy of DPH incorporated into microsomes (FIG. 2). The decrease in fluorescence anisotropy observed in this study indicates that PGs increase the microsomal membrane fluidity. PGE_2 and PGA_2 decreased the temperature of the lipid-phase separation from $28.3 \pm 1.4°$ to $17.5 \pm 1.2°$ and $16.1 \pm 0.9°$, respectively, and this is matched by a similar effect on the Ca^{2+}-stimulated ATPase activity. Such a depression in the temperature presumably reflects the interaction of the PGs with lipid components of the microsomal membranes. The "hairpin" conformation hypothesis implied a multipoint interaction of a prostaglandin molecule with its binding site regulated, in part, by the configuration of the side chains. It is probable that ionic forces also contribute to the association of PGE_2 and PGA_2 with microsomes.[11] Modulations of tumor cell membrane fluidity by PGE_2 and PGA_2 could affect the activity of a variety of other membrane-bound enzymes, binding of growth factors, and conformation of receptors.[12] It may thus be suggested that the increase in membrane fluidity is an early key event in tumor growth inhibition induced by PGE_2 and PGA_2.

REFERENCES

1. BREGMAN, M. D. *et al.* 1982. Biochem. Biophys. Res. Commun. **104:** 1080–1086.
2. OHNO, K. *et al.* 1986. Biochem. Biophys. Res. Commun. **139:** 808–815.
3. JOHNSON, G. S. & I. PASTAN. 1971. J. Natl. Cancer Inst. **47:** 1357–1364.
4. RESINK, T. J. *et al.* 1986. Eur. J. Biochem. **154:** 451–456.
5. DELICONSTANTINOS, G. & S. FOTIOU. 1986. J. Endocrinol. **110:** 395–404.
6. DELICONSTANTINOS, G. 1986. Cell. Mol. Biol. **32:** 113–119.
7. DELICONSTANTINOS, G. *et al.* 1983. Biomed. Pharmacother. **37:** 339–343.
8. DELICONSTANTINOS, G. & G. RAMANTANIS. 1983. Biochem. J. **212:** 445–452.
9. SHINITZKY, M. & Y. BARENHOLZ. 1978. Biochim. Biophys. Acta **515:** 367–394.
10. DELICONSTANTINOS, G. 1988. Comp. Biochem. Physiol. **89B:** 585–594.
11. ANDERSEN, N. H. *et al.* 1976. Adv. Prost. Thromb. Res. **1:** 271–289.
12. DELICONSTANTINOS, G. 1987. Anticancer Res. **7:** 1011–1022.

Active Ca^{2+} Accumulation in the Endoplasmic Reticulum of Different Hepatomas: Stimulation by Phosphates and Ca^{2+}-Releasing Effect of IP_3[a]

A. ROMANI,[b] R. FULCERI,[b] A. POMPELLA,[b] M. FERRO,[c]
AND A. BENEDETTI[b,d]

[b]Istituto di Patologia Generale della Università di Siena
Siena, Italy

[c]Istituto di Patologia Generale della Università di Genova
Genova, Italy

Some properties of the endoplasmic reticulum (ER) compartment with respect to active Ca^{2+} accumulation and Ca^{2+} mobilization by inositol 1,4,5-trisphosphate (IP_3) have been studied in two cultured hepatoma cell lines—MH_1C_1 and HTC cells—and in isolated microsomes obtained from a solid hepatoma—Morris hepatoma 3924A.

The Ca^{2+}-accumulating capacity of ER *in situ* was studied in cultured hepatoma cells whose plasma membrane was made permeable with digitonin.[1] Microsomes were obtained from Morris hepatoma as previously reported.[2] The MgATP-dependent Ca^{2+} accumulation, and its variations by inorganic phosphates (P_i) and/or IP_3, were evaluated by monitoring the free Ca^{2+} concentration in the incubates with a Ca^{2+}-electrode.[3] In some experiments, the uptake of $^{45}Ca^{2+}$, added to incubations, was also measured in order to determine the net amount of Ca^{2+} released by IP_3 from microsomes and cells. Isolated microsomes and permeabilized cells were incubated at 37°C in a medium with the following composition (mM): KCl 100, NaCl 20, $MgCl_2$ 5, NaN_3 (as mitochondrial inhibitor) 5, MOPS 10 (pH 7.2), in the presence of 2 mM ATP plus an ATP-regenerating system (10 mM phosphocreatine and 10 μunit/ml creatine phsophokinase). Total Ca^{2+} in the incubation systems was determined by atomic absorption spectroscopy.

STIMULATION OF ER CA^{2+} ACCUMULATION BY INORGANIC PHOSPHATES

It was previously shown by Bygrave and Anderson[4] that P_i stimulate the initial rate of Ca^{2+} transport in microsomes from ascites-sarcoma 180/TG cells. On this basis, the

[a]This work was supported by Grant No. 86.00370.44 (Project: Oncology) from the Consiglio Nazionale delle Ricerche (Italy), M. Comporti, Principal Investigator. Additional funds were received from the Association for International Cancer Research (U.K.).

[d]Address for correspondence: Dr. Angelo Benedetti, Istituto di Patologia Generale, Strada del Laterino, 8, 53100 Siena, Italy.

possibility that inorganic phosphates may favor reticular Ca^{2+} accumulation in hepatomas was tested. The addition of inorganic phosphates in cytosol-like concentration (5 mM)[5] stimulated Ca^{2+} accumulation by both Morris microsomes and ER *in situ* of the two hepatomas (FIG. 1). This allowed the buffering (down to approximately 0.2 μM) of relatively high concentrations of Ca^{2+} (FIG. 1). Such an effect of P_i is minor in liver microsomes.[1,3] The observation that P_i stimulate Ca^{2+} accumulation, besides ascites-sarcoma cells, in three different hepatomas suggests that cytosolic phosphates might play a role in modulation of Ca^{2+} exchanges between ER and cytosol of tumor cells.

EFFECT OF IP$_3$ ON ER CA^{2+} ACCUMULATED UNDER P$_i$ STIMULATION

Because inorganic phosphates stimulate ER Ca^{2+} accumulation in hepatoma cells, we have verified whether this results in an enlargement of the reticular Ca^{2+} pool available for mobilization by IP$_3$. The addition of IP$_3$ (5 μM) caused a rapid release of the Ca^{2+} previously accumulated in absence of P_i stimulation, from both Morris microsomes and the nonmitochondrial pool of HTC (FIG. 2) and MH_1C_1 (not shown)

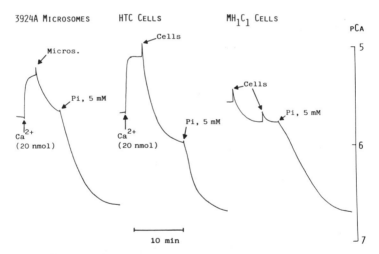

FIGURE 1. Ca^{2+} accumulation by isolated Morris 3924A hepatoma microsomes, and by digitonin-permeabilized HTC cells and MH_1C_1 cells, and its stimulation by inorganic phosphates (P$_i$). Where indicated, microsomes (1 mg protein) and cells (HTC: 3 mg protein; MH_1C_1: 4.8 mg protein) were added to 1 ml 100 mM KCl, 20 mM NaCl, 5 mM $MgCl_2$, 5 mM NaN_3, 10 mM MOPS (pH 7.2), containing 2 mM ATP plus an ATP-regenerating system (10 mM phosphocreatine and 10 μunit/ml creatine phosphokinase), in a "thermostated" (37°C) plastic vessel.[3] P$_i$ were added where indicated as small volumes of a potassium phosphate buffer, pH 7.2. Ca^{2+} accumulation was estimated by monitoring the free Ca^{2+} concentration in the incubates with a Ca^{2+} electrode.[3] Total Ca^{2+} content of the incubation medium prior to the addition of microsomes/cells was (μM) 31, 31 and 11, for 3924A microsomes, HTC cells, and MH_1C_1 cells, respectively, as determined by atomic absorption spectroscopy. A typical experiment out of four is reported.

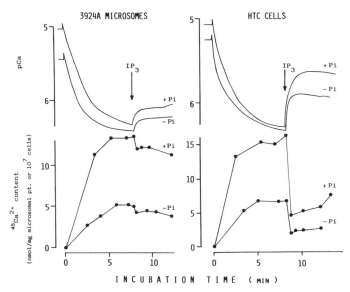

FIGURE 2. Ca^{2+}-releasing effect of IP_3 added to isolated Morris 3924A microsomes and digitonin-permeabilized HTC cells which had previously accumulated Ca^{2+} in the absence or presence of 5 mM P_i. Incubation mixtures (1.5 ml) included 100 mM KCl, 20 mM NaCl, 5 mM $MgCl_2$, 5 mM NaN_3, 10 mM MOPS (pH 7.2), plus trace amounts of $^{45}CaCl_2$ (1 μCi), and 2 mg (3924A microsomes) or 3 mg protein/ml (HTC cells). When the incubations were carried out in the presence of P_i, $CaCl_2$ was also added (3924A microsomes: 20 μM; HTC cells: 10 μM). Ca^{2+} present in the incubation mixtures prior to Ca^{2+} additions was 13 μM. Ca^{2+} accumulation was started by the addition (50 μl) of ATP plus an ATP-regenerating system. At various intervals, aliquots (50 μl) of the incubation mixtures were drawn, in order to measure the $^{45}Ca^{2+}$ content of microsomes and cells.[2] Simultaneously, the free Ca^{2+} concentration in the incubation medium was monitored by means of a Ca^{2+} electrode.[3] Where indicated, 5 μM (final concentration) IP_3 was added. Ca^{2+} nonspecifically bound to microsomes and cells prior to the addition of ATP (lower than 10% of the maximum accumulation value, in any instance) was subtracted in order to have active Ca^{2+} accumulation. Two typical experiments out of four are shown.

hepatoma cells. Such an effect was more marked in HTC cells than in MH_1C_1 cells and Morris microsomes. As can be seen in FIGURE 2, Ca^{2+}-loading of microsomes—as well as of the nonmitochondrial pool of HTC permeabilized cells—under P_i stimulation results in a higher net Ca^{2+} release upon addition of IP_3. Thus, the higher ER Ca^{2+} contents after stimulation by inorganic phosphates appeared to produce an enlargement of the reticular IP_3-sensitive Ca^{2+} pool, which might in turn result in an enhancement of the IP_3-induced, Ca^{2+}-mediated cell responses.

ACKNOWLEDGMENT

Professor T. Galeotti (Universitá Cattolica del Sacro Cuore, Rome) is gratefully acknowledged for supplying Morris 3924A hepatoma rats.

REFERENCES

1. FISKUM, G. 1985. Cell Calcium **6:** 25–37.
2. BENEDETTI, A., R. FULCERI & M. COMPORTI. 1985. Biochim. Biophys. Acta **816:** 267–277.
3. SIEGEL, E. & H. AFFOLTER. 1987. Meth. Enzymol. **141:** 25–36.
4. BYGRAVE, F. L. & T. A. ANDERSON. 1981. Biochem. J. **200:** 343–348.
5. WEINHOUSE, S. 1955. Adv. Cancer Res. **3:** 269–325.

Ca^{2+}-Transport-Mediated Regulation of Metabolism in Hepatoma Mitochondria

ANNE N. MURPHY AND GARY FISKUM

Department of Biochemistry
George Washington University
School of Medicine
Washington, D.C. 20037

In models of cellular response to various extracellular stimuli, Ca^{2+} plays a pivotal role in both activating cytosolic processes associated with energy utilization and stimulating mitochondrial ATP production to meet cellular energy demands.[1] The increase in cytosolic [Ca^{2+}] resulting from agonist stimulation is believed to be followed by an increase in mitochondrial Ca^{2+} uptake and matrix free [Ca^{2+}], which activates key regulatory dehydrogenases (NAD$^+$-isocitrate, 2-oxoglutarate, and pyruvate dehydrogenase). In tumor cells, both mitochondrial Ca^{2+} transport properties and patterns of metabolism are altered in comparison to the cells of their tissue of origin.[2] Given that the response of tumor cells to growth factors can involve an increase in cytosolic [Ca^{2+}][3] this study specifically addresses the role of extramitochondrial Ca^{2+} in the control of mitochondrial energy metabolism in the highly malignant rat AS-30D ascites hepatoma cell line.

Results using extracts of isolated mitochondria indicate that the enzyme kinetics and Ca^{2+} sensitivities of AS-30D hepatoma and normal liver NAD$^+$-isocitrate and 2-oxoglutarate dehydrogenase do not significantly differ[4]; however, the maximal activity of both Ca^{2+}-sensitive NAD$^+$-isocitrate and Ca^{2+}-insensitive NADP$^+$-isocitrate dehydrogenase are 7- and 3.5-fold higher in the tumor than rat liver extracts.[4] These findings indicate that the potential exists for control of the tricarboxylic acid cycle by Ca^{2+} in tumor mitochondria, as has been demonstrated for mitochondria from several normal tissues.[5] Furthermore, measurements of oxygen consumption by isolated hepatoma mitochondria suspended in media containing isocitrate, glutamate, or 2-oxoglutarate in the presence of buffered free Ca^{2+} concentrations from 0.2 μM to 2.0 μM have confirmed the Ca^{2+}-sensitivity of mitochondrial respiration under physiologically realistic conditions.[6]

Respiratory measurements similar to those employing isolated mitochondria have been extended to include the use of digitonin-permeabilized hepatoma cells.[7] FIGURE 1 describes parallel measurements of the medium free [Ca^{2+}], the oxidation-reduction state of pyridine nucleotides (NAD(P)H), and the consumption of O$_2$ by digitonin-permeabilized AS-30D hepatoma cells suspended in the presence of 5 mM glutamate and 0.1 mM malate as oxidizable substrates. Under these conditions, glutamate is transaminated in the presence of malate to form 2-oxoglutarate within the mitochondrion. Addition of cells to the medium resulted in a rapid decline in the [Ca^{2+}] to a steady-state level of $\approx$0.2 μM due to ATP-dependent endoplasmic reticulum Ca^{2+} uptake.[8] Under these conditions, the pyridine nucleotide redox state remained relatively oxidized (low fluorescence) because of rapid ATP turnover. The addition of

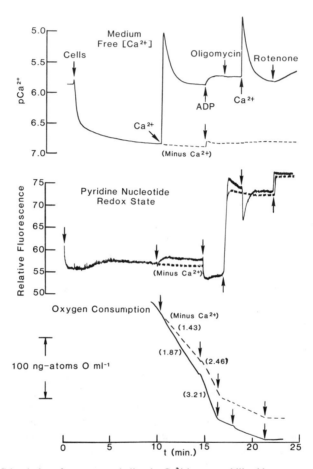

FIGURE 1. Stimulation of energy metabolism by Ca^{2+} in permeabilized hepatoma cells. AS-30D rat ascites hepatoma cells were maintained and washed free of erythrocytes as previously described[7] and added at a concentration of 12.5×10^6 ml^{-1} to 2.3 ml of medium (37°C) containing 125 mM KCl, 2 mM K_2HPO_4, 15 mM NaCl, 4.3 mM $MgCl_2$, 3.4 mM Na_2 ATP, 20 mM HEPES (pH 7.0), 5 mM glutamate, 0.1 mM malate, and 0.004% digitonin. The concentrations of added substances were $CaCl_2$ (24 μM), ADP (0.4 mM), oligomycin (4.3 μg ml^{-1}), and rotenone (4 μM). The medium free $[Ca^{2+}]$ was monitored with an Ionetics Calcium Selectrode calibrated with a Ca-EGTA buffer system. Pyridine nucleotide redox state was measured spectrofluorometrically using 352-nm excitation and 464-nm emission wavelengths. O_2 consumption was measured in a closed system using a Clark-type O_2 electrode. Numbers in parentheses refer to rates of respiration in ng-atoms $O/min/10^6$ cells.

exogenous $CaCl_2$ (1.9 nmol/10^6 cells) produced a rapid increase in free Ca^{2+}, followed by a return to a steady-state level of ≈ 1.5 μM. This reponse indicates that the Ca^{2+}-uptake capacity of the nonmitochondrial organelles was reached and that the mitochondria accumulated and "buffered" the majority of the added Ca^{2+}. These events were accompanied by a small but consistent net reduction in the pyridine nucleotide redox state (increased fluorescence) as well as a 30% stimulation of respiration. The Ca^{2+}-dependent elevation of respiration was also evident after the addition of ADP, which induced net pyridine nucleotide oxidation, as expected. Inhibition of the mitochondrial ATP synthetase by oligomycin slowed respiration to a resting level and brought the pyridine nucleotides to a near maximal level of reduction. A subsequent addition of Ca^{2+} was rapidly accumulated by the mitochondria, as indicated by the return of the free $[Ca^{2+}]$ to the mitochondrial "buffer-point" and by the corresponding transient pyridine nucleotide oxidation (observable in the absence of ATP synthesis). Respiration was terminated by the addition of the electron transport inhibitor, rotenone, which brought the pyridine nucleotides to a fully reduced state and induced the net release of Ca^{2+} from the mitochondria into the medium.

The most plausible explanation for the Ca^{2+}-dependent net reduction of pyridine nucleotides and stimulation of glutamate-dependent respiration is the activation of 2-oxoglutarate dehydrogenase by an increase in the mitochondrial matrix free $[Ca^{2+}]$. These findings are consistent with similar observations for mitochondria isolated from normal tissues[1,5] and strongly suggest that Ca^{2+}-activated mitochondrial energy metabolism can occur in tumor cells *in vivo*.

REFERENCES

1. DENTON, R. M. & J. G. MCCORMACK. 1985. Am. J. Physiol. **249**: E543–E554.
2. FISKUM, G. 1986. *In* Mitochondrial Physiology and Pathology. G. Fiskum, Ed :180–201. Van Nostrand Reinhold. New York.
3. MOODY, T. W., A. MURPHY, S. MAHMOUD & G. FISKUM. 1987. Biochem. Biophys. Res. Commun. **147**(1): 189–195.
4. MURPHY, A. N. & G. FISKUM. 1988. FASEB J. **2**: A338.
5. HANSFORD, R. G. 1985. Rev. Physiol. Pharmacol. **102**: 1–72.
6. MURPHY, A. N. & G. FISKUM. Unpublished results.
7. MOREADITH, R. W. & G. FISKUM. 1984. Anal. Biochem. **137**: 360–367.
8. BECKER, G. L., G. FISKUM & A. L. LEHNINGER. 1980. J. Biol. Chem. **255**(19): 9009–9012.

The Localization of an Antiorganelle Monoclonal Antibody (10A8) in the Golgi Apparatus of Rat Cells[a]

SPYROS G. E. MEZITIS, C. HARKER RHODES,
ANNA STIEBER, NICHOLAS K. GONATAS,
JACQUELINE O. GONATAS, AND [b]BECCA FLEISCHER

University of Pennsylvania
School of Medicine
Philadelphia, Pennsylvania 19104
[b]*Vanderbilt University*
Nashville, Tennessee

Several intrinsic membrane or resident proteins of the Golgi apparatus have been isolated and characterized[1-5]; these are the mannose-6-phosphate receptor, an N-acetylglucosaminyl transferase, an α-mannosidase II, and galactosyl- and sialytrans-ferases.[1-5] It is generally believed that enzymes involved in the glycosylation of polypeptides processed through the Golgi apparatus are arranged in a stereotype, *cis* to *trans* gradient.[6] However, recent evidence indicates that the distributions of enzymes and the mannose-6-phosphate receptor through the cisternae of the Golgi apparatus are cell-specific.[7,8]

Monoclonal antibodies are useful morphologic markers and tools for the isolation of proteins that may be difficult to identify and isolate with other methods. Furthermore, since Louvard and collaborators have shown the feasibility of preparing antiorganelle antisera, monoclonal or polyclonal antibodies become increasingly useful in studies of organelles.[9]

In this study, we have raised a monoclonal antibody (10A8) against fractions of neuronal Golgi apparatus from rat brain.[10] These fractions were extracted with carbonate according to Howell and Palade[11]; this treatment removes most of the soluble proteins trapped in the vesicles of Golgi fractions.[11]

The antibody, which was of the IgG2b class, was localized by ultrastructural immunocytochemistry in medial cisternae of the Golgi apparatus of rat brain neurons, glia, pituitary, cultured rat pheochromocytoma (PC 12), and endocrine pancreas[12] (FIGS. 1 and 2). The antibody did not react with mouse, rabbit, bovine, and human brain. Immunoblotting of enriched fractions of Golgi apparatus from rat brain neurons and liver showed a single band of 160 kD in apparent molecular mass. Paradoxically, the antibody did not stain rat liver and exocrine pancreas, although it stained strongly the adjacent endocrine pancreas. These findings suggest that the antigenic deter-

[a]This study was supported in part by National Science Foundation Grant DCB 8402370 (to B. F.) and by National Institutes of Health Grants NS 05572 and 32TNS07064 (to N. K. G.). A segment of this study was performed in partial fulfillment of the requirements for the doctoral thesis of S. G. E. M.

minant(s) in neurons, glia, pituitary and PC 12 cells reacting with the monoclonal antibody are not accessible to it in liver and exocrine pancreas.

Ongoing studies to isolate and further characterize the 160-kD polypeptide may clarify its structure and function and provide insights into the organization of the Golgi apparatus of neurons.

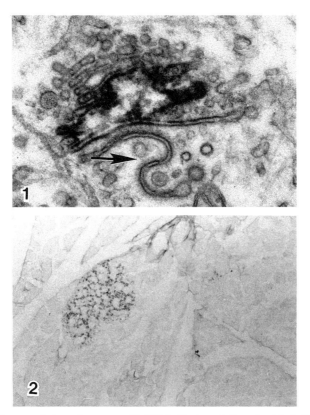

FIGURE 1. Immunoelectronmicrogram with 10A8 of rat cerebellum fixed in 2.5% glutaraldehyde and treated with 0.05% saponin. Note immunoperoxidase staining of medial cisternae of the Golgi apparatus of a Purkinje neuron. *Arrow* shows unstained coated cisternae at the *trans* face of the Golgi apparatus. Magnification ×59,000.

FIGURE 2. Immunocytochemically obtained image of rat pancreas with 10A8. Note intensely stained cells in an islet and unstained exocrine pancreas. Magnification ×525.

ACKNOWLEDGMENT

We would like to express our appreciation to Judith A. Westley for typing this manuscript.

REFERENCES

1. DAHMS, N. M., P. LOBEL, J. BREITMEYER, J. M. CHIRGWIN & S. KORNFELD. 1987. Cell **50:** 181–192.
2. DUNPHY, W. G., R. BRANDS & J. E. ROTHMAN. 1985. Cell **40:** 463–472.
3. NOVIKOFF, P. M., D. R. P. TULSIANI, O. TOUSTER & A. B. NOVIKOFF. 1983. Proc. Natl. Acad. Sci. USA **80:** 4364–4368.
4. ROTH, J. & E. G. BERGER. 1982. J. Cell Biol. **93:** 223–229.
5. ROTH, J., D. J. TAATGES, J. WEINSTEIN, J. C. PAULSON, P. GREENWELL & W. M. WATKINS. 1986. J. Biol. Chem. **261:** 14307–14312.
6. FARQUHAR, M. G. 1985. Annu. Rev. Cell Biol. **1:** 447–488.
7. ROTH, J., D. J. TAATGES, J. WEINSTEIN, J. C. PAULSON, P. GREENWELL & W. M. WATKINS. 1986. J. Biol. Chem.
8. BROWN, W. J. & M. G. FARQUHAR. 1987. Proc. Natl. Acad. Sci. USA **84:** 9001–9005.
9. LOUVARD, D., H. REGGIO & G. WARREN. 1982. Antibodies to the Golgi complex and the rough endoplasmic reticulum. J. Cell Biol. **92:** 92–107.
10. GONATAS, J. O., N. K. GONATAS, A. STIEBER & B. FLEISCHER. 1985. J. Neurochem. **45:** 497–507.
11. HOWELL, K. E. & G. E. PALADE. 1982. J. Cell Biol. **92:** 822–832.
12. GONATAS, J. O., N. K. GONATAS, A. STIEBER & D. LOUVARD. 1987. J. Neurochem. **49:** 1498–1506.

Fructose 2,6-Bisphosphate and Glycolysis in Ascites Tumor Cells[a]

FEDERICA I. WOLF[b] AND ACHILLE CITTADINI[c]

[b]Institute of General Pathology
Catholic University
Rome, Italy

[c]Department of Cell Biology and Physiology
University of L'Aquila
L'Aquila, Italy

Cancer cell energy metabolism is characterized by several features, among which high aerobic glycolysis and low Pasteur effect are the most peculiar. These phenomena are not fully understood, although they can be accounted for by qualitative and quantitative changes in glycolytic enzyme activity, hexose transport, and deranged intracellular compartmentalization. Glycolysis is regulated by a fine balance of numerous effectors (substrates, adenine nucleotides, hormones, and ions) on key enzymes, among which phosphofructo 1-kinase (PFK1) is probably the most important. In the last decade, fructose 2,6-bisphosphate (F2,6-P_2) has been identified as the most potent stimulator of PFK1. F2,6-P_2 is an acid-labile compound, found in all mammalian cells, whose metabolism has been widely studied (see Ref. 1 for a review). In the normal cell, the effects of F2,6-P_2 on PFK1 are particularly determinant for full stimulation of glycolysis when glucose is abundant and ATP concentration inhibitory.[2]

Information regarding the role of F2,6-P_2 in determining the glycolytic characteristics of tumor cells is fragmentary. In this work we studied the role of F2,6-P_2 in the regulation of energy metabolism of Ehrlich ascites tumor cells (ATC). The F2,6-P_2 content of ATC incubated at 37°C under O_2 is 5.7 pmol/mg protein (TABLE 1) with 5 mM glucose and it remains stable if glucose is increased. Time-course experiments show that the maximal level of F2,6-P_2 is achieved at 8 minutes. Unlike the case in normal cells,[2] F2,6-P_2 increases more than three times in anaerobic conditions (namely, with rotenone, N_2 and NaCN [TABLE 1]), but it does not change in the presence of the uncoupler TTFB. It must be underlined that these four conditions are characterized by an increase of lactate accumulation (TABLE 1), whereas ATP content is selectively decreased only in the presence of TTFB. These observations suggest that in the tumor cells studied F2,6-P_2 is responsible for the increase of glycolysis under anaerobic condition, probably by removing Mg-ATP inhibition of PFK1. When ATP decreases, as in the case of treatment with TTFB, high glycolysis is due to PFK1 stimulation by ADP and P_i.

Our data, together with other observations on the capability of tumor promoters and oncogenic virus of enhancing glycolytic flux and F2,6-P_2 concentration,[3,4] emphasize the importance of fructose 2,6-P_2 in determining glycolytic features of tumor cells.

[a]This work was supported by MPI and by an AIRC grant to A. C.

TABLE 1. Effect of Different Metabolic Conditions on Fructose 2,6-Bisphosphate, Lactate and ATP Levels and Glucose Consumption in Ehrlich Ascites Tumor Cells

	$F2,6-P_2$ (pmol/mg protein)	Lactate	ATP (nmol/mg protein)	Glucose
O_2	5.7 ± 1.2 (5)	193.2 ± 6.5 (3)	7.6 ± 1.0 (3)	120.0 (2)
Rotenone				
(6.7 μM)	15.7 ± 2.3 (7)	338.4 ± 29.8 (5)	10.5 ± 0.8 (5)	310.0 (2)
N_2	19.7 (2)	319.9 ± 19.1 (3)	8.8 (2)	349.0 (2)
NaCN				
(1.0 mM)	22.6 (2)	332.6 (2)	8.6 (2)	366.0 (2)
TTFB				
(8.0 μM)	4.9 (2)	391.8 (2)	1.1 (2)	452.0 (2)

NOTE: Ascites tumor cell suspensions were incubated at 37°C with 5.0 mM glucose and metabolic inhibitors for 15 minutes.; $F2,6-P_2$ was assayed spectrophotometrically on alkaline extracts using PP_i-PFK.

Further work is needed to elucidate the intimate mechanisms of these phenomena, taking into account that $F2,6P_2$ may interfere with other biochemical reactions essential for uncontrolled cancer cell proliferation.

REFERENCES

1. VAN SCHAFTINGER, E. 1987. Fructose 2,6-bisphosphate. Adv. Enzymol. **59:** 315–395.
2. HUE, L. & M. H. RIDER. 1987. Role of fructose 2,6-bisphosphate in the control of glycolysis in mammalian tissues. Biochem. J. **245:** 313–324.
3. BOSCA, L., G. G. ROUSSEAU & L. HUE. 1985. Phorbol 12-myristate 13-acetate and insulin increase the concentration of fructose 2,6-bisphosphate and stimulate glycolysis in chicken embryo fibroblasts. Proc. Natl. Acad. Sci. USA **82:** 6440–6444.
4. BOSCA, L., M. MOJENA, J. GHYSDAEL, G. G. ROUSSEAU & L. HUE. 1986. Expression of the v-src or v-fps oncogene increases fructose 2,6-biphosphate in chick-embryo fibroblasts. Biochem. J. **236:** 595–599.

The Effect of Magnesium on Glucose Utilization in Ascites Tumor Cells

DONATELLA BOSSI, FEDERICA I. WOLF,
GABRIELLA CALVIELLO, AND [a]ACHILLE CITTADINI

Institute of General Pathology
Catholic University
Rome, Italy

[a]*Department of Cell Biology and Physiology*
University of L'Aquila
L'Aquila, Italy

Magnesium, the most abundant intracellular divalent cation, is deeply involved in the modulation of a variety of biochemical and biological properties of mammalian cells. The role of intracellular magnesium is particularly important in the regulation of enzymes involved in transphosphorylation and other reactions concerned with cell division and proliferation; the alteration of cell Mg^{2+} homeostasis has been proposed as playing a major role in the pathogenesis of cancer.[1]

After our previous work on the regulation of Ca^{2+} and Mg^{2+} metabolism in cancer cells,[2,3] we undertook this study on the effect of magnesium on the energy metabolism of Ehrlich ascites tumor cells (ATC).

Ascites tumor cells, supplemented with increasing (0.0–5.0 mM) Mg^{2+} concentration, were treated with 3 mM glucose. The addition of glucose provokes a sudden decrease of endogenous ATP content, proportional to the amount of Mg^{2+}, and within 5 minutes the whole adenine nucleotide pool is exhausted. Under the same condition glucose 6-P is also reduced (FIG. 1), while lactate accumulation decreases to 46.2 from 94.6 nmol/mg dry weight. Under the same experimental condition glucose-supplemented respiration is inhibited by about 35 percent. To exclude the possibility of damage as a cause rather than attributing a specific metabolic effect to Mg^{2+} we ran a trypan blue exclusion test and assayed LDH released from the cells. We found no signs of cell damage (more than 95% were viable cells). The effect of Mg^{2+} is strongly dependent on the extent of the preincubation with Mg^{2+} (FIG. 1C). When Mg^{2+} is added after glucose no effect can be detected in either ATP or glycolytic intermediates.

The treatment with Mg^{2+} seems to make ATP more available for utilization, and glucose is an adequate substrate for phosphorylation even though glucose 6-P and lactate production were significantly reduced. Glucose utilization indicates that Mg^{2+} stimulates glucose consumption, particularly during the first 3 minutes (FIG. 1D). Besides glycolysis, phosphorylated glucose can be utilized via either glycogen synthesis or pentose shunt. Preliminary data on glucose 1-P and 6-P gluconate do not yet give indication about the effect of Mg^{2+} on these metabolic pathways.

Mg^{2+} could modulate the rate of glucose transport in ATC. To study this possibility we ran experiments with cytochalasin B, which inhibits glucose transport. The results are reported in FIGURE 2. Ten micromolar cytochalasin B completely

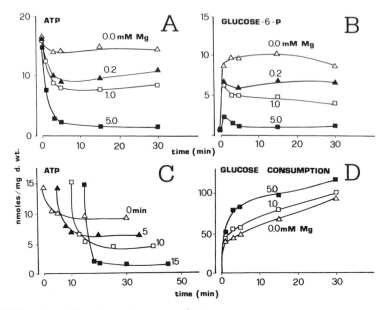

FIGURE 1. (**A** and **B**) Preincubation with Mg^{2+} decreases ATP and glucose 6-P content of ATC upon addition of glucose (time 0). (**C**) The effect is proportional to the preincubation time with 5mM Mg^{2+}. (**D**) Mg^{2+} increases glucose consumption.

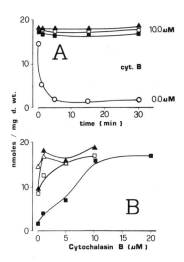

FIGURE 2. ATP content of ATC treated with glucose in the presence of cytochalasin B and pretreated with Mg^{2+}. (**A**) Cytochalasin B prevents decrease of ATP due to Mg^{2+}. (**B**) The amount of cytochalasin B required to achieve the effect increases with Mg^2 concentration.

annuls the effect of Mg^{2+} on ATP and on all other variables. The amount of cytochalasin B required to counteract the effect of Mg^{2+} increases with increasing cation concentration.

In conclusion, our data indicate that Mg^{2+} exerts a clear effect upon Ehrlich ascites tumor cell energy metabolism and suggest that this divalent cation certainly plays a primary role in metabolic regulation of the cell, although its exact function in the entire process is still difficult to assess.

REFERENCES

1. RUBIN, H. 1982. Cancer Res. **42:** 1761–1768.
2. CITTADINI, A. & A. SCARPA. 1983. Arch. Biochem. Biophys. **227:** 202–209.
3. CALVIELLO, G., D. BOSSI & A. CITTADINI. 1987. Arch. Biochem. Biophys. **259:** 38–45.

Bidirectional Cross-Talk of Intracellular Signaling Systems in Splenocytes from Athymic Nude Mice

ERIK WIENER AND ANTONIO SCARPA

Department of Physiology/Biophysics
School of Medicine
Case Western Reserve University
Cleveland, Ohio 44106.

Peptide growth factors are multifunctional. They can either stimulate or inhibit cellular proliferation depending on the other growth factors present in the medium.[1] One mechanism for this multifunctionality is cross-talk between the different receptor-coupled effector systems. B lymphocytes offer a model system for the study of such

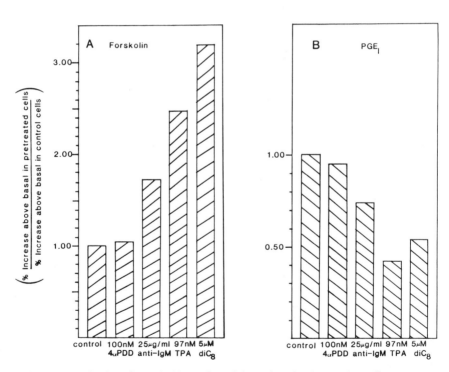

FIGURE 1. Activation of protein kinase C modulates the adenylate cyclase effector system. Splenocytes were incubated for 10 min at 37°C prior to (**A**) a 10-min incubation with 114 μM forskolin or (**B**) a 2-min incubation with 50 nM PGE_1. The protein kinase C activators indicated in the figure were added 3 min prior to the cAMP agonists.

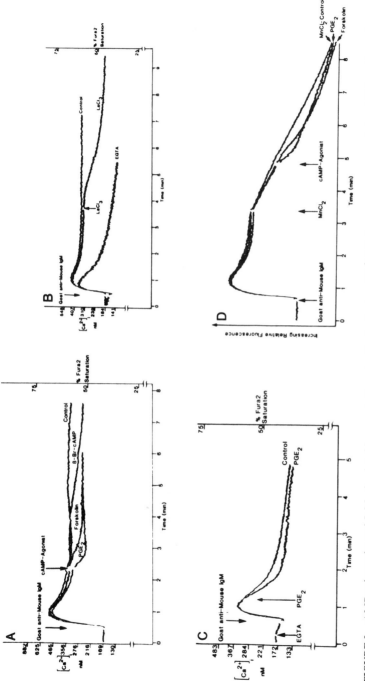

FIGURE 2. cAMP reduces the anti-IgM-stimulated increase in the cytosolic steady-state [Ca²⁺]. Splenocytes were suspended at 6.125×10^5 cells/ml and stimulated with 10 μg/ml goat anti-mouse IgM. (**A**) After stimulation with anti-Ig, 560 nM PGE_2, 106 μM forskolin, 3.3 mM 8-Bromo-cAMP, or 0.1% EtOH as vehicle (control) was added. (**B**) 3 mM EGTA was added prior to the anti-Ig; 30 μM $LaCl_3$ was added after the anti-Ig; or only anti-Ig was added. (**C**) EGTA (3.0 mM) was added prior to the anti-Ig. After stimulation with anti-Ig either 560nM PGE_2 or 0.1% EtOH as vehicle was added. (**D**) After stimulation with anti-Ig, 300 μM $MnCl_2$ was added. After $MnCl_2$ either 560 nM PGE_2, 106 μM forskolin, or 0.1% EtOH as vehicle was added.

interactions, since cytokines and prostaglandins secreted by other leukocytes mediate the response of B cells to antigen. We have examined whether signals from the inositol phospholipid signaling cascade modulate the adenylate cyclase system and whether the activation of adenylate cyclase modulates the signals generated by the phosphoinositidase C effector system in splenocytes from athymic nude mice.

Single-cell suspensions were prepared as previously described.[2] However, in experiments examining changes in the $[Ca^{2+}]_i$, only a balanced salt solution containing bovine serum albumin (BSA) and 10 mM glucose was used. cAMP was extracted and measured by radioimmunoassay, and $[Ca^{2+}]_i$ was measured by the fluorescent indicator Fura 2, as previously described.[2,3]

We report here that products from the two effector systems modulate one another in splenocytes from athymic nude mice. Activation of protein kinase C by three different reagents potentiated the forskolin- but decreased the PGE_1- induced increase in cAMP. That is, pretreating the splenocytes with tumor-promoting phorbol esters potentiated the forskolin-induced rise in cAMP from 1.7 ± 0.1 to 4.3 ± 0.6 pmol cAMP/10^6 cells (FIG. 1A), but decreased the PGE_1 response from 0.98 ± 0.06 to 0.51 ± 0.03 pmol cAMP/10^6 cells (FIG. 1B). Similarly, pretreating the cells with 5 μM diC$_8$ increased the forskolin response from 1.7 ± 0.1 to 5.1 ± 0.2 pmol cAMP/10^6 cells (FIG. 1A), but it decreased the PGE_1 response from 1.15 ± 0.03 to 0.75 ± 0.04 pmol cAMP/10^6 cells (FIG. 1B). The inactive phorbol ester 4α-phorbol-12,13-didecanoate had no effect on either the forskolin or PGE_1 response. Goat anti-mouse IgM potentiated the forskolin-induced increase by 76% (FIG. 1A), but it decreased the response to PGE_1 by 30% (FIG. 1B). These results are consistent with the hypothesis that activation of an enzyme involved in the inositol phospholipid signaling cascade, protein kinase C, modulates the adenylate cyclase system in mouse splenocytes.

We next examined whether activation of the adenylate cyclase effector system could modulate the inositol phospholipid signaling system. Treating splenocytes with PGE_2, forskolin, or 8-Br cAMP after stimulation with anti-Ig decreased the steady-state level of the cytosolic free $[Ca^{2+}]$ (FIG. 2A). The anti-Ig-stimulated increase in the steady-state Ca^{2+} is partially sustained by an influx of extracellular Ca^{2+}. Either lowering the extracellular $[Ca^{2+}]$ with EGTA or blocking influx with La^{3+} resulted in much lower stimulated steady-state $[Ca^{2+}]$ (Fig. 2B). However, increasing the cellular cAMP levels in the presence of EGTA also increased the rate of decline of the anti-Ig-stimulated Ca^{2+} transient (FIG. 2C). Furthermore, increasing the cAMP levels did not reduce Mn^{2+} influx as measured by the Mn^{2+}-induced quenching of Fura 2 (FIG. 2D). These last two results imply that cAMP is not lowering the steady-state $[Ca^{2+}]$ by reducing receptor-stimulated Ca^{2+} influx.

In conclusion, activation of protein kinase C results in the dual modulation of the adenylate cyclase system, and activation of adenylate cyclase lowers the steady-state $[Ca^{2+}]_i$ induced by anti-Ig in splenocytes from athymic nude mice.

REFERENCES

1. SPORN, M. B. & A. B. ROBERTS. 1988. Nature **332:** 217–219.
2. WIENER, E. C. & A. SCARPA. 1987. FEBS Lett. **224:** 33–37.
3. WIENER, E. C., D. LEBMAN, J. CEBRA & A. SCARPA. 1987. Arch. Biochem. Biophys. **254:** 462–471.

Shape Change Leading to Cell Death and Ca^{2+} Entry in Yoshida Hepatoma Cells

M. A. RUSSO,[a] D. BOSSI,[b] M. OSTI,[a] G. CALVIELLO,[b]
AND A. CITTADINI[c]

[a]Department of Experimental Medicine
Università "La Sapienza"
Rome, Italy

[b]Institute of General Pathology
Catholic University
Rome, Italy

[c]Department of Cell Biology and Physiology
University of L'Aquila
L'Aquila, Italy

INTRODUCTION

Cytosolic Ca^{2+} overload has been suggested to be a common step in the pathogenesis of cell degeneration and death.[1-4] We have suggested that Ca^{2+}-dependent

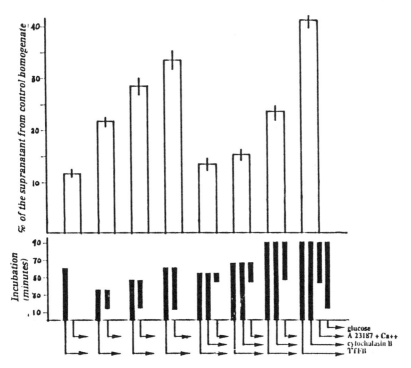

FIGURE 1. LDH release from YHC incubated under various metabolic conditions.

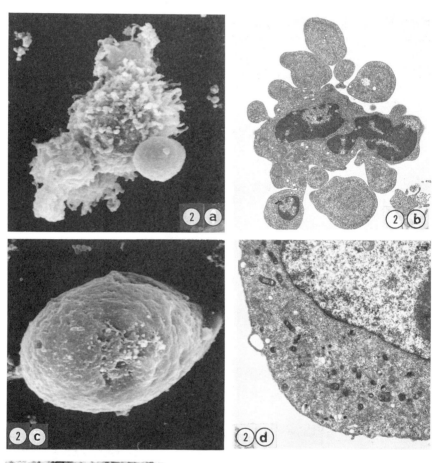

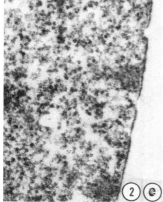

FIGURE 2. SEM (**a**) and TEM (**b**) photomicrographs of YHC treated with A 23187 + Ca^{2+}; pretreatment with CB partially prevents bleb formation, as observed by SEM (**c**) and TEM (**d**); cells in advanced necrosis show holes in the plasma membrane (**e**).

cytoskeletal functions play an important role in the changes leading to necrosis and, in particular, that bleb formation may be a common feature in the necrobiotic sequence.[2,3] The present data show that the same occurs in cancer cells.

Yoshida hepatoma cells (YHC) were preincubated at 37°C for 15 min in the presence of TTFB (3 μM), with or without 10 μg/ml cytochalasin B (CB), in a Ca^{2+}-free medium. After 40 min Ca^{2+} and Ca^{2+} ionophore (CI) A23187 were added, and the effect of 20 mM glucose was also tested. Samples were taken after 20, 30, and 60 minutes for scanning (SEM) and transmission electron microscopy (TEM). LDH released in the medium was measured as a quantitative index for necrosis (FIG. 1).

RESULTS AND CONCLUSIONS

Short-Term Incubation. Yoshida hepatoma cells incubated 20–30 min with Ca^{2+} under a condition of exhaustive energy depletion in the presence of the CI A23187 show deep modification of shape (FIG. 2a and b), which can be partly prevented by addition of glucose and/or by pretreatment with CB (FIG. 2c and d).

Long-Term Incubation. Incubation, for 60–80 min under the same condition as above, induces severe, mostly necrotic alterations (FIG. 2e). Also under these conditions, pretreatment with CB partly prevents necrosis; on the contrary, addition of glucose enhances cell disorganization (blebs) and necrosis (increased LDH release; FIG. 1).

In conclusion: (1) Cytosolic Ca^{2+} overload causes sublethal cytoskeletal changes, which in late phases leads to cell disorganization and necrosis. (2) Blebs and cell disorganization can be prevented by CB, indicating that actin filaments are involved in the pathogenesis of such modifications. (3) In the early phases of incubation, energy supply can likely control cytosolic Ca^{2+} homeostasis and then prevent, at least partly, blebs and LDH release. In late phases, Ca^{2+} homeostasis is impaired; Ca^{2+}-dependent cell functions, especially cytoskeletal contractility, become irreversibly activated. Energy supply to the cell (glucose addition) favors cytoskeletal contraction and hence enhances cell damage. (4) Excessive cytoskeletal contraction induces bleb formation, cytoskeletal disorganization, plasma membrane disruption and cell necrosis.

REFERENCES

1. JEWELL, S. A., G. BELLOMO, H. THOR & S. ORRENIUS. 1982. Science **217:** 1257–1259.
2. RUSSO, M. A., A. CITTADINI, A. M. DANI, G. INESI & T. TERRANOVA. 1981. J. Mol. Cell. Cardiol. **13:** 265–279.
3. RUSSO, M. A., A. B. KANE, E. E. YOUNG & J. L. FARBER. 1982. Am. J. Pathol. **109:** 133–143.
4. SCHANNE, F. A., A. B. KANE, E. E. YOUNG & J. L. FARBER. 1979. Science **202:** 700–702.

Lonidamine-Induced Membrane Permeability and the Effect of Adriamycin on the Energy Metabolism of Ehrlich Ascites Tumor Cells[a]

ARISTIDE FLORIDI, ANNA BAGNATO, CARLO BIANCHI,
MAURIZIO FANCIULLI, BRUNO SILVESTRINI,[b] AND
ANTONIO CAPUTO

Regina Elena Institute for Cancer Research
Rome, Italy
[b]*Institute of Pharmacology and Pharmacognosy*
University of Rome "La Sapienza"
Rome, Italy

It has been demonstrated that lonidamine (LND) potentiates the effect of adriamycin (ADM) by increasing membrane permeability of Ehrlich ascites tumor cells, thus allowing low ADM concentration (18 μM) to interfere also with oxidative metabolism.[1] In order to establish whether the enhanced ADM susceptibility of LND-treated cells might be ascribed to a modified mitochondrial permeability, the effect of LND on cytochrome c oxygen reductase, a marker of the extent of outer mitochondrial membrane permeabilization, was evaluated.

LND does not significantly inhibit (<3% at 200 μM) cytochrome oxidase,[2] which, on the contrary, is strongly affected by ADM.[3] Therefore, the decrease in oxygen consumption at the third energy-conserving site by the association of ADM + LND cannot be ascribed to LND.

FIGURE 1A shows the concentration-dependent effect of ADM on N',N',N',N'-tetramethyl-p-phenyleudiamine (TMPD) oxidation in Ehrlich ascites tumor mitochondria. Adriamycin, up to 150 μM, does not affect the rate of oxygen consumption, but, with higher drug concentrations, a sharp decline of the respiratory rate occurs.

When mitochondria are preincubated with LND (1 min) (FIG. 1B), the addition of 150 μM ADM decreases the respiration rate and the extent of inhibition increases linearly with LND concentration. Because LND does not affect the electron transfer in the cytochrome $c \rightarrow$ oxygen segment of the respiratory chain,[2] this inhibition might depend on a greater interaction of ADM molecules with their inhibitory sites. In order to verify this hypothesis, we evaluated the capacity of ADM (FIG. 1C) and LND (FIG. 1D) to influence the accessibility of exogenous ferrocytochrome c to inner-membrane reduction sites. FIGURE 1C shows that ADM does not modify the rate of electron

[a]This work was supported by Grant PFO 87.01532.44 from the Consiglio Nazionale Ricerche (CNR) and by Associazione Italiana Ricerche sul Cancro (AIRC) 88.

transfer from exogenous reduced cytochrome c to respiratory chain, thus indicating that the drug does not increase the outer membrane permeability. When the oxidation of externally added ferrocytochrome c is evaluated in LND-treated mitochondria, oxygen consumption is markedly stimulated, which must be related to an increased permeability of the outer mitochondrial membrane achieved by LND.

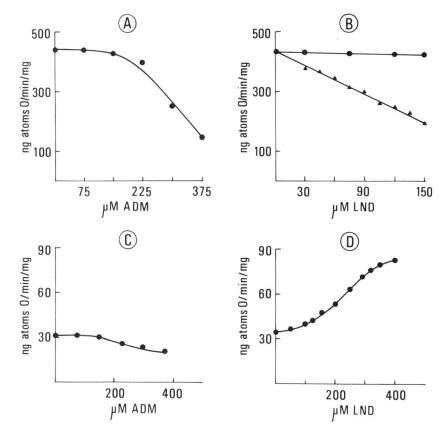

FIGURE 1. Effect of **(A)** ADM and **(B)** LND [●] and ADM + LND [▲] on energy consumption at the energy-conserving site 3 of the respiratory chain in Ehrlich ascites tumor mitochondria. The concentration of ADM in **(B)** was 150 μM. Effect of **(C)** ADM and **(D)** LND on the cytochrome c:oxygen oxido-reductase activity. Each point was averaged from five different mitochondrial preparations and yielded reproducible results ($\pm 4\%$).

These results emphasize the role of intracellular membranes in eliciting a response to cytotoxic drugs and indicate that the cell sensitivity also depends on the physical status of the membranes. Furthermore, the data reported here strongly suggest the possibility of obtaining and/or enhancing the cytotoxicity of low ADM concentrations, thus reducing the toxic side effects of this drug.

REFERENCES

1. FLORIDI, A., A. GAMBACURTA, A. BAGNATO, C. BIANCHI, M. G. PAGGI, B. SILVESTRINI, A. CAPUTO. 1988. Modulation of adriamycin uptake by lonidamine in Ehrlich ascites tumor cells. Exp. Mol. Pathol. In press.
2. FLORIDI, A. & A. L. LEHNINGER. 1983. Action of the antitumor and antispermatogenic agent lonidamine on electron transport in Ehrlich ascites tumor mitochondria. Arch. Biochem. Biophys. **226:** 73–83.
3. BIANCHI, C., A. BAGNATO, M. G. PAGGI & A. FLORIDI. 1987. Effect of adriamycin on electron transport in rat heart, liver and tumor mitochondria. Exp. Mol. Pathol. **45:** 123–135.

Flow Cytometric Measurements of Cytosolic [Ca^{2+}] in Normal and Leukemic Progenitor Cells

DANIEL S. COWEN,[a] HILLARD M. LAZARUS,[b] AND
GEORGE R. DUBYAK[a,c]

Department of [a]Pharmacology, [b]Medicine, and [c]Physiology and
Biophysics, and the Ireland Cancer Center

[c]Case Western Reserve University
School of Medicine
Cleveland, Ohio 44106

Well-characterized activators of neutrophils and monocytes, including formylated chemotactic peptides (FMLP), leukotriene B$_4$ (LTB$_4$), and platelet-activating factor (PAF) bind to surface receptors coupled to the phosphoinositide-specific phospholipase-C (phosphoinositidase).[1-3] In previous studies we have demonstrated that extracellular ATP can similarly stimulate phosphoinositidase, and consequent mobilization of intracellular Ca^{2+} stores.[4] Our present studies are concerned with determining when, developmentally, neutrophil and monocyte precursor cells begin expressing the various receptors coupled to phosphoinisotidase. Agonist-induced Ca^{2+} mobilization was monitored with the fluorescent Ca^{2+} indicators Fura-2 and Indo-1. Fura-2-loaded cell suspensions were used for measurements of mean cytosolic [Ca^{2+}] changes occurring in particular cell populations, while Indo-1 loaded cells were monitored by flow cytometry for calculations of the percentage of cells capable of responding to specific agonists.

Mean population changes in cytosolic [Ca^{2+}] elicited in Fura-2-loaded cell supensions are difficult to interpret for heterogeneous cell populations. For example, in Fura-2-loaded mononuclear blood cells isolated from a patient with leukemia classified as FAB-M4, containing 90% blasts (myeloblasts and promonocytes) and 10% lymphocytes, 100 μM ATP elicited only a three-fold rise in cytosolic [Ca^{2+}] from a basal level of 137 nM. This less than maximal rise could be interpreted as showing either: (1) that most cells responded to ATP by mobilizing only a small fraction of their Ca^{2+} stores and/or (2) that only a subpopulation of cells was responsive. However, flow cytometric analysis of Indo-1-loaded cells demonstrated that 76% of the mononuclear cells were capable of responding to ATP with small Ca^{2+} transients (FIG. 1). This was in contrast to the less than 1% of cells capable of responding to 30 μM FMLP.

Of interest, cells from the established HL-60 promyelocyte line showed heterogeneity in receptor expression. Ten (10) μM ATP elicited in Fura-2-loaded cells greater than 10-fold mean increases in cytosolic [Ca^{2+}] from a basal level of 180 nM. In contrast, 30 μM FMLP produced only 1.5-fold mean elevations. Flow cytometric analysis of Indo-1-loaded cells revealed that whereas 94 $\pm$ 1% ($n = 3$) of undifferentiated HL-60 cells responded to ATP with large Ca^{2+} transients, FMLP elicited only small transients in 69 $\pm$ 7% ($n = 3$) of cells (FIG. 2). In contrast, differentiation of

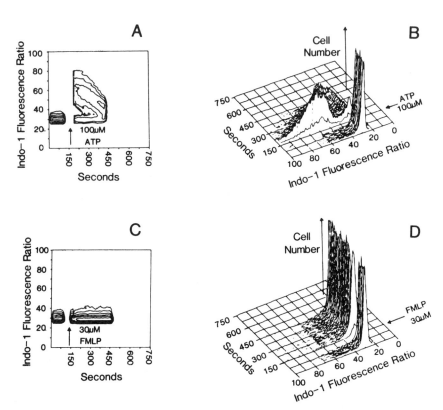

FIGURE 1. Flow cytometric measurements of ATP- or FMLP-induced changes in cytosolic $[Ca^{2+}]$ in individual cells isolated from a patient with FAB M4 leukemia. Indo-1-loaded blood mononuclear cells isolated from a patient with myelomonocytic leukemia (FAB M4), comprising 90% blasts (myeloblasts and promonocytes) and 10% lymphocytes, were stimulated at the indicated times with ATP (**A** and **B**) or FMLP (**C** and **D**) and examined by flow cytometry. In **A** and **C** data are displayed as contour plots of time versus Indo-1 ratio of violet/blue emission (which is proportional to cytosolic $[Ca^{2+}]$). In **B** and **D** data are displayed as time versus Indo-1 ratio of violet/blue emission vs. number of cells.

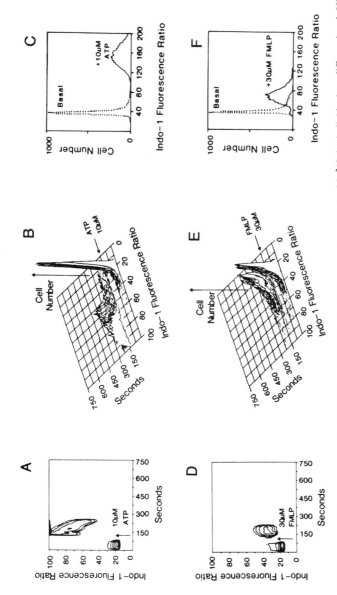

FIGURE 2. Flow cytometric measurements of ATP- or FMLP-induced changes in cytosolic [Ca^{2+}] in individual undifferentiated HL-60 cells. Indo-1-loaded HL-60 cells were stimulated at the indicated times with ATP (A–C) or FMLP (D–F) and examined by flow cytometry. In A and D data are displayed as contour plots of time versus Indo-1 ratio of violet/blue emission (which is proportional to cytosolic [Ca^{2+}]). In B and E data are displayed as time versus Indo-1 ratio of violet/blue emission versus number of cells. C and F display histograms of the ration of Indo-1 violet/blue emission versus number of cells for HL-60 cells prior to stimulation (*dotted lines*) and for cells during the period 35–60 seconds after addition of agonist (*solid lines*).

HL-60 cells after 48 hours[5] of treatment with 500 μM dibutyryl cAMP, a membrane-permeable analogue of cyclic AMP, caused nearly all cells to respond to ATP and FMLP equally, with large Ca^{2+} transients.

REFERENCES

1. CONRAD, G. W. & T. J. RINK. 1986. J. Cell. Biol. **103:** 439–450.
2. NACCACHE, P. H., T. F. P. MOLSKI, P. BORGEAT, & R. I. SHA'AFI, 1985. J. Cell. Physiol. **122:** 273–280.
3. LEW, P. D., A. MONOD, F. A. WOLDVOGEL, & T. POZZAN. 1987. Eur. J. Biochem. **162:** 161–168.
4. COWEN, D. S., H. L. LAZARUS, S. E. STOLL, S. B. SHURIN, & G. R. DUBYAK. Submitted for publication.
5. CHAPLINSKI, T. J. & J. E. NIEDEL. 1982. J. Clin. Invest. **70:** 953–964.

Bicarbonate Abolishes Intracellular Alkalinization in Mitogen-Stimulated NIH 3T3 Cells

BENJAMIN S. SZWERGOLD, TRUMAN R. BROWN, AND
JEROME J. FREED

Fox Chase Cancer Center
Philadelphia, Pennsylvania 19111

Regulation of intracellular pH (pH_i) is an important aspect of the process of intracellular homeostasis.[1] Recently a number of studies have shown that a prompt and persistent increase in pH_i occurs after mitogenic stimulation of cells. This intracellular alkalinization has been proposed to be a part of the process of cellular response to stimulation.[2-4]

We have tested this hypothesis on NIH 3T3 cells by ^{31}P NMR in a system in which

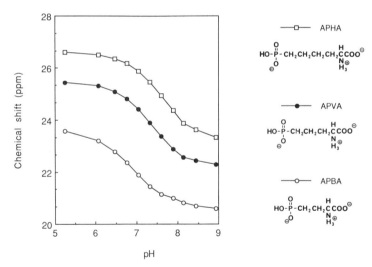

FIGURE 1. Structures and titration curves of the phosphonoamino acid ^{31}P NMR pH indicators: 2-amino-6-phosphono hexanoic acid (APHA), $pK_a = 7.60$; 2-amino-5-phosphono valeric acid (APVA), $pK_a = 7.35$; 2-amino-4-phosphono butyric acid (APBA), $pK_a = 6.90$

pH_i can be monitored continously over many hours as the cells are stimulated by mitogen and go through the cell cycle. Using novel amino-acyl phosphonate pH_i indicators (FIG. 1), we have determined that in nonphysiological, bicarbonate-free media, stimulation of cells with serum results in a transient (up to 1 hour) alkalinization of 0.16 pH units (FIG. 2B). This finding is in agreement with previous reports, most of which were carried out in the absence of bicarbonate. In contrast, when an identical experiment is performed in physiological bicarbonate-buffered media, mitogenic stimulation of cells has no effect on pH_i (FIG. 2C). Not only is pH_i clearly steady at 7.25 during the first hour after exposure to serum, but also preliminary experiments show that it remains constant at that value as these cells progress through the cell cycle. It should be noted that in this system we can observe the proliferation of the stimulated cells (FIG. 3), and therefore we conclude that an elevation in pH_i is not necessary for mitogenic stimulation of NIH 3T3 cells.

Our results, while strictly applicable only to the NIH 3T3 cells, are in agreement with previous reports on other cell types such as human lymphocytes[5] and the A431 epidermoid carcinoma cells.[6] The cumulative impact of our studies and the prior reports refutes the hypothesis that an increase in pH_i is a signal of any sort in the process of mitogenic stimulation.

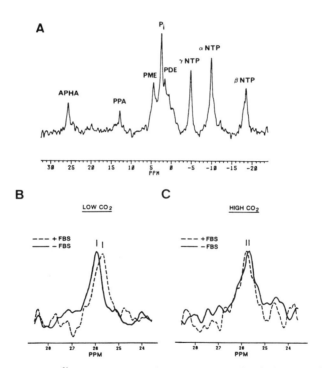

FIGURE 2. (A) A typical ^{31}P NMR spectrum of NIH 3T3 cells perfused with serum-free DMEM at pH 7.0 and 37°C. Peaks are identified as follows: APHA, intracellular pH indicator 2-amino-6-phosphono-hexanoic acid; PPA, extracellular pH indicator phenyl phosphonic acid; PME, phosphate monoesters; P_i, inorganic phosphate; PDE, phosphodiesters; NTP, nucleoside triphosphates. The P_i signal observed is due to extracellular phosphate in the perfusion medium. (B) Spectra of intracellular APHA in a low bicarbonate medium ($pCO_2 = 0.8$ Torr, $pH_e = 6.95$) before (60 minutes, *solid line*) and after (40 minutes, *dashed line*) the addition of 15% FBS. (C) Spectra of intracellular APHA in bicarbonate-buffered medium before (90 minutes, *solid line*) and after (50 minutes, *dashed line*) addition of 15% FBS.

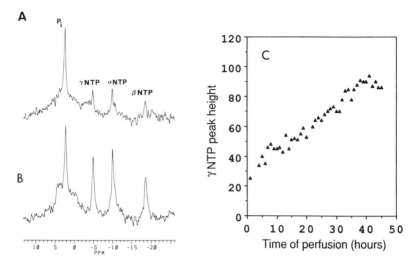

FIGURE 3. (A) ^{31}P NMR spectrum of NIH 3T3 cells at the beginning of perfusion. Cell density 4×10^6 cells/ml beads. (B) ^{31}P NMR spectrum of the same cell sample after 42 hours of perfusion in the NMR spectrometer. Cell density 12×10^6 cells/ml beads. Acquisition parameters were identical to those used in (A). (C) Time course of increase in the peak height of the γ NTP during the course of the experiment.

REFERENCES

1. MADHUS, I. H. 1988. Regulation of intracellular pH in eukaryotic cells. Biochem. J. **250:** 1–8.
2. MOOLENAAR, W. H. 1986. Effects of growth factors on intracellular pH regulation. Annu. Rev. Physiol. **48:** 363–376.
3. ROZENGURT, E. 1986. Early signals in the mitogenic response. Science **234:** 161–166.
4. LAGARDE, A. E. & J. M. POUYSSÉGUR. 1986. The Na$^+$:H$^+$ antiport in cancer. Can. Biochem. Biophys. **9:** 1–14.
5. DEUTSCH, C., J. S. TAYLOR & M. Price. 1984. pH homeostasis in human lymphocytes: Modulation by ions and mitogens. J. Cell Biol. **98:** 885–894.
6. CASSEL, D. *et al.* 1985. Mitogen independent activation of Na$^+$/H$^+$ exchange in human epidermoid carcinoma A431 cells: Regulation by medium osmolarity. J. Cell Physiol. **122:** 178–186.

Neuropeptides Elevate Cytosolic Calcium in Small-Cell Lung Cancer Cells[a]

T. W. MOODY, S. MAHMOUD, J. STALEY,
AND G. FISKUM

Department of Biochemistry
The George Washington University
School of Medicine and Health Sciences
Washington, D.C. 20037

Small-cell lung cancer (SCLC) is a neuroendocrine tumor which is enriched in its neuroendocrine properties.[1] SCLC cells have high levels of neuronal enzymes and polypeptide hormones including bombesin (BN)-like peptides and neurotensin (NT). Upon secretion from the SCLC cells the neuropeptides may diffuse and bind to receptors present on the cell surface. SCLC cells have receptors that bind BN and the structurally related gastrin-releasing peptide (GRP) with high affinity.[2] Also, SCLC cells have receptors for NT, vasoactive intestinal polypeptide (VIP), and somatostatin (SRIF).[3-5] VIP and SRIF receptors interact with adenylate cyclase positively and negatively, respectively.[5,6] BN-like peptides stimulate phosphatidylinositol turnover and the resulting inositol-1,4,5-trisphosphate released elevates cytosolic Ca^{2+} levels.[7-9] Here we investigated whether neuropeptides alter the cytosolic Ca^{2+} levels in SCLC cells.

SCLC cell line NCI-H345 was cultured in SIT medium (RPMI-1640 containing 3×10^{-8} M Na_2SeO_3, 5 µg/ml insulin and 10 µg/ml transferrin) supplemented with 2.5% heat-inactivated fetal calf serum, and was then harvested. The cells (2.5×10^6/ml) were incubated with 5 µM fura 2 AM and unloaded fura 2 was removed by centrifugation at $150 \times g$ for 3 minutes. Cells were resuspended in SIT containing 10 mM HEPES (pH 7.4) and after addition of the peptides, the fluorescence intensity was determined using an excitation wavelength of 340 nm and an emission wavelength of 510 nm.

Previously, we showed that BN or GRP elevated cytosolic Ca^{2+} levels in a dose-dependent manner. The half-maximal effective dose (ED_{50}) for BN or GRP was 1 nM and BN-like peptides released Ca^{2+} from intracellular stores. The C-terminal but not the N-terminal of BN or GRP was essential for biological activity. Also, substance P (SP) analogues such as (D-Arg^1, D-Pro^2, D-Trp^{7,9}, Leu^{11})SP (10 µM) functioned as BN receptor antagonists in that they inhibited the increase in cytosolic Ca^{2+} caused by 10 nM BN.[10] Because the SP analogues inhibited the increase in the clonal growth of SCLC caused by BN, they may function as useful agents to disrupt the BN-induced autocrine growth cycle of SCLC.[10]

Here we investigated the ability of neuropeptides to elevate the cytosolic Ca^{2+} levels

[a]This research is supported by Grants CA33767 and CA42306 (to T. W. M.) and Grant CA32946 (to G. F.) from the National Cancer Institute.

in SCLC cells. TABLE 1 shows that at a $1-\mu M$ dose BN or GRP elevated the cytosolic Ca^{2+} levels from 150 to 200 mM using cell line NCI-H345 after approximately 15 seconds. There was then a slow decline in the cytosolic Ca^{2+} levels, which may be due to receptor desensitization. After 4 minutes the Ca^{2+} levels returned to basal levels. Similarly, NT elevated the cytosolic Ca^{2+} levels from 150 to 250 nM. NT increased the cytosolic Ca^{2+} levels to maximal values after 15 seconds, and then the cytosolic Ca^{2+} levels slowly declined and after 2 minutes returned to basal levels. Because EGTA (5 mM) had no effect on the ability of NT to elevate cytosolic Ca^{2+} levels, NT may release Ca^{2+} from intracellular stores. If NT was added 4 minutes after the initial addition of NT, the cytosolic Ca^{2+} levels were unchanged because all the NT receptors were occupied. If BN was added after NT administration, however, the cytosolic Ca^{2+} levels increased identical to those of BN alone. Similarly, if NT was added after BN administration, the cytosolic Ca^{2+} levels increased identical to that of NT alone. These data indicate that NT or BN do not deplete the intracellular Ca^{2+} stores. Whereas 1 μM VIP increased the intracellular cAMP levels 10-fold, it had no effect on the cytosolic Ca^{2+} levels. If BN or NT were added after VIP there was an increase in the

TABLE 1. Ability of NT Analogues to Elevate Cytosolic Ca^{2+}

Peptide	Ca^{2+} Response
BN	Strong
GRP	Strong
NT	Strong
VIP	Inactive
SRIF	Inactive

NOTE: The ability of various neuropeptides at a $1-\mu M$ dose to elevate cytosolic Ca^{2+} levels in SCLC cell line NCI-H345 was determined. BN caused a transient elevation in the cytosolic Ca^{2+} within 30 seconds after addition from 150 to 200 nM, whereas NT increased the cytosolic Ca^{2+} from 150 to 270 nM.

cytosolic Ca^{2+} levels identical to that of BN or NT alone. These data indicate cyclic AMP-dependent protein kinases have no effect on the ability of BN or NT to elevate cytosolic Ca^{2+} levels.

In summary, NT and BN but not VIP or SRIF elevate cytosolic Ca^{2+} levels in SCLC cells. BN and NT receptors may desensitize after administration of high neuropeptide doses. BN and NT elevate the cytosolic Ca^{2+} levels through distinct neuropeptide receptors.

REFERENCES

1. CARNEY, D. N., A. F. GAZDAR, G. BEFPLER, J. G. GUCION, P. J. MARANGOS, T. W. MOODY, M. H. ZWEIG & J. D. MINNA. 1985. Cancer Res. 45: 2913–2923.
2. MOODY, T. W., D. N. CARNEY, F. CUTTITA, K. QUATTROCCHI & J. D. MINNA. 1985. Life Sci. 36: 105–113.
3. SCHAFFER, M. M., D. N. CARNEY, L. Y. KORMAN, G. S. LEBOVIC & T. W. MOODY. 1987. Peptides 8: 1101–1108.
4. ALLEN, A. E., D. N. CARNEY & T. W. MOODY. 1988. Peptides 9: 57–61.

5. KEE, K., L. Y. KORMAN & T. W. MOODY. 1988. Peptides **9:** 257–261.
6. KORMAN, L. Y., D. CARNEY, M. CITRON & T. W. MOODY. 1986. Cancer Res. **46:** 1214–1218.
7. MOODY, T. W., A. MURPHY, S. MAHMOUD & G. FISKUM. 1987. Biochem. Biophys. Res Commun. **147:** 189–195.
8. TREPEL, J. B., J. MOYER, R. HEIKKILLA, L. M. NECKERS & E. A. SAUSVILLE. 1987. Reg. Peptides **19:** 141.
9. HEIKKILA, R., J. B. TREPEL, F. CUTTITTA, L. M. NECKERS & E. A. SAUSVILLE. 1987. J. Biol. Chem. **262:** 16456–16460.
10. MOODY, T. W., S. MAHMOUD, A. KOROS, F. CUTTITTA, J. WILLEY, M. ROTSCH, U. ZEYMER & G. BEPLER. 1986. Fed. Proc. **46:** 2201.

Regulation of the Expression of Cell Cycle Genes

RENATO BASERGA, BRUNO CALABRETTA,
SALVATORE TRAVALI, DARIUSZ JASKULSKI,
KENNETH E. LIPSON, AND J. KIM DeRIEL

*Department of Pathology and
Fels Research Institute
Temple University Medical School
Philadelphia, Pennsylvania 19140*

INTRODUCTION

It is generally acknowledged that the study of cell proliferation in animal cells can be reduced to a very simple proposition, namely, that the extent of proliferation depends on (1) the environmental signals (that is, the stimulatory and inhibitory growth factors in the environment that regulate cell proliferation) and (2) the genes and gene products that interact with and respond to the growth factors. In our laboratory we have been mostly concerned with the second part of this proposition. In fact, it was in our laboratory that we first provided the formal demonstration that unique-copy gene transcription is necessary for the transition of cells from a resting (G_0) to a growing stage (for a review, see Baserga[1]). This demonstration served as the basis for the search for genes that regulate the proliferation of animal cells in culture and, by extrapolation, in the living animal. There are essentially four approaches to the identification of such genes, namely: (1) the identification of proto-oncogenes (i.e., the cellular equivalents of retroviral transforming genes); (2) the isolation and molecular cloning of genes that complement defects in temperature-sensitive mutants of the cell cycle; (3) the identification of growth factors and growth factor receptors; and (4) the identification of sequences that are growth-regulated (i.e., that are preferentially expressed in a specific phase of the cell cycle or are induced by growth factors). This last approach is based on the differential screening of cDNA libraries. The present discussion deals largely with growth-regulated genes.

Growth-regulated genes can be defined in general as genes that are not expressed in quiescent nonproliferating cells, but that become expressed when cells are induced to proliferate by the addition of appropriate growth factors. The proliferation of BALB/c3T3 cells (and of other cells like human diploid fibroblasts) is exquisitely regulated by growth factors.[2] A picture is emerging that indicates that in general two growth factors are necessary and sufficient to induce the transition of cells from a G_0 state to the S phase.[1,2] For instance, in BALB/c3T3 cells the two growth factors that are necessary for the induction of cellular DNA synthesis are platelet-derived growth factor (PDGF) and insulin-like growth factor 1 (IGF1). Some growth-regulated genes are inducible by PDFG only, whereas other growth-regulated genes require both PDGF and IGF1.

Although not well established, it is widely believed that the most important and critical events for the transition of cells from G_0 back into the cell cycle occur very early after stimulation. It is a reasonable assumption which has led us in the past to attribute great importance to events occurring in the first one to two hours after stimulation of quiescent cells with growth factors. More recently this assumption has been reinforced by the finding that the expression of certain cellular oncogenes is markedly increased in the first hours after G_0 cells are stimulated.[3] However, it has also become evident that these same oncogenes are stimulated by agents that not only do not induce cell proliferation, but actually induce cell differentiation and decrease cell proliferation.[3] This has led us to the conclusion that early events after stimulation of G_0 cells by growth factors are part of a common response of cells to a variety of environmental signals (which include proliferative stimuli), while the events occurring in late G_1 are specific for cell proliferation. It is, though, an important part of this conclusion that under ordinary circumstances the proliferation specific events cannot take place unless the cell has been previously primed by its initial response, that is, by the early events.

On this basis, we have, in our laboratory, recently focused on the regulation of those genes that require both PDGF and IGF1 for inducibility. PDGF-inducible genes (also called early genes or competence genes) are certainly important and include well known oncogenes like c-*fos*, and c-*myc*. Critical controls of cell proliferation, though, probably occur sometime later in the cell cycle, presumably in late G_1 or at the G_1-S boundary. On this assumption we have investigated the regulation of genes that are part of the DNA-synthesizing machinery.

The proteins of the DNA-synthesizing machinery constitute a group of gene products that are generally expressed co-ordinately at the G_1-S boundary of the cell cycle. We have investigated how growth factors regulate the expression of two of these genes: the one coding for the proliferating cell nuclear antigen, PCNA, and the one for thymidine kinase, TK. PCNA is a nuclear protein that has been recently identified as the auxiliary protein of DNA polymerase-delta.[4,5] TK is a member of a group of enzymes involved in the synthesis of cellular DNA. TK enzyme activity,[6] and TK mRNA steady-state levels,[7-9] increase sharply when cells enter S phase. Because of their close association with DNA synthesis, PCNA and TK can be used to study the regulation of expression of an important group of genes that require both PDGF and IGF1 for inducibility. Genes of this group could act as primary agents in the control of cell cycle progression and, therefore, cell proliferation.

RESULTS

Growth Factor Regulation of the PCNA and TK Genes

These studies were carried out in BALB/c3T3 cells which, as mentioned above, are exquisitely sensitive to growth factors. Steady-state mRNA levels were determined in G_0 cells and in cells stimulated by serum or growth factors. TABLE 1 summarizes the effect of serum or plasma on the mRNA levels of PCNA and TK. It is clear that neither is truly inducible by platelet-poor plasma, indicating that both genes require both PDGF and IGF1 for inducibility. The modest expression of TK at 16 and 24 hours after stimulation with platelet-poor plasma is due to a small fraction of cycling cells that are always present in every quiescent population. This is confirmed by the fact that the same Northern blots indicated that at 16 and 24 hours after stimulation with platelet-poor plasma there was a small amount of expression of histone mRNA (TABLE 1). The cells that respond to platelet-poor plasma are probably cells that have already been primed by PDGF or have, at least, responded already with the expression of early genes and can be induced to enter S phase by the second growth factor only.

With individual growth factors the results were as follows[10]: (*a*) PDGF alone, or in combination with EGF and insulin, increased PCNA mRNA levels; (*b*) PDGF alone did not affect TK mRNA levels, which increased only when a combination of growth factors caused the cells to enter S phase; (*c*) PCNA, but not TK, was induced by EGF alone. However, the induction of PCNA by EGF was not sustained and the cells did not enter S phase. It should be noted that growth factors induced a sustained increase in either PCNA or TK only when stimulation of DNA synthesis also occurred. We conclude that for the full expression of PCNA and TK, both growth factors, PDGF and IGF1, are required.

Coppock and Pardee[8] have shown that low concentrations of cycloheximide (100 ng/ml) completely suppressed induction of TK mRNA by serum. The same conditions also inhibit the increase in PCNA mRNA levels that occurs 24 hours after serum stimulation. This is significant because all of the early genes that have been tested,

TABLE 1. Steady-State mRNA Levels of PCNA and TK in BALB/c 3T3 Cells

	G_0	8 Hours	16 Hours	24 Hours
PCNA				
(S)	+	+ +	+ + +	+ + + +
(P)		+	−	−
TK				
(S)	−	−	+ +	+ + +
(P)		+	+	+
Histone H3				
(S)	−	−	+	+ + + +
(P)	−	−	−	+

NOTE: G_0 cells were incubated with either 10% fetal calf serum (S) or platelet-poor plasma (P). RNA was extracted at the times indicated and the amounts were estimated on Northern blots.

including c-*myc,* c-*fos* and others, are cycloheximide-insensitive (that is, the levels of mRNA still increase after growth factor stimulation, even in the presence of very high concentrations of cycloheximide that cause a 99% inhibition of protein synthesis). The inability of cycloheximide to inhibit the inducibility of early genes by growth factors has been interpreted as signifying that the expression of these genes does not require *de novo* protein synthesis, and specifically that it does not require products of other growth-factor-inducible genes. This is obviously not true in the case of PCNA and TK.

Expression of PCNA and TK mRNAs in a G_1 Specific ts *Mutant of the Cell Cycle*

We used as the *ts* mutant *ts*13 cells that arrest in G_1 at the restrictive temperature.[11] In these cells PCNA and TK are similarly regulated. Both of them are induced by serum at the permissive temperature, but no induction occurs if the quiescent cells are stimulated at the restrictive temperature.[7,10] These experiments, and those with cycloheximide, clearly indicate that these genes behave in a different way than do the early genes and that they require for full induction the product of another growth-factor-inducible gene.

The Role of the Promoter in the Regulation of the Expression of the TK Gene

Four reports have appeared in the literature stating that cell cycle regulation of the TK gene is determined by its coding sequence, and not by the promoter or other regulatory regions.[9,12–14] To further test this hypothesis we have established cell lines carrying different chimeric constructs of the TK gene.[15] The results are summarized in TABLE 2. Clearly, regulation of expression of the TK gene, in terms of TK mRNA levels, depends strongly on the type of promoter used. If a G_1 promoter is used, like that of calcyclin,[16] the TK mRNA is expressed very strongly in G_1 and decreases in S phase. If the promoter of the heat-shock protein HSP70 is used to drive the TK cDNA, the induction of TK mRNA depends exclusively on the heat shock, regardless of the

TABLE 2. Regulatory Elements of the TK Gene

Constructs		
Promoter	Coding Sequence	Maximal Expression of TK mRNA
SV40	TKcDNA	S phase
Calcyclin	TKcDNA	G_1 phase
HSP-70	TKcDNA	Any phase, but only after heat shock
Human TK	CAT	S phase

NOTE: The appropriate constructs were transfected into BALB/c3T3 cells, and cell lines were established. Gene expression was determined by RNA blots.

position of the cell in the cell cycle. Conversely, if a human TK promoter is used to drive the CAT gene, CAT expression is limited to S phase. The contention of Stewart et al.[9] that the cDNA of TK regulates its cell-cycle-dependent expression was based on the fact that they could not see any difference in the levels of mRNA when the TK cDNA was placed either under its own promoter or the promoter of SV40. It seems, though, that while the promoter of SV40 responds to growth factors, it is not growth-regulated like the TK promoter. This is illustrated in FIGURE 1. In this experiment BALB/c3T3 cells were stimulated with serum with or without the addition of cycloheximide. In the presence of cycloheximide (100 ng/ml) 8 hours after serum stimulation, the induction of TK mRNA is suppressed as is the induction of histone H3 mRNA). However, in cells that carry, in addition to their endogenous TK gene, an SV40 TK cDNA construct the results are quite surprising. Although the endogenous TK gene is regulated as usual (that is, inhibited by cycloheximide), the expression of the SV40 TK cDNA construct is actually increased by cycloheximide. These experiments conclusively demonstrate that the SV40 TK cDNA construct is regulated in a totally different way from the endogenous TK gene.

FIGURE 1. Effect of cycloheximide on the expression of TK genes. (A) BALB/c3T3 cells. (B) 3T3KL cells (that is, BALB/c3T3 carrying a construct in which the human TK cDNA is driven by the SV40 early promoter). RNA was extracted from cells, blotted and hybridized to probes for TK and histone H3. *Lane 1:* G_0 cells; *lane 2:* cells stimulated for 24 hours with serum; cycloheximide (100 ng/ml) was added 8 hours after stimulation; *lane 3:* cells serum-stimulated for 24 hours in the absence of cycloheximide.

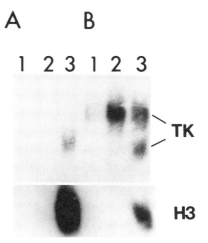

Anti-Sense Oligodeoxynucleotides to PCNA Inhibit Cell Proliferation

The TK gene codes for a dispensable protein. Cells in culture, and some animals, have no TK and yet they not only survive but can also proliferate very vigorously. We investigated whether PCNA is necessary for DNA synthesis and cell cycle progression. For this purpose we synthesized sense and anti-sense oligodeoxynucleotides of the first 18 nucleotides of PCNA immediately after the AUG codon. The oligodeoxynucleotides were added to exponentially growing BALB/c3T3 cells. Anti-sense, but not sense, oligodeoxynucleotides effectively inhibited cellular DNA synthesis and cellular proliferation (TABLE 3). The anti-sense oligodeoxynucleotides also inhibited the expression of the PCNA protein.[17]

CONCLUSIONS

The three components of the cell cycle that are present in all animal cells are: doubling of cell size, DNA replication, and mitosis.[1] Genes inducible by growth factors can be divided into two large groups: those whose expression is cycloheximide-resistant

TABLE 3. Effect of Anti-Sense Oligodeoxynucleotides to PCNA on the Growth of BALB/c3T3 Cells

Treatment	Percent of Labeled Cells	Mitoses/1,000
Anti-sense oligodeoxynucleotides	<1	0
Sense oligodeoxynucleotides	48	16
Controls	47	10

NOTE: Both percentage of labeled cells and mitotic index were determined 24 hours after addition of either anti-sense or sense oligodeoxynucleotides of PCNA. Labeling with [^{3}H]thymidine was done for 30 minutes prior to fixation.

(early genes, competence genes) and those whose expression is inhibited by cyclohex-imide. Of the early genes, one can say that (*a*) there are many; (*b*) there is a lot of redundancy; and (*c*) many of them are inducible by agents that do not induce cellular DNA synthesis. Evidence is rapidly accumulating that early G_1 events are nonspecific for cell proliferation and that the critical steps for the regulation of cell cycle progression occur at the G_1/S boundary, when the proteins of the DNA synthesizing machinery make their appearance. It is reasonable to assume that the regulation of the expression of the genes coding for these proteins holds the key to our understanding of how cell proliferation is regulated. Our investigations are directed at identifying the regulatory elements in two of these genes, the PCNA and TK genes, whose inducibility requires two growth factors, PDGF and IGF1. Our hypothesis is that, in animal cells, the rate-limiting event for cell proliferation is in the protein or proteins that activate the genes of the DNA-synthesizing machinery.

REFERENCES

1. BASERGA R. 1985. The Biology of Cell Reproduction. Harvard University Press. Cambridge, MA.
2. STILES, C. D., G. T. CAPONE, C. D. SCHER, H. N. ANTONIADES, J. J. VAN WYK & W. J. PLEDGER. 1979. Dual control of cell growth by somatomedins and platelet-derived growth factor. Proc. Natl. Acad. Sci. USA **76:** 1279–1283.
3. BASERGA, R. & S. FERRARI. 1987. The molecular basis of cell reproduction: Have we been looking in the wrong place? Haematologica **72:** 1–4.
4. BRAVO, R., R. FRANK, P. A. BLUNDELL & H. MACDONALD-BRAVO. 1987. Cyclin-PCNA is the auxiliary protein of DNA polymerase delta. Nature **326:** 515–517.
5. PRELICH, G., C. K. TAN, M. KOSTURA, M. B. MATHEWS, A. G. SO, K. M. DOWNEY & B. STILL. 1987. Functional identity of proliferating cell nuclear antigen and a DNA polymerase delta auxiliary protein. Nature **316:** 517–520.
6. BRENT, T. P., J. A. V. BUTLER & A. R. CRATHORN. 1965. Variations in phosphokinase activities during the cell cycle in synchronous populations of HeLa cells. Nature **207:** 176–177.
7. LIU, H. T., C. W. GIBSON, R. R. HIRSCHHORN, S. RITTLING, R. BASERGA & W. E. MERCER. 1985. Expression of thymidine kinase and dihydrofolate reductase genes in mammalian ts mutants of the cell cycle. J. Biol. Chem. **260:** 3269–3274.
8. COPPOCK, D. L. & A. B. PARDEE. 1987. Control of thymidine kinase mRNA during the cell cycle. Mol. Cell Biol. **7:** 2925–2932.
9. STEWART, C. J., M. ITO & S. E. CONRAD. 1987. Evidence for transcriptional and post-transcriptional control of the cellular thymidine kinase gene. Mol. Cell. Biol. **7:** 1156–1163.
10. JASKULSKI, D., C. GATTI, S. TRAVALI, B. CALABRETTA & R. BASERGA. 1988. Regulation of the PCNA-cyclin and thymidine kinase mRNA levels by growth factors. J. Biol. Chem. **263:** 10175–10179.
11. TALAVERA, A. & C. BASILICO. 1977. Temperature-sensitive mutants of BHK cells affected in cell cycle progression. J. Cell Phys. **92:** 425–436.
12. MERRILL, G. F., S. D. HAUSCHKA & S. L. MCKNIGHT. 1984. TK enzyme expression in differentiating muscle cells is regulated through an internal segment of the cellular TK gene. Mol. Cell. Biol. **4:**1777–1784.
13. HOFBAUER, R., E. MULLER, C. SEISER & E. WINTERSBERGER. 1987. Cell cycle regulated synthesis of stable mouse thymidine kinase mRNA is mediated by a sequence within the cDNA. Nucleic Acids Res. **15:** 741–751.
14. LEWIS, J. A. & T. A. MATKOVITCH. 1986. Genetic determinants of growth phase-dependent and adenovirus V responsive expression of the Chinese hamster thymidine kinase gene are contained within thymidine kinase mRNA sequences. Mol. Cell. Biol. **6:** 2262–2266.

15.   TRAVALI, S., K. E. LIPSON, D. JASKULSKI, E. LAURET & R. BASERGA. 1987. The role of the promoter in the regulation of the thymidine kinase gene. Mol. Cell. Biol. **8:** 1551–1557.
16.   FERRARI, S., B. CALABRETTA, J. K. DERIEL, R. BATTINI, F. GHEZZO, E. LAURET, C. GRIFFIN, B. S. EMANUEL, F. GURRIERI & R. BASERGA. 1987. Structural and functional analysis of a growth-regulated gene, the human calcyclin. J. Biol. Chem. **262:**8325–8332.
17.   JASKULSKI, D., J. KIM DE RIEL, W. E. MERCER, B. CALABRETTA & R. BASERGA. 1988. Inhibition of cellular proliferation by anti-sense oligodeoxynucleotides to PCNA cyclin. Science. **240:** 1544–1546.

DISCUSSION

R. LOTAN (*University of Texas, Houston, Texas*): Deciding what time is the right time to construct a cDNA after growth inhibition or stimulation is a problem scientists face. Would you suggest, in the context of proliferation-regulating genes, making a cDNA library at mid-to-late part of G_1 stage for different cell types?

R. BASERGA (*Temple University, Philadelphia, Pennsylvania*): Any time is right; it depends on what you want.

G. GUIDOTTI (*University of Parma, Italy*): You mentioned as a necessary gene for the cell to enter the S-phase a co-factor of delta-polymerase. Can you comment on its function at the molecular level?

BASERGA: It is the auxiliary factor of DNA polymerase delta.

Transforming Growth Factor-β: Multifunctional Regulator of Cell Growth and Phenotype

LALAGE M. WAKEFIELD,

NANCY L. THOMPSON, KATHLEEN C. FLANDERS,

MAUREEN D. O'CONNOR-McCOURT,[a]

AND MICHAEL B. SPORN

Laboratory of Chemoprevention
National Cancer Institute
Bethesda, Maryland 20892

[a]*Biotechnology Research Institute*
Montreal, Quebec, H4P 2R2, Canada

INTRODUCTION

Transforming growth factor-β (TGF-β) is the founder member of a growing family of structurally related peptides that are involved in the regulation of cell growth and development in organisms as phylogenetically distant as flies and man. TGF-β itself is a multifunctional molecule whose biological effects are highly context-dependent. Thus, depending on the cell type and the cell environment, TGF-β *in vitro* can stimulate or inhibit proliferation, promote or block differentiation, and modulate cellular function.[1,2] Two closely related forms of TGF-β have been identified. Type 1 and type 2 TGF-β share 70% sequence identity and are indistinguishable in most biological assay systems. However, activities unique to each type are emerging,[2] and it can be anticipated that the two types may be differentially regulated. Structurally, the active form is a disulfide-linked homodimer of 25 kDa and the monomeric unit is encoded as the C-terminal 112 amino acids of a 390-residue precursor.[3] With 2 mg extractable TGF-β/kg, platelets are the most concentrated source of TGF-β 1 in the body, probably reflecting an important role for the molecule in wound healing.[4] Bone is also a major source (0.1 mg/kg), and all other tissues examined have detectable, though lower (0.01–0.03 mg/kg) TGF-β levels. Thus, although TGF-β was initially discovered and named in the context of a transformation assay, the ubiquitous distribution of the molecule suggests that it must play a fundamental regulatory role in normal cell physiology.

CONTROL OF EPITHELIAL CELL GROWTH

In contrast to the situation with mesenchymal cells, where TGF-β may either stimulate or inhibit cell growth depending on the context, TGF-β appears to be invariably inhibitory for epithelial cells growing *in vitro*. Thus it has been shown to inhibit the growth of liver, lung, intestine and kidney epithelial cells and keratinocytes (references cited in Ref. 2). Generally the ED_{50} for inhibition is subpicomolar, making TGF-β the most potent known inhibitor of epithelial cell growth. In most cases, these epithelial cells have been shown to secrete TGF-β in culture and all have functional TGF-β receptors, raising the possibility that TGF-β might function as an endogenous autocrine inhibitor of epithelial cell growth *in vivo*.

The act of culturing cells can induce expression of high levels of TGF-β mRNA that are not observed in the same cells *in vivo*.[5] Therefore in order to determine whether TGF-β protein is expressed by epithelial cells *in vivo*, we did immunohistochemical analyses using a polyclonal antiserum against a synthetic peptide corresponding to residues 1–30 of the N-terminal of TGF-β. The micrographs in FIGURE 1 show that there is detectable staining for TGF-β in the epithelial cells of the lung bronchioles and the villi of the small intestine in adult mice, consistent with a possible role for the molecule in *in vivo* regulation of epithelial cell growth. Other epithelial cells that show staining for TGF-β include kidney, liver and skin.[6] *In situ* hybridization techniques are being developed to determine whether the TGF-β is indeed being synthesized in these cells, as opposed to taken up from external sources. Recent work on regenerating liver has suggested that negative regulation by TGF-β may involve paracrine rather than autocrine mechanisms in some systems.[7] Ultimately proof of an involvement of TGF-β in regulation of epithelial cell growth *in vivo* will require the demonstration that agents that antagonize TGF-β action will disrupt the regulatory loop and cause uncontrolled proliferation in the epithelial cell compartment.

Transformation of epithelial cells *in vitro* frequently results in loss of sensitivity to TGF-β. Thus hepatocytes transformed by aflatoxin[8] and normal bronchial epithelial cells[9] or hepatoctyes[10] transfected with the H-*ras* oncogene are no longer growth-inhibited by TGF-β. Since the initiating event in chemical carcinogenesis is frequently a *ras* mutation, this suggests that loss of negative regulation by TGF-β may be an important event in development of epithelial malignancies. One would therefore predict that carcinoma cells from naturally occurring tumors might be defective in TGF-β production or response. The data in TABLE 1 show that we now have examples of human carcinoma lines exhibiting lesions at each step in the TGF-β inhibitory loop. Thus cell growth appears to be regulated by a dynamic interplay of opposing positive and negative effectors, and any impairment in the negative arm of this balanced system may give rise to uncontrolled cell proliferation and contribute to carcinogenesis. For the hormone-dependent breast cancer line MCF-7, treatment with the anti-estrogen tamoxifen appears to reinduce the TGF-β inhibitory loop in these cells,[11] suggesting that pharmacologic agents that enhance TGF-β action may be therapeutically useful in the prevention or treatment of epithelial malignancies.

REGULATION OF TGF-β ACTION

TGF-β mRNA is expressed in virtually every tissue of the adult animal,[6] and detectable amounts of TGF-β protein can be extracted from nearly all tissues. Similarly, receptors for TGF-β have been found on every normal cell type examined so far, and the receptor seems to be relatively unmodulated.[12] This raises the question of how the action of this very potent bioeffector might be regulated, since it is presumably not constitutively active throughout the organism. One major candidate for a regulatory step is suggested by the observation that TGF-β is secreted by virtually all cell

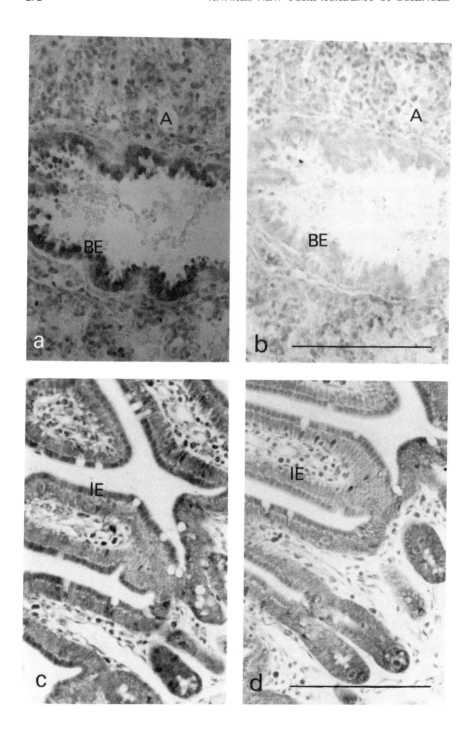

TABLE 1. Carcinoma Cells with Lesions in the TGF-β Autocrine Loop

Human Cell Designation and Type		Lesion
A2380	Pancreatic carcinoma	No TGF-β secretion[19]
A549	Lung adenocarcinoma	Inability to activate latent form[19]
SSC-25	Squamous skin carcinoma	No TGF-β receptors[26]
CaLu 1	Squamous lung carcinoma	Postreceptor defect[19]
HuT 292	Mucoepidermoid lung carcinoma	Postreceptor defect[19]
SW900	Undifferentiated lung carcinoma	Postreceptor defect[19]

types in a biologically latent form.[12] In this form, TGF-β cannot bind to its cellular receptor, nor is it recognized by polyclonal antisera to TGF-β, suggesting that the receptor binding site and other epitopes are somehow masked.[13]

By Western blot analysis and other techniques, we have identified three distinct latent forms of TGF-β, and models for these are shown in FIGURE 2. The latent form of TGF-β secreted by platelets and the majority of cells in culture is a three-component complex with a molecular weight of ~235 kDa.[13] This consists of mature, active TGF-β (25-kDa dimer), noncovalently associated with the remainder of the TGF-β precursor sequence (75-kDa dimer), and a third unidentified component (135 kDa). The precursor sequences alone seem sufficient to confer latency, since the TGF-β complex secreted by recombinant constructs is latent but lacks the third 135-kDa component.[14] Thus the remainder of the precursor is the TGF-β-binding protein, and the 135-kDa component probably has a modulatory role in targeting or activation of the latent form.

The third latent form of TGF-β that we have identified consists of TGF-β bound to alpha$_2$-macroglobulin (α_2M), a major serum protein involved in scavenging and clearing proteases from the circulation.[15] This is the predominant latent form of TGF-β in serum. We have further shown that the latent form of TGF-β secreted by cells and platelets cannot bind to α_2M unless the cell/platelet latent TGF-β is first activated.[13] We therefore propose that the latent forms of TGF-β secreted by cells and platelets represent "delivery" complexes that probably function to extend the half-life of TGF-β in the extracellular milieu, possibly allowing paracrine action at more distant sites, and to restrict the target range of TGF-β to cells that possess a specific activating

---

◄**FIGURE 1.** Immunohistochemical staining for TGF-β 1 in normal murine tissues. Five-day-old neonatal lung (**a-b**) and 6-week-old adult mouse small intestinal (**c-d**) tissue were fixed sequentially in neutral buffered formalin and Bouin's solution and paraffin-embedded. Six-micron sections were dewaxed, hydrated, and pretreated sequentially with hydrogen peroxide to block endogenous peroxidase, with hyaluronidase to permeabilize tissue, and with excess protein to block nonspecific antibody absorption. Sections were then incubated overnight at 4°C with either a polyclonal rabbit IgG raised to a synthetic peptide corresponding to the N-terminal 30-amino-acid sequence of TGF-β 1 (anti-LC(1–30)) (**a,c**), or with an equivalent concentration of normal rabbit IgG (**b,d**). The avidin-biotin-peroxidase detection system was used to localize immunoreactive TGF-β 1. Sections were counterstained with Giemsa and May-Grunwald. (**a**) Anti-LC(1–30) localizes to cytoplasm of epithelial cells (BE) in the terminal bronchioles but not to cells lining the alveoli (A). (**c**) Anti-LC(1–30) localizes to the cytoplasm of intestinal epithelial cells (IE) uniformly along sides and apex of villi with little or no staining of cells in the crypt (C). No staining was observed in control sections incubated with normal rabbit IgG in either lung (**b**) or intestine (**d**). Bar = 100 μm.

mechanism. By contrast, the TGF-β/α_2M complex found in serum probably represents a clearance complex, with α_2M binding any excess active TGF-β, thus keeping its effects local to the activating site. This proposal is consistent with *in vivo* work showing that TGF-β is cleared from the circulation very rapidly, predominantly through the liver.[16]

Chaotropic agents and extremes of pH will activate latent TGF-β *in vitro,* presumably by disrupting the quaternary structure of the complex.[13,17] However, recent work showing that the proteases plasmin and cathepsin D can activate latent TGF-β[18] suggests that activation *in vivo* could involve controlled proteolysis. Since the TGF-β receptor seems to be essentially universally and constitutively expressed,[12] it is likely that activation of TGF-β from its latent form constitutes the major extracellular

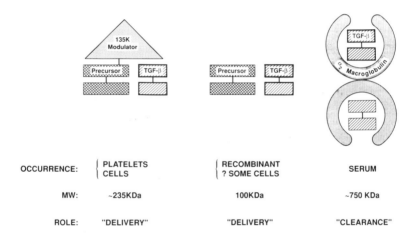

OCCURRENCE:	PLATELETS CELLS	RECOMBINANT ? SOME CELLS	SERUM
MW:	~235KDa	100KDa	~750 KDa
ROLE:	"DELIVERY"	"DELIVERY"	"CLEARANCE"

FIGURE 2. Models for the various biologically latent forms of TGF-β. The subunit designated "precursor" is the sequence that remains when the signal peptide and the mature TGF-β (residues 279–390) are cleaved from the 390-amino-acid biosynthetic precursor. This subunit is glycosylated and exists as a 75-kDa dimer, while mature TGF-β is an unglycosylated 25-kDa dimer. Vertical lines linking subunits indicate disulfide bonds. For further details see the text.

site of regulation in TGF-β action. Thus target tissues for TGF-β would be determined by the ability of a cell type to activate the latent form, or its proximity to other cells with that ability. Unlike the parent bronchial epithelial cell, the human A549 lung adenocarcinoma line continues to proliferate in the presence of the high levels of latent TGF-β that it secretes, although it is potently inhibited by active TGF-β.[19] This suggests that loss of the activating mechanism may have been a contributing step in the malignant progression of these cells. The majority of cell lines in tissue culture appear to be unable to activate latent TGF-β[20] possibly because of negative selection resulting from growth in the presence of serum, which contains latent forms of TGF-β. Work with primary cultures should help to determine which cell types would be able to activate latent TGF-β *in vivo.*

MECHANISM OF ACTION OF TGF-β

Essentially nothing is known about the signal transduction mechanism for TGF-β. Unlike receptors for many other peptide growth factors, the TGF-β receptor family has no intrinsic tyrosine kinase activity and no other enzymic activity has yet been ascribed to it. In systems where TGF-β acts to antagonize the mitogenic effects of other growth factors, it has no effect on all the early responses of the cell to the mitogen, such as phosphorylation, ion fluxes, activation of protein kinase C, or breakdown of phospho-inositides.[21] Thus the pathway of transduction of the TGF-β signal from the extracellular surface to the nucleus is unclear.

At a more distal level, however, it is clear that TGF-β modulates the expression of many proteins that are critically involved in growth and differentiation. Most strikingly, it has pronounced effects on the elaboration of extracellular matrix in many target cell types. Thus it promotes the expression of extracellular matrix proteins and their receptors, while decreasing the expression of proteases involved in matrix degradation or increasing expression of the corresponding protease inhibitors, the net effect being an increase in newly formed extracellular matrix (Ref. 2 and references therein). Many of the biological actions of TGF-β can be explained by its effects on matrix proteins. For instance, peptide antagonists of fibronectin binding will inhibit the phenotypic transformation of normal rat kidney fibroblasts induced by TGF-β.[22] This suggests that the "transforming" activity of TGF-β that gave the molecule its name is due to the induction of fibronectin expression, allowing the cells to modulate their extracellular environment in such a way as to permit anchorage-independent growth.

OTHER PROPERTIES OF TGF-β RELEVANT TO TUMORIGENESIS

Recent work has shown TGF-β to be a potent immunomodulator. It inhibits the proliferation of T- and B-lymphocytes, inhibits antibody production by B-lymphocytes, depresses cytolytic activity of natural killer (NK) cells, and inhibits generation of lymphokine-activated killer (LAK) cells and cytotoxic T cells (Ref. 23 and refs. therein). Studies on patients with glioblastomas suggest that immunosuppression by TGF-β also occurs in vivo. These patients have depressed cell-mediated immunity that is restored to normal when the tumor is removed. When the immunosuppressing agent was purified from conditioned medium of tumor-derived lines, it was found to be TGF-β 2 (Ref. 24). This suggests that tumors that secrete high levels of TGF-β may be better able to escape the normal immune surveillance process.

In contrast to its inhibitory effects on other cells of the immune system, TGF-β is chemotactic for monocytes and stimulates these cells to secrete angiogenic factors and growth factors for fibroblasts.[25] TGF-β injected in vivo promotes fibrosis and angiogenesis, probably mediated in part by monocytes, and immunohistochemical studies show TGF-β protein localized in mouse embryos in areas of active angiogenesis, also indicating an important role for the molecule in blood vessel development (reviewed in Ref. 1). Thus secretion of TGF-β by tumors may promote angiogenesis and the development of a nurturing tumor stroma.

CONCLUSIONS

TGF-β is the most potent natural inhibitor of epithelial cell growth discovered to date. Immunohistochemical studies showing the localization of TGF-β in epithelial tissues such as the lung and intestine confirm that TGF-β, acting in an autocrine or paracrine fashion, may play an important regulatory role in epithelial homeostasis *in vivo*. Activation of TGF-β from its biologically latent form constitutes a critical control point in this process. The demonstration that many carcinoma cell lines exhibit lesions in the TGF-β negative autocrine loop further suggests that TGF-β is normally important in regulating epithelial cell proliferation. This raises the possibility that agents such as the antiestrogens, which induce TGF-β secretion, may be useful in the development of novel chemopreventive or therapeutic strategies for treatment of epithelial malignancies. However, while direct growth-inhibitory effects of TGF-β on epithelial cells would presumably retard malignant progression, inhibitory effects of TGF-β on the immune surveillance system and stimulatory effects on angiogenesis may create an environment that is actually more permissive for tumor development. Currently, little is known about the immediate mechanism of transduction of the TGF-β signal from the cell surface to the nucleus, but at a more distal level, many of the biological activities of TGF-β appear to be due to the ability of TGF-β to modulate the composition and quantity of extracellular matrix. The elucidation of other functional gene families regulated by TGF-β and mechanistically involved in its action will be critical for the rational development of the molecule as a therapeutic tool.

REFERENCES

1. ROBERTS, A. B., K. C. FLANDERS, P. KONDAIAH, N. L. THOMPSON, E. VAN OBBERGHEN-SCHILLING, L. M. WAKEFIELD, P. ROSSI, B. DE CROMBRUGGHE, U. HEINE & M. B. SPORN. 1988. Rec. Prog. Hormone Res. **44:** 157–197.
2. SPORN, M. B. & A. B. ROBERTS. 1988. Biofactors **1:** 89–93.
3. DERYNCK, R., J. A. JARRETT, E. Y. CHEN, D. H. EATON, J. R. BELL, R. K. ASSOIAN, A. B. ROBERTS, M. B. SPORN, D. V. GOEDDEL. 1985. Nature **316:** 701–705.
4. ASSOIAN, R. K., A. KOMORIYA, C. A. MEYERS, D. M. MILLER, & M. B. SPORN. 1983. J. Biol. Chem. **258:** 7155–7160.
5. JAKOWLEW, S. B., P. KONDAIAH, K. C. FLANDERS, N. L. THOMPSON, P. J. DILLARD, M. B. SPORN & A. B. ROBERTS. 1988. Oncogene Res *2:* 135–148.
6. THOMPSON, N. Submitted for publication.
7. BRAUN, L., J. E. MEAD, M. PANZICA, R. MIKUMO, G. I. BELL & N. FAUSTO. 1988. Proc. Natl. Acad. Sci. USA **85:** 1539–1543.
8. MCMAHON, J. B., W. L. RICHARDS, A. A. DEL CAMPO, M-K. H. SONG & S. S. THORGEIRSSON. 1986. Cancer Res. **46:** 4665–4671.
9. MASUI, T., L. M. WAKEFIELD, J. F. LECHNER, M. A. LA VECK, M. B. SPORN & C. C. HARRIS. 1986. Proc. Natl. Acad. Sci. USA **83:** 2438–2442.
10. HOUCK, K. A., S. C. STROM & G. K. MICHALOPOULOS. 1987. Proc. Am. Assoc. Cancer Res. **28:** 64.
11. KNABBE, C., M. E. LIPPMAN, L. M. WAKEFIELD, K. C. FLANDERS, A. KASID, R. DERYNCK & R. B. DICKSON. 1987. Cell **48:** 417–428.
12. WAKEFIELD, L. M., D. M. SMITH, T. MASUI, C. C. HARRIS & M. B. SPORN. 1987. J. Cell. Biol. **105:** 965–975.
13. WAKEFIELD, L. M., D. M. SMITH, K. C. FLANDERS & M. B. SPORN. 1988. J. Biol. Chem. **263:** 7646–7654.
14. WAKEFIELD, L. & A. LEVINSON. Unpublished material.

15. O'CONNOR-MCCOURT, M. D. & L. M. WAKEFIELD. 1987. J. Biol. Chem. **262:** 14090–14099.
16. COFFEY, R. J., JR., L. J. KOST, R. M. LYONS, H. L. MOSES & N. F. LARUSSO. 1987. J. Clin. Invest. **80:** 750–757.
17. PIRCHER, R., P. JULLIEN & D. A. LAWRENCE. 1986. Biochem. Biophys. Res. Commun. **136:** 30–37.
18. KESKI-OJA, J., R. M. LYONS, & H. L. MOSES. 1987. J. Cell. Biochem. Suppl. **11a:** 60.
19. WAKEFIELD, L. & T. MASUI. Unpublished material.
20. WAKEFIELD, L. Unpublished material.
21. CHAMBARD, J-C. & J. POUYSSEGUR. 1988. J. Cell. Physiol. **135:** 101–107.
22. IGNOTZ, R. A. & J. MASSAGUÉ. 1986. J. Biol. Chem. **261:** 4337–4345.
23. MULÉ, J. J., S. L. SCHWARZ, A. B. ROBERTS, M. B. SPORN & S. A. ROSENBERG. 1988. Cancer Immunol. Immunother. **26:** 95–100.
24. WRANN, M., S. BODMER, R. DE MARTIN, C. SIEPL, R. HOFER-WARBINEK, K. FREI, E. HOFER & A. FONTANA. 1987. EMBO J. **6:** 1633–1636.
25. WAHL, S. M., D. A. HUNT, L. M. WAKEFIELD, N. MCCARTNEY-FRANCIS, L. M. WAHL, A. B. ROBERTS & M. B. SPORN. 1987. Proc. Natl. Acad. Sci. U.S.A. **84:** 5788–5792.
26. SHIPLEY, G. D., M. R. PITTELKOW, J. J. WILLE, JR., R. E. SCOTT & H. L. MOSES. 1986. Cancer Res. **46:** 2068–2071.

DISCUSSION

C. BOREK (*Columbia University, New York, New York*): Agents that act as a double-edged sword usually exact their various effects in a dose-related manner. Are the inhibitory effects of TGF-β and the stimulatory effects on various cells exerted at different doses of TGF-β?

L. WAKEFIELD (*National Institutes of Health, Bethesda, Maryland*): In cells for which TGF-β is growth-stimulatory, such as osteoblasts, the dose-response curve is indeed biphasic, with growth inhibition occurring at higher TGF-β concentrations. However, this is not the whole story. For instance, Anita Roberts and coworkers have shown that in fibroblasts transfected with the *myc* oncogene, TGF-β inhibits colony formation in soft agar when the cells are grown in the presence of EGF, whereas the same concentration of TGF-β is stimulatory if the cells are grown in the presence of PDGF. In this case, the direction of the response is determined not by the concentration of the TGF-β, but by the spectrum of other growth factors acting on the cell at the same time.

L. CANTLEY (*Tufts University, Boston, Massachusetts*): You showed that TGF-β enhances fibronectin secretion in fibroblasts. Does addition of fibronectin to early embryo fibroblast cultures mimic the growth-stimulatory effect of TGF-β?

WAKEFIELD: Massagué and his coworkers have shown that fibronectin will substitute for TGF-β in stimulating the growth of normal rat kidney fibroblasts in soft agar. This suggests that the growth-stimulatory effect of TGF-β on NRK cells is due to its ability to induce fibronectin synthesis, and thus allow the cell to modify its microenvironment in such a way that this becomes permissive for growth. It is quite possible that the same phenomenon would occur in early embryo fibroblasts, but there are no published data for these cells.

G. GUIDOTTI (*University of Parma, Italy*): Is there any information on changes in

TGF-β secretion in psoriasis, a disease in which the transit time of cells from basal to spinous layer is markedly accelerated and final differentiation of cells is inhibited.

WAKEFIELD: No information is available so far, but it will be a very interesting system to examine for a possible involvement of TGF-β.

R. LOTAN (*University of Texas, Houston, Texas*): Is there any cell type where retinoic acid growth-inhibitory action may be explained by increasing TGF-β production?

WAKEFIELD: In the B6F$_{10}$ melanoma, which is one of the cell types that is most sensitive to the inhibitory action of retinoids, unpublished work from our laboratory suggests that retinoic acid is not acting via an induction of TGF-β. However, we do anticipate that in some systems there will be interaction between the retinoid and TGF-β growth-regulatory systems, possibly at the level of TGF-β receptor induction or activation of latent forms of TGF-β, not necessarily just at the level of TGF-β expression.

B. SZWERGOLD (*Fox Chase Cancer Center, Philadelphia, Pennsylvania*): What is the chromosomal location of the TGF-β gene (or genes)?

WAKEFIELD: Rik Derynck and his coworkers have shown that the gene for type 1 TGF-β is located on human chromosome 19 on the long arm in the region q13.1-13.3.

Purification of a Phosphotyrosine Phosphatase That Dephosphorylates the Epidermal Growth Factor Receptor Autophosphorylation Sites

C. J. PALLEN,[a] G. N. PANAYOTOU, L. SAHLIN,[b]
AND M. D. WATERFIELD

Ludwig Institute for Cancer Research
London W1P 8BT, England

Investigation of the control of normal growth and development and of the abnormal growth of cancer cells has, mainly through the study of the regulation of cells in culture, led to the isolation and characterization of a diverse series of polypeptide growth regulators.[1] Through the use of rDNA techniques and extremely sensitive assays, the groups of growth regulators isolated and defined originally through characterization of bioactive proteins have been extended to include many peptides previously only recognized by their biological activities. The recognition that growth factors can, in many cases, act as both positive and negative regulators of cell growth has led to the accepted use of the term growth regulator rather than simply growth factor for these polypeptides. Studies of the mechanism of action of epidermal growth factor, first purified and characterized by S. Cohen,[2] led to the recognition that the biological effects of the regulators are mediated by an initial interaction with a cell surface receptor, which must be displayed on the surface of all target cells. In the case of the EGF receptor the target cells can be derived from all three germ layers, although only certain cell lineages display such receptors and respond mitogenically to EGF.[3] Studies of EGF and its receptor have laid the groundwork for investigation of many other growth regulator-receptor systems, and such studies continue to provide a basis for understanding the mechanism of action of other growth signal transduction mediators.[4]

The initial event in the triggering of a growth response to EGF is its interaction with a specific 170,000-kD glycoprotein cell surface receptor. Knowledge of the complete amino acid sequence of the receptor, information derived from biochemical studies, the application of cell physiology, and extrapolation from the known structure of such membrane proteins as bacterorhodopsin have all led to the notion that the EGF receptor can be divided into three distinct domains.[5] EGF binding is mediated by an external domain of 115 kD, which is linked to a single hydrophobic transmembrane domain that is in turn joined to a cytoplasmic domain that has a ligand-stimulatable

[a]Present address: Institute of Molecular and Cell Biology, National University of Singapore, 10 Kent Ridge Crescent, Singapore 0511, Republic of Singapore.
[b]Present address: FEFE, Kliniska Forsknings Laboratoriet, Karolinska Sjukhuset, S-10401 Stockholm, Sweden.

tyrosine protein kinase activity. The mechanism by which ligand binding activates the kinase activity is still not well understood. Since the single receptor transmembrane domain may not be able to directly transduce a signal, it has been suggested that receptor dimerization may be involved.[6] Thus, the ligand-induced juxtaposition of two external domains could bring together the two cytoplasmic domains and induce the conformational change needed to activate the kinase. An alternative mechanism involving a monomeric receptor would require a push-pull, rotational, or other physical change, perhaps resulting in an altered membrane interaction that induces the conformational change necessary for kinase activation. Whatever the mechanism involved, the kinase activity itself is probably essential for signal transduction since the ablation of kinase activity by chemical modification or mutagenesis of a lysine residue essential for ATP binding results in loss of the mitogenic signal transduction function.[7,8]

Activation of the tyrosine protein kinase activity leads to phosphorylation of the receptor cytoplasmic domain at predominantly a single site in intact cells and at two to three additional sites in vitro.[9] These sites—P1, P2 and P3—are located in a distinct subdomain which is C-terminal in the receptor and susceptible to cleavage by calpain to generate a 150,000-kD receptor sequence.[10] The phosphorylation is an intramolecular event in solubilized[11] receptors, but could be intermolecular in intact cells. *In vitro* studies have suggested that P1 (or P2 and P3) phosphorylation has no effect[12] or may result in activation of the kinase itself.[13] Recent mutagenesis studies have revealed a role for P1 in receptor kinase activation at low substrate concentrations and have demonstrated a reduction (50%), but not a complete inhibition, of the mitogenic effect of EGF as a consequence of conversion of P1 tyrosine to phenylalanine.[14]

The importance of the protein tyrosine kinase activity and of receptor phosphorylation in the EGF receptor functions remains unclear, and because such modifications can activate the tyrosine kinase of the insulin receptor[15] or regulate the tyrosine kinase of the pp60[c-src] protein,[16] it is important to continue to search for the role of tyrosine phosphorylation in the function of the EGF receptor and other related tyrosine kinases.

A particularly important feature of this family of enzymes, which now includes a number of growth regulator receptors, is their activation through mutation. First discovered was the pp60[v-src] oncogene, which is now known to have eight or nine related homologous genes capable of activation as oncogenes,[17] and a series of growth regulator receptors which include the EGF, *erb*-b2 (or *neu* receptor-like protein), CSF-1, c-*kit*, MET, TRK, *ret* and c-*ros* proteins. Oncogenic conversion of the EGF receptor by the avian erythroblastosis virus (AEV), observed as the acquisition of a truncated EGF receptor, was the first reported example of oncogenic subversion of growth regulator receptor function.[18] For the CSF-1 (*fms*),[19] *kit*,[20] MET,[21] TRK,[22] *ret*,[23] and *ros*[24] proteins, a diverse series of mutational events have been shown to induce expression of transforming proteins which are subverted receptors.

In most cases, the alteration of kinase regulation, either through mutation of ligand-binding regions or removal of tyrosine kinase phosphorylation sites, is thought to be important in the generation of the transforming protein. Against this background, the importance of tyrosine phosphorylation is evident and for this reason we have undertaken the purification and characterization of tyrosine phosphatases, which could be important in regulating the removal of tyrosine residues from the EGF receptor and *src* family proteins.

That tyrosine phosphorylation of the EGF receptor is reversible is evident from the work of Brautigan *et al.*,[25] who showed that the receptor in membranes could be dephosphorylated by an endogenous membrane phosphatase. Several tyrosine phosphatase activities have been either partially or completely purified from a diverse series of tissues, the enzymes having molecular weights of 23–95 kD.[26-30] The enzymes are distinguishable from other types of phosphatases since they are stimulated or unaffected by chelating agents such as EDTA, inhibited by magnesium or manganese, and strongly inhibited by zinc and vanadate.

Because the tyrosine kinase family of receptors and the *src* kinases are membrane-associated we have chosen to focus on tyrosine phosphatases that can be purified from membranes. Similar activities have been detected in several membrane preparations, and recently Tonks *et al.*[31] have reported, in outline, the partial characterization of two phosphatases from placental membranes.

In this paper we briefly describe the purification of a tyrosine phosphatase (PTP1) from placental membranes. A detailed description of the purification of this enzyme and other tyrosine phosphatases from the same tissue will be published elsewhere.[38] The activity of the purified phosphatase on the EGF receptor has been examined, and the enzyme is shown to be able to remove phosphate from tyrosine residues P1, P2 and P3.

METHODS

Materials

A-431 membranes were prepared by the method of Thom *et al.*[32] EGF purified from mouse submaxilliary glands was obtained from Dr. J. Hsuan and casein kinase II and the catalytic subunit of the cyclic AMP-dependent protein kinase from Dr. P. Parker. Histone type II-AS, *p*-nitrophenylphosphate (PNPP), phosphorylase kinase, and phosphorylase *b* were from Sigma.

Substrate Phosphorylation and Purification

A synthetic peptide (RR-*src*) similar to the phosphorylation site of pp60*src* (RRLIEDAEYAARG) was synthesized (ABI peptide synthesizer and f-*moc* chemistry) and phosphorylated by the EGF receptor kinase. A 1-ml reaction mixture containing 120 μg A431 membranes, 0.2 μM EGF, 50 mM HEPES (pH 7.4), 150 mM NaCl, 0.02% Triton X-100, 5% glycerol, 2 mM $MnCl_2$, 100 μM Na_3VO_4 1.4 mM RR-*src* and 250 μM ATP (2000–3000 cpm/pmol) was pre-incubated for 1 hour at 20°C, and ATP was added for a further 16 hrs at 30°C. The reaction was terminated with 1 ml of 1 N acetic acid and centrifuged and the supernatant was applied to an anion exchange column equilibrated in 0.5 N acetic acid (1.5 × 47 cm Biorad AG-1X2). The phosphopeptide was eluted with 1 N acetic acid. Phospho- and dephosphopeptides were separated from reaction mixtures using an antiphosphotyrosine monoclonal antibody column (1 ml) equilibrated in 0.05 M ammonium acetate pH 6.8 and eluted with a ten-fold concentration of the same buffer. Eluted radioactive fractions were counted and phosphopeptide concentrations were calculated on the basis of ATP specific activities.

Phosphorylation of Protein Substrates

Phosphoseryl casein and phosphoseryl histone were prepared using casein kinase II (0.5 mg/ml casein, 3 U/ml enzyme) and cAMP-dependent protein kinase (1 mg/ml casein, 10 U/ml enzyme), respectively. Reaction was for 30 minutes at 30°C, when it was stopped with 20% TCA. Centrifuged pellets were washed with 20% TCA (twice), dissolved in 40 μl of NaOH, and diluted to 0.5 ml with 20 mM Tris (pH 7.5) and 10 mM 2-mercaptoethanol. Phosphotyrosyl casein or histone was prepared using 0.5 mg casein or 1 mg/ml histone, 125 μg A431 membranes, 2 μg/ml EGF, 12 mM MgCl$_2$, 2 mM MnCl$_2$, 20 mM HEPES (pH 7.4) and [γ-^{32}P]ATP (100 μM, 1000 cpm/pmol). Incubation was for 60 min at 30°C when the reaction was stopped with 15 mM EDTA, and membranes were removed by centrifugation. Proteins were dialyzed into 20 mM Tris (pH 7.5) and 10 mM 2-mercaptoethanol. Phosphorylase a was prepared using phosphorylase kinase.[33]

Assay of Phosphatase Activity

Phosphosubstrate (in 50 mM MES [pH 6], 0.1 M NaCl, 2 mM EDTA, 1 mg/ml BSA, 0.5 mM DTT and 0.01% CHAPS) was incubated with enzyme at 30°C. For peptide substrates the reaction was stopped with 5% TCA and for proteins with 20% TCA (with 0.25 mg/ml BSA as carrier) and the ^{32}P released was quantitated by the method of Chan et al.[34]

The EGF receptor was purified by immunoprecipitation from A431 lysates using the monoclonal antibody R1. Autophosphorylation of the receptor and tryptic phosphopeptide mapping were performed as previously described.[9]

RESULTS

Purification of Tyrosine Phosphatases

An outline of the purification process is shown diagramatically in FIGURE 1.

Placental Membranes. These were prepared from placentas obtained at normal delivery or cesarean section, stored on ice, and cut into small pieces. After homogenization in a Waring blender at 4°C in buffer A (25 mM HEPES [pH 7.6], 0.15 M NaCl, 2 mM EGTA, 0.1 mM PMSF, 10 μg/ml aprotinin, 500 ml/placenta) the homogenate was centrifuged (5700× g, for 20 min) and the supernatant spun at 32000 g for 60 min, resuspended in buffer A, and recentrifuged. The membrane pellet was then resuspended in buffer B (50 mM Tris-HCl [pH 7.6], 20% Triton X-100, 0.1 M NaCl, 2 mM EDTA, 0.05 mM PMSF and 10 μg/ml aprotinin), homogenized, stirred for 60 min, and centrifuged at 100,000 g for 60 min; it could then be stored at −70°C for 1 week.

Heparin Sepharose Chromatography. Detergent-solubilized membranes were diluted 1:4 in buffer B and fractionated on a heparin Sepharose column. Protein was eluted with a gradient of NaCl from 0 to 0.4 M.

FPLC Gel Filtration Chromatography. Active fractions (PTP-1) from heparin Sepharose were brought to 0.01% with CHAPS, concentrated, and applied to two linked Superose-12 columns. A single peak of phosphatase activity was eluted (approximate MW 50,000).

FPLC Ion-Exchange Chromatography. The gel-filtered enzyme (PTP-1) was brought to 0.1% CHAPS and dialyzed to remove salt. A major peak of phosphatase

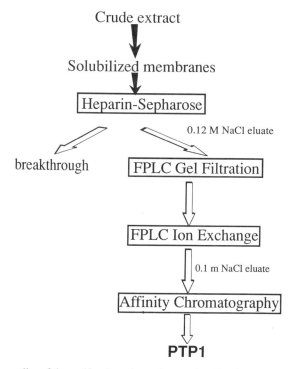

FIGURE 1. An outline of the purification scheme for tyrosine phosphatase PTP1 from placental tissue.

activity was eluted from a Mono Q FPLC column at 0.1 M NaCl. This enzyme could be stored at −70°C (after dialysis into buffer D: 50 mM Tris [pH 7.5], 0.1 M NaCl, 50% glycerol, 0.5 mM DTT, and 0.01% CHAPS).

Affinity Chromatography. Immobilized *p*-aminobenzyl phosphonic acid (PABP) was obtained from Pierce. The eluate from the Mono Q FPLC column (see above) was applied to the PABP column and the phosphatase activity was eluted by a salt gradient. The active fractions were pooled, dialyzed to remove salt, and re-passed over the affinity column. Results are shown in FIGURE 2.

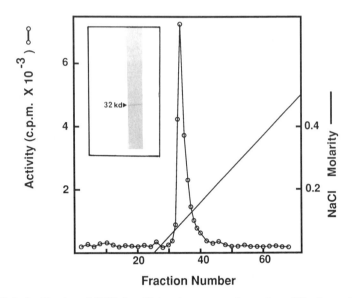

FIGURE 2. Purification of PTP1 by affinity chromatography on immobilized *p*-aminobenzyl phosphonic acid. Fractions were assayed for phosphatase activity as described in the METHODS section. Inset shows a Coomassie blue staining gel track of a peak fraction analyzed by SDS-PAGE.

A major protein of apparent MW of 32–33,000 could be visualized by SDS-PAGE analysis together with minor contaminating bands of molecular weight 40–60,000.

Reverse-Phase Chromatography. The affinity-purified phosphatase activity was analyzed by reverse-phase HPLC using an Aquapore RP-300 column. A protein of 32,000 kD was eluted by 0.1% trifluoroacetic acid buffers from the column at an acetonitrile concentration of 60%. Since this purified protein was denatured and inactive as eluted from the column its identification as a phosphatase awaits reconstitution of enzyme activity.

Characterization of Protein Tyrosine Phosphatase 1

We analyzed both the susceptibility of protein tyrosine phosphatase 1 (PTP 1) to various established phosphatase activators and inhibitors as well as substrate specificity. These analyses were made with enzyme prior to its purification by affinity chromatography.

Activity Requirements. The pH optimum for the synthetic *src* substrate was 6.0, although the enzyme was 40% as active at pH 7.5 as it was at pH 6. The enzyme was inhibited by manganese and magnesium, unaffected by calcium, inhibited by Ca^{2+} plus calmodulin, and stimulated by EDTA. The enzyme was inhibited by pyrophosphate, ADP and ATP and only partially inhibited by sodium fluoride (28%). The phosphatase was strongly inhibited by vanadate and zinc (TABLE 1).

Substrate Specificity. The purification of the enzyme was monitored using the synthetic phosphorylated *src* peptide substrate. Since this substrate could only be phosphorylated to a level of 5%, it was necessary to develop a technique for purification of the fully phosphorylated peptide. This was achieved by using a monoclonal antibody to phosphotyrosine as an affinity matrix. By use of the fully phosphorylated peptide as a substrate, the kinetics of dephosphorylation were analyzed and values of 4 μM for the K_m and 70 nmol/min per mg for the V_{max} were obtained. A higher affinity and V_{max} may be achieved using the affinity-purified enzyme. The enzyme was inactive towards free phosphoamino acids.

The activity of PTP1 towards *p*-nitrophenyl phosphate was compared with that of alkaline phosphatase,[35] which is known to remove phosphate from this substrate. No activity was observed at pH 9 and at pH 5.8 the activity was more than 1000-fold less than that of placental alkaline phosphatase (135 nmol/min per mg compared to 324 μmol/min per mg).[36] The specificity of the enzyme for phosphotyrosine protein substrates was measured using phosphorylase phosphorylated on serine residues and casein either phosphorylated on serine or on tyrosine residues. In each case only the phosphotyrosyl-containing substrates were dephosphorylated. The specificity of the PTP1 enzyme towards the phosphotyrosine residues of the autophosphorylated EGF receptor was measured by analysis of the rate of dephosphorylation of the P1, P2, P3 sites. (FIG. 3). The removal of phosphate at each site was measured after tryptic digestion of the receptor and separation of phosphorylated peptides by reverse-phase HPLC. The enzyme was able to remove phosphate from the three sites examined. The rate of removal of phosphate from sites P1 and P2 was slightly faster than that from P3.

DISCUSSION

The characterization of tyrosine phosphatase activities in placental membranes described here has revealed a number of distinct enzyme activities that can be

TABLE 1. Effects of Various Compounds on Tyrosine Phosphatase (PTP1) Activity

Effector	Concentration	Percentage of Phosphatase Activity
None	—	100
EDTA	2 mM	155
EGTA	2 mM	168
MnCl$_2$	2 mM	59
MgCl$_2$	20 mM	47
CaCl$_2$	1 mM	91
Calmodulin		5
Calmodulin + CaCl$_2$		19
ATP	1 mM	33
ADP	1 mM	33
PP$_i$	2 mM	0.4
50 mM NaF + 2 mM EDTA		72

separated by sequential chromatography and heparin Sepharose, gel filtration, FPLC ion-exchange, and affinity chromatography. The purification and characterization have been monitored with a phosphorylated *src* peptide analogue as a substrate, and the purification of a major membrane-associated enzyme, PTP-1, is briefly described. Clearly several cytosolic activities and more membrane-associated enzymes remain to be purified and their relationship to the PTP1 enzyme described here defined. The precise relation of PTP1 to previously characterized cytosolic and membrane-associated activities, particularly those purified by Tonks *et al.*[31] will require structural

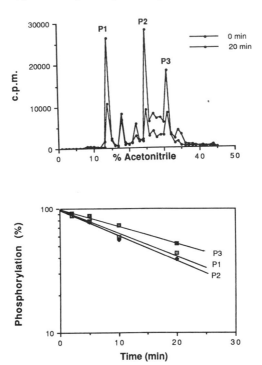

FIGURE 3. Dephosphorylation of tyrosine residues of the EGF receptor by PTP1. *Top:* Tryptic phosphopeptide map of purified EGF receptor from A431 cells after incubation with PTP1 for 0 and 20 minutes. *Bottom:* Time course of dephosphorylation of the EGF receptor. Autophosphorylated receptor was incubated with PTP1 for the indicated times. The reaction was stopped by quickly washing the immune complexes in assay buffer and incubating at 100°C for 2 minutes followed by HPLC analysis of the tryptic phosphopeptides.

analysis, which is currently in progress using the HPLC-purified material. Rosenberg and Brautingan[37] have suggested that the cytosolic enzymes may be derived from the membrane-associated forms by proteolysis, and thus the nature of any sequences or modifications that confer membrane association will be particularly interesting. Analysis of PTP-1 activators and inhibitors reveals a pattern of phosphatase activity similar to that reported previously. Thus the stimulation of activity by EDTA, inhibition by manganese and magnesium, and inhibition by μM amounts of zinc and vanadate conform to previous paradigms for this family of phosphatases. These responses, together with the demonstrated specificity of PTP1 for tyrosine phosphate and lack of effect on serine phosphate, distinguish the enzyme from the phosphoseryl/ threonyl protein phosphatases types 2A and 2B, which have been shown to posses tyrosine phosphatase activity.[27,34,35] The inhibition of activity by magnesium, absence of activity at pH 9, and inability to dephosphorylate histone or casein show that PTP-1 is not an alkaline phosphatase. The bone and prostate acid phosphatase also differ from PTP-1, which is not inhibited by fluoride, cannot hydrolyze free phosphotyrosine, and is inhibited by zinc. In addition, the effects of EDTA and manganese upon PTP-1 activity are exactly opposite to those of protein phosphatase 2A. Furthermore since ATP and PPi stimulate the phosphotyrosine phosphatase activity of the type 2A enzymes and inhibit PTP-1, the enzyme must be fundamentally distinct from these proteins. The results obtained thus far show that PTP-1 has a restricted specificity confined to phosphotyrosyl residues on peptides or proteins. Within this constraint, the enzyme appears relatively unspecific, being able to remove phosphate from the tyrosine residues of casein, histone, the EGF receptor, and the *src* peptide. The analysis of dephosphorylation of the EGF receptor phosphorylation sites reveals slightly different rates for phosphatase activity, and further analysis may establish that certain substrates could show distinct specificity. Since the enzyme may act at the membrane level, the specificity could be restricted in intact cells; furthermore, a more detailed examination of a variety of important tyrosine kinase substrates could reveal an important role in regulation of their activity.

REFERENCES

1. GOUSTIN, A. S., E. B. LEOF, G. D. SHIPLEY & H. L. MOSES. 1986. Cancer Res. **46:** 1015–1029.
2. COHEN, S. 1962. J. Biol. Chem. **237:** 1555–1552.
3. ADAMSON, E. D. & A. R. REES. 1981. Mol. Cell. Biochem. **34:** 129–152.
4. YARDEN, Y. & A. ULLRICH, 1988. Biochemistry **27:** 3113–3119.
5. CARPENTER, G. 1987. Annu. Rev. Biochem. **56:** 881–
6. YARDEN, Y. & J. SCHLESSINGER. 1987. Biochemistry **26:** 1443–1437.
7. HONEGGER, A. M., T. J. DULL, S. FELDER, E. VAN OBBERGHEN, F. BELLOT, D. SZAPARY, A. SCHMIDT, A. ULLRICH & J. SCHLESSINGER. 1987. Cell. **57:** 199–209.
8. HONEGGER, A. M., D. SZAPARY, A. SCHMIOT, R. LYALL, E. VAN OBBERGHEN, T. J. DULL, A. ULLRICH & J. SCHLESSINGER. 1987. Mol. Cell Biol. **7:** 4568–4571.
9. DOWNWARD, J., P. PARKER & M. D. WATERFIELD. 1984. Nature **311:** 483–485.
10. GATES, A. E. & L. E. KING. 1985. Biochemistry **24:** 5209–5215.
11. WEBER, W., P. J. BERTICS & G. N. GILL. 1984. J. Biol. Chem. 14631–14636.
12. DOWNWARD, J., M. D. WATEFIELD & P. J. PARKER. 1985. J. Biol. Chem. **260:** 14538–14546.
13. BERTICS, P. J. & G. N. GILL. 1985. J. Biol. Chem. **260:** 14642–14647.
14. BERTICS, P. J., W. S. CHEN, L. HUBLER, C. S. LAZAR, M. G. ROSENFIELD & G. N. GILL. 1988. J. Biol. Chem. **263:** 3610–3617.
15. YU, K.-T. & M. P. CZECH. 1984. J. Biol. Chem. **259:** 5277–5286.
16. KMIECIK, T. E. & D. SHALLOWAY. 1987. Cell **49:** 65–73.
17. HUNTER, T. 1987. Cell **49:** 1–4.
18. DOWNWARD, J., Y. YARDEN, E. MAYES, G. SCRACE, N. TOTTY, P. STOCKWELL, A. ULLRICH, J. SCHLESSINGER & M. D. WATERFIELD. 1984. Nature **307:** 521–527.
19. SHERR, C. J., C. W. RETTENMIER, R. SACCA, M. F. ROUSSEL, A. T. LOOK & E. R. STANLEY. 1985. Cell **41:** 665–676.
20. QIU, F., R. PRADIR, K. BROWN, P. E. BARKER, S. JHANWAR, F. H. RUDDLE & P. BESMER. 1988. EMBO J. **7:** 1003–1011.

21. DEAN, M., M. PARK, M. M. LeBEAU, T. S. ROBINS, M. O. DIAZ, J. D. ROWLEY, D. G. BLAIR & G. F. VANDE WOUDE. 1985. Nature **318**: 385–388.
22. MARTIN-ZANCA, D., S. H. HUGHES & M. BARBACID. 1986. Nature **319**: 743–748.
23. TAKAHASHI, M. & G. M. COOPER. 1987. Mol. Cell. Biol. **7**: 1378–1385.
24. ULLRICH, A., J. R. BELL, E. Y. CHEN, R. HERRERA, L. M. PETRUZZELLI, T. J. DULL, A. GRAY, L. COUSSENS, Y.-C. LIAO, M. TSUBOKAWA, A. MASON, P. H. SEEBURG, C. GRUNFIELD, O. M. ROSEN & J. RAMACHANDRAN. 1985. Nature **313**:736–761.
25. BRAUTIGAN, D. L., P. BORNSTEIN & B. GALLIS. 1981. J. Biol. Chem. **256**: 6519–6524.
26. HORLEIN, D., B. GALLIS, D. L. BRAUTIGAN & P. BORNSTEIN. 1982. Biochemistry **21**: 5577–5382.
27. FOULKES, J. G., E. ERIKSON & R. L. ERIKSON. 1983. J. Biol. Chem. **258**: 431–438.
28. NELSON, R. L. & P. E. BRANTON. 1984. Mol. Cell. Biol. **4**: 1003–1007.
29. SHRINER, C. L. & D. L. BRAUTIGAN. 1984. J. Biol. Chem. **259**: 11383–11389.
30. TUNG, H. Y. L. & L. J. REED. 1987. Anal Biochem. **161**: 412–418.
31. TONKS, N. K., C. D. DILTZ & E. H. FISCHER. 1987. Adv. Protein Phosphatases **4**: 431.
32. THOM, D., A. J. POWELL, C. W. LLOYD & D. A. REES. 1977. Biochem. J. **168**: 187–194.
33. KREBS, E. G., A. B. KENT & E. H. FISCHER. 1958. J. Biol. Chem. **231**: 73–83.
34. CHAN, C. P., B. GALLIS, D. K. BLUMENTHAL, C. J. PALLEN, J. H. WANG & E. G. KREBS. 1986. J. Biol. Chem. **261**: 9890–9898.
35. SWARUP, G., S. COHEN & D. L. GARBERS. 1981. J. Biol. Chem. **256**: 8197–8204.
36. HARKNESS, D. R. 1968. Arch. Biochem. Biophys. **126**: 513–518.
37. ROSENBERG, S. A. & D. L. BRAUTIGAN. 1987. Biochem. J. **243**: 747–751.
38. PALLEN, C. J., L. SAHLIN & M. D. WATERFIELD. Manuscript in preparation.

Growth Factor and Oncogene Influences on Cell Growth Regulation

LESLIE A. SERUNIAN AND LEWIS C. CANTLEY

Department of Physiology
Tufts University School of Medicine
Boston, Massachusetts 02111

INTRODUCTION

Efforts to understand the mechanisms that control cell growth have, in a variety of scientific fields, stimulated research in which different cell types and proliferative agents such as retroviruses and mitogens are used. Progress in unraveling the complexities of growth regulation has been facilitated by studies of growth factor receptor structure, viral oncogenes, and their cellular homologues called proto-oncogenes. Moreover, the realization that cell transformation involves the same biochemical processes that regulate nontransformed cell growth, but with loss of the normal restraints, has provided new insights into the mechanisms of cell proliferation.

Recent biochemical and molecular information about growth factors and their receptors has shed light on their relationships to both normal and abnormal cell growth. Several peptide growth factor receptors have now been purified and, with a few notable exceptions, constitute a family of homologous proteins with growth-factor-stimulated tyrosine kinase activities. These tryosine kinases undergo autophosphorylation, which appears to be critical for their regulation; other physiologically relevant targets of these tyrosine kinases have not yet been determined. Furthermore, the peptide growth factor receptors that have been purified are structurally very distinct from the hormone receptors whose signals are known to be transduced by GTP-binding proteins.

In spite of the structural similarities among growth factor receptors, the primary responses to different growth factors are quite distinct. For example, whereas at least some of the primary responses to platelet-derived growth factor (PDGF) and bombesin are due to elevation of second messengers produced from turnover of products in the phosphatidylinositol (PI) pathway, many other growth factors do not appear to involve the classical PI pathway. Recent evidence that insulin stimulates the production of a novel second messenger derived from glycosylated PI raises the interesting possibility that other growth factors may act through other novel second-messenger molecules derived from PI. A link between elevation of second messengers and the growth response is based primarily on activators of protein kinase C that affect cell proliferation. However, this activation only partially explains growth responses to some factors and is irrelevant to growth responses to other factors.

Like certain growth factor receptors, may oncoproteins are enzymes with protein-tyrosine kinase activity toward as yet unidentified target molecules which, in turn, convey a growth signal to the cell nucleus. In contrast, a distinct subset of oncoproteins target directly to the nucleus, where at least some are thought to act as specific transactivators and regulators of RNA and DNA synthesis. Although many cellular and nuclear oncogenes have been isolated and characterized, the specific biochemical functions associated with their products in the regulation of gene expression and cell growth remain elusive.

This paper will describe recent findings which advance our understanding of the biochemical mechanisms by which growth factors and oncogenes propagate signals necessary for cell division. Since this area of research is still in its infancy, we expect the future discovery of new signal transduction mechanisms to explain many of the still unresolved questions. With this in mind, the strengths and inadequacies of known signal transduction systems as mediators of cell proliferation are discussed.

GROWTH FACTOR RECEPTORS

It has long been known that nontransformed cells in primary culture require metal ions, nutrients, and certain growth factors to proliferate. Although these protein factors do not usually affect the same cells that synthesize them, in some cases, a given cell produces its own growth-stimulating factors, resulting in continuous autocrine proliferation. An extreme example of this phenomenon is evident in transformed cells and is associated with uncontrolled growth (discussed below).

In the last few years, the techniques of molecular biology have been used to determine the primary structures of a variety of growth factors and growth factor receptors. One striking observation to emerge from this work is that many of the growth factor receptors are remarkably similar. For example, the receptors for epidermal growth factor (EGF), PDGF, colony-stimulating factor-1 (CSF-1), and insulin all have in common a three-domain structure.[1] The amino terminal half of the protein forms the extracellular domain and contains the binding site for the specific growth factor; a single stretch of hydrophobic amino acids forms a transmembrane domain; and the carboxy terminal half of the protein forms the cytosolic domain, which has the largest degree of similarity among all of these receptors. For the growth factors mentioned above, the cytosolic domain has been shown to have an enzymatic activity which catalyzes the transfer of the gamma phosphate of ATP to a protein tyrosine residue. This activity is stimulated by the binding of growth factor to the extracellular domain and appears to be essential for the propagation of the growth response.[2]

SECOND MESSENGERS OF GROWTH FACTOR RESPONSES

The observation that growth-factor-stimulated tyrosine kinase activity is intrinsic to growth factor receptors has provided the basis for a testable hypothesis for the biochemical mechanism of growth factor responses. This activity is stimulated within seconds after growth factor binding to the receptor and may explain the rapid series of events occurring at the plasma membrane very early after activation.[3-5] For example, within the first minute after PDGF stimulation of quiescent fibroblasts, the hydrolysis of phosphatidylinositol bisphosphate (PIP_2) produces both inositol trisphosphate and diacylglycerol (DAG). The products of growth-factor-stimulated PIP_2 breakdown have been shown to act as potent second messengers: the inositol trisphosphate isomer,

(1,4,5)IP$_3$, mediates calcium release from intracellular stores by interacting with a specific receptor in the endoplasmic reticulum[4] and DAG activates a calcium- and phospholipid-dependent protein kinase called protein kinase C.[6]

In the same manner, a variety of other events observed at the plasma membrane in response to PDGF addition are mediated either directly or indirectly by these two second messengers. For example, PDGF stimulates both Na$^+$ influx and cytosolic alkalinization as a consequence of the activation of the plasma membrane Na$^+$/H$^+$ exchange system. The Na$^+$/H$^+$ pump is, in turn, mediated by both the elevation of Ca^{2+} and the activation of protein kinase C, thus generating a cascade of events initiated by the formation of the PIP$_2$-derived second messengers.[5] PDGF activation of the (Na$^+$/K$^+$) pump is a consequence of elevated cytosolic Na$^+$'s saturating the pumping site for this ion.[5] Since PIP$_2$ breakdown explains many of the early responses to PDGF, it is attractive to speculate that the PDGF receptor directly affects a PIP$_2$-specific phospholipase C.

The structural and functional homologies of the cytosolic domains of the peptide growth factor receptors might suggest that they evoke similar primary responses. However, even in well-controlled systems where multiple types of receptors are examined in a single, clonal cell line, each growth factor elicits a unique cellular response. This observation indicates that the primary targets of the various receptors differ. For example, while PDGF stimulates polyphosphoinositide breakdown, EGF seems to have only a minor effect on this system. CSF-1, fibroblast growth factor (FGF), and insulin appear not to stimulate polyphosphoinositide breakdown at all, at least when each is added alone. It seems likely that these other growth factors stimulate the production of second messengers distinct from (1,4,5)IP$_3$ and DAG. In fact, recent evidence strongly suggests that other inositol phosphate isomers such as (1,3,4)IP$_3$ and novel polyphosphoinositides may be involved in the growth responses of cells to certain oncogenes as well as to PDGF (Ref. 7 and unpublished observations).

Although cAMP and cGMP have been implicated as second messengers of proliferative responses after growth factor stimulation,[5] the activation of these molecules also fails to explain the diverse cellular responses to most peptide growth factors. One very intriguing possibility is that the variety of responses is due to the existence of additional PI-derived second messengers which have just begun to be characterized.[7]

Recently, Saltiel and coworkers have found evidence for a glucosamine-containing phosphoinositide whose breakdown is regulated by insulin.[8] A phospholipase C specific for this molecule is stimulated in response to insulin and the soluble breakdown product is proposed to act as a second messenger for insulin action. A series of parallel PI-derived second-messenger systems, each responding to different peptide growth factor receptors, provides an attractive model for the evolution of the diverse growth factor responses in complex organisms.

An alternative explanation for the differences in growth factor responses is that different factors affect different steps in the PI turnover pathway. For example, recent evidence indicates that inositol-1,3,4,5-tetrakisphosphate [(1,3,4,5)IP$_4$], produced by phosphorylation of (1,4,5)IP$_3$, also acts as a second messenger.[9] Thus, differential responses to growth factors may be a consequence of the different effects of receptors on the enzyme that converts (1,4,5)IP$_3$ to (1,3,4,5)IP$_4$. In addition, (1,4,5)IP$_3$ is produced only by PIP$_2$ breakdown, while DAG can be produced by breakdown of any of the phosphoinositides. Thus, the ratio of these two second messengers could be regulated by the kinases that phosphorylate the D-4 and D-5 positions of phosphatidyl-inositol.[10] Some of the synergistic effects of growth factors might also be explained by different receptors acting at different steps in the same pathway.

SIGNAL TRANSDUCTION FROM GROWTH FACTOR
TO SECOND MESSENGER

Although certain growth factor receptors clearly stimulate PI turnover and other responses at the plasma membrane, the biochemical mechanisms by which these responses are activated are unknown. Despite extensive research by many laboratories, the relevant targets for growth factor receptor tyrosine kinases have remained elusive. In fact, the only proteins that have been shown to be stoichiometrically phosphorylated on tyrosine under physiological conditions are the growth factor receptors themselves. Several explanations for the failure to pinpoint the critical targets have been advanced. One is that very few copies of the target proteins make them difficult to detect. A second possibility is that the phosphorylation site of the target protein turns over very rapidly even under conditions designed to inhibit phosphoprotein phosphatases. A third possibility is that the critical substrate is the receptor itself: autophosphorylation (or transphosphorylation between subunits of receptor clusters) could induce a conformational change that affects a noncovalent interaction with a target protein. This third possibility suggests a model similar to that which has been elucidated for the beta-adrenergic receptor.

Signal transduction by the beta-adrenergic receptor is mediated by a nucleotide-binding protein.[11] The unoccupied receptor exists in a tight complex with a GTP-binding protein which transduces the response to the target enzyme, adenylate cyclase. The GTP-binding protein in this system is homologous to GTP-binding proteins that transduce signals from alpha-2 receptors, muscarinic acetylcholine receptors, and vertebrate rhodopsin: the protein consists of an alpha subunit (approximately 40,000 Da) and two smaller subunits (beta and gamma). Upon hormone binding to the receptor, a conformational interaction causes GTP to replace GDP, which is released from the nucleotide binding site of the alpha subunit. This substitution stimulates the dissociation of the beta and gamma subunits from the alpha subunit. The interaction between alpha and the target enzyme propagates the signal. In the case of beta-receptors, the target is adenylate cyclase. GTP hydrolysis to GDP converts the alpha subunit back to a form that re-associates into the original complex to terminate the signal. Several signal-transducing GTP-binding proteins have recently been purified and cDNA clones have been sequenced. It is clear that a large family of these transducing proteins exists, perhaps one for each type of hormone receptor.

At the present time no solid evidence supports the involvement of a GTP-binding protein in peptide growth factor responses. However, the effects of PDGF on PI turnover in fibroblasts can be mimicked in permeabilized cells by the addition of GTP and nonhydrolyzable GTP analogues.[10] Although it is unclear whether or not GTP is directly activating the same stimulatory pathway as PDGF, or a parallel pathway, GTP-binding proteins have been shown to mediate hormone-stimulated PI turnover in a variety of cells. The characteristics of these responses suggest that the PIP_2-specific phospholipase C is regulated by a mechanism analogous to that of adenylate cyclase. The use of pertussis toxin and cholera toxin as specific modifiers of GTP-binding proteins indicates that different transducing proteins are used to activate phospholipase C in different systems. Further work is needed to determine what GTP binding proteins are involved in this pathway and which peptide growth factor receptors interact with GTP binding proteins.

From a comparison of their primary structures, it would appear that peptide growth factor receptors do not belong to the family of hormone receptors that interacts with GTP-binding proteins. The beta-adrenergic and muscarinic acetylcholine receptors have recently been sequenced; these proteins do not have tyrosine kinase domains, but rather have a gross structural homology to rhodopsin. It is clear that these receptors, unlike the peptide growth factor receptors, are not protein kinases. Moreover, their interactions with GTP-binding proteins are through direct contact rather than covalent modification. However, these receptors are substrates for other cytosolic protein kinases and phosphorylation appears to modulate their responses to hormones.[11] Thus, the possibility would not be too unexpected that the peptide growth factor receptors have combined a domain for binding a transducing protein with an intrinsic tyrosine kinase activity to modify that domain. Of interest, support for a signal transduction model involving conformational interactions with transducing proteins is provided by the sequence of the receptor for nerve growth factor (NGF), the first peptide growth factor to be purified. This receptor has a gross structure that is distinct from both the tyrosine kinase family of growth factor receptors and the beta agonist type of receptor.[12,13] Like other peptide growth factor receptors, the amino terminus of this protein forms an extracellular peptide binding domain which is linked to the cytosolic domain by a single stretch of hydrophobic residues. However, the cytosolic domain of the NGF receptor does not have tyrosine kinase activity and more closely resembles the carboxy terminal cytosolic domain of the beta-adrenergic receptor. This domain appears to be phosphorylated by cytosolic protein kinases and perhaps also interacts with GTP-binding proteins.[12,13]

The discussion above is not intended to imply that other important targets for tyrosine kinases besides the receptors themselves will not be found. A particularly interesting substrate for the EGF receptor tyrosine kinase has been identified in a transformed cell line which overexpresses this receptor. This substrate protein has a molecular weight of 35,000 Da and binds to the cytosolic surface of the plasma membrane.[14] Its primary sequence reveals homology to a large family of calcium-dependent membrane-binding proteins.[15] The 35,000-Da protein has been called lipocortin because of its induction by corticosteroids and its association with lipids. Lipocortin inhibits phospholipase A_2 activity *in vitro*.[16] A related protein of 36,000 Da is phosphorylated on tyrosine in Rous sarcoma virus–transformed cells and has been called calpactin for its ability to bind actin in a calcium-dependent manner.[17] However, it is not clear whether either the phospholipase A_2 inhibitory activity of lipocortin or the actin-binding activity of calpactin is relevant *in vivo*. In fact, stoichiometric phosphorylations of these proteins in response to growth factors have not been demonstrated in nontransformed cells.

A potential candidate for a substrate of PDGF receptor tyrosine kinase is a distinct isomer of the enzyme that phosphorylates phosphatidylinositol (PI) to phosphatidylinositol-phosphate (PIP). Fibroblasts contain two distinct PI kinases and one of these enzymes correlates with an 85,000-Da protein that is immunoprecipitated with antibodies directed against phosphorylated tyrosine residues from cells stimulated with PDGF but not from quiescent cells.[18] Interestingly, this 85,000-Da PI kinase appears to be unlike the major PI kinase in red cells, liver, and fibroblasts in that it phosphorylates the D-3 position of the inositol ring and not the D-4 position,[7] thus producing a PI-3-P molecule. This novel PI-3-kinase is not an intermediate in the inositol 1,4,5-trisphosphate signalling pathway and may be activated when phosphorylated by growth factor receptors such as the PDGF receptor. Since in fibroblasts, anti-phosphotyrosine antibodies appear to immunoprecipitate the PI-3-kinase in a complex with the PDGF receptor, recruitment of this PI-3-kinase to the PDGF receptor may be more critical than its phosphorylation. In any case, this association could contribute to the activation of a novel PI turnover pathway by PDGF. That is, an increased PI-3-kinase activity in the region of the membrane where PI turnover is stimulated would feed additional substrates for polyphosphoinositide-specific phospholipases C^{10} or for the production of novel polyphosphoinositides that are critical in growth regulation (unpublished observations).

From the information presented in this section, it is clear that additional work on the purification and characterization of proteins associated with the growth factor receptors will be necessary to understand the intricacies of the various signal transduction mechanisms.

DO THE EARLY RESPONSES TO GROWTH FACTORS MEDIATE PROLIFERATION?

It is not at all clear that DNA synthesis and cell division are consequences of the same signal transduction pathway as the early responses observed at the plasma membrane. Whereas the second messengers that are produced within seconds of growth-factor-binding elicit early changes in cell physiology and morphology, the critical trigger for cell division may involve a completely different signal which occurs when a portion of the receptor or growth factor itself enters the nucleus. However, the fact that antibodies against receptors can often activate cells makes it unlikely that growth factor internalization is critical for responses. It is more difficult to eliminate receptor internalization as an essential signal for cell division.

In general, growth factors must be continuously present for many hours before a cell divides. This observation suggests that during many hours of continuous incubation with growth factor, some cellular molecule slowly builds up to a critical level, but is rapidly degraded in the absence of the factor. Second messengers such as (1,4,5) IP_3 and cAMP generally peak and decline within a minute after cell stimulation, whereas DAG builds up more slowly but probably reaches a steady state within 10 to 20 minutes. Clearly, these changes are not rate-limiting for cell division; however, continuous activation of these messages could be necessary for a slow build-up of another factor (RNA or protein) that influences cell division.

Although it is difficult to prove that the second messengers produced at early times mediate the later proliferative responses, there is some evidence to support this view.[1,3] The strongest support comes from studies with tumor-promoting phorbol esters— molecules that specifically activate protein kinase C by replacing its normal regulator, DAG. When added to quiescent cells, phorbol esters partially mimic growth factors by stimulating DNA synthesis, thus suggesting that the DAG elevation, which occurs in response to growth factors, contributes to proliferation.

Addition of calcium ionophores, which mimic the effect of $(1,4,5)IP_3$ by elevating cytosolic calcium levels, appears to enhance the response to phorbol esters in lymphocyte activation. Phorbol esters and calcium ionophores also stimulate the expression of c-*fos,* a proto-oncogene that is normally expressed transiently within 20 minutes after growth factor addition to quiescent cells. These results support the notion that the second messengers of PI turnover can regulate gene transcription. Since both calcium elevation and activation of protein kinase C cause phosphorylation of a number of cellular proteins on serine and threonine residues, it is plausible that these phosphorylation events contribute to the changes that occur in the nucleus. However, downregulation of protein kinase C only partially inhibits c-*fos* induction by bombesin (a potent stimulator of PI turnover) and has no effect on c-*fos* expression by EGF (a relatively poor stimulator of PI turnover in most cells).[19] These results not only raise the possibility that branching, parallel pathways are involved in propagating signals to the nucleus, but also suggest that alternative signaling pathways probably exist to mediate growth responses.

An obvious problem that comes to mind in proposing that such commonly used second messengers as $(1,4,5)IP_3$, DAG, or cAMP contribute to the proliferative signal is the question of why hormonal elevation of these molecules in terminally differen-tiated tissues does not result in cell growth. One possible answer is that in most terminally differentiated tissues, irreversible DNA rearrangement prevents response to the normal signals for proliferation. In tissues such as liver or lymphocytes, which are capable of switching from quiescent to proliferative states, this explanation is not sufficient. Instead, in such tissues, growth-inhibitory signals may be produced to counteract stimulatory signals. An alternative explanation is that the critical signals for growth are not the conventional second messengers, but instead are novel molecules such as those in the PI-3-P pathway discussed above.

ONCOGENES AND CELL GROWTH

A major contribution to our understanding of the mechanisms of cell growth regulation has evolved from the study of oncogenes. Oncogenes are characterized by their ability to transform a normal, contact-inhibited cell into a cell that grows beyond confluence in culture in low or no serum and that is capable of forming tumors in animals.[20,21] Such genes were originally observed in certain retroviruses that caused tumors in birds and rodents. It is now clear that retroviral oncogenes originated in the genomes of host animals and were transduced by the viruses from normal, cellular DNA sequences called proto-oncogenes. Many retroviruses have incorporated into their genomes aberrant versions of the cellular genes encoding growth factors, growth factor receptors, or transducers (either cytosolic or nuclear) of growth responses. Thus, it is often either overexpression or mutation of a normal host gene that provokes uncontrolled cell growth. In a sense, these viruses have already "cloned out" many of the genes that are critical to growth regulation. More recently it has been shown that chemical mutagens induce tumors through mutation or overexpression of some of the same oncogenes that are carried by retroviruses.[21]

In general, four different mechanisms can account for the ways in which oncogenes confer growth autonomy to a cell[3,21]: (1) by directly coding for or indirectly stimulating the production of a growth factor whose receptor is synthesized in the same cell type (autocrine stimulation); (2) by encoding a mutated growth factor receptor that is continuously activated in the absence of the growth factor; (3) by encoding a protein that transduces signals from the receptor to cytosolic second messengers; and (4) by coding for a protein that communicates second-messenger responses to nuclear targets. There are excellent examples of the first two types of oncogenes as well as examples of oncogenes that fit into the latter two classes (see below).

The first solid link between oncogenes and growth factors was the discovery that the simian sarcoma virus oncogene, v-*sis*, encodes a protein that is homologous to the beta subunit of PDGF.[22,23] Functionally, the PDGF-like molecule mimics normal PDGF by binding to the receptor and stimulating cell growth. However, since this factor is constitutively produced in the same cell that synthesizes its receptor, the normal mechanisms of terminating the growth response are circumvented. It was recently shown that overproduction of EGF by cells containing EGF receptors also results in cell transformation.[24]

One example of the second class of oncogenes is the *erbB* oncogene of avian erythroblastosis virus. This gene codes for a truncated verson of the EGF receptor,[25] which lacks the EGF-binding domain but contains a continuously-active tyrosine kinase domain. Other examples of oncogenes that are derived from cellular genes encoding growth factor receptors include v-*fms* (the CSF-1 receptor), v-*ros* (a receptor for uncharacterized growth factor), and *neu* (a receptor with extensive homology to the EGF receptor).[1] From studies using molecular gene constructs of various cloned receptors, it appears that mutations that eliminate the amino terminal growth factor–binding domain often produce spontaneously active tyrosine kinases and result in cell transformation.

A number of candidates exist for the class of oncogenes that act as cytosolic transducers of growth factor responses. For example, v-*src*, v-*abl*, and v-*fps* are viral oncogenes that encode protein tyrosine kinases like the growth factor receptors, but the normal cellular homologues of these proteins are not growth factor receptors.[2] These proteins associate with the plasma membrane but have no transmembrane or extracellular domain. It has been suggested that these tyrosine kinases are regulated by growth factor receptor–tyrosine kinases in a cascade reaction.[26] Of interest, cells transformed with v-*src*, v-*abl*, v-*yes*, and polyoma virus middle T antigen (which stimulates the tyrosine kinase activity of $pp60^{c\text{-}src}$) have increased levels of intermediates of phosphatidylinositol turnover. These results suggest that these tyrosine kinases either directly or indirectly regulate a PI pathway.[27–31] Moreover, the same novel PI-3-kinase activity that associates with the PDGF receptor also tightly associates with both $pp60^{v\text{-}src}$ and the middle $T/pp60^{c\text{-}src}$ complex, thus implicating a direct interaction between $pp60^{src}$ and the PI-3-kinase.[7,27]

A second group of oncogenes that may transduce growth factor receptor responses is the *ras* family. These genes code for small proteins (approximately 21,000 Da) which bind and slowly hydrolyze GTP to GDP. *ras* proteins have sequence homology to the family of GTP binding proteins that transduces hormone and vision responses discussed earlier. It is tempting to speculate that these proteins play a role in transducing responses from growth factor receptors to their target molecules. Like polyoma middle $T/pp60^{c\text{-}src}$ transformed cells, cells transformed with mutated *ras* genes show elevated levels of second messengers of the PI turnover pathway. Although data are consistent with the constitutive activation of phospholipase C[32], it has not yet been possible to prove such an interaction in reconstitution experiments.

The final class of oncogenes is represented by a growing family that includes SV40 large T, polyoma large T, E1A, E1B, *myc, fos, ski, myb,* and *jun.*[21,33] The protein products of these genes appear in the nucleus and in many cases may either directly or indirectly regulate cellular genes involved in transcription and the mitotic program.[21] Recent evidence indicates that the proto-oncogene product of human c-*jun* has both structural and functional properties of transcription factor AP-1[33,34] and that the former may be a member of a family of genes encoding related transcriptional regulatory proteins.[33] Thus, while it is itself affected by cellular growth signals, the nuclear c-*jun* product may function to control directly the expression and transcription of specific target cellular genes by binding to DNA enhancers and promotor sequences.

Some of these oncogenes mediate nuclear events that are associated with growth factor stimulation. For example, after the addition of growth factors to quiescent cells, both c-*fos* and c-*myc* messages are elevated within 20 minutes and 2 hours, respectively. As discussed earlier, expression of these same genes can also be induced by agents which mimic the second messengers of the PI turnover pathway.

Much of what we know about the biochemistry and molecular biology of growth regulation has been gleaned from the studies of oncogenes. On the basis of our present knowledge of the functions of retroviral oncogenes, it is surmised that newly discovered as well as the remaining oncogenes may code for either cytoplasmic transducers of growth factor receptor signals or nuclear targets that regulate the genes of mitosis. Further evaluation of the functions of these oncogene families is likely to provide a clearer picture of the biochemical machinery that leads from growth factor receptor stimulation to cell division.

REFERENCES

1. YARDEN, Y., J. A. ESCOBEDO, W.-J. KUANG, T. L. YANG-FENG, T. O. DANIEL, P. M. TREMBLE, E. Y. CHEN, M. E. ANDO, R. N. HARKIN, U. FRANCKE, V. A. FRIED, A. ULLRICH & L. T. WILLIAMS. 1986. Structure of the receptor for platelet-derived growth factor helps define a family of closely related growth factor receptors. Nature (London) **323:** 226–232.
2. HUNTER, T. & J. A. COOPER. 1985. Protein-tyrosine kinases. Annu. Rev. Biochem. **54:** 897–930.
3. WHITMAN, M., L. FLEISCHMAN, S. B. CHAHWALA, L. CANTLEY & P. ROSOFF. 1986. Phosphoinositides, mitogenesis, and oncogenesis. *In* Receptor Biochemistry and Methodology. J. W. Putney, Jr. Ed.: 197–217. Alan R. Liss. New York.
4. BERRIDGE, M. J. & R. F. IRVINE. 1984. Inositol trisphosphate, a novel second messenger in cellular signal transduction. Nature (London) **312:** 315–321.
5. ROZENGURT, E. 1986. Early signals in the mitogenic response. Science **234:** 161–166.
6. NISHIZUKA, Y. 1984. The role of protein kinase C in cell surface signal transduction and tumor promotion. Nature (London) **308:** 693–698.
7. WHITMAN, M., C. P. DOWNES, M. KEELER, T. KELLER & L. CANTLEY. 1988. Type I phosphatidylinositol kinase makes a novel inositol phospholipid, phosphatidylinositol-3-phosphate. Nature (London) **332:** 644–646.
8. SALTIEL, A. R., P. SHERLINE & J. A. FOX. 1987. Insulin-stimulated diacylglycerol

production results from the hydrolysis of a novel phosphatidylinositol glycan. J. Biol. Chem. **262:** 1116–1121.

9. IRVINE, R. F. & R. M. MOOR. 1986. Micro-injection of inositol 1,3,4,5-tetrakisphosphate activates sea urchin eggs by a mechanism dependent on external Ca^{2+}. Biochem. J. **240:** 917–920.

10. CHAHWALA, S. B., L. FLEISCHMAN & L. CANTLEY. 1987. Kinetic analysis of GTP-gamma-S effects on phosphatidylinositol turnover in NRK cell homogenates. Biochemistry **26:** 612–622.

11. STRYER, L. & H. R. BOURNE. 1986. Annu. Rev. Cell Biol. **2:** 391–419.

12. JOHNSON, D., A. LANAHAN, C. R. BUCK, A. SEHGAL, C. MORGAN, E. MERCER, M. BOTHWELL & M. CHAO. 1986. Expression and structure of the human NGF receptor. Cell **47:** 545–554.

13. RADKE, M. J., T. P. MISKO, C. HSU, L. A. HERZENBERG & E. M. SHOOTER. 1987. Gene transfer and molecular cloning of the rat nerve growth factor receptor. Nature (London) **325:** 593–597.

14. FAVA, R. A. & S. COHEN. 1984. Isolation of a calcium-dependent 35-kilodalton substrate for the epidermal growth factor receptor/kinase from A-431 cells. J. Biol. Chem. **259:** 2636–2645.

15. KRETSINGER, R. H. & C. E. CREUTZ. 1986. Consensus in exocytosis. Nature (London). **320:** 573.

16. HUANG, K.-S., B. P. WALLNER, R. J. MATTALIANO, R. TIZARD, C. BURNE, A. FREY, C. HESSION, P. MCGRAY, L. K. SINCLAIR, E. P. CHOW, J. L. BROWNING, K. L. RAMACHANDRAN, J. TANG, J. E. SMART & R. B. PEPINSKY. 1986. Two human 35 kd inhibitors of phospholipase A_2 are related to substrates of $pp60^{v-src}$ and of the epidermal growth factor receptor/kinase. Cell **46:** 191–199.

17. SARIS, C. J. M., B. F. TACK, T. KRISTENSEN, J. R. GLENNEY, JR. & T. HUNTER. 1986. The cDNA sequence of the protein-tyrosine kinase substrate p36 (calpactin I heavy chain) reveals a multidomain protein with internal repeats. Cell **46:** 201–212.

18. WHITMAN, M., D. KAPLAN, T. ROBERTS & L. CANTLEY. 1987. Evidence for two distinct phosphatidylinositol kinases in fibroblasts: Implications for cellular regulation. Biochemical J. **247:** 165–174.

19. MCCAFFREY, P., W. RAN, J. CAMPISI & M. R. ROSNER. 1987. Two independent growth factor-generated signals regulate c-fos and c-myc mRNA levels in Swiss 3T3 cells. J. Biol. Chem. **262:** 1446–1448.

20. BISHOP, J. M. 1985. Viral Oncogenes. Cell **42:** 23–38.

21. WEINBERG, R. A. 1985. The action of oncogenes in the cytoplasm and nucleus. Science **230:** 770–776.

22. WATERFIELD, M. D., G. T. SCRACE, N. WHITTLE, P. STROOBANT, A. JOHNSSON, A. WASTESON, D. WESTERMARK, C.-H. HELDIN, J. S. HUANG & T. F. DEUEL. 1983. Platelet-drived growth factor is structurally related to the putative transforming protein p28 sis of simian sarcoma virus. Nature (London) **304:** 35–38.

23. DOOLITTLE, R. F., M. W. HUNKAPILLER, L. H. HOOD, S. G. DEVARE, K. C. ROBBINS, S. A. AARONSON & H. N. ANTONIADES. 1983. Simian sarcoma virus onc gene, v-sis, is derived from the gene (or genes) encoding a platelet derived growth factor. Science **221:** 275–280.

24. STERN, D. F., D. L. HARE, M. A. CECCHINI & R. A. WEINBERG. 1987. Construction of a novel oncogene based on synthetic sequences encoding epidermal growth factor. Science **235:** 321–324.

25. DOWNWARD, J., Y. YARDEN, E. MAYES, G. SCRACE, N. TOTTY, P. STOCKWELL, A. ULLRICH, J. SCHLESSINGER & M. D. WATERFIELD. 1984. Close similarity of epidermal growth factor receptor and v-erbB oncogene product sequences. Nature (London) **307:** 521–527.

26. RALSTON, R. & J. M. BISHOP. 1985. The product of the proto-oncogene c-src is modified during the cellular response to platelet-derived growth factor. Proc. Natl. Acad. Sci. USA **82:** 7845–7849.

27. KAPLAN, D., M. WHITMAN, B. SCHAFFHAUSEN, L. RAPTIS, R. L. GARCEA, T. M. ROBERTS & L. CANTLEY. 1986. Phosphatidylinositol metabolism and polyoma-mediated transformation. Proc Natl. Acad. Sci. USA **83:** 3624–3628.

28. DIRINGER, H. & R. R. FRIIS. 1977. Changes in phosphatidylinositol metabolism correlated to growth state of normal and Rous sarcoma virus-transformed Japanese quail cells. Cancer Res. **37:** 2979–2984.
29. SUGIMOTO, Y., M. WHITMAN, L. C. CANTLEY & R. L. ERIKSON. 1984. Evidence that the Rous sarcoma transforming gene product phosphorylates phosphatidylinositol and diacylglycerol. Proc. Natl. Acad. Sci. USA **81:** 2117–2121.
30. WHITMAN, M., D. KAPLAN, B. SCHAFFHAUSEN, L. CANTLEY & T. ROBERTS. 1985. A phosphatidylinositol kinase activity is associated with Polyoma middle T competent for transformation. Nature (London) **315:** 239–242.
31. FRY, M. J., A. GEBHARDT, P. J. PARKER & G. FOULKES. 1985. Phosphatidylinositol turnover and transformation of cells by Abelson murine leukemia virus. EMBO J. **4:** 3173–3178.
32. FLEISCHMAN, L. F., S. B. CHAHWALA & L. CANTLEY. 1986. *ras*-transformed cells have altered steady state levels of phosphatidylinositol-4,5-bisphosphate and catabolites. Science **231:** 407–410.
33. VARMUS, H. 1987. Oncogenes and transcriptional control. Science **238:** 1337–1339.
34. BOHMANN, D., T. J. BOS, A. ADMON, T. NISHIMURA, P. K. VOGT & R. TJIAN. 1987. Human proto-oncogene c-*jun* encodes a DNA binding protein with structural and functional properties of transcription factor AP-1. Science **238:** 1386–1392.

DISCUSSION

V. CHAIRUGI (*Florence, Italy*): Could one propose that the PDGF activity on inositol lipid metabolism is c-*src*-dependent in analogy of what happens with the middle T antigen?

L. CANTLEY (*Tufts University, Boston, Massachusetts*): We investigated that possibility by immunoprecipitating pp60^{c-src} from PDGF-stimulated cells and assaying for associated PI-3 kinase activity. No significant activity was found.

G. DUBYAK (*Case Western Reserve University, Cleveland, Ohio*): Does activation of PIP$_2$-phospholipase C in PDGF-treated or polyoma-transfected cells result in altered production of inositol polyphosphate products consistent with the novel substrate PIP$_2$ isomers?

CANTLEY: Kurt Auger and Leslie Serunian in my laboratory find prolonged elevation of Ins-1,3,4-P$_3$ in response to PDGF. This could possibly be derived from PI-3,4-P$_2$, but might also be explained by the previously described pathway in which Ins-1,4,5-P$_3$ is converted to Ins-1,3,4-P$_3$ via the Ins-1,3,4,5-P$_4$ intermediate.

Mechanisms by Which Genes Encoding Growth Factors and Growth Factor Receptors Contribute to Malignant Transformation

MATTHIAS H. KRAUS, JACALYN H. PIERCE,
TIMOTHY P. FLEMING, KEITH C. ROBBINS,
PIER PAOLO DI FIORE, AND STUART A. AARONSON

National Cancer Institute
Laboratory of Cellular and Molecular Biology
Bethesda, Maryland 20892

Investigations of genetic alterations associated with neoplasia have identified a limited set of cellular genes, termed proto-oncogenes, which are highly conserved in vertebrate evolution. Acute transforming retroviruses have substituted viral genes essential for replication with these discrete segments of host genetic information. When incorporated within the retroviral genome, such transduced sequences acquire the ability to induce neoplastic transformation.[1,2] Analysis of products of viral oncogenes has revealed important insights concerning the physiological roles of their normal cellular counterparts, designated proto-oncogenes. In particular, recent evidence has established a direct link between oncogene products and either growth factors or growth factor receptors. The v-*sis* oncogene is derived from the gene encoding the B chain of the platelet-derived growth factor,[3,4] while the v-*erb* B and v-*fms* oncogenes are derived from the genes encoding epidermal growth factor (EGF)[5] or colony-stimulating factor-1 (CSF-1)[6] receptors, respectively. An important consequence of these findings has been the ability to construct specific, testable models concerning the molecular basis of oncogenesis.

The present review summarizes our current understanding of the mechanisms involved in transformation by genes normally encoding growth factors or growth factor receptors. We have studied the transforming properties of the v-*sis* oncogene in model systems and the role of the *sis* proto-oncogene in the development of human malignancies. We also investigated the ability of the gene for another factor, transforming growth factor (TGFα), to act as an oncogene in cells expressing the appropriate receptors. Finally, we have investigated the transforming potential for two growth factor receptor genes, epidermal growth factor receptor (EGFR) gene and the closely related *erb* B-2 gene, and have identified aberrations of these genes associated with naturally occurring malignancies. Our findings have implications concerning the conditions under which the expression of growth factors and growth factor receptors may play a role in tumor initiation or progression.

Oncogenes and Growth Factors: The sis *Oncogene*

The first direct link between oncogenes and growth-factor-mediated proliferative pathways was established when it was determined that the amino acid sequence of one of the two major polypeptide chains of human PDGF is highly similar to that predicted for the transforming protein encoded by the v-*sis* oncogene.[3,4] Molecular cloning and nucleotide sequence analysis of the c-*sis* proto-oncogene from the normal human cellular genome demonstrated that it encodes the PDGF-B or -2 chain.[7] Abnormal expression of the c-*sis* gene was found in some spontaneous human malignancies of connective tissue.[8] This suggested the hypothesis that the *sis*/PDGF-B may very well play a role in the neoplastic process and supplied the rationale to test whether the transcriptional activation of the gene coding for this normal growth factor can transform a cell responsive to its growth-promoting action.

Activation of the Normal Human c-sis/PDGF-B Gene as an Oncogene

The strategy for assessing the transforming potential of the normal human-*sis*/PDGF-2 proto-oncogene involved introduction of the PDGF-B coding sequences into a cell that was susceptible to SSV-induced malignant transformation. Transfection of NIH/3T3 cells with the entire PDGF-B coding sequence had no effect on the morphology or growth of these cells. However, when the normal PDGF-B coding sequence was positioned downstream of a retroviral transcriptional promoter capable of inducing mRNA expression, the molecular construct transformed NIH/3T3 cells with an efficiency comparable to that of the v-*sis* oncogene.[22] These findings established that activation of expression of a normal human growth factor gene can cause it to acquire oncogenic properties. Moreover, when captured by a retrovirus, the v-*sis*/PDGF-2 transforming gene has been shown to induce fibrosarcomas and glioblastomas,[9,10] which are derived from cell types that express PDGF receptors. Human tumors, fibrosarcomas and glioblastomas, often express *sis*/PDGF-2 transcripts not detectable in normal human fibroblasts or glial cells.[8] Thus, transcriptional activation of this gene in a cell susceptible to the growth-promoting action of PDGF may be an important step in the induction of certain naturally occurring human tumors.

Identification and Localization of the PDGF-B Gene Product in Human Tumor Cells

As a further test of this hypothesis, we screened a variety of tumor cell lines to compare their levels of the c-*sis*/PDGF-B transcript with those of v-*sis* transcripts in SSV-transformed NRK or NIH/3T3 cell lines. The latter express moderate or barely detectable levels of the v-*sis* gene product, respectively. Northern blot analysis using a v-*sis* probe revealed variable levels of PDGF-2 mRNA among the human tumor cell lines examined. The A172 glioblastoma cell line contained levels of PDGF-2 mRNA comparable to those found in SSV-transformed NIH/3T3 cells. Thus, we utilized this tumor cell line in efforts to detect human PDGF-2 gene products. In SSV transformants, the v-*sis* product has been shown to be predominantly membrane-associated.[11] In fact, reproducible detection of p28[v-*sis*] in SSV-transformed NIH/3T3 cells requires its enrichment in cell membrane fractions. Efforts to detect human PDGF-B polypeptides in whole cell extracts of A172 tumor cells were not successful.

It was difficult to detect c-*sis*/PDGF-B gene products in A172 glioblastoma cells. However, low levels of p26[c-*sis*] were found in the membrane fraction of these cells. The protein was identical in size to p26[c-*sis*] observed in COS cells transfected with the PDGF-B cDNA construct. All of these findings establish that human tumor cells can express PDGF-B products analogous to those in cells induced to express functional PDGF-B.[12] Our detection of PDGF-B homodimeric molecules in a human glioblastoma cell line strongly supports the concept that expression of this mitogen can play a role in the development of naturally occurring tumors of connective tissue origin.[12]

Other Growth Factors and the Malignant Process

Transforming growth factor α, TGFα, is a 50-amino-acid peptide which, like epidermal growth factor (EGF), is a potent mitogen for cells possessing the EGF receptor. The secretion of TGFα by transformed but apparently not by normal cells,[13-15] as well as the ability of this growth factor to induce anchorage-independent growth of certain cells,[16] raised speculation concerning its role in tumor development.[17] The molecular cloning of TGFα cDNA[18] provided the opportunity to directly test the effects of expression of this growth factor in cells possessing its receptor and to compare its biological activity with v-*sis*/PDGF-B and the activated form of the cognate receptor for TGFα, the v-*erb*/EGF receptor.

Effects of TGFα Expression Vectors on the Phenotype of NIH/3T3 Cells

To test the biological activity of human TGFα, the coding region of the TGFα cDNA was cloned into two different expression vectors and introduced into NIH/3T3 cells. Therefore, the TGFα cDNA sequence encoding the human TGFα precursor protein[18] was cloned into Moloney murine leukemia virus (Mo-MuLV)-based vector pZip NeoSV(X)1.[19] In this vector (pZipTGFα), TGFα should be translated from an mRNA transcribed under the control of the LTR. We also placed the TGFα coding sequence under the control of the metallothionein promoter,[20] which can be induced by heavy metals. To assess the transforming potential of these human TGFα expression vectors, we transfected the DNA of each onto NIH/3T3 cells. Neither induced detectable transformed foci with as much as 10μg DNA added per plate.[21] Moreover, induction of the metallothionein promoter by inclusion of 80 mM $ZnCl_2$ in medium of cells transfected with pMTTGFα DNA did not result in the appearance of foci. In contrast, both the v-*sis* oncogene of simian sarcoma virus and the v-*erb* oncogene of avian erythroblastosis virus induced readily observable transformed foci at titers of more that 10^3 foci/μg DNA under identical conditions.[22,23]

Although transfection of TGFα recombinant plasmids onto NIH/3T3 cells did not produce any observable altered foci, clonal lines expressing high levels of TGFα exhibited a morphology readily distinguishable from control NIH/3T3 lines. These cells were refractile and grew in multiple layers without a regular growth pattern, whereas the control cells grew as a flat contact-inhibited monolayer. Moreover, the saturation density achieved in 10% calf serum by each of the TGFα-expressing lines was at least 2- to 4-fold higher than that of control cell lines and comparable to that of a v-*erb* B transformant.[21]

Mechanism Accounting for the Overgrowth of TGFα Expressing NIH/3T3 Cells

There was an apparent paradox between the lack of focus formation induced by the TGFα expression vectors upon primary transfection of NIH/3T3 and the densely growing nature of clonal lines selected after transfection as TGFα-expressing cells. It was possible that introduction of the TGFα coding sequence required more cell generations for expression of a genetically determined transformed phenotype than was required for the viral oncogenes v-*sis* or v-*erb* B. Alternatively, the growth-promoting effects of TGFα might not be exerted specifically on the cell synthesizing this growth factor, but instead might act on the entire cell population after secretion into the medium. By the former hypothesis, stable TGFα-transformed cells might be expected to form densely growing colonies when plated as single cells on monolayers of contact inhibited NIH/3T3, an assay for transformed cells.[24,25] If the latter were the case, the number of TGFα-secreting cells plated might not secrete a sufficient quantity of TGFα into the medium to cause the culture to grow in a transformed manner. V-*sis* and v-*erb* B transformants efficiently formed readily detectable colonies when plated on NIH/3T3 monolayers. In striking contrast, as many as 10^4 cells of a representative TGFα-producing subline, MTTGF-15-2, failed to yield visible colonies under these same conditions. This lack of growth did not represent random clonal variation since none of eight TGFα sublines analyzed yielded detectable colonies in the monolayer assay. These results strongly implied that the overgrowth potential of the TGFα-expressing sublines was not an intrinsic genetic property.[21]

To determine whether added TGFα could induce the growth alterations observed in dense cultures of clonal TGFα-expressing NIH/3T3 cells, we added different TGFα concentrations to control NIH/3T3 cells. At a TGFα concentrations of 5 ng/ml, comparable to that released by dense TGFα sublines, the cells grew to 2- to 4-fold increased saturation density and exhibited morphologic alterations indistinguishable from those of clonally derived TGFα-producing NIH/3T3 sublines. To directly assess the effect of TGFα expression on the *in vivo* growth properties of the cells, transfectants containing pZipTGFα or pMTTGFα, as well as v-*sis* or v-*erb* B, were inoculated subcutaneously in nude mice. As shown in TABLE 1, v-*sis* and v-*erb* transfectants rapidly formed tumors with as few as 10^4 cells inoculated. In contrast several TGFα-expressing NIH/3T3 lines, which secreted large quantities of TGFα, failed to induce tumors even when as many as 10^6 cells were inoculated. All of the above results strongly argue that TGFα exerts little or no direct effect on the growth properties of the individual NIH/3T3 cell from which it is secreted.[21] TGFα is known to be actively secreted with less than 1% of the amount synthesized remaining cell-associated. In contrast, the *sis*/PDGF-B gene product is processed through the secretory pathway but remains tightly cell-associated with less than 5% released into culture fluids. This presumably makes available a high concentration of this growth factor available specifically to the cell producing it and may account at least in part for the much greater potency of *sis*/PDGF-B compared to TGFα as a transforming gene for

TABLE 1. *In Vivo* Growth Properties of TGFα-Expressing NIH/3T3 Lines

Cell Line	'No. of Tumors/No. of Animals[a]		
	10^6 Cells	10^5 Cells	10^4 Cells
neo			
SV2 neo-7-1	0/4	0/4	0/4
SV2 neo-9M	0/4	0/4	0/4
TGFα			
MTTGF-16-7	0/4	0/4	0/4
MTTGF-15-2	0/4	0/4	0/4
ZipTGF-IIM	0/4	0/4	0/4
ZipTGF-12M	0/4	0/4	0/4
erbB			
ERB-21M	4/4	4/4	1/4
sis			
sis-13M	4/4	3/4	1/4

[a]NFR nude mice were inoculated subcutaneously with the indicated number of cells from individual cell lines. The animals were observed twice weekly for 6 weeks for the formation of tumors greater than 2 mm in diameter.

NIH/3T3 cells. Thus, the mode of processing of a growth factor as well as the availability of cell surface receptors with which to interact may play an important role in determining its ability to drive the cell expressing it along pathways toward malignancy.

Growth Factor Receptors and Oncogenes: The EGF Receptor

There is accumulating evidence that genes encoding growth factor receptors can also become oncogenes. For example, activation of the EGF receptor (EGFR) gene (c-*erb*B locus) as an oncogene involves an NH_2-terminal truncation of the protein together with other modifications.[26] More subtle alterations appear to have been

TABLE 2. Growth Properties of EGF Receptor-Transfected NIH/3T3

Transfected DNA	Transforming Activity[a] (ffu/pM)		Growth in Soft Agar[b] (%)	
	−EGF	+EGF	−EGF	+EGF
LTR/EFGR	$<10^0$	2×10^2	0.4	19.7
LTR/v-*erb*B	3×10^2	3×10^2	12.7	15.4
pSV$_2$/gpt	$<10^0$	$<10^0$	<0.01	<0.01

[a]Transfection was performed with 40 μg of calf thymus DNA as a carrier by means of the calcium phosphate precipitation technique. Eighteen days after the transfection, EGF at 20 ng/ml was added to the indicated plates and focus formation was scored 8 days after EGF addition on duplicate plates. The specific activity is adjusted to focus-forming units per picomoles of cloned DNA added on basis of the relative molecular weights of the respective plasmids.

[b]Single-cell suspensions were plated at 10-fold serial dilution in 0.45% sea plaque agarose medium plus 10% calf serum. Visible colonies were scored at 14 days. The results represent the mean values of three independent experiments.

responsible for the oncogenic conversion of v-*fms*.[27,28] The *neu* oncogene,[29] which also encodes a growth factor receptor-like protein,[30,31] has been shown to be activated in chemically induced rat neuroblastomas[29] by a single amino acid substitution in its transmembrane domain.[32] In human malignancies, abnormalities involving growth factor receptors most commonly appear to involve gene amplification and/or overexpression of apparently normal RNAs and proteins.[33–40] These findings have suggested that overexpression of a normal growth factor receptor may also provide a selective growth advantage to a cell progressing along the malignant pathway. The EGF receptor gene is often amplified in human tumors.[33,35,37,39] We, therefore, sought to investigate the selective advantage, if any, that overexpression of the normal EGF receptor might confer to cells in a model system.

Overexpression of the EGF Receptor Confers a Conditional Growth Advantage to NIH/3T3 Cells

In an effort to directly assess whether overexpression of the normal EGF receptor could alter cell growth properties, we engineered an eukaryotic expression vector in which the EGFR cDNA was put under the transcriptional control of the Mo-MLV LTR.[41] This vector (LTR/EGFR) was transfected onto NIH/3T3 cells, which normally express a low number of functional EGF receptors (approximately 3×10^3 receptors per cell). As shown in TABLE 2, the LTR/EGFR DNA construct failed to induce morphologically altered foci in NIH/3T3 cells in repeated experiments. In contrast, LTR/v-*erb*B, which encodes an oncogenically activated form of the avian EGF receptor[23] engineered in an identical expression vector, induced transformed foci with efficiencies of 3×10^2 focus-forming units per picomole of DNA (TABLE 2). We reasoned that if increased number of EGF receptors were to confer a selective growth advantage to NIH/3T3 cells, this advantage might be unmasked by the addition of EGF to the medium. Indeed, when the transfection assay was performed in medium supplemented with 20 ng/ml EGF, the same LTR/EGFR construct displayed readily detectable transforming activity of around 2×10^2 focus-forming units per picomole of DNA (TABLE 2). The conditional nature of the transforming activity associated with LTR/EGFR transfection strongly argues that the normal EGFR gene was responsible for transforming activity observed in the presence of EGF. The specificity of the EGF effect was further demonstrated by the lack of a detectable increase in the transforming potential of the LTR/v-*erb*B DNAs upon EGF addition to the culture medium (TABLE 2).[41]

As an independent approach toward assessing the conditional nature of the growth alterations induced by the LTR/EGFR construct, we investigated the ability of transfected NIH/3T3 cells to display anchorage-independent growth, a property known to correlate well with the malignant phenotype.[42] After transfection and marker selection, a mass population, designated NIH-EGFR cells, was suspended in semi-solid medium. In the absence of EGF supplement, the cells displayed only a low colony-forming ability of around 9.4% (TABLE 2). However, upon addition to EGF to the agar medium, we observed a dramatic increase in colony formation (TABLE 2) with a shift towards large, progressively growing colonies (>0.20 mm in diameter). The optimal EGF concentration for induction of this effect was 20 ng/ml (data not shown). In comparison, NIH/3T3 cells transfected with the v-*erb*B gene displayed a high

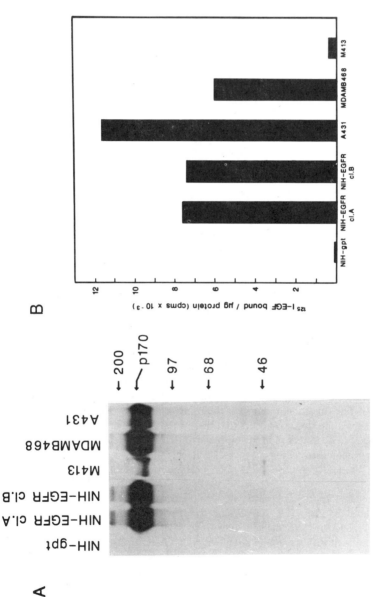

FIGURE 1. Comparison of the levels of EGF receptor in LTR/EGFR-transfected clonal populations of NIH/3T3 cells and human tumor cell lines. (A)Immunoprecipitation and electrophoretic analysis. 3.0×10^7 TCA-precipitable cpm of metabolically labelled extracts from two clones of agar-selected NIH-EGFR cells (NIH-EGFR cl. A and cl. B) and two human tumor cell lines (A431 and MDAMB468) that overexpress the EGF receptor were immunoprecipitated with the EGF receptor-specific antibody, RK2. NIH-gpt are NIH/3T3 cells transfected with the control plasmid pSV2/gpt and selected in killer HAT medium. M413 cells are normal human fibroblasts. (B) ^{125}I-EGF binding. Saturation binding of ^{125}I-EGF on the indicated cell lines was measured in the presence of ^{125}I-EGF at a concentration of 200 nM (specific activity, 97 mCi/mg). Results are expressed as cpm of ^{125}I-EGF specifically bound after subtraction of nonspecific binding, determined in the presence of a 100-fold excess of unlabeled EGF.

clonogenic capability in semi-solid medium that was not enhanced by EGF addition (TABLE 2).[41]

Some Human Tumor Cells Overexpress EGF Receptor at Levels that Confer Conditional Growth Advantage to NIH/3T3 Cells

The EGF receptor gene has been reported to be amplified and/or overexpressed in a wide array of human malignancies.[33,35,37,39] Hence, we sought to compare the levels of overexpression of the EGFR protein in human tumor cell lines possessing amplified EGFR genes with that of clonal NIH/3T3 cell lines containing the EGFR coding sequence and exhibiting an EGF-dependent transformed phenotype.[41] Human tumor cell lines A431 and MDAMB468 have been shown to exhibit EGFR gene amplification[26,39,43–45] accompanied by overexpression of the EGFR protein.[45–47] By immunologic analysis, the levels of the 170-kD EGFR protein in A431 and MDAMB468 cells were markedly elevated over that in M413 human embryo lung fibroblasts, which contain the unamplified EGF receptor gene (FIG. 1). As shown in FIGURE 1 the elevated levels of EGFR protein detected in the tumor lines were similar to those observed in NIH-EGFR CI A and CI B cell lines, as detected both by immunoprecipitation of the EGFR with a specific antibody (FIG. 1A) and by a saturation binding assay of ^{125}I-EGF (FIG. 1B). Thus, human tumor cell lines that overexpressed the EGFR gene demonstrated levels of EGFR similar to those capable of conferring the EGF-dependent transformed phenotype in a model system.[41]

Identification of a Novel EGF Receptor-Related Gene Amplified in a Human Mammary Adenocarcinoma

As a further approach to identify activated oncogenes which encode receptor-related proteins, we subjected DNAs of mammary tumor cell lines and tissues to Southern blot analysis utilizing v-*erb*B as a probe.[34] We did so because of the homology of v-*erb*B to the EGF receptor[5] and the observation that certain mammary tumor cell lines exhibit high EGF-binding levels, comparable to those observed in the epidermoid carcinoma cell line A431, which displays EGF receptor gene amplification and rearrangement.[26,33] In an effort to identify genes that might be candidates for new receptor coding sequences of this gene family, we employed hybridization conditions of moderate stringency under which proto-oncogenes related to other viral oncogenes of the tyrosine kinase family did not hybridize (data not shown). Thus, any gene detected might be expected to have a closer relationship to v-*erb*B than to other members of the tyrosine kinase family. DNA prepared from a human mammary carcinoma, MAC117, showed a pattern of hybridization differing both from that observed with DNA of normal human placenta and the A431 squamous cell carcinoma line. In A431 DNA, we observed four *Eco*R1 fragments that had increased signal intensities compared to those of corresponding fragments in placental DNA. In contrast, MAC117 DNA contained a 6-kilobase pair (kbp) fragment, which appeared to be amplified compared to corresponding fragments observed in both A431 and placental DNAs. These findings were consistent with the possibility that the MAC117 tumor contained an

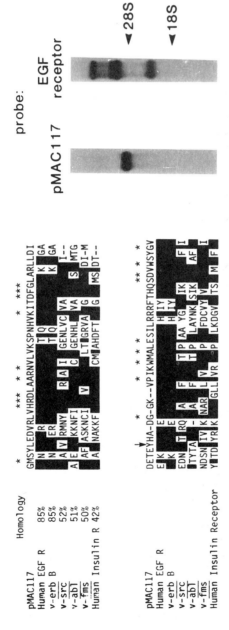

FIGURE 2. (*Left*) Comparison of the putative encoded amino acid sequence in pMAC117 with known tyrosine kinase sequence. Black regions represent homologous amino acids. Differing amino acids are shown in one-letter code. Amino acid positions conserved in all sequences are denoted by asterisks. The tyrosine homologous to that autophosphorylated by the *v-src* protein is shown by an *arrow*. The *v-abl* sequence contains a tyrosine residue in this region displaced by two positions. The amino acid sequences of human EGF receptor, *v-erb*B, *v-src*, *v-abl*, *v-fms*, and human insulin receptor were aligned by a computer program. The homology observed with the predicted amino acid sequences of *v-yes* and *v-fms* was 51 percent and 48 percent, respectively. (*Right*) Detection of distinct messenger RNA species derived from the pMAC117 gene and the human EGF receptor gene.

amplified DNA sequence related to, but distinct from, the cellular *erb*B proto-oncogene.[34]

To define its structure, we undertook the molecular cloning of the 6-kbp *Eco*R1 fragment and we determined its nucleotide sequence in the region most homologous to v-*erb*B. This sequence contained two regions of nucleotide sequence homology to v-*erb*B separated by 122 nucleotides. The predicted amino acid sequence (FIG. 2) was 85% homologous to two regions that are contiguous in the EGF receptor sequence.[26] By comparison of the predicted amino acid sequence of the clones designated pMAC117 with corresponding sequences of several members of the tyrosine kinase family, the most striking homology was observed with the human EGF receptor and v-*erb*B. However, we observed 42% to 52% homology with the predicted amino acid sequences of other tyrosine kinase encoding genes as well.

The availability of cloned probes of the gene made it possible to investigate its expression in a variety of cell types. The probe detected a single 5-kb transcript in A431 cells (FIG. 2). Under the stringent conditions of hybridization utilized, this probe did not detect any of the three RNA species recognized by EGF receptor complementary DNA. Thus, the gene, designated *erb*B-2, represented a new functional gene within the tyrosine kinase family, closely related to, but distinct from the gene encoding the EGF receptor.[34]

Overexpression of the erbB-2 Gene in Human Mammary Tumor Cell Lines by Different Molecular Mechanisms

The initial identification of *erb*B-2 gene amplification in tissue from a primary mammary adenocarcinoma (MAC 117) suggested the possibility that *erb*B-2 overexpression might contribute to neoplastic growth in this tumor type. To assess the role of *erb*B-2 in human mammary neoplasia, we compared mRNAs of 16 mammary tumor cell lines, normal M413 human fibroblasts, and a human mammary epithelial cell line, HBL100.[39] Increased expression of an apparently normal size 5-kb transcription was detected in 8 of 16 tumor cell lines, when total cellular RNA was subjected to Northern blot analysis (data not shown). An aberrantly sized *erb*B-2 mRNA was not detected in any of the cell lines analyzed.[39] The amount of *erb*B-2 transcript in the eight cell lines overexpressing *erb*B-2 was measured by dot-blot analysis (TABLE 3).

To investigate alterations of the *erb*B-2 gene associated with its overexpression, we examined the *erb*B-2 gene by Southern blot analysis in these cell lines. The normal restriction pattern was detected in all DNA samples tested, indicating that gross rearrangements in the proximity of the *erb*B-2 coding region did not occur in these cell lines. When compared with normal human fibroblast DNA, the *erb*B-2-specific restriction fragments appeared amplified in several cell lines, including SK-BR-3, BT474 and MDA-MB361.[39] Quantitation of *erb*B-2 gene copy number was accomplished using DNA dot-blot analysis. These studies revealed a 4-to 8-fold *erb*B-2 gene amplification in SK-BR-3 and BT474 relative to normal human DNA and a 2- to 4-fold *erb*B-2 gene amplification in the MDA-MB453 and MDA-MB361 cell lines. Thus, gene amplification was associated with overexpression in the four tumor cell lines with the highest levels of *erb*B-2 mRNA (TABLE 3). In contrast, gene amplification could not be detected by Southern blot analysis or DNA dot-blot analysis in four tumor cell lines in which the *erb*B-2 transcript was increased to intermediate levels.[39]

TABLE 3. Overexpression and Gene Amplification of erbB-2 in Human Mammary
Neoplasia

Source	Overexpression of mRNA[a]	Gene Amplification
M413	1	1
HBL100	1	1
MCF-7	1	1
SK-BR-3	128	4–8
BT474	128	4–8
MDA-MB361	64	2–4
MDA-MB453	64	2
ZR-75-1	8	1
ZR-75-30	4	<1
MDA-MB175	8	1
BT483	8	<1

[a]Overexpression above normal fibroblast and HBL 100.

In chemically induced rat neuroblastomas, a point mutation within the transmembrane domain activates neu, the rat homologoue of erbB-2 to acquire transforming activity in the NIH/3T3 transfection assay.[32] The lack of transforming activity in the NIH/3T3 focus-forming assay of a large group of mammary tumors and tumor cell lines,[48] including those that exhibited erbB-2 amplification and/or overexpression in the absence of aberrant transcript size, suggested that a structurally normal erbB-2 coding sequence was overexpressed in some human mammary tumor cell lines.[39]

The erbB-2 Gene is a Potent Oncogene When Overexpressed in NIH/3T3 Cells

To directly assess the effects of erbB-2 overexpression on cell growth properties, expression vectors based on the transcriptional initiation sequences of either the Moloney murine leukemia virus long terminal repeat (MuLV LTR) or the SV40 early promoter were constructed in an attempt to express the erbB-2 cDNA at different levels in NIH/3T3 cells.[49,50]

To assess the biologic activity of our human erbB-2 vectors, we transfected NIH/3T3 cells with serial dilutions of each DNA. As shown in TABLE 4, two different LTR-based erbB-2 expression vectors, LTR-1/erbB-2 and LTR-2/erbB-2, induced transformed foci at high efficiencies of 4.1×10^4 and 2.0×10^4 focus-forming units per picomole of DNA (ffu/pM), respectively.[50] In striking contrast, the SV40/erbB-2 construct failed to induce any detectable morphologic alteration of NIH/3T3 cells transfected under identical assay conditions (TABLE 4). Since the SV40/erbB-2 construct lacked transforming activity, these results demonstrated that the higher levels of erbB-2 expression under LTR influence correlated with its ability to exert transforming activity.[50]

To compare the growth properties of NIH/3T3 cells transfected by these genes, we analyzed the transfectants for anchorage-independent growth in culture, a property of many transformed cells.[42] The colony-forming efficiency of the LTR-1/erbB-2 transformant was very high and comparable to that of cells transformed by LTR-driven v-H-ras and v-erbB (TABLE 4). Moreover, the LTR-1/erbB-2 transfectants were as malignant in vivo as cells transformed by the highly potent v-H-ras oncogene and

50-fold more tumorigenic than cells tranfected with v-*erb*B. In contrast, SV40/*erb*B-2 transfectants failed to display anchorage-independent growth *in vitro* and did not grow as tumors in nude mice even when 10^6 cells were injected (TABLE 4).[50]

While the predicted *erb*B-2 protein bears structural similarity to the EGF receptor, there is evidence that EGF is not the ligand for the *erb*B-2 product.[51,52] In fact, the normal ligand for this receptor-like protein has yet to be identified. In present in serum, this ligand might be responsible for stimulating the overexpressed *erb*B-2 product and triggering its transforming ability. To address this possibility, we investigated whether *erb*B-2 transformed cells maintained their altered phenotype when cultured in medium lacking serum.[50] NIH/3T3 cells grow in a chemically defined medium which contains EGF, PDGF, or FGF, and high concentrations of insulin (W. Taylor, O. Segatto, S.A. Aaronson, unpublished). These growth factors were excluded as possible exogenous ligands for the *erb*B-2 gene product.[51,52] In medium lacking EGF, PDGF or FGF, LTR-1/*erb*B-2-transfected cells continued to exhibit a stable transformed phenotype by growing as foci of densely packed cells.[50] These findings demonstrate that neither EGF nor any factors present in serum are required for maintaining the transformed phenotype of NIH/3T3 cells overexpressing *erb*B-2.

Human Mammary Tumors with Amplified erbB-2 Genes Express the erbB-2 Protein at High Levels Comparable to LTR/erbB-2 NIH/3T3 Transformants

In order to assess the relevance of *erb*B-2 protein levels inducing *in vitro* transformation for *erb*B-2 overexpression in mammary neoplasia, we sought to compare the level of overexpression of the *erb*B-2-encoded 185-kD protein in human mammary tumor cell lines possessing amplified *erb*B-2 genes with that of NIH/3T3 cells transformed by the *erb*B-2 coding sequence.[50] An anti-*erb*B-2 peptide serum detected several discrete protein species ranging in size from 150 to 185 kD in extracts of MDA-MB361 and SK-BR-3 mammary tumor cell lines, as well as in LTR/*erb*B-2

TABLE 4. Transformed Phenotype of *erb*B-2 Transfectants

DNA Transfectant[a]	Specific Transforming Activity[b] (ffu/pM)	Colony-Forming Efficiency in Agar[c] (%)	Cell Number Required for 50% Tumor Incidence[d]
LTR-1/*erb*B-2	4.1×10^4	45	10^3
SV40/*erb*B-2	$<10^0$	<0.01	$>10^6$
LTR/*erb*B	5.0×10^2	20	5×10^4
LTR/*ras*	3.6×10^4	35	10^3
pSV2/gpt	$<10^0$	<0.01	$>10^6$

[a]All transfectants were isolated from plates that received 1 μg cloned DNA and were selected by their ability to grow in the presence of killer HAT medium.
[b]Focus-forming units were adjusted to ffu/pM of cloned DNA added on the basis of relative molecular weights of the respective plasmids.
[c]Cells were plated at 10-fold serial dilutions in 0.33% soft agar medium containing 10% calf serum. Visible colonies comprising >100 cells were scored at 14 days.
[d]NFR nude mice were inoculated subcutaneously with each cell line. Ten mice were tested at cell concentrations ranging from 10^6 to 10^3 cells/mouse. Tumor formation was monitored at least twice weekly for up to 30 days.

NIH/3T3 transformants (FIG. 3). The relative levels of the 185-kD *erb*B-2 product were similar in each of the cell lines and markedly elevated over that expressed by MCF-7 cells, where the 185-kD *erb*B-2 protein was not detectable under these assay conditions (FIG. 3). Thus, human mammary tumor cells which overexpressed the *erb*B-2 gene demonstrated levels of the *erb*B-2 gene product capable of inducing malignant transformation in a model system.

The Role of Growth Factors and Growth Factor Receptors in the Malignant Process

While knowledge of the chain of events leading a normal cell to become malignant is still incomplete, there is considerable evidence that activated oncogenes often

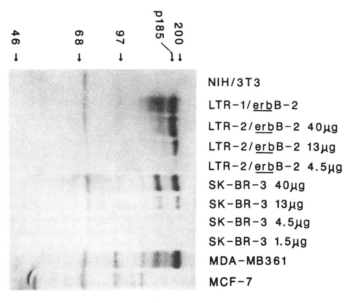

FIGURE 3. Immunoblot analysis of *erb*B-2 proteins in LTR/*erb*B-2-transformed NIH/3T3 cells and human tumor cell lines.

subvert the pathways normally regulating cell proliferation. In the present review, we have provided evidence that quantitative alterations in the expression of growth factors and growth factor receptors are capable of transforming cells under controlled *in vitro* conditions in model systems. Moreover, we have summarized evidence that related alterations can be detected in a significant fraction of naturally occurring human malignancies.

On the basis of evidence summarized here, the ability of a growth factor to induce autonomous *in vitro* or *in vivo* growth of a cell in which it is synthesized is dependent upon more than expression of cognate receptors by the same cell. After its synthesis, the v-*sis*/PDGF-2 gene product is processed along the same pathway as the PDGF receptor.[11] Moreover, the v-*sis* product is not actively secreted and remains tightly cell-associated. Whether this growth factor interacts with its receptor within the cell or at a surface location, it is capable of triggering the autonomous growth of the single cell in which it is synthesized.[22] In contrast, our results demonstrate that growth stimulation by secreted TGFα must be mediated indirectly by its "feeding effect" on the entire culture with little, if any, direct action on the cell in which it is synthesized. Thus, the mode of growth factor synthesis and secretion may play critical roles in determining its actions. If, for example, the growth factor was processed through the cell as an inactive precursor or in a different cellular compartment from its receptor and/or was rapidly secreted and diffused, it may exert little or no direct effect on growth of that cell.

By analogy with what has been established for the transforming activity of the PDGF2/*sis* gene, our studies on the oncogenic activation of growth factor receptors also demonstrate that the simple overexpression of an otherwise unaltered molecule, involved in the control of normal cell proliferation, can lead to the acquisition of the malignant phenotype.[40,50] We showed, in fact, that the human *erb*B-2 gene can be activated as an oncogene by its overexpression in NIH/3T3 cells.[50] The level of the *erb*B-2 product was shown to be critical in determining its transforming ability. An SV40-driven *erb*B-2 cDNA construct lacked detectable focus-forming ability despite the fact that NIH/3T3 cells containing this construct exhibited readily detectable levels of the *erb*B-2 protein. When the same *erb*B-2 protein. When the same *erb*B-2 cDNA was placed under LTR control, a further 5- to ten-fold increase in expression of the *erb*B-2 product was associated with the acquisition of transforming properties by the gene. Similarly, introduction of an EGFR expression vector induced the appearance of a ligand-dependent transformed phenotype in NIH/3T3 cells.[41] These *in vitro* observations were paralleled by the *in vivo* finding of overexpression of the *erb*B-2 gene in a number of human tumors including a significant percentage of human mammary adenocarcinomas.

Interesting differences emerged from the analysis of the transformed phenotypes induced by the EGFR and the *erb*B-2 protein. In the former case, the overexpression of EGFR was capable of inducing malignant transformation of NIH/3T3 cells only when the cells were grown in the presence of the physiological ligand (EGF) for the receptor.[41] In the latter case, transformation seemed to be ligand-independent.[50] In fact, transfected NIH/3T3 cells overexpressing the *erb*B-2 product were still capable of altered growth in chemically defined medium supplemented with EGF, FGF, and insulin, all of which have been excluded as exogenous ligands for this receptor-like protein. This raises intriguing questions about the mechanisms of transformation by these two related receptors.

We demonstrated increased ligand-sensitivity for NIH/3T3 cells expressing high levels of EGFR.[41] Control NIH/3T3 cells showed a 4- to 5-fold increase in DNA synthesis upon EGF addition. However, NIH/3T3 clones that overexpressed EGF receptors at levels 500- to 1000-fold over that of control cells demonstrated a markedly amplified response to EGF, exhibiting 80- to 100-fold increases in DNA synthesis. They also demonstrated responsiveness to the ligand at EGF concentrations that failed to stimulate control NIH/3T3 cells. Thus, our findings support the concept that EGF receptor overexpression amplifies normal EGF signal transduction. Since the ligand for the *erb*B-2 protein has yet to be identified, it is not possible to exclude that *erb*B-2 transformed cells themselves might produce the ligand for this receptor protein. Alternatively, an increased number of receptors may cause transformation either by raising the level of constitutive tyrosine kinase activity to a threshold required for growth stimulation or by facilitating receptor-receptor interactions that may be a prerequisite for their activation.

We also demonstrated that EGF receptor gene overexpression was much less effective than normal *erb*B-2 overexpression in inducing the transformed phenotype. These findings support the hypothesis that the *erb*B-2 gene product is coupled to a distinct and more potent growth-signaling pathway than the *erb*B/EGF receptor in NIH/3T3 cells. Investigation of recombinants between the EGF-receptor and *erb*B-2 coding sequences should make it possible to identify those regions that confer more potent transforming activity to the *erb*B-2 protein.

REFERENCES

1. WEISS, R. A., N. TEICH, H. VARMUS, & J. COFFIN, Eds. 1984. Molecular Biology of Tumor Viruses, 2nd ed. Cold Spring Harbor Laboratory. Cold Spring Harbor, NY.
2. BISHOP, J. M. 1983. Annu. Rev. Biochem. **52:** 201–254.
3. DOOLITTLE, R. F., M. W. HUNKAPILLER, L. E. HOOD, S. G. DEVARE, K. C. ROBBINS, S. A. AARONSON, & H. N. ANTONIADES. 1983. Science **221:** 275–277.
4. WATERFIELD, M. D., G. T. SCRACE, N. WHITTLE, P. STROOBANT, A. JOHNSSON, A. WASTESON, B. WESTERMARK, C. H. HELDIN, J. S. HUANG, & T. D. DEUEL. 1983. Nature **304:** 35–39.
5. DOWNWARD, J., Y. YARDEN, E. MAYES, G. SCRACE, N. TOTTY, P. STOCKWELL, A. ULLRICH, J. SCHLESSINGER, & M. D. WATERFIELD. 1984. Nature **307:** 521–527.
6. SHERR, C. J., C. W. RETTENMIER, R. SACCA, M. F. ROUSSEL, A. T. LOOK, & E. R. STANLEY. 1985. Cell **41:** 665–676.
7. IGARASHI, H., A. GAZIT, I.-M. CHIU, A. SRINIVASAN, A. YANIV, S. R. TRONICK, K. C. ROBBINS, & S. A. AARONSON. 1985. *In* Cancer Cells 3/Growth Factors and Transformation. J. Feramisco, B. Ozanne & C. Stiles, Eds.: 159–166. Cold Spring Harbor Laboratories, Cold Spring Harbor, NY.
8. EVA, A., K. C. ROBBINS, P. R. ANDERSEN, A. SRINIVASAN, S. R. TRONICK, E. P. REDDY, N. W. ELLMORE, A. T. GALEN, J. A. LAUTENBERGER, T. S. PAPAS, E. H. WESTIN, F. WONG-STAAL, R. C. GALLO, & S. A. AARONSON. 1982. Nature **295:** 116–119.
9. WOLFE, L. G., F. DEINHARDT, G. J. THEILEN, H. RABIN, T. KAWAKAMI, & L. K. BUSTAD. 1971. J. Natl. Cancer Inst. **47:** 1115–1120.
10. WOLFE, L. G., R. K. SMITH, & R. DEINHARDT. 1972. J. Natl. Cancer Inst. **48:** 1905–1907.
11. ROBBINS, K. C., F. LEAL, J. H. PIERCE, & S. A. AARONSON. 1985. EMBO J. **4:** 1783–1792.
12. IGARASHI, H., C. D. RAO, M. SIROFF, F. LEAL, K. C. ROBBINS, & S. A. AARONSON. 1987. Oncogene **1:** 79–85.
13. TODARO, G. J., J. E. DE LARCO, & S. COHEN. 1976. Nature **264:** 26–31.
14. TWARDZIK, D. R., G. J. TODARO, H. MARQUARDT, F. H. REYNOLDS, & J. R. STEPHENSON. 1982. Science **216:** 894–897.
15. TWARDZIK, D. R., G. J. TODARO, F. H. REYNOLDS, & J. R. STEPHENSON. 1983. Virology **124:** 201–207.
16. DE LARCO, J. E., & G. J. TODARO. 1978. Proc. Natl. Acad. Sci. USA **75:** 4001–4005.
17. SPORN, M. B. & G. J. TODARO. 1980. N. Engl. J. Med. **303:** 878–880.
18. DERYNCK, R., A. B. ROBERTS, M. E. WINKLER, E. Y. CHEN, & D. V. GOEDDEL. 1984. Cell **38:** 287–297.
19. CEPKO, C. L., B. E. ROBERTS, & R. C. MULLIGAN. 1984. Cell **137:** 1053–1062.
20. PALMITER, R. D., H. Y. CHEN, & R. L. BRINSTER. 1982. Cell **29:** 701–710.
21. FINZI, E., T. FLEMING, O. SEGATTO, C. Y. PENNINGTON, T. S. BRINGHMAN, R. DERYNCK, & S. A. AARONSON. 1987. Proc. Natl. Acad. Sci. USA **84:** 3733–3737.
22. GAZIT, A., H. IGARASHI, I.-M. CHUI, A. SRINIVASAN, A. YANIV, S. R. TRONICK, K. C., ROBBINS, & S. A. AARONSON. 1984. Cell **39:** 89–97.

23. GAZIT, A., J. H. PIERCE, M. H. KRAUS, P. P. DiFIORE, C. Y. PENNINGTON, & S. A. AARONSON. 1986. J. Virol **60:** 19–28.
24. AARONSON, S. A., & G. J. TODARO. 1968. Science **162:** 1024–1026.
25. AARONSON, S. A., G. J. TODARO, & A. E. FREEMAN. 1970. Exp. Cell Res. **61:** 1–5.
26. ULLRICH, A., L. COUSSENS, J. S. HAYFLICK, T. J. DULL, A. GRAY, A. W. TAM, J. LEE, Y. YARDEN, T. A. LIBERMANN, & J. SCHLESSINGER *et al.* 1984. Nature **309:** 418–425.
27. COUSSENS, L., C. VAN BEVEREN, D. SMITH, E. CHEN, R. L. MITCHELL, C. M. ISACKE, C. M. I. M. VERMA, & A. ULLRICH. 1986. Nature **320:** 277–280.
28. ROUSSEL, M. F., T. J. DULL, C. W. RETTENMIER, P. RALPH, A. ULLRICH, & C. J. SHERR. 1987. Nature **325:** 549–552.
29. SHIH, C., L. C. PADHY, M. MURRAY, & R. A. WEINBERG. 1981. Nature **290:** 261–264.
30. PADHY, L. C., C. SHIH, D. COWING, R. FINKELSTEIN, & R. A. WEINBERG. 1982. Cell **28:** 865–871.
31. SCHECHTER, A. L., D. F. STERN, L. VAIDYANATHAN, S. J. DECKER, J. A. DREBIN, M. I. GREENE, & R. A. WEINBERG. 1984. Nature **312:** 513–516.
32. BARGMAN, C. I., M. C. HUNG, & R. A. WEINBERG. 1986. Cell **45:** 649–657.
33. XU, Y. H., N. RICHERT, S. ITO, G. T. MERLINO, & I. PASTAN. 1984. Proc. Natl. Acad. Sci. USA **81:** 7308–7312.
34. KING, C. R., M. H. KRAUS, & S. A. AARONSON. 1985. Science **229:** 974–976.
35. KING, C. R., M. H. KRAUS, L. T. WILLIAMS, G. T. MERLINO, I. H. PASTAN, & S. A. AARONSON. 1985. Nucleic Acid Res **13:** 8447–8486.
36. SEMBA, K., N. KAMATA, K. TOYOSHIMA, & T. YAMAMOTO. 1985. Proc. Natl. Acad. Sci. USA **82:** 6497–6501.
37. LIBERMANN, T. A., H. R. NUSBAUM, N. RAZON, R. KRIS, I. LAX, H. SOREQ, N. WHITTLE, M. D. WATERFIELD, A. ULLRICH, & J. SCHLESSINGER. 1985. Nature **313:** 144–147.
38. FUKUSHIGE, S., K. MATSUBARA, M. YOSHIDA, M. SASAKI, T. SUZUKI, K. SEMBA, K. TOYOSHIMA, & T. YAMAMOTO. 1986. Mol. Cell. Biol. **6:** 955–958.
39. KRAUS, M. H., N. C. POPESCU, S. C. AMSBAUGH, & C. R. KING. 1987. EMBO J. **6:** 605–610.
40. SLAMON, D. J., G. M. CLARK, S. G. WONG, W. J. LEVIN, A. ULLRICH, & W. L. McGUIRE. 1987. Science **235:** 177–182.
41. DiFIORE, P. P., J. H. PIERCE, T. P. FLEMING, R. HAZAN, A. ULLRICH, C. R. KING, J. SCHLESSINGER, & S. A. AARONSON. 1987. Cell **51:** 1063–1070.
42. MACPHERSON, I. & L. MONTAGNIER. 1964. Virology **23:** 291–294.
43. LIN, C. R., W. S. CHEN, W. KRUIGER, L. S. STOLARSKY, W. WEBER, R. M. EVANS, I. M. VERMA, G. N. GILL, & M. G. ROSENFELD. 1984. Science **224:** 843–848.
44. MERLINO, G. T., T. G. XU, S. ISHII, A. J. CLARK, K. SEMBA, K. TOYOSHIMA, T. YAMAMOTO, & I. PASTAN. 1984. Science **224:** 417–419.
45. FILMUS, J., M. N. POLLAK, R. CAILLEAU, & R. N. BUICK. 1985. Biochem. Biophys. Res. Commun. **128:** 898–905.
46. FABRICANT, R. N., J. E. DE LARCO & G. J. TODARO. 1977. Proc. Natl. Acad. Sci. USA **74:** 565–569.
47. WRANN, M. M. & C. F. FOX. 1979. J. Biol. Chem. **254:** 8083–8086.
48. KRAUS, M. H., Y. YUASA, & S. A. AARONSON. 1984. Proc. Natl. Acad. Sci. USA **81:** 5384–5388.
49. GORMAN, C. M., G. T. MERLINO, M. C. WILLINGHAM, I. PASTAN, B. H. HOWARD. 1982. Proc. Natl. Acad. Sci. USA **79:** 6777–6781.
50. DI FIORE, P. P., J. H. PIERCE, M. H. KRAUS, O. SEGATTO, C. R. KING, & S. A. AARONSON. 1987. Science **237:** 178–182.
51. AKIYAMA, T., C. SUDO, H. OGAWARA, K. TOYOSHIMA, & T. YAMAMOTO. 1986. Science **232:** 1644–1646.
52. STERN, D. F., P. A. HEFFERNAN, & R. A. WEINBERG. 1986. Mol. Cell. Biol. **5:** 1729–740.

DISCUSSION

L. CANTLEY (*Tufts University, Boston, Massachusetts*): Does introduction of oncogenes into the keratinocyte cell line eliminate the requirement for ethanolamine in the defined growth medium?

S. AARONSON (*National Institutes of Health, Bethesda, Maryland*): In those transformants that we have analyzed, there still seems to be an ethanolamine requirement.

Regulation of Intercellular Communication and Growth by the Cellular *src* Gene[a]

WERNER R. LOEWENSTEIN AND ROOBIK AZARNIA

Department of Physiology and Biophysics
University of Miami School of Medicine
Miami, Florida 33136

We deal with the action of a cellular oncogene on a membrane channel, a channel present in the junctions between cells. The channel is a collaborative product: each cell in a junction contributes a symmetric half. The halves are tightly joined, forming a continuous, direct aqueous passageway between the cell interiors.[1] This cell-to-cell channel is an old friend; one of us ushered it in 22 years ago at a meeting, such as the one today, the first biomembranes conference of the New York Academy of Sciences.[2] The channel is a basic conduit for intercellular communication. It has a diameter of 16–20 angstroms,[3] wide enough to pass a range of cytoplasmic molecules, and growth-regulatory signals may be among them.[4]

We had learned before about two physiological regulators of this channel: the calcium ion, which is a key element in the control of the channel open state, and cyclic AMP-dependent protein kinase, which controls the channel formation process.[4,5] Here we give an account of a regulation that is ruled by the cellular *src* gene, which was discovered 12 years ago.[6] This gene encodes the 60,000-dalton membrane-bound protein, tyrosine kinase, pp60$^{c\text{-}src}$.[7]

The gene is present in all animal species and is highly conserved. Its functions were heretofore unknown.

C-*src* OVEREXPRESSION REDUCES JUNCTIONAL COMMUNICATION

Our approach to function was straightforward. In collaboration with T.E. Kmiecik, S. Reddy and D. Shalloway, we raised the gene expression level by incorporating highly transcribed c-*src* recombinant DNAs into mammalian cells, and examined the junctional transfer and growth.[8] The expression vectors contained the chicken c-*src* sequences ligated to retroviral long terminal repeats providing promoter/enhancer sequences and polyadenylation sites (FIG 1). Mouse NIH 3T3 cells were transfected with these plasmids, and cell lines were cloned from foci (suffix *foc*) or co-selected for expression of *Eco-gpt* (suffix *cos*) or *neo*. Thus, we obtained cell clones whose pp60$^{c\text{-}src}$ levels were elevated 10 to 20 times over those of the controls.

The effect of this overexpression on junctional communication was striking: whereas normally the NIH 3T3 cells transferred Lucifer Yellow or lissamine rhodamine B–labeled glutamatic acid (LRB-Glu) extensively to first- and higher-order

[a]The research work reported in this paper was supported by Research Grant CA14464 from the National Cancer Institute, U.S. National Institutes of Health.

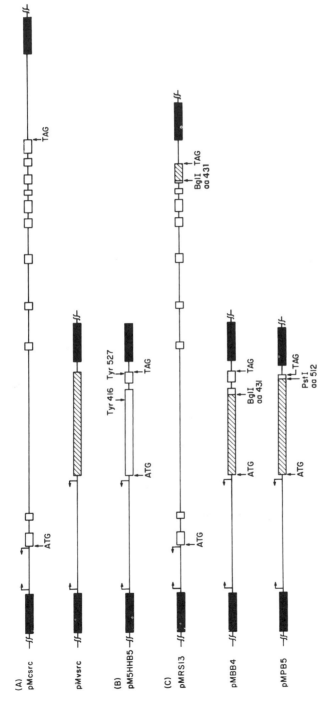

FIGURE 1. The c-src and v-src expression plasmids. The plasmids contain the chicken c-src, v-src, mutant and chimeric src genes ligated to transcriptional enhancers, promoters and polyadenylation signals provided by flanking Moloney murine leukemia virus long terminal repeats (solid boxes) from the vector pEVX. The locations of Moloney splice donors and first src splice acceptors are denoted by bent arrows. Coding regions are indicated by open (c-src) or hatched boxes (v-src) and the locations of the initiating (ATG) and terminating (TAG) codons are marked. Only the functionally relevant eukaryotic regions of the plasmids are shown. (A) pmMcsrc, the plasmid used in the experiments on c-src overexpression, and the structurally similar v-src expression plasmid pMvsrc; (B) pM5HHB5, the plasmid used for site-directed mutagenesis to produce Tyr→ Phe codon exchange at amino acids 416, 527 or both, generating plasmids pcsrc416, pcsrc527 and pc416-527. (C) Plasmids pMRS13 and pMBB4 are mirror-image chimeric constructions with junctions at site BglI at amino acid (aa) codon 431. pMPB5 is similar to pmBB4, except that the junction is at PstI site at amino acid 512. (From Azarnia et al.[8] Reproduced by permission.)

neighbors, the overexpressor cells transferred these tracers rarely and only to first-order neighbors. The incidence of cell interfaces permeable to the tracers was reduced by 57 to 73%, the cells with the highest pp60$^{c\text{-}src}$ expression level exhibiting the greatest reduction (TABLE 1).

Nonjunctional membrane permeability was not affected by the overexpression; the rates of tracer loss from the cells were negligible and not different from those of the control cells.[8]

TABLE 1. Junctional Communication of c-src Overexpressor Cells[a]

Cells	pp60^{src} Expression Level[b]	in Vivo Phospho-tyrosine[c]	Permeable Interfaces[d]	
			Lucifer Yellow	LRB-Glu
NIH 3T3 (control)	1	1.0	1.0 ± 0.08 (32)	1.0 ± 0.03 (20)
NIH (pMcsrc/cos)A	10	1.0	0.43 ± 0.08 (38)	
NIH (pMcsrc/foc)A	14		0.36 ± 0.03 (50)	
NIH (pMcsrc/foc)B	20	2.5	0.27 ± 0.05 (55)	0.31 ± 0.01 (40)
NIH (pMvsrc/cos)A	2	3.4	0.05 ± 0.03 (50)	
NIH (pMvsrc/foc)A	2	5.3	0.11 ± 0.05 (16)	0.09 ± 0.01 (40)

[a]From Azarnia et al.[8]

[b]Relative steady-state pp60^{src} expression was determined from immunoprecipitates (mAb327) of cell lysates labeled to equilibrium with [^{35}S]methionine for 48 hours, normalized to the expression level of NIH 3T3 control cells. Standard fractional error $<^{\times}_{+}1.2$; the low endogenous level of NIH 3T3 may have additional background subtraction error.

[c]Relative phosphotyrosine in vivo from ^{32}P-labeled phosphoamino acid analysis of total cell proteins. Labeled phosphoamino acids were separated by two-dimensional electrophoresis, and the proportion of radioactivity in phosphotyrosine was calculated as a fraction of radioactivity in all cell phosphorylhydroxyamino acids and normalized to the phosphotyrosine fraction in NIH 3T3 controls (0.015%).

[d]Relative incidence of interfaces permeable to Lucifer Yellow or LRB-Glu. The values are arithmetic means ±SE of the incidences determined on individual cells, normalized to the mean incidence of NIH 3T3 controls (the absolute means of the controls were 0.37 ± 0.03, Lucifer Yellow; 0.86 ± 0.03, LRB-Glu). The number of independent measurements, that is, the number of cells injected, is given in parentheses. Each incidence value of the src-transfected cells differed from controls at a significance level $p < 0.00005$ (t test with Bonferroni's correction for multiple comparisons).

POST-TRANSLATIONAL MODIFICATION OF c-src PROTEIN REDUCES COMMUNICATION

The answer provided by the preceding experiments was clear: the cellular src gene, by way of its product pp60$^{c\text{-}src}$, downregulates junctional permeability. How might the gene circuitry be controlled physiologically? One possibility, of course, is control of the gene transcription rate; this flows immediately from the preceding results. But other possibilities present themselves beyond transcription and translation, through interactions between the gene product and cellular proteins—interactions of a sort physiologists and biochemists are more at home with.

Our guidepost was the finding of an interaction between pp60$^{c\text{-}src}$ and the middle T antigen, a protein encoded by polyoma virus. This protein binds to pp60$^{c\text{-}src}$ (Ref. 9) and enhances the protein kinase activity of pp60$^{c\text{-}src}$ (Ref. 10). We studied the action of

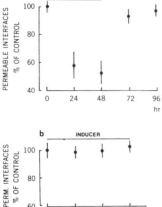

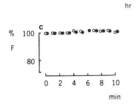

FIGURE 2. (a) Reversible reduction of junctional LRB-Glu transfer in F-111 cells containing a glucocorticoid-sensitive middle T recombinant DNA (mT). Dexamethasone (1 μM), the inducer, was applied at time zero and removed at 48 hr. Mean incidence of permeable interfaces (±SE) was normalized with respect to uninduced controls (100%). (b) A control run in which dexamethasone (1 μM) was applied to F-111 cells not containing mT. (c) Photometric measurement of intracellular LRB-Glu fluorescence loss from cells containing mT induced with (1 μM) dexamethasone (●) and uninduced (O). (From Azarnia and Loewenstein.[11] Reproduced by permission.)

middle T on junctional communication, using an expression vector constructed by Raptis, Lamfrom and Benjamin,[19] in which the middle T gene was ligated to a glucocorticoid-sensitive promoter. The vector was incorporated into rat F-111 cells. Transcription of the gene could be conveniently turned on and off by adding the hormone (dexamethasone) to the cell medium and removing it.

The effect of middle T was as pronounced as that of c-*src* overexpression: junctional permeability was reversibly reduced (FIG 2) and the reduction correlated with an increase in protein tryosine kinase activity.[11]

So, we have here a case of a regulation of junctional communication through posttranslational control of c-*src*. The interacting protein, in this instance, is encoded by a virus, but it would not be surprising if the viral trick of pp60^c-*src* activation turned out to be a variation, albeit extreme, of older physiological themes.

TYR^527 AND TYR^419 ARE REGULATORY SITES OF COMMUNICATION

Tyr^527, a site near the carboxyl terminus of the c-*src* protein, is phosphorylated *in vivo*.[12] Several lines of evidence indicated that this phosphorylation inhibits the enzyme and transforming activity of pp60^c-*src* (Refs. 13–15), and so we centered on this site.

In one kind of experiment we played the evolutionary game in reverse, using chimeric genes in which the c-*src* sequences downstream of amino acid 431 had been replaced with the corresponding carboxyl terminal sequences of the evolutionary successor of c-*src*, the viral *src* gene (v-*src*), which lacks Tyr^527. In another kind of experiment we used c-*src* genes with a point mutation at Tyr^527.

The c-*src*/v-*src* chimeras induced marked losses of junctional permeability, even at low gene expression levels (TABLE 2). The losses were comparable to those induced by the full viral gene.[16,17] Conversely, mirror-image chimeric constructions (v-*src*/c-*src*) with the c-*src* carboxyl terminal sequences spliced downstream of the amino acid codon 431 of v-*src*, or chimeric constructions with such a splice at codon 512, produced no significant loss of junctional permeability (TABLE 2).

For point mutation we used a site-directed mutant c-*src* vector (pcsrc527; FIG. 1) that encodes a substitution of Phe (which cannot be phosphorylated) for Tyr^{527}. This mutation was nearly as effective as the c-*src*/v-*src* chimeras; most cells exhibited no junctional transfer at all. The incidence of permeable interfaces in the cells containing the Tyr^{527} mutation was much lower than that of the controls containing the unmodified c-*src* vector (pM5HHB5), even though the pp60$^{c\text{-}src}$ expression level was lower than in the controls (TABLE 3). The mutation evidently had caused a reduction of junctional permeability on its own.

As in the case of c-*src* overexpression, the expression of Tyr^{527} mutants or c-*src*/v-*src* chimeras did not sensibly affect the permeability of nonjunctional cell membrane.

Another site of interest lies 110 amino acids upstream: Tyr.^{416} This site is phosphorylated (autophosphorylated) in v-*src* protein, but not in c-*src* protein.[7,12] However, it becomes phosphorylated when Tyr^{527} is mutated in c-*src*.[13] We examined the effect on communication of a Phe substitution for Tyr^{416}, made coordinately with a Phe substitution for Tyr^{527}.

The coordinate substitution led to suppression of the downregulatory effect on communication of the Tyr^{527} mutation. Cells containing genes with the Tyr^{416} mutation in addition to that at Tyr^{527} (pcsrc416.527) had junctional permeabilities similar to those of the overexpressor controls (pM5HHB5) (TABLE 3).[8]

TABLE 2. Junctional Communication of Cells Expressing *src* Chimeras[a]

Cells	pp60^{src} Expression Level[b]	Permeable Interfaces[c]	p^{d}
Control			
NIH 3T3	1	1.0 ± 0.08 (32)	
c-*src*/v-*src*			
NIH (pMRS13/foc)C	2	0.13 ± 0.06 (50)	0.000007
NIH (pMRS13/foc)D	2	0.04 ± 0.02 (50)	0.000007
v-*src*/c-*src*			
NIH (pMBB4/foc)E	3	0.8 ± 0.2 (50)	NS
NIH (pMBB4/foc)I	6	0.9 ± 0.2 (32)	NS
NIH (pMBB4/foc)H	2	0.5 ± 0.2 (50)	NS (0.07)
NIH (pMPB5/foc)B	7	1.3 ± 0.2 (50)	NS
NIH (pMPB5/foc)E	1	1.0 ± 0.2 (50)	NS

[a]From Azarnia *et al.*[8]

[b]Values are geometric means of two measurements; the fractional standard errors ranged from $\times 1.06$ to 1.3; methods as in Table 1.

[c]Probings with Lucifer Yellow. Arithmetic means ± SE; the number of cells injected is in parentheses.

[d]Statistical significance level of the difference between the various mean incidence values of permeable interfaces and the mean value of the controls (*t* test with Bonferroni's correction); NS, not significant.

TABLE 3. Tyr^{527} and Tyr^{416} Mutants^a

Cells	c-src Expression or Mutation	pp60^{src} Expression Level*^{b}	Specific Enzyme Activity^c	Permeable Interfaces^d	p_1^e	p_2^e
NIH 3T3	Endogenous pp60^{c-src}	1		1.00 ± 0.08 (32)	NS	0.04
NIH (pSV2neo)	Endogenous pp60^{c-src}	1		0.92 ± 0.08 (50)		0.02
NIH (pM5HHB5/neo)	Overexpressed pp60^{c-src}	13	1	0.73 ± 0.08 (50)	0.02	
NIH (pcsrc527/neo)	Tyr^{527} → Phe^{527}	8	13	0.28 ± 0.05 (82)	0.000005	0.00004
NIH (pcsrc416/neo)	Tyr^{416} → Phe^{416}	10	1	0.93 ± 0.04 (30)	NS	NS
NIH (pcsrc416·527/neo)	Tyr^{527} → Phe^{527}; Tyr^{416} → Phe^{416}	20	7	0.66 ± 0.08 (59)	0.04	NS

^{a}From Azarnia et al.[8]

^{b}Geometric-mean values; the fractional standard errors ranged from ×1.03 to 1.2; The low NIH 3T3 and NIH (pSV2neo) expression level may have additional background subtraction error. The methods were as in TABLE 1, but the [^{35}S]methionine labeling for pp60^{src} was 40 to 44 hours. Values for pp60^{src} expression are not comparable to those of TABLES 1 and 2.

^{c}In vitro specific tyrosine kinase activity was determined by incubating monoclonal antibody-bound [^{35}S]methionine-labeled pp60^{src} with [γ-^{32}P]ATP and comparing the amount of phosphate transferred to α-enolase with the amount of pp60^{src} present as determined by double-channel scintillation counting. Values are geometric means of three measurements; the fractional errors ranged from ×1.5 to 1.6.

^{d}Arithmetic means ± SE; the number of cells injected are in parentheses.

$^e P_1$ and P_2 are the statistical significance levels of the differences between the various mean incidence values of permeable interfaces and the mean value of the NIH(pSV2neo) controls or the NIH(pMHHB5/neo) controls, respectively (t test with Bonferroni's correction).

When Tyr^{416} was substituted alone, the effect of c-*src* overexpression on communication was suppressed. Cells containing the single Tyr^{416} mutation (pcsrc416) and expressing pp60$^{c\text{-}src}$ at a level ten times the endogenous one, exhibited junctional permeabilities not significantly different from those of the controls (pSV2neo) containing only endogenous pp60$^{c\text{-}src}$ (TABLE 3).

Thus, in sum, both Tyr^{527} and Tyr^{416} are important in the regulation of junctional communication. Their states of phosphorylation seem to be coupled, phosphorylation of Tyr^{527} leading to dephosphorylation of Tyr^{416} and vice versa.[13] In control-systems

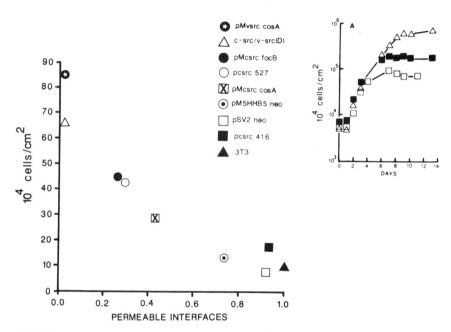

FIGURE 3. Mean saturation density versus mean incidence of permeable interfaces in cells containing endogenous c-*src*, overexpressed c-*src*, and mutations at the c-*src* carboxyl terminal coding region (key of symbols given in *inset*). *Inset:* three sample growth curves—cell density (log scale) versus time—representative of cells with the highest, intermediate, and lowest saturation densities.

terms this would amount to a latch-switch; the switching in the direction of downregulation of communication would require phosphorylation of Tyr.^{416}

REGULATION OF GROWTH

The changes in junctional permeability caused by the various manipulations of c-*src* correlated with changes in saturation density, the density at which the cells stopped growing. FIGURE 3 shows a plot of saturation density versus junctional transfer, a summary of the data of all conditions studied. In general, the larger the loss of communication, the higher the saturation density.

This, precisely, is the kind of relationship expected if growth-regulating signals are transmitted through the cell-to-cell channels and must diffuse some distance through the cell community to exert their regulatory effects. Because junctional permeability determines the extent of the diffusion, the growth regulation would be ruled by that permeability in an interplay with signal degradation and leakage. A basic model predicts that a decrease of junctional permeability would lead to an increase of growth.[18]

The data from the c-*src* overexpressor cells all fell in the middle range of the curve, where modest losses of communication were associated with modest rises of saturation density (FIG. 3). Unlike the cells expressing the v-*src* gene, the c-*src*/v-*src* chimeras and the Tyr[527] mutants, which exhibited the largest losses of communication, the largest increases of saturation density, and were transformed, the c-*src* overexpressors were not transformed. The cellular *src* gene thus seems to rule the middle range of communication, regulating growth without the pathogenic excesses of its viral counterpart.

REFERENCES

1. LOEWENSTEIN, W. R. 1981. Junctional intercellular communication. The cell-to-cell membrane channel. Physiol. Rev. **61:** 829–913.
2. LOEWENSTEIN, W. R. 1966. Permeability of membrane junctions. Ann. N.Y. Acad. Sci. **137:** 441–472.
3. SCHWARTZMANN, G. O. H., H. WIEGANDT, B. ROSE, A ZIMMERMAN, D. BEN-HAIM & W. R. LOEWENSTEIN. 1981. The diameter of the cell-to-cell junctional membrane channels, as probed with neutral molecules. Science **213:** 551–553.
4. LOEWENSTEIN, W. R. 1987. The cell-to-cell channel of gap junctions. Cell **48:** 725–726.
5. LOEWENSTEIN, W. R. 1985. Regulation of cell-to-cell communication by phosphorylation. Biochem. Soc. Symp. London **50:** 43–58.
6. STEHELIN, D., H. E. VARMUS, J. M. BISHOP & P. K. VOGT. 1976. DNA related to the transforming gene(s) of avian sarcoma viruses is present in normal avian DNA. Nature **260:** 170–173.
7. HUNTER, T. & J. COOPER. 1985. Protein-tyrosine kinases. Annu. Rev. Biochem. **54:** 897–930.
8. AZARNIA, R., S. REDDY, T. E. KMIECIK, D. SHALLOWAY & W. R. LOEWENSTEIN. 1988. The cellular *src* gene product regulates junctional cell-to-cell communication. Science **239:** 398–401.
9. COURTNEIDGE, S. A. & A. E. SMITH. 1983. Polyoma virus transforming protein associates with the product of the c-*src* cellular gene. Nature **303:** 435–439.
10. BOLEN, J. B., C. J. THIELE, M. A. ISRAEL, W. YONEMOTO, L. A. LIPSICH & J. S. BRUGGE. 1984. Enhancement of cellular *src* gene product associated tyrosyl kinase activity following polyoma virus infection and transformation. Cell **38:** 767–777.
11. AZARNIA, R. & W. R. LOEWENSTEIN. 1987. Polyoma virus middle T antigen downregulates junctional cell-to-cell communication. Mol. Cell. Biol. **7:** 946–950.
12. COOPER, J. A., K. L. GOULD, C. A. CARTWRIGHT & T. HUNTER. 1986. Tyr[527] is phosphorylated in pp60[c-src]: Implications for regulation. Science **231:** 1431–1434.
13. KMIECIK, T. E. & D. SHALLOWAY. 1987. Activation and suppression of pp60[c-src] transforming ability by mutation of its primary sites of tyrosine phosphorylation. Cell **49:** 65–73.
14. PIWNICA-WORMS, H., K. B. SAUNDERS, T. M. ROBERTS, A. E. SMITH & S. H. CHENG. 1987. Tyrosine phosphorylation regulates the biochemical and biological properties of pp60[c-src]. Cell **49:** 75–82.

15. CARTWRIGHT, C. A., W. ECKHART, S. SIMON & P. L. KAPLAN. 1987. Cell transformation by pp60$^{c\text{-}src}$ mutated in the carboxyl-terminal regulatory domain. Cell **49:** 83–91.
16. ATKINSON, M. M., A. S. MENKO, R. G. JOHNSON, J. R. SHEPPARD, J. D. SHERIDAN. 1981. Rapid and reversible reduction of junctional permeability in cells infected with a temperature-sensitive mutant of avian sarcoma virus. J. Cell Biol. **91:** 573–578.
17. AZARNIA, R. & W. R. LOEWENSTEIN. 1984. Intercellular communication and the control of growth. X. Alteration of junctional permeability by the *src* gene. A study with temperature-sensitive mutant Rous sarcoma virus. J. Membrane Biol. **82:** 191–205.
18. LOEWENSTEIN, W. R. 1979. Junctional intercellular communication and the control of growth. Biochim. Biophys. Acta Cancer Rev. **560:** 1–65.
19. RAPTIS, L. H., H. LAMFROM & T. L. BENJAMIN. 1985. Regulation of cellular phenotype and expression of polyoma virus middle T antigen in rat fibroblasts. Mol. Cell. Biol. **5:** 2467–2485.

DISCUSSION

L. WAKEFIELD (*National Cancer Institute, Bethesda, Maryland*): Does the inverse relationship between cell saturation density and junction permeability suggest that it is a negative growth signal that normally passes through the junction? And if so, do you have any evidence as to the identity of this negative signal?

W. LOEWENSTEIN: To answer your last question first: no, the identity of the signal is not known. All we can say is that the signal molecule must be small enough to fit through the channel, which sets an upper limit of about 2000 daltons. The experiments with c-*src* give no clue on signal sign. Our earlier experiments—work with Parmender Mehta and John Bertram (1981. Cell **44:**187–196)—is equally pertinent here. That work showed that the growth of transformed cells in mixture with normal cells is inhibited by junctional communication with normal cells. At first glance, this might suggest that an inhibitory signal is transmitted from the normal cells to the transformed cells. But the result is just as compatible with a stimulatory signal from the transformed cells' being diluted out by the volume of the connected cell population. The dependence of growth, that is, saturation density, on junctional permeability is the expected behavior if growth-regulating signals are disseminated by the cell-to-cell channel through the cell community and must diffuse some distance through the community to exert their regulatory effect. But one can model the dependence with negative as well as positive signals (Loewenstein. 1979. Biochim. Biophys. Acta Cancer Res. **560:**1–65). In the case of positive signals the permeability-growth relationship may not be strictly monotonic, but the important point is that from a certain value onwards, it always is in the right direction: it gives an inverse relationship.

G. GUIDOTTI (*University of Parma, Italy*): Is any correlation known between the ability to form channels and appearance of their c-*src* product-mediated regulation in low eukaryotes such as sponges?

LOEWENSTEIN: Junctional communication is present in sponge cells. That much is known. But we have no information on whether c-*src* works there as it does in mammalian cells.

G. NERI (*University "D'Annunzio", Italy*): Is decreased communication among cells due to a decreased number of channels?

LOEWENSTEIN: The decrease of communication caused by *src* protein probably reflects a decrease in the number of open channels.

B. SZWERGOLD (*Fox Chase Cancer Center, Philadelphia, Pennsylvania*): Do the cell-cell channels show any charge selectivity in addition to their size selectivity?

LOEWENSTEIN: Yes, the channels are charge-selective. There is not much charge discrimination for the small inorganic ions; these readily zip through. But if one probes the channels with molecules of a size approaching the channel diameter, the channel selectivity is brought out; the channel discriminates against negatively charged permeants.

T. CHAN (*Purdue University, Lafayette, Indiana*): You said that the reduced cell-cell communication in the *src*-transformant is the result of decreased opening of the channels rather than a reduction of channel number. How did you measure actual numbers of channels on the transformants?

LOEWENSTEIN: No, I said a decrease in the number of *open* channels, that is, channels in the open state. The number of channels in a junction can be estimated from the number of particles of gap junctions detected by freeze-fracture electronmicroscopy. Each particle embodies a channel half which may be open or closed. There is no evidence of a change in the particle number by action of *src* protein. On the other hand, in another kind of modification of junctional permeability, a modification induced by cyclic AMP-dependent protein kinase, we see changes in the particle number.

Lymphocyte Activation: Modulatory Effects Mediated by Interactions between Cell Adhesion Molecules

P. ANDERSON, C. MORIMOTO, M. L. BLUE, M. STREULI,
H. SAITO, AND S. F. SCHLOSSMAN

Division of Tumor Immunology
Dana-Farber Cancer Institute
Boston, Massachusetts 02115

INTRODUCTION

In recent years, it has become increasingly clear that the aberrant expression of one or more of a panel of genes involved in cellular proliferation can contribute directly to the process of malignant transformation.[1] The overproduction or mutation of soluble growth factors or their receptors can, under experimental conditions, result in uncontrolled proliferation characteristic of the transformed phenotype.[2] An understanding of cellular transformation is therefore predicated on an understanding of normal processes of cell growth. The activation of T lymphocytes proceeds through an ordered cell cycle progression that is characteristic of all dividing cells. Unlike many cell types, lymphocyte activation is triggered initially through cell:cell interactions mediated by intercellular adhesion molecules. Lymphocytes and their hematopoietic accessory cells express a panel of structurally related molecules capable of mediating cell adhesion.[3,4] Lymphocyte-specific structures such as CD2 (T11), CD4 (T4), and CD8 (T8) bind specifically to the accessory cell surface molecules LFA-3, class II MHC and class I MHC, respectively. Such interactions differ from soluble ligand receptor associations in that only molecules present in the area of cell contact can participate. Furthermore, the particulate nature of the interacting cells affords a support that tends to aggregate those ligands involved in cell-cell adhesion. As a consequence of these constraints, the repertoire of adhesion molecules expressed on antigen-presenting cells will dictate a group of lymphocyte surface molecules that will co-aggregate in regions of cell-cell contact.[4] We have examined the possibility that the co-aggregation of specific lymphocyte surface structures might contribute to the regulation of T-cell activation.

CO-AGGREGATION OF CD3:TCR WITH INDIVIDUAL ACCESSORY MOLECULES MODULATES T-CELL ACTIVATION

Resting T cells were cultured with Sepharose-immobilized antibodies for 4 days in the presence or absence of IL-2.[5,6] Lymphocyte proliferation was quantitated using [³H]thymidine incorporation during a 16-hr pulse. As shown in FIGURE 1, Sepharose-immobilized antibody failed to induce the proliferation of accessory cell-depleted T cells in the absence of IL-2. In the presence of recombinant IL-2 (5 U/ml), proliferation was observed in the presence of immobilized anti-CD3, but not in the

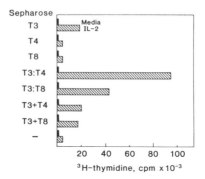

FIGURE 1. Proliferative response of T cells to immobilized antibodies. T lymphocytes were cultured in triplicate samples in round-bottom 96-well microtiter plates (Intermed, Denmark) containing 4×10^4 cells/well in RPMI containing 10% human serum (Pel-Freeze) at 37°C in a humidified CO_2 incubator. Sepharose beads were added at a concentration of 1 ml packed beads per well. When combinations of immobilized antibodies were used, each was added at a concentration of 0.5 μl packed beads per well. Where indicated, human recombinant IL-2 (Roche Labs, Nutley, NJ) was added at a concentration of 5 U/ml. Cultures were continued for 4 days, at which point they were pulsed with [³H]thymidine (0.2 μCi/well) (New England Nuclear, Boston, MA) for 16 hr. Cells were then harvested using a Mash II apparatus, and [³H]thymidine incorporation was measured using a Packard Liquid scintillation counter. Reported results represent the means of triplicate determinations in which standard errors were <15%.

presence of immobilized anti-CD4 or anti-CD8. In the presence of IL-2, Sepharose beads conjugated to both anti-CD3 and anti-CD4 induced a proliferative response that was significantly greater than that induced by an equal density of anti-CD3 alone (stimulation index = 3.2 ± 1.1-fold, $n = 7$). The combination of Sepharose (CD3) and Sepharose (CD4) failed to enhance proliferation in a similar manner. Sepharose (CD3:CD8) also induced more proliferation than Sepharose (CD3) or Sepharose (CD3) + Sepharose (CD8). These results suggest that crosslinking of CD3 with either CD4 or CD8 enhances the proliferation induced by immobilized anti-CD3 alone. This effect was not observed by co-stimulation using identical antibody combinations immobilized on separate beads.

The specificity of the proliferative enhancement produced by Sepharose (CD3:CD4) was tested using purified T-cell subpopulations. As shown in FIGURE 2, Sepharose (CD3:CD4) enhanced the proliferation of CD4+ cells over that produced

FIGURE 2. Proliferative response of T-cell sub-populations to immobilized antibodies. Designated populations of T cells were stimulated by immobilized antibody in the presence of IL-2, 5 U/ml as described by the legend to FIGURE 1.

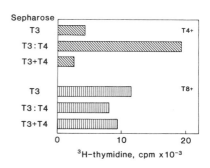

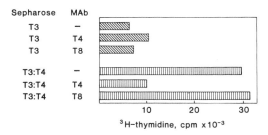

FIGURE 3. Effect of soluble antibody on proliferation induced by immobilized antibody. Designated populations of T cells were stimulated by immobilized antibody in the presence of IL-2, 5 U/ml as described in the legend for FIGURE 1. At the initiation of culture, soluble antisera were added as a 1/5000 dilution of ascites. In all cases, Sepharose-immobilized antibodies were preincubated in media containing 10% human serum to block unoccupied protein A sites prior to the addition of soluble antisera.

by Sepharose (CD3). Whereas CD8+ cells proliferated in response to Sepharose (CD3), there was no enhancement produced by Sepharose (CD3:CD4).

The effect of soluble anti-CD4 on proliferation induced by immobilized antibodies is shown in FIGURE 3. Sepharose (CD3)-induced proliferation in unseparated T cells was essentially unaffected by soluble anti-CD4 or anti-CD8. On the other hand, the enhanced proliferation induced by Sepharose (CD3:CD4) was abrogated in the presence of soluble anti-CD4, but not in the presence of soluble anti-CD8. These results suggest that interference with co-crosslinking between CD3 and CD4 eliminates the proliferative enhancement.

LYMPHOCYTE SUBPOPULATIONS

Recently, monoclonal antibodies reactive with subpopulations of human (2H4+4B4− and 2H4−4B4+; Refs. 7, 8) and rat (OX22− and OX22+; Refs. 9, 10) CD4+ lymphocytes have been shown to define distinct functional subsets. The 2H4+4B4− population selectively induces suppression of immunoglobulin production in a pokeweed-mitogen-stimulated system, resulting in its designation as a suppressor-inducer population.[8] The reciprocal population, which is phenotypically 2H4−4B4+, functions as a helper cell in immunoglobulin production in the pokeweed mitogen

TABLE 1. Proliferative Response of CD4+ Lymphocyte Subsets to Various Stimuli

	Proliferative Response	
Stimulus	CD4+2H4+	CD4+2H4−
Sepharose (CD3)	+	+
Mitogen	+++	+++
Allo-MLR	+++	+++
Soluble antigen	±	+++
Sepharose (CD3:CD4)	+	+++
Auto-MLR	+++	+
Anti-T11$_2$ + Anti-T11$_3$	+++	+

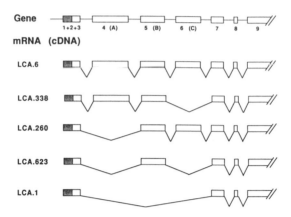

FIGURE 4. Genomic organization of human LCA and putative patterns of alternative splicing resulting in mRNAs corresponding to individual LCA isoforms.

system.[7] These populations possess distinctive functional programs and differ in their proliferative responses to various stimuli (TABLE 1).[11]

The 2H4 antibody recognizes a restricted epitope on the leukocyte common antigen (LCA) family of glycoproteins.[12] The human LCA gene has been cloned and its genomic organization characterized.[13] As schematized in FIGURE 4, the 5' region of the LCA gene contains 3 exons that can be joined in various combinations, by alternative splicing. As a result, individual isoforms of LCAs differing in their extracellular domains are produced. Two of the largest LCA isoforms (M_r: 205 kD and 220 kD) include exon 4, which has been shown to encode the 2H4 epitope.[14] CD4 + lymphocytes expressing these high molecular weight LCA isoforms constitute a functionally distinct lymphocyte subpopulation.

The differential responsiveness of these lymphocyte subpopulations to soluble antigen presented in the context of self-MHC was of particular interest. It has been postulated that antigen presentation by accessory cells involves the formation of a quaternary complex involving antigen, class II-MHC, CD4 and CD3-Ti.[15,16] If CD3:CD4 cross-linking by Sepharose (CD3:CD4) enhances proliferation by mimicking such a complex, this enhancement might be confined to the T-cell population which normally responds to soluble antigen.

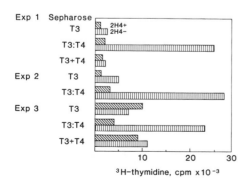

FIGURE 5. Proliferative response of T-cell subpopulations to immobilized antibodies. CD4 + lymphocytes were separated into 2H4 + and 2H4 − populations by adherence to anti-2H4-coated dishes. The proliferative response to Sepharose-immobilized antibody was measured in the presence of IL-2, 5 U/ml.

To test this hypothesis, CD4+ lymphocytes were separated into 2H4+ and 2H4− populations by panning. Each of these populations was tested for its proliferative response to immobilized antibodies as shown in FIGURE 5. Results from three independent experiments are shown to point out the donor variability in these experiments. Whereas both populations proliferated in response to Sepharose (CD3), enhanced proliferation induced by Sepharose (CD3:CD4) was demonstrated preferentially in the CD4+2H4− helper population. The average stimulation index of Sepharose (CD3:CD4) was 1.5 ± 0.9 ($n = 5$) in the CD4+2H4+ population and 4.5 ± 2.4 ($n = 5$) in the CD4+2H4− population. These results suggest that the proliferative enhancement induced by CD3:CD4 crosslinking reflects a preferential activation of the CD4+2H4− helper population.

One interpretation of these results is shown schematically in FIGURE 6. In this view, a heterobivalent interaction between the antigen:MHC complex on accessory cells and both CD4 and CD3:TCR on responder cells is postulated to induce the aggregation of

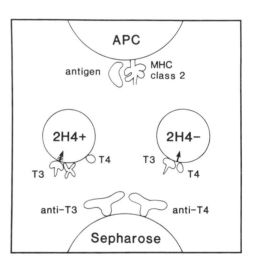

FIGURE 6. Hypothetical interactions between lymphocyte surface molecules induced by interactions with antigen:MHC structures on antigen presenting cells (*top*) versus Sepharose-immobilized antibodies (*bottom*).

CD4 with CD3:TCR. This effect is mimicked by Sepharose (CD3:CD4). On 2H4− lymphocytes, this co-aggregation of CD4 and CD3:TCR results in an activation signal. In 2H4+ lymphocytes, either the co-aggregation of CD4 and CD3:TCR is precluded from occurring, or the signal transduced by this association is not delivered.

CONCLUSIONS

The T-cell receptor complex is known to consist of a disulfide-linked heterodimer (Ti) associated noncovalently with three to four transmembrane proteins collectively known as CD3. The aggregation of this complex by immobilized antibodies reactive with either Ti or CD3 is sufficient to induce the proliferation of resting T lymphocytes.[17] Physiologically, antigen recognition requires specific interactions between T cells and accessory cells bearing self major histocompatibility molecules.

T lymphocytes expressing CD4 recognize antigen in the context of class II MHC, whereas CD8-positive lymphocytes are restricted to class I MHC-bearing accessory cells.[18,19] These relationships have been postulated to result from a specific recognition of class II MHC by CD4 and class I MHC by CD8. Such interactions have been further postulated to stabilize T cell:accessory cell adhesion, allowing antigen recognition to occur.[20] Although such interactions probably play some role in antigen-specific T-cell activation, it is becoming increasingly clear that CD4 and CD8 function as more than just cell adhesion molecules. Perturbation of CD4 by monoclonal antibodies is capable of inhibiting lectin or antibody-induced lymphocyte proliferation.[21] Furthermore, our results suggest that co-aggregation of CD4 and the TCR complex enhances T-lymphocyte activation. The potential for cell surface molecules to interact with one another in the modulation of cellular activation and proliferation may provide an added level of control over lymphocyte growth.[22]

REFERENCES

1. BISHOP, J. M. 1983. Annu. Rev. Biochem. **52:** 350–354.
2. DOWNWARD, J., Y. YARDEN, E. MAYES, G. SCRACE, N. TOTTY, P. STOCKWELL, A. ULLRICH & M. D. WATERFIELD. 1984. Nature **307:** 521–527.
3. WILLIAMS, A. F. 1987. Immunol. Today **8:** 298–303.
4. ANDERSON, P., C. MORIMOTO, J. B. BREITMEYER & S. F. SCHLOSSMAN. 1988. Immunol. Today. **9:** 199–203.
5. ANDERSON, P., M. L. BLUE, C. MORIMOTO & S. F. SCHLOSSMAN. 1987. J. Immunol. **139:** 678–682.
6. ANDERSON, P., M. L. BLUE, C. MORIMOTO & S. F. SCHLOSSMAN. 1988. Cellular Immunol. **115:** 246–256.
7. MORIMOTO, C., N. L. LETVIN, A. W. BOYD, et al. 1985. J. Immunol. **134:** 3762–3769.
8. MORIMOTO, C., N. L. LETVIN, J. A. DISTASO, W. R. ALDRICH & S. F. SCHLOSSMAN. 1985. J. Immunol. **134:** 1508–1515.
9. ARTHUR, R. P. & D. MASON. 1986. J. Exp. Med. **163:** 774–786.
10. BARCLAY, A. W., D. I. JACKSON, A. C. WILLIS & A. F. WILLIAMS. 1987. EMBO J. **6:** 1259–1264.
11. MATSUYAMA, T., P. ANDERSON, J. F. DALEY, S. F. SCHLOSSMAN & C. MORIMOTO. 1988. Eur. J. Immunol. **18:** 1473–1476.
12. RUDD, C. E., C. MORIMOTO, L. L. WONG & S. F. SCHLOSSMAN. 1987. J. Exp. Med. **166:** 1758–1773.
13. STREULI, M., L. R. HALL, Y. SAGA, S. F. SCHLOSSMAN & H. SAITO. 1987. J. Exp. Med. **166:** 1548–1566.
14. STREULI, M., T. MATSUYAMA, C. MORIMOTO, S. F. SCHLOSSMAN & H. SAITO. 1987. J. Exp. Med. **166:** 1567–1572.
15. REINHERZ, E. L., S. C. MEUER & S. F. SCHLOSSMAN. 1983. Immunol. Rev. **74:** 83–112.
16. SCHWARTZ, R. H. 1985. Annu. Rev. Immunol. **3:** 237–251.
17. MEUER, S. C., J. C. HODGDON, R. E. HUSSEY, J. P. PROTENTIS, S. F. SCHLOSSMAN, & E. L. REINHERZ. 1983. J. Exp. Med. **158:** 988–996.
18. SWAIN, S. L. 1981. Proc. Natl. Acad. Sci. USA **78:** 7101–7105.
19. MEUER, S. C., S. F. SCHLOSSMAN & E. L. REINHERZ. 1982. Proc. Natl. Acad. Sci. USA **79:** 4395–4399.
20. GREENSTEIN, J. L., B. MALISSEN & S. J. BURAKOFF. 1985. J. Exp. Med. **162:** 369–374.
21. BANK, I. & L. CHESS. 1985. J. Exp. Med. **162:** 1294–1301.
22. ANDERSON, P., M. L. BLUE & S. F. SCHLOSSMAN. 1988. J. Immunol. **140:** 1732–1737.

DISCUSSION

N. SHARON (*Weizmann Institute, Rehovot, Israel*): Must the monoclonal antibodies against lymphocyte surface antigens be presented for activation on cells or beads, or will they act also if they are just in polymerized form?

P. ANDERSON (*Harvard Medical School, Boston, Massachusetts*): Recently, two groups have reported results employing heterodimeric antibody conjugates consisting of anti-CD3 coupled to anti-CD4. In soluble form, such conjugates markedly enhanced calcium influx induced by T-cell receptor aggregation when compared to anti-CD3 alone.

Effects of Transformation by p21 $^{N\text{-}ras}$ upon the Inositol Phospholipid and Adenylate Cyclase Signaling Pathways in NIH 3T3 Cells

SHIREEN-ANNE DAVIES, MILES D. HOUSLAY,
AND MICHAEL J.O. WAKELAM

Molecular Pharmacology Group
Department of Biochemistry
University of Glasgow
Glasgow G12 8QQ, Scotland

The product of the *ras* oncogene, p21^{ras}, has been shown to transform murine fibroblasts and is involved in the proliferation and differentiation of PC12 pheochromocytoma cells. The mechanisms by which *ras* induces transformation are not known, but it possibly occurs through growth-factor-mediated events at the cell surface involving the *ras* gene product.

The cell line we used to investigate the effects of *ras* transformation on second messenger systems was the T15 cell line. This is an NIH 3T3 cell line transfected with the human N-*ras* proto-oncogene under control of a dexamethasone-inducible promoter.[1] When the cells are grown in the absence of dexamethasone (T15−) they are morphologically similar to NIH 3T3 cells, but when grown in the presence of dexamethasone (T15+) there is overexpression of p21^{ras} resulting in cell proliferation and transformation.

Hydrolysis of phosphatidylinositol 4,5, bisphosphate (PIP$_2$) results in the production of two second messengers, inositol 1,4,5 trisphosphate and 1,2-diacylglycerol,[2] both of which are associated with cell proliferation. We have already shown in N-*ras*-transformed cells that the inositol phosphate response to bombesin is amplified and that this is linked to p21$^{N\text{-}ras}$ expression.[3,4]

We have also shown that this effect is due to the hydrolysis of PIP$_2$ resulting in inositol 1,4,5 trisphosphate production and that bombesin stimulation of T15+ cells but not T15− cells results in an increase in intracellular Ca^{2+} (in preparation).

The effects of p21$^{N\text{-}ras}$ transformation on the adenylate cyclase system were also examined by determining basal and stimulated cAMP levels in this cell line.

METHODS

T15, NIH 3T3, and *sis*771 cells were grown on 24-well plates in DMEM/10% calf serum and 2 μM dexamethasone as required. For inositol phosphate experiments the cells were labeled with 10 μCi · ml^{-1} [^{3}H] inositol in serum and inositol-free DMEM for 16 hours. After the labeling period the cells were washed and preincubated in Hanks buffered saline solution with 10 mM lithium chloride, 10 mM glucose, and 1%

TABLE 1. Results of Inositol Phosphate and Cyclic AMP Experiments[a]

Cell Type	[3H] Inositol 1,4,5 Trisphosphate (dpm/well)				[3H] cAMP (dpm/well)[b]	
	Basal	Stimulated		Basal	Isoproterenol	Adrenaline
	Basal					
T15−	320 ± 21	258 ± 18		2988 ± 1	39137 ± 2934	38426 ± 2706
T15+	241 ± 23	536 ± 18		3441 ± 475	23139 ± 2518	21109 ± 2745
NIH 3T3	ND	ND		2939 ± 424	12985 ± 1231	12526 ± 1950
NIH 3T3 + dexamethasone	ND	ND		2877 ± 342	12549 ± 843	11294 ± 268
sis 771	ND	ND		4376 ± 123	28800 ± 498	37717 ± 171

[a]Results are from one experiment (*n* = 3) and are representative of at least 4–6 separate experiments. ND indicates that only total inositol phosphate values have been determined.
[b]Results are combined data from two separate experiments, *n* = 3 in each.

BSA for 5 minutes. The cells were then stimulated with 2.5 μM bombesin for various time intervals and the reactions quenched with 10% (v/v) PCA. The samples were neutralized and added to 1 ml of 5 mM $NaBO_4/0.5$ mM EDTA before being applied to Dowex Formate columns. Inositol phosphates were eluted off as previously described.[5]

For the cAMP measurements, the cells were labeled in serum-free DMEM containing 2 μCi [^{3}H]adenine per well for 90 minutes. After this period, they were washed and stimulated with the appropriate ligand at a concentration of 10^{-5} M for 10 minutes in the presence of 1mM IBMX. Reactions were quenched with 10% (w/v) TCA and cAMP content per well was estimated by the method of Salomon et al.[6]

RESULTS AND DISCUSSION

Table 1 shows the effect of p21[N-ras] expression on bombesin-stimulated inositol 1,4,5 trisphosphate production in T15 cells. The data show values obtained at a 10-second stimulation with bombesin and is the peak of inositol 1,4,5 trisphosphate production in these cells. The stimulated production of inositol 1,4,5 trisphosphate is not due to the action of dexamethasone since NIH 3T3 cells treated with dexamethasone show levels of basal and stimulated total inositol phosphates similar to those of untreated cells.

The results suggest that the amplification of the inositol phosphate response in T15+ cells is due to hydrolysis of PIP_2 as a consequence of receptor occupation. This effect seems specific to ras transformation since sis-transformed cells show a negligible stimulated inositol phosphate response to bombesin. This suggests that p21[N-ras] may interact with growth factor-phospholipase C coupling in this cell line.

The effect of p21[ras] expression on the adenylate cyclase system is not well understood, but we have shown that p21[N-ras] expression results in a decrease in stimulated cAMP levels. If the cells are grown for longer than two passages in dexamethasone, they exhibit a transformed morphology and at this stage there is increased downregulation of the stimulated cAMP response as well as desensitization of the forskolin response (in preparation). These results are not due to dexamethasone treatment since control and dexamethasone-treated NIH 3T3 cells have the same basal and stimulated cAMP levels. These effects seem specific to p21[N-ras] expression since sis-transformed cells have basal and stimulated cAMP levels similar to those of T15− cells. We have shown that p21[N-ras] overexpression in NIH 3T3 cells can cause an increased bombesin-stimulated production of inositol 1,4,5 trisphosphate associated with cell proliferation. Also, the overexpression of p21[N-ras] can result in an inhibition of stimulated cAMP levels in this cell line. This may be a consequence of β-adrenergic receptor phosphorylation by protein kinase C, activated as a result of PIP_2 hydrolysis.[7] It has been shown recently that cAMP analogues can reverse morphologic transformation[8] and it appears that an inhibition of stimulated cAMP levels and cell proliferation are closely associated.

These changes in the production of these two second messengers are associated with cell proliferation and may be integral to the mechanisms by which uncontrolled cell proliferation as a result of p21[N-ras] overexpression can occur in the presence of growth factors.

REFERENCES

1. McKay, I. A., C. J. Marshall, C. Cales & A. Hall. 1986. EMBO J. **5:** 2617–2621.
2. Berridge, M. J. 1987. Biochim. Biophys. Acta **907:** 33–45.
3. Wakelam, M. J. O., S. A. Davies, M. D. Houslay, I. McKay, C. J. Marshall & A. Hall. 1986. Nature **323:** 173–176.
4. Wakelam, M. J. O., M. D. Houslay, S. A. Davies, C. J. Marshall & A. Hall. 1987. Biochem. Soc. Trans. **15:** 45–47.
5. Berridge, M. J., C. P. Downes & M. R. Hanley. 1982. Biochem J. **206:** 587–595.
6. Salomon, Y., C. Londos & M. Rodbell. 1974. Anal. Biochem. **58:** 541–548.
7. Bouvier, M., L. M. Leeb-Lundberg, J. L. Benovie, M. G. Caron & R. J. Lekfowitz. 1987. J. Biol. Chem. **262:** 3106–3113.
8. Tagliaferri, P., D. Katsaros, T. Clair, L. Neckers, R. K. Robins & Y. S. Cho-Chung. 1988. J. Biol. Chem. **263:** 409–416.

The Role of Mitochondrial Hexokinase in Neoplastic Phenotype and Its Sensitivity to Lonidamine[a]

MARCO G. PAGGI,[b] MAURIZIO FANCIULLI,[b] NICOLA
PERROTTI,[b] ARISTIDE FLORIDI,[b]MASSIMO ZEULI,[b]
BRUNO SILVESTRINI,[c] AND ANTONIO CAPUTO[b]

[b]Regina Elena Institute for Cancer Research
Rome, Italy

[c]Institute of Pharmacology and Pharmacognosy
"La Sapienza" University
Rome, Italy

Hexokinase (ATP:D-hexose 6-phosphotransferase, EC 2.7.1.1.), the key enzyme in glucose catabolism, controls many important metabolic pathways necessary for cell growth and replication. Its product, glucose-6-phosphate, is in fact involved in energy-yielding processes and in the biosynthesis of many compounds such as nucleic acids, lipids, proteins, and reducing equivalents. Hexokinase activity was enhanced in tumors in a way strictly related to neoplastic progression in experimental murine tumors as well as in human neoplasms. Hexokinase activity is also increased in tumor cells by the binding of the protein to the outer mitochondrial membrane.[1]

Lonidamine [1, (2,4 dichlorobenzyl)-1-H-indazol-3-carboxylic acid] (LND) is able to inhibit mitochondrially bound hexokinase activity, thus reducing the high aerobic glycolysis peculiar to the neoplastic phenotype. The drug is effective not only in murine systems, but also in human tumors (TABLE 1). Particularly in human malignant gliomas the inhibitory effect of the drug has been demonstrated to be strictly related to the degree of malignancy, so that the more a tumor displays elevated aerobic glycolysis and hexokinase activity, the greater the extent of the inhibition.[2]

Tumor hexokinase II has been purified to homogeneity from the highly malignant, highly glycolytic AS-30D rat hepatoma cell line and polyclonal antibodies were raised against it.[3] These antibodies had an extensive cross-reaction with hexokinases from normal rat tissues[3] and from a human grade III astrocytoma,[4] thus indicating in some way a degree of homology among these hexokinases from different tissues.

Since the N-terminal of the purified mitochondrial hexokinase was blocked, it was impossible to sequence it. Screening of a cDNA library, constructed in λgt11 using mRNA from a human hepatoma, with the polyclonal antibody as a probe is currently under way, bringing to identification 11 positive clones. Contemporarily, the purified enzyme has been digested with trypsin, and peptides were separated in reverse-phase HPLC to be sequenced by automated Edman degradation. The subsequent construction of a synthetic cDNA probe aids in screening of the human hepatoma library.

[a]This work was supported by C.N.R. PFO Grants 87.01532.44 and 87.0550.44 and by AIRC/86.

TABLE 1. Effect of LND on the Utilization of [¹⁴C]Glucose by Cultured Human Astrocytoma Cells

Label in Glucose	LND	Glucose Utilized (μmol)	[¹⁴C]lactate			[¹⁴C]O$_2$		Lipids		RNA		DNA		Supporting Structures	
			μmol	cpm	GE	cpm	GE	cpm	GE	cpm	GE	cpm	GE	cpm	GE
1	−	8.23	12.73	432,521	4.04	13,424	.125	5794	.054	8408	.078	1619	.015	3618	.034
6	−	8.12	13.32	520,875	4.42	6590	.056	5485	.046	5584	.047	1200	.010	3669	.031
U	−	8.69	13.58	483,325	4.07	14,449	.122	4873	.041	6348	.054	1601	.014	3568	.030
1	+	5.27	8.55	337,091	3.15	9053	.085	2829	.026	3935	.037	1101	.010	1464	.014
6	+	5.22	7.30	321,074	2.72	4879	.041	3378	.029	3332	.028	800	.007	1472	.012
U	+	5.30	7.30	297,595	2.51	11,734	.099	3279	.028	4918	.041	1063	.009	2220	.019

Note: Values are expressed for 2×10^7 cells/2 hr of incubation. The initial specific activities of 1-¹⁴C, 6-¹⁴C and U-¹⁴C were 107,060 cpm/μmol, 117,893 cpm/μmol and 118,652 cpm/μmol, respectively. The final concentration of LND was 0.2 mM. Values are averaged from three different cell preparations. GE: glucose equivalents (μmoles).

REFERENCES

1. PEDERSEN, P. L. 1978. Tumor mitochondria and the bioenergetics of the cancer cell. Prog. Exp. Tumor Res. **22:** 190–274.
2. PAGGI, M. G. & A. CAPUTO. 1987. Clinical significance of a rearranged hexokinase isozyme expression in neoplastic phenotype: The astrocytoma model. *In* Human Tumor Markers. Cimino *et al.,* Eds. Walter De Gruyter. Berlin.
3. NAKASHIMA, R. A., M. G. PAGGI, L. J. SCOTT & P. L. PEDERSEN. 1988. Purification and characterization of a bindable form of mitochondrial bound hexokinase from the highly glycolytic AS-30D rat hepatoma cell line. Cancer Res. 48: 913–919.
4. PAGGI, M. G. Unpublished data.

Harvey-*ras*, but Not Kirsten or N-*ras*, Inhibits the Induction of C-*fos* Expression

ARNALDO CARBONE,[a] GABRIELE L. GUSELLA,[b]
DANUTA RADZIOCH,[b] AND LUIGI VARESIO[b]

[a]*Istituto di Anatomia Patologica*
Università Cattolica del Sacro Cuore
00168 Rome, Italy

[b]*Laboratory of Molecular Immunoregulation*
Frederick, Maryland 21701

The oncogenes of the *ras* family comprise Harvey-*ras* (H-*ras*) and Kirsten-*ras* (K-*ras*), which are related to acutely transforming murine sarcoma retroviruses, and N-*ras*, which was isolated from a neuroblastoma.[1] The *ras* oncogenes encode membrane-associated, G-like proteins: the *ras* p21s. Mutated *ras* p21s or overex-

TABLE 1. TPA-Induced Expression of c-*fos* and TPA Receptors in Normal and Transformed NIH 3T3 Fibroblasts

Transforming Gene[a]	c-*fos* Induction		TPA Receptors[d] (sites/cell ×10⁻⁵)
	mRNA[b]	Protein[c]	
None	100%	100%	7.5
Mutated H-*ras*	8%	15%	8.0
Overexpressed H-*ras*	10%	5%	7.7
N-*ras*	30%	Not determined	5.0
K-*ras*	115%	91%	7.1
trk	92%	Not determined	6.5
dbl	103%	Not determined	6.8

[a]NIH 3T3 cells transformed by H-*ras* and *trk* genes were provided by Dr. M. Barbacid (Developmental Oncology Section, LBI); NIH 3T3 cells transformed by N-, K-*ras* and *dbl* were kindly provided by Dr. A. Eva (National Cancer Institute, NIH).

[b]Fifteen micrograms of total RNA were analyzed by the Northern blot technique for the presence of c-*fos* mRNA. The c-*fos* mRNA levels were quantified by densitometry analysis of the X-ray films. The results are expressed as percent of c-*fos* mRNA induction by TPA in transformed NIH 3T3 cells relative to normal NIH 3T3.

[c]After metabolic labeling, proteins were extracted, immunoprecipitated with a specific anti-*fos* antibody, and analyzed by SDS-PAGE analysis, followed by autoradiography. The relative c-*fos* protein levels were quantified as above.

[d]TPA-binding capacity of these cells was measured by incubating intact cells with increasing amounts of ³H-labeled phorbol-12,13-dibutyrate in the presence or absence of specific competitors (Scatchard analysis).

pressed normal H-*ras* p21 induces transformation of NIH-3T3 fibroblasts, probably by deregulating the transmission of extracellular stimuli.[2] *ras* p21s oncogenes differ substantially in the amino acidic composition of the C-terminus, and this biochemical difference renders K-*ras*, but not H-*ras* p21, susceptible to phosphorylation by protein kinase C (PKC).[3] On the bases of these data, we reasoned that the biological response of *ras*-transformed cells to PKC activators could be different, depending upon the nature of the transforming *ras*. To examine this possibility, we studied the expression of c-*fos* proto-oncogene mRNA and protein in response to the tumor promoter 12-*O*-tetradecanoyl-phorbol-13-acetate (TPA), a potent PKC agonist, in normal or *ras*-transformed fibroblasts. TPA, in fibroblasts, induces expression of the proto-oncogene c-*fos*, which encodes a nuclear phosphoprotein that may be involved in the orderly progression of fibroblasts throughout the cell cycle.[4]

RESULTS

We found that H-*ras* has the unique ability to inhibit c-*fos* induction by TPA (TABLE 1). In contrast, normal c-*fos* expression was induced by TPA in fibroblasts transformed by N-or K-*ras* or by the *ras*-unrelated oncogenes *dbl* or *trk* (onc-D). The inhibition of c-*fos* induction by H-*ras* was not due to alteration in the binding of TPA to the transformed cells (TABLE 1) or to selection of idiosyncratic clones (data not shown).

CONCLUSIONS

These results provide the first evidence that H-*ras* is functionally different from K- or N-*ras*, and suggest that H-*ras* could interfere with the PKC-dependent cascade of signals leading to the induction of c-*fos* expression.

REFERENCES

1. BARBACID, M. 1987. Ras genes. Annu. Rev. Biochem. **56:** 779–827.
2. LEVINSON, A. D. 1986. The transforming activity of ras oncogenes. Trends in Genetics. **2:** 81–95.
3. BALLESTER, R., M. E. FURTH & O. M. ROSEN. 1987. Phorbol ester and protein kinase c-mediated phosphorylation of the cellular Kirsten ras gene product. J. Biol. Chem. **262:** 2688–2695.
4. SAMBUCETTI, L. A. & T. CURRAN. 1986. The fos protein complex is associated with DNA in isolated nuclei and binds to DNA cellulose. Science **234:** 1417–1419.

Functional Characterization of a Potential Receptor for Growth Factor of Human B and Reed-Sternberg Cells

ASHRAF IMAM, S. STATHOPOULOS,
SUZAN L. HOLLAND, AND CLIVE R. TAYLOR

Department of Pathology, and
Comprehensive Cancer Center
University of Southern California
Los Angeles, California 90033

A developmentally related antigen expressed on the plasma membrane of B lymphocytes and on Reed-Sternberg cells was identified using a monoclonal antibody produced by immunization of a Balb/c mouse with a Hodgkin's cell line (HDLM-3F). The antibody was termed anti-BLA-36 (B lymphocytes antigen-36) to indicate its predominant reactivity and the molecular weight of the corresponding antigen. When immunoperoxidase techniques were used, anti-BLA-36 reacted strongly with the Hodgkin's cell line that served as immunogen (FIG. 1), and to a lesser degree with pre-B and B cell lines, but showed no detectable binding activity with other hematopoietic cell lines. In normal tissues, BLA-36 antigen was detectable predominantly on cells in the germinal center and mantle zone of reactive follicles in lymph nodes and spleen. In hematopoietic malignancy, BLA-36 antigen was detectable on the surface of Reed-Sternberg cells, mononuclear Hodgkin's cells, and on malignant cells of B-cell lineage. Under these conditions, T lymphocytes, histiocytes, granulocytes, macrophages, and stromal cells in lymphoid tissue were consistently negative for the expression of the antigen. The findings in lymphomas exactly mirrored the patterns of staining observed in the cell line panels that were examined. T-cell lymphomas and diffuse histiocytic lymphomas were consistently nonreactive. B-cell lymphomas (Raji and Daudi) and the so-called pre-B cell line (SUAMB-1 and SUAMB-2) were, by contrast, clearly reactive, as were examples of lymphoblastoid and undifferentiated lymphoma (BL-1 and Nu-LB-1), all of which show some features of B-cell differentiation. Furthermore, acute lymphoblastic leukemia of B-cell derivation (BALL-1 and BALM-2) clearly showed positivity with the antibody. The results suggest that anti-BLA-36, unlike most other antileukocyte antibodies, retains its immunoreactivity in paraffin-embedded tissue sections, and it distinguishes Reed-Sternberg cells and B-cell lymphomas from all other malignant cells. Therefore, anti-BLA-36 appears to have diagnostic utility. As a consequence of the observed reactivity with "pre-B cells," sections of liver were examined from fetuses ranging in age from 10–24 weeks. A small number of round mononuclear cells consistent in appearance with lymphoblasts were identified in fetal liver, increasing in number up to 17 or 18 weeks of gestation. In fetuses of 22 or more weeks of age, the number of these cells leveled off or even decreased. Specimens of bone marrow derived from fetuses ranging in age from 17 to 22 weeks also showed a small proportion of cells that were positive for the expression of BLA-36. In frozen sections double-staining methods revealed that BLA-36-positive

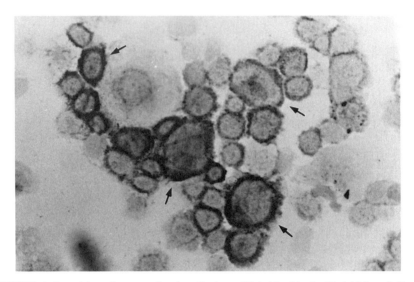

FIGURE 1. Reactivity of a monoclonal antibody to BLA-36 with the Hodgkin's cell lines. Cytocentrifuge preparations of cell lines were made, fixed with cold acetone, and stained with the antibody. Hodgkin's cell line (HDLM-3) showed strong reactivity mainly with the cell surface, as indicated by *arrows*. Original magnification × 200.

FIGURE 2. The effects of antibody to BLA-36 on cell growth and cell viability were assayed by adding 2.0 μg of purified specific antibody (■) or an irrelevant antibody of the same immunoglobulin class (□) on culture containing 1.6×10^5 Hodgkin's cell line (HDLM-3) per well of a 24-well tissue culture plate as indicated by *arrow*. The cells were removed at 24-hr intervals and were counted with a hemacytometer. The viability of cells was 99% as determined by trypan-blue dye exclusion.

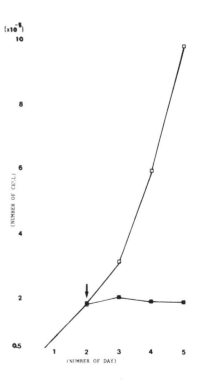

cells also contained immunoglobulin mu heavy chain, suggesting that these cells represented an early stage in the B-cell differentiation pathway. These observations indicate that BLA-36 is expressed during an early stage of B-cell development, perhaps even prior to the expression of mu-chain. Biochemical and immunologic analyses indicate that BLA-36 is distinct from previously identified antigens of hematopoietic cell lineage with respect to molecular weight, immunologic recognition, and resistance (for antibody-binding activity) to treatment with organic solvents. Finally, in order to explore the effect of anti-BLA-36 on cell growth, Hodgkin's, B-, T-, and large-cell lymphoma cell lines were grown in the presence of various concentrations of the antibody. Antibody concentrations in the range of 1 to 2 μg/ml completely inhibited the growth of the Hodgkin's and B-cell lines (FIG. 2). In addition, an irrelevant monoclonal antibody of the same immunoglobulin class in the above concentration range exhibited no inhibition of growth under the same conditions (FIG. 2). No such effect was observed when the antibody in the above concentration range was incubated with antigen-negative cell lines (CEM, MOLT-4, U-937, and SU-DHL-1). Furthermore, anti-BLA-36 markedly (60%) inhibited [^{3}H]thymidine incorporation, whereas equivalent concentrations of control antibody had no effect. These aspects are the subject of continued study.

Receptor for Epidermal Growth Factor in Neoplastic and Non-Neoplastic Human Thymus

F. BATTAGLIA,[a] G. SCAMBIA,[a] C. PROVENZANO,[b]
S. ROSSI,[c] P. BENEDETTI PANICI,[a] G. FERRANDINA,[a]
E. BARTOCCIONI,[b]F. CRUCITTI,[c] AND S. MANCUSO[a]

[a]Department of Obstetrics and Gynecology
[b]Institute of General Pathology
Rome, Italy

[c]Department of Surgical Pathology
Catholic University
Rome, Italy

It has been recently reported that epidermal growth factor (EGF) modulates several biological functions of thymic epithelial cells *in vitro*.[1] In particular, it increases the secretion of intrathymic mediators and stimulates the proliferation of small thymic epithelial cells (TECs), which express a high-affinity EGF receptor (EGF-R). We have studied EGF-R expression in the normal and pathologic human thymus.

MATERIALS AND METHODS

Twenty-four normal and pathologic thymic tissue specimens were frozen on dry ice shortly after surgical removal and stored at $-80°C$ until receptor assay. A representative section of specimen was retained for histopathologic examination and the quantitative distribution of epithelial and lymphoid components of thymus specimens was evaluated. The receptor assay was performed as described elsewhere.[2] Tissue weight permitting Scatchard analysis was carried out with concentrations of ^{125}I-labeled EGF ranging from 0.24 to 2.6 nM either alone or in the presence of unlabeled EGF ($1\mu g$). Results were expressed as fmoles per milligram of membrane protein. Statistical analysis was performed using the Wilcoxon rank sum test.

RESULTS AND DISCUSSION

As shown in TABLE I EGF-R is present in both the neoplastic and the non-neoplastic human thymus. It is worth noting that in thymomas that were analyzed EGF-R levels seem to be related to the epithelial cell content, suggesting that EGF-R is mainly expressed by the epithelial cells. EGF-R levels were found to be higher in thymomas (median = 14.1; range = 3.5–58.44 fmol/mg protein) than in hyperplastic thymuses (median = 6.9; range = 0–11.16 fmol/mg protein) ($p < 0.05$). Although the number of cases analyzed was small, the EGF-R levels were higher in male than in

TABLE 1. Epithelial Growth Factor Receptor in Neoplastic and Non-neoplastic Thymuses

Case No.	Sex of Patient	Epithelial/Lymphoid Ratio	EGF-R (fmol/mg protein)
Thymoma			
5	F	+/+++	3.5
2	F	+/++	3.79
3	F	+/++	19.11
4	F	++/++	7.8
1	F	++/+	14.1
6	M	+++/+	18.29
8	F	+++/−	12.6
7	M	+++/−	43.06
9	M	+++/−	58.44
Thymic hyperplasia			
19	F		0
17	M		1.06
12	F		2.48
16	M		2.81
18	M		3.7
11	F		6.9
10	F		7.65
14	F		9.2
20	F		11.16
13	F		11.78
15	M		16.25
Involuted thymus			
24	M		0
22	F		0
23	F		7.2
21	M		11.7

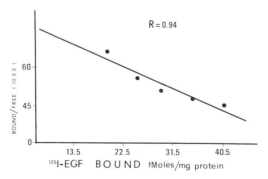

FIGURE 1. Scatchard analysis of case 9; (kD 1.012 nM; number of binding sites is 58.44 fmol/mg proteins).

female patients with thymoma ($p < 0.05$), suggesting a possible role of sex hormones in modulating thymic EGF-R expression. FIGURE 1 depicts Scatchard analysis of one pure epithelial thymoma showing a high-affinity EGF-R (Kd = 1.012 nM), with a number of binding sites of 58.44 fmol/mg protein. This finding is in agreement with reports by Nieburgs et al.,[1] who found high-affinity EGF-R in TECs in vitro.

Our results, showing EGF-R expression both in normal and neoplastic human thymuses, support the hypothesis of intrathymic lymphoid cell regulation by "nonimmunologic" mediators (such as EGF) acting on thymic epithelial cells.[1] Moreover, Johnson et al.[3] found that EGF was capable of completely replacing helper cell requirement for IFN-a production with a mechanism not related to the proliferative effect on cells. Also, EGF increases the in vitro production of PGE2 by epithelial thymus cells.[1] The release of PGE2 into the thymic microenvironment seems to modulate the T-lymphocyte response to influences of stimulating factors, such as IL-2[4] or other mitogenic factors.[5] Finally, an abnormal microenvironment of epithelial cells has been described in human thymic neoplasms.[6] We can hypothesize that EGF or EGF-like substances may contribute indirectly to the disturbance of T-cell "education" which is associated with thymus diseases and is expressed as autoimmune disorders.

REFERENCES

1. NIEBURGS, A. C., J. K. KORN, P. T. PICCIANO & S. COHEN. 1987. Thymic epithelium in vitro. Regulation of growth and mediator production by epidermal growth factor cell. Immunol. **108:** 396–404.
2. BATTAGLIA, F., G. POLIZZI, G. SCAMBIA, S. ROSSI, P. BENNEDETTI PANICI, S. IACOBELLI, F. CRUCITTI & S. MANCUSO. 1988. Receptor for epidermal growth factor and steroid hormones in human breast cancer. Oncology. **45:** 424–427.
3. JOHNSON, H. M. & B. A. TORRES. 1985. Peptide growth factors, PDGF, EG and FGF regulate interferon-gamma production. J. Immunol. **134:** 2824–2826.
4. RAPPAPORT, R. S. & G. R. DODGE. 1982. Prostaglandin E inhibits the production of human interleukin 2. J. Exp. Med. **155:** 943–948.
5. TOMAR, R. H., T. L. DARROW & P. A. JOHN. 1981. Response to and production of prostaglandin by murine thymus, spleen, bone marrow and lymph node cells. Cell Immunol. **60:** 335–346.
6. HAYNES, B. F. 1984. The human thymic microenvironment. Adv. Immunol. **36:** 87–141.

Protein Kinase C in Cell Proliferation and Differentiation[a]

SERGIO ADAMO,[b,c] CLARA NERVI,[d] ROBERTA CECI,[d]
LUCIANA DE ANGELIS,[d] AND MARIO MOLINARO[d]

[c]Department of Experimental Medicine
University of L'Aquila
L'Aquila, Italy

[d]Institute of Histology and General Embryology
University of Rome
Rome, Italy

Calcium-, phospholipid-dependent protein kinase (PKC), an amphitropic single-chain protein of 79 kDalton, is a key enzyme in the signal transduction system based on phosphoinositide hydrolysis. PKC may be found either in association with the membrane lipids or soluble in the cytoplasm, its calcium-mediated specific association with phosphatidylserine (PS) being essential for the enzyme activity. Diacylglycerol (DG) produced during phosphoinositide hydrolysis increases PKC affinity for calcium and PS, thereby activating the enzyme.[1] Since PKC activators such as DG and TPA (a potent tumor promoter which mimics DG in its ability to activate PKC) affect cell proliferation and differentiation, it is tempting to speculate that changes in PKC activity and subcellular localization may occur along with spontaneous changes of the cell proliferative and differentiative condition.

Human fibroblasts (HF) seeded at widely different initial densities (e.g., 20% and 120% confluency) and cultured in medium containing 10% fetal calf serum allow the comparison between actively proliferating and mitotically quiescent cells exposed to the same external environment. PKC activity was extracted and measured as previously described.[2] While in proliferating cells 70% of PKC activity is membrane-associated (and can be solubilized by the use of cation chelators), membrane-associated PKC drops to 24% in mitotically quiescent cells (FIG. 1). We have shown that this shift in the ratio of membrane-associated versus soluble PKC is not dependent on the cell density but on the proliferation rate.[2]

Terminal differentiation of mononucleated myoblasts (MB) into multinucleated striated myotubes (MT), and the expression of muscle-specific products in MT, is selectively inhibited by TPA.[3] The subcellular distribution of PKC was measured in MB and in MT, representing cells of the same lineage exposed to the same environment but distinct in terms of stage of differentiation. Although in 24-hr MB 81% of PKC activity is membrane-associated, only 26% of 96-hr MT PKC activity is membrane-associated. Although in our experimental conditions the vast majority of 24-hr MB is

[a]This work was supported by C.N.R. Special Project "Oncology" Grant 87.01150.44.
[b]Address for correspondence: Sergio Adamo, Dipartimento di Medicina Sperimentale, Collemaggio, 67100 L'Aquila, Italy.

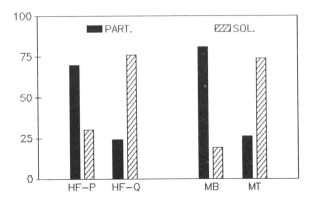

FIGURE 1. Subcellular distribution of PKC activity in human fibroblasts and myogenic cells. The data are expressed as percent of the total activity. HF-P: proliferating HF; HF-Q: mitotically quiescent HF; MB: 24-hr MB; MT: 96-hr MT.

postmitotic, as judged by autoradiography after [³H]thymidine labeling, it may still be argued that the different subcellular distribution of PKC between MB and MT is related to differences in cell proliferation rather than differentiation. However, we have recently shown that a correlation exists between cell differentiation and decrease of membrane-associated PKC in cultured rat Sertoli cells, in the absence of any change in the mitotic activity.[4]

We have investigated the possible role of some of the factors that could be responsible for the described difference in PKC subcellular distribution, a parameter strictly related to the actual activity of the enzyme in the intact cell. Modifications of the phospholipid composition of the membrane (and, in particular, of the PS to phosphatidylcholine [PC] ratio) could play a role in modifications of the subcellular distribution of PKC. Indirect evidence for this is provided by experiments indicating

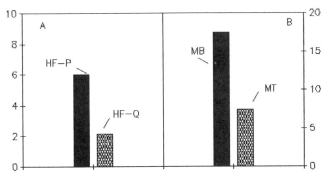

FIGURE 2. Basal rate of IPs production (cpm/μg protein) in human fibroblasts (**A**) and myogenic cells (**B**). IPs were extracted and measured as described in Ref. 7. Abbreviations are the same as in FIGURE 1.

that PKC-mediated responses are modified by pretreatment of the cells with PS- or PC-containing liposomes.[5,6]

Also, the rate of phosphoinositide hydrolysis (which determines the rate of DG production) is likely to influence the binding to the membrane and the activation of PKC. We have measured the production of inositol phosphates (IPs) under basal conditions (in the absence of specific stimulators of phosphoinositide hydrolysis) both in proliferating versus quiescent HF and in MB versus MT.

When cells prelabeled with [^{3}H]inositol to isotopic equilibrium with respect to phosphatidylinositol labeling are used, the production of IPs is constant during the time period examined (0–60′). However, the constant level of IP1 produced by proliferating HF is approximately 2.5-fold higher than that of quiescent HF; and 24-hr MB IP1 production is approximately 3-fold higher than that of 96-hr MT (FIG. 2).

These data are of interest in interpreting the spontaneous modulation of PKC subcellular distribution occurring along with spontaneous changes in the cell proliferative or differentiative condition.

REFERENCES

1. NISHIZUKA, Y. 1986. Science **233**: 305–312.
2. ADAMO, S., C. CAPORALE & S. AGUANNO. 1986. FEBS Lett. **195**: 352–356.
3. COSSU, G., M. PACIFICI, S. ADAMO, M. BOUCHÈ & M. MOLINARO. 1982. Differentiation **21**: 62–65.
4. GALDIERI, M., C. CAPORALE & S. ADAMO. 1986. Mol. Cell. Endocrinol **48**: 213–220.
5. ADAMO, S., C. CAPORALE, S. AGUANNO & M. MOLINARO. 1986. Cell Biol. Int. Rep. **10**: 215.
6. COSSU, G., S. ADAMO, M. I. SENNI, C. CAPORALE & M. MOLINARO. 1986. Biochem. Biophys. Res. Commun. **137**: 752–758.
7. ADAMO, S., B. M. ZANI, C. NERVI, M. I. SENNI, M. MOLINARO & F. EUSEBI. 1985. FEBS Lett. **190**:161–164.

Inhibitory Action of Transforming Growth Factor Beta on Thyroid Cells[a]

G. COLLETTA, A. M. CIRAFICI, M. IMBRIACO,
AND G. VECCHIO

Dipartimento di Biologia e
Patologia Cellulare e Molecolare
Centro di Endocrinologia e
Oncologia Sperimentale del CNR
II Facoltà di Medicina e Chirurgia
Naples, Italy

The transforming growth factor beta (TGF-beta) is a prototype of a family of polypeptide growth factors that regulate cell growth and differentiation. This factor consists of a 25-kD disulfide-linked homodimer originally found in transformed fibroblasts.[1]

In the present work we focused our attention on the effects of this transforming growth factor on rat thyroid cells in culture, the FRTL-5 clone. These cells are epithelial cells differentiated in culture that maintain properties from the original tissue such as the capability of synthesizing and secreting thyroglobulin (TG) in the culture medium, the ability to take up the iodide, and dependence on thyrotropic hormone (TSH) for growing.[2] Rat thyroid cells grow in the presence of calf serum and require a mixture of six hormones (6H): insulin, transferrin, TSH, hydrocortisone, somatostatin, and glycyl-hystidyl-lysine acetate. But these cells are particularly dependent upon the addition of TSH to the medium. Upon withdrawal of this hormone the cells cease to divide, but maintain their viability and become hypersensitive to RSH with respect to cAMP production.[3]

When TGF-beta (TGF-Beta Type 1 from B & D System) is added to the culture medium, the thyroid cells acquire a different morphologic pattern: they appear enlarged and more adherent to the plate and their growth potential is reduced. A clear inhibition of cellular proliferation is noticed when cells are treated with only 5 ng/ml of TGF-beta in the presence of calf serum and 6H.

Since TSH is the major growth factor in this epithelial system, [³H]thymidine uptake experiments were performed to verify the direct effect of TGF-beta on TSH-induced DNA synthesis. Thyroid cells deprived of the hormone mixture can be made quiescent even in the presence of 5% calf serum, and under these conditions only background levels of [³H]thymidine were observed. However, to exclude any interference by serum protein in the action of growth factors we carried out experiments in serum-free medium. TABLE 1 reports results obtained by stimulating quiescent thyroid cells with pure TSH and TGF-beta. TSH induces the maximal stimulation at concentration of 1 mU per milliliter after 18 hours of treatment.[4] TGF-beta alone is

[a]This work was supported by Progetto Finalizzato Oncologia CNR and by the Associazione Italiana Ricerca sul Cancro.

TABLE 1. FRTL5 [^{3}H]Thymidine Uptake after TSH and TGFB Treatment

Factors Added	cpm^a	Stimulation Index^b
None	1115	1
6H	458,000	41
5% calf serum	3250	2.9
TSH 0.001 mU/ml	1902	1.7
TSH 0.01 mU/ml	18,500	16.5
TSH 0.1 mU/ml	27,650	27.8
TSH 1 mU/ml	33,100	29.5
TGFB 0.1 ng/ml	1120	1
TGBF 1 ng/ml	1200	1
TGFB 10 ng/ml	980	1
TGFB 5 ng/ml^c	890	1
TGFB 5 ng/ml + TSH 0.001 mU/ml	1350	1.5
TGFB 5 ng/ml + TSH 0.01 mU/ml	3900	4.3
TGFB 5 ng/ml + TSH 0.1 mU/ml	11,400	12.8
TGFB 5 ng/ml + TSH 1 mU/ml	16,720	18.7

^{a}Count per minute of [^{3}H]thymidine incorporated into trichloroacetic-acid-insoluble material (average of two assays).

^{b}Stimulation index: ratio between counts incorporated into growth-factor-stimulated cultures relative to counts incorporated into identically treated unstimulated cultures.

^{c}Quiescent cells were pretreated with TGFB for 24 hours and then stimulated with TSH.

unable to produce any effect on thyroid cells. Instead, TSH-induced DNA synthesis is clearly reduced when thyroid cells are pretreated with TGF-beta, 5 ng/ml, for 24 hours.

The results presented here indicate that in the rat thyroid cell system, TGF-beta is able to interact with growth factors and modulate the proliferative potential of cell culture, altering the thyroid cells' ability to respond to TSH as a mitogenic signal.

REFERENCES

1. SPORN, M. B., *et al.* 1987. J. Cell Biol. **105:** 1039–1045.
2. AMBESI-IMPIOMBATO, F. S., *et al.* 1980. Proc. Natl. Acad. Sci. USA **79:** 6680–6685.
3. VALENTE, W. A., *et al.* 1983. Endocrinology **112:** 71–79.
4. COLLETTA, G., *et al.* 1986. Science **233:** 458–460.

Amino Acid and Sugar Transport in Mouse 3T3 Cells Expressing Activated *ras* and *neu* Oncogenes[a]

NICOLA LONGO, RENATA FRANCHI-GAZZOLA,
OVIDIO BUSSOLATI, VALERIA DALL'ASTA,
FRANCA A. NUCCI, AND GIAN C. GAZZOLA

Istituto di Patologia Generale
Università di Parma
43100 Parma, Italy

An enhancement of sugar transport is among the most characteristic biochemical markers of cellular transformation.[1] In the case of amino acid transport, an increased uptake has been reported for virus- and chemically transformed cell lines, but not for cells derived from spontaneously occurring tumors.[2] In any case, it was not possible to correlate variations in amino acid transport to the expression of a specific oncogene. Here we report amino acid and sugar transport in mouse 3T3 cells transformed by *ras* or *neu* oncogenes.

Oncogenes were integrated in the genomic DNA by transforming 3T3 mouse cells with Harvey sarcoma virus (line XHT), or with DNA fragments from human bladder carcinoma[3] (line EJ) and from rat neuroblastoma cells[4] (line B104). The activity of amino acid and glucose transport systems was evaluated by measuring the initial rate of entry of specific substrates[5-7] (TABLE 1). Solute uptake was normalized to intracellular water.[7]

FIGURE 1 shows confluent cultures of 3T3 mouse fibroblasts before (A) and after transformation with activated *ras*[3] (B, C) or *neu*[4] (D) oncogenes. Transformed cells could grow in multilayers and their multiplication was not contact-inhibited. The growth rate of tumor cells was similar to that of control fibroblasts in sparse cultures, but remained very high in confluent cultures, when 3T3 cells ceased to proliferate. Amino acid and sugar uptake were measured either in sparse or in confluent fibroblasts. TABLE 1 reports the influx of amino acids and sugar in normal and transformed 3T3 mouse fibroblasts. Both in sparse and in confluent cultures, glucose transport was increased in tumor as compared to nontransformed cells. The activity of amino acid transport systems A, ASC, L, and y^+ was not grossly different among the four cell lines. The activity of NA^+-independent system x_C^- (L-glutamic acid transport) was increased in confluent transformed cells. By contrast, the uptake of L-aspartic acid via Na^+-dependent system X_{AG}^- was very low both in sparse and in confluent transformed cells. As previously observed in cultured human fibroblasts,[6] system X_{AG}^- activity increased and system x_C^- activity decreased in confluent, as compared to

[a]This work was supported by M.P.I. and C.N.R., Rome, Italy.

374

TABLE 1. Amino Acid and Sugar Transport in Normal and Transformed 3T3 Fibroblasts

System Substrate Concentration (mM)		Amino Acid Transport (nmol · ml⁻¹ of intracellular water · min⁻¹)						Sugar Transport (nmol · ml⁻¹ · sec⁻¹)
		A MeAIB 0.1	ASC L-Ser 0.05	L L-Leu 0.01	X_{AG}^- L-Asp 0.01	x_C^- L-Glu 0.02	y^+ L-Arg 0.02	Glucose OMG 1
3T3	S	123 ± 4	139 ± 50	66 ± 1	44 ± 1	37 ± 1	161 ± 20	28 ± 4
	C	112 ± 12	118 ± 4	45 ± 1	89 ± 14	26 ± 2	162 ± 18	19 ± 3
EJ (*ras*)	S	97 ± 14	166 ± 34	42 ± 4	3 ± 1	45 ± 3	102 ± 4	48 ± 3
	C	86 ± 9	76 ± 7	35 ± 1	3 ± 1	61 ± 4	118 ± 9	45 ± 4
XHT (*ras*)	S	89 ± 15	198 ± 9	58 ± 13	3 ± 1	34 ± 3	129 ± 2	80 ± 9
	C	104 ± 2	98 ± 4	44 ± 1	2 ± 1	89 ± 11	106 ± 2	57 ± 2
B104 (*neu*)	S	113 ± 25	187 ± 22	60 ± 3	4 ± 1	79 ± 3	207 ± 5	57 ± 8
	C	96 ± 2	80 ± 4	49 ± 10	3 ± 1	177 ± 11	144 ± 7	54 ± 3

NOTE: Data represent the mean value of three independent determinations ± SD. The activity of Na^+-independent amino acid transport systems, L, x_C^-, and y^+ was measured in a Na^+-free medium, in which choline replaced sodium in Earle's solution. MeAIB: α-(methylamino)isobutyric acid; OMG: 3-O-methyl-D-glucose; S = sparse cells; C = confluent cells.

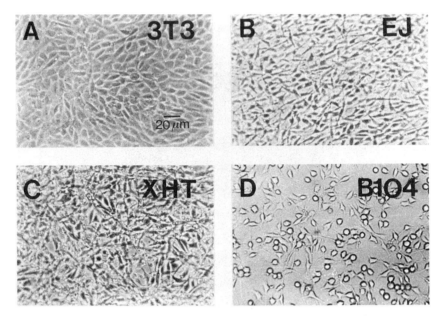

FIGURE 1. Photomicrograph of normal 3T3 mouse fibroblasts before (**A**) and after transfection with the transforming gene of a human bladder carcinoma cell line (*ras*) (**B**), with Harvey sarcoma virus (*ras*) (**C**), or with the *neu* oncogene (**D**) derived from a rat neuroblastoma cell line.

actively growing 3T3 cells. This suggests that both the suppression of system X_{AG}^- and the increase in system x_C^- activity in tumor cells may be somehow related to their growth characteristics. The study of other tumor cells will establish whether variations in anionic amino acid transport represent a novel biochemical marker of cell transformation.

ACKNOWLEDGMENTS

We thank Dr. Richard A. Weinberg, Whitehead Institute, Cambridge, Massachusetts, for kindly supplying transformed 3T3 cells.

REFERENCES

1.   FLIER, J. S., M. M. MUECKLER, P. USHER & H. F. LODISH. 1987. Science **235:** 1492–1495.
2.   GAZZOLA, G. C., V. DALL'ASTA, R. FRANCHI-GAZZOLA, O. BUSSOLATI, N. LONGO & G. G. GUIDOTTI. 1985. *In* Cell Membranes and Cancer. T. Galeotti, A. Cittadini, G. Neri, S. Papa & L. A. Smets, Eds.: 169–174. Elsevier. Amsterdam.
3.   TABIN, C. J., S. M. BRADLEY, C. I. BARGMANN, R. A. WEINBERG, A. G. PAPAGEORGE, E. M. SCOLNICK, R. DHAR, D. R. LOWY & E. H. CHANG. 1982. Nature **300:** 143–149.
4.   SCHECHTER, A. L., D. F. STERN, L. VAIDYANATHAN, S. J. DECKER, J. A. DREBIN, M. I. GREENE & R. A. WEINBERG. Nature **312:** 513–516.

5. GAZZOLA, G. C., V. DALL'ASTA & G. G. GUIDOTTI. 1980. J. Biol. Chem. **255:** 929–936.
6. DALL'ASTA, V., G. C. GAZZOLA, R. FRANCHI-GAZZOLA, O. BUSSOLATI, N. LONGO & G. G. GUIDOTTI. 1983. J. Biol. Chem. **258:** 6371–6379.
7. BUSSOLATI, O., P. C. LARIS, F. A. NUCCI, V. DALL'ASTA, N. LONGO, G. G. GUIDOTTI & G. C. GAZZOLA. 1987. Am. J. Physiol. **253:** C391–C397.

Membrane Structures Involved in the Proliferation and Differentiation of T-Cell Precursors

L. M. LAROCCA, N. MAGGIANO, A. CARBONE,
F. O. RANELLETTI,[a] M. PIANTELLI, AND A. CAPELLI

Pathology and [a]Histology Departments
Università Cattolica
00168 Rome, Italy

The thymus represents the major anatomic site where immunocompetent T cells are generated. Direct contact of developing thymocytes with nonlymphoid components of the thymus and subsequent proliferation are important for normal T-cell maturation. To evaluate what kind of signals are involved in the proliferation and differentiation of T-cell precursors, immature, double-negative CD3-CD1⁻ thymocytes were co-cultured with thymic epithelial cell (TE) or with accessory cells. TE were cultured as previously reported.[1] An enriched culture medium containing cholera toxin, insulin, hydrocortisone, adenine, and epidermal growth factor was used along with a feeder layer of mitomycin-C-treated mouse 3T3 fibroblasts. CD3-CD1⁻ cells, obtained as described,[2] bound to *in vitro* cultured TE and proliferated with a peak response at day 3. The addition of exogenous IL2 (5 U/ml) enhanced this proliferative response and sustained it for 3 additional days. Anti-CD2 monoclonal antibody (MoAb) inhibited almost completely both the binding of CD3-CD1⁻ thymocytes to TE and their proliferation. Moreover, CD3-CD1⁻ cells did not proliferate when co-cultured with TE in the same culture chamber, but separated by a porous membrane bottom (Transwell 3415, Costar, Cambridge, MA). Anti-class I and II, −CD4 and −CD8 MoAbs did not modify CD3-CD1⁻ proliferative response to TE, whereas anti-CD25 produced a partial inhibition (TABLE 1). Phenotypically 35 ± 6% SE and 53 ± 8% SE ($n = 3$) of nonadherent thymocytes expressed CD3 antigen after 48 and 72 hours of culture, respectively.

CD3-CD1⁻ thymocytes co-cultured with thymic accessory cells or with peripheral blood monocytes require for proliferation the addition of Rec-IL2 (5 U/ml). In this case the maximal proliferative response was observed after 7 days of culture (TABLE 1) and was dependent on direct contact between cells. Accessory cells, pretreated with anti-class II MoAb, were unable to sustain CD3-CD1⁻ cell proliferation. Unlike anti-CD4 and -CD8, anti-CD25 and -CD2 MoAbs almost completely inhibited the accessory-cell-dependent proliferation. Phenotypically, CD3⁺ cells constituted 36 ± 9% SE and 82 ± 13% SE ($n = 3$) of the total cell population at 4 and 8 days of culture, respectively.

Our data indicate that direct contact between immature T-cell precursors and thymic stromal cells is necessary for proliferation of thymocyte precursors. The membrane structures involved are CD2, CD25, LFA3, and class II antigens. Interactions between TE and immature thymocytes are sufficient for the production of growth

TABLE 1. ^{3}H-TDR Incorporation (cpm/well) of CD3-CD1$^-$ Thymocytes Co-cultured with TE or Accessory Cells^a

		Days						
Cells	IL2	2	3	4	5	6	7	8
TE	−	4230	11,560	5520	1210	470	<300	<300
TE	+	8090	10,280	11,100	9320	8760	5120	1990
M0	−	ND	ND	<300	<300	<300	800	<300
M0	+	ND	<300	450	2980	7700	13,810	9130

^{a}CD3-CD1$^-$ thymocytes (3 × 10^4 cells/well) were co-cultured with mitomycin-treated epithelial (TE) or accessory cells (M0) (both at 1 × 10^4 cells/well) in the absence or in the presence of 5U/ml of recombinant IL2. Cells were pulsed for 5 hr with 0.6 μCi/well of ^{3}H-labeled methyl thymidine, spec. act. 2 Ci/mmol. This table shows results of a typical experiment of three performed. ND: not done.

factors able to sustain the proliferation and/or the maturation of double-negative cells. On the contrary, the accessory-cell-triggered proliferation is class II and exogenous-IL2-dependent and appeared kinetically slower than that induced by TE. These findings may be explained by two different hypotheses: (*a*) the immature thymocytes interacting with TE are functionally different from those interacting with accessory cells; (*b*) the same cells can be triggered by different thymic stromal cells throughout different activation pathways.

REFERENCES

1. RHEINWALD, J. & H. GREEN. 1975. Serial cultivation of strains of human epidermal keratinocytes: The formation of keratinizing colonies from a single cell. Cell **6**: 331–339.
2. LAROCCA, L. M., M. PIANTELLI, N. MAGGIANO & P. MUSIANI. 1987. T cell surface markers expression by immature human thymocytes in in vitro culture: Role of Ia+ accessory cells. J. Immunol. **138**: 2410–2416.

Molecular Cloning of PTC, a New Oncogene Found Activated in Human Thyroid Papillary Carcinomas and Their Lymph Node Metastases[a]

M. GRIECO,[b] M. SANTORO,[b] M. T. BERLINGIERI,[b]
R. DONGHI,[c] M. A. PIEROTTI,[c] G. DELLA PORTA,[c]
A. FUSCO,[b] AND G. VECCHIO[b]

[b]Centro di Endocrinologia e Oncologia Sperimentale del C.N.R.
and
Dipartimento di Biologia e Patologia Cellulare e Molecolare
II Facoltà di Medicina e Chirurgia
Naples, Italy

[c]Istituto Nazionale dei Tumori
Milan, Italy

In our laboratory we have recently analyzed human thyroid papillary carcinomas for the presence of activated oncogenes by means of the transfection assay of NIH/3T3 cells.[1] This study has led us to conclude that a common oncogene was activated in thyroid papillary carcinomas in five of twelve patients. Furthermore, this oncogene did not cross-hybridize with a broad range of cloned viral and human oncogenes, indicating the activation of a new transforming gene in these carcinomas. This new oncogene, which we designate PTC for papillary thyroid carcinoma, was found to be activated frequently in thyroid papillary carcinomas as well as in primary tumors and lymph node metastases in the same patients.

In order to molecularly clone the gene, genomic libraries have been constructed by inserting size-fractionated *Eco*RI and *Mbo*I partial digests of DNAs from tertiary transfectants of a primary tumor into EMBL 4 lambda phage. Several overlapping phages that possessed human repetitive sequences of the Alu family were isolated. They contained the same Alu-positive restitution fragments that were detected in the transfectants, and covered about 60 kbp of the gene. Comparison of restriction sites again showed no homology to known oncogenes. Alu-free fragments were isolated from these phages and hybridized with DNAs from transfectants of the five positive patients showing that the same gene was present in all of them.

Two Alu-free fragments were shown to be able to detect transcripts of about 4, 2.6 and 1.8 kbp of human origin. No mRNAs were detected with these probes in NIH/3T3 cells.

Preliminary results of *in situ* hybridizations and somatic cell hybrid experiments

[a]This work was supported by Progetto Finalizzato Oncologia, C.N.R., and Associazione Italiana Ricerca sul Cancro.

TABLE 1. Transformation of NIH/3T3 Cells with DNAs from Thyroid Papillary Carcinomas and Lymph Node Metastases

DNA Code	Source of DNA	Primary Foci	Tumor Induction	Secondary Foci	Tertiary Foci
T32	Tumor of patient 1	8/4	4/4	200/4	180/4
M32	Lymph node metastases of patient 1	10/4	4/4	160/4	167/4
T28	Tumor of patient 2	80/4	4/4	148/4	159/4
M28	Lymph node metastases of patient 2	20/4	4/4	160/4	166/4
T18	Tumor of patient 3	4/4	4/4	93/4	101/4
T22	Tumor of patient 4	6/4	4/4	68/2	73/2
T36	Tumor of patient 5	10/4	4/4	50/2	64/4
—	Normal thyroids and lymphocytes of patients 1, 2, 3 and 4	0/20	—	—	—
I.E.	Tumor of patient I.E.	85/4	6/6	90/2	N.T.

performed with two Alu-free fragments as a probe show that this oncogene is located on chromosome 10. No oncogene has been located so far on this chromosome.

Work is in progress to isolate a full-length cDNA of PTC as well as to obtain sequence data of the coding regions that have been identified.

Finally, we are molecularly cloning another activated oncogene detected in a sixth patient harboring a papillary thyroid carcinoma. This gene is not homologous to PTC and does not cross-hybridize with the *ras* gene family.

REFERENCE

1. FUSCO, A., M. GRIECO, M. SANTORO, M. T. BERLINGIERI, S. PILOTTI, M. A. PIEROTTI, G. DELLA PORTA & G. VECCHIO. 1987. A new oncogene in human thyroid papillary carcinomas and their lymph-nodal metastases. Nature **328**: 170–172.

CRGF: An Autocrine Growth Factor Associated with Colorectal Carcinomas

G. POMMIER, J. M. CULOUSCOU, F. GARROUSTE,
AND M. REMACLE-BONNET

Laboratoire d'Immunopathologie
Faculté de Médecine
13385 Marseille Cedex 5, France

INTRODUCTION

Recent advances have accumulated showing several structural and functional relationships between numerous oncogene products and growth factors or their cellular receptors.[1] Among many hypotheses, the concept of autocrine stimulation has been advanced to explain the constitutive alterations in cellular growth control which play a fundamental role in the pathogenesis of neoplastic transformation.[1] Suramin, a polyanionic compound able to dissociate a large spectrum of receptor-bound growth factors and to inhibit their binding to their receptors, has been recently used to show the autocrine hypothesis.[2] In previous work done in our laboratory[3] we showed that the HT-29 human colonic carcinoma cell line releases, among others, a growth factor (colorectum-associated growth factor [CRGF], $M_r \sim 25,000$) that is immunologically related to, but distinct from, epidermal growth factor (EGF). Here, we show that CRGF can function as an autocrine growth factor.

MATERIALS AND METHODS

Cell lines, cell culture conditions, and preparation of CRGF and EGF radioreceptor assays were as previously described.[3] Suramin was a gift from Specia, Paris, France.

RESULTS

The ability of CRGF to displace ^{125}I-EGF binding at 4°C on FR3T3 fibroblasts showed a linear relationship with that of authentic human EGF. The concentration of CRGF was therefore converted to equivalent nanograms of human EGF. Scatchard analysis of binding data of ^{125}I-EGF to confluent rat (FR3T3, NRK-clone 49F) or human (HSF2) fibroblasts and breast adenocarcinoma cell line (MCF-7) in the presence or not of 10 ng EGF-equivalent of CRGF showed that the decreased EGF binding induced by CRGF was always due to an increase in the dissociation constant of EGF receptors (K_d). In contrast, the number of EGF receptors remained unchanged. When tested on a wide variety of different cell lines, CRGF produced a significant inhibition (32.9–70.1%) of ^{125}I-EGF binding whatever the cell type assayed (FIG. 1).

These data suggest that CRGF receptors are present on cells of epithelial and mesenchymal origin, on normal and tumor cells, on cells derived from adult and embryonic tissues, and on cells from different species. In contrast, as shown in FIGURE 1, exogenously supplied CRGF did not alter EGF binding to HT-29, HT-29-clone D4 and CAL-14 colonic adenocarcinoma cell lines as it did not to the HRT-18 rectal tumor cell line. Neither the number (24×10^3 receptors/cell) or the affinity ($K_d = 0.5$ nM) of EGF receptors on HT-29 cells was altered by exogenous CRGF. It follows from these results that CRGF receptors are either absent or occupied by endogenous CRGF

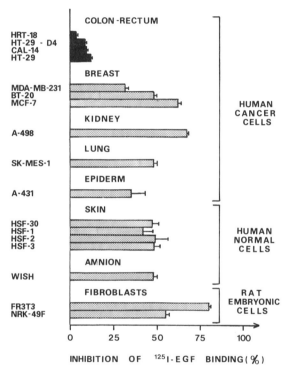

FIGURE 1. Inhibitory potential of CRGF on various normal and tumor cell lines. Results are expressed as percent of inhibition of control specific binding obtained without CRGF.

at the cell surface of colorectal cell lines. To further investigate this hypothesis HT-29 cells were cultured overnight at 37°C in suramin-containing medium. FIGURE 2 shows that this treatment allowed CRGF to induce 65% inhibition of EGF binding on HT-29 cells. A similar result was obtained when HT-29-clone D4, CAL-14 and HRT-18 cell lines were exposed to suramin. These data strengthen the suggestion that autocrine secretion of CRGF may be a common feature for at least these four colorectal carcinoma cell lines, explaining the apparent lack of CRGF receptors on these cell lines. Suramin-treated HT-29 cells expressed an unaffected number of EGF receptors as compared to untreated cells. However, in the presence of exogenously added CRGF,

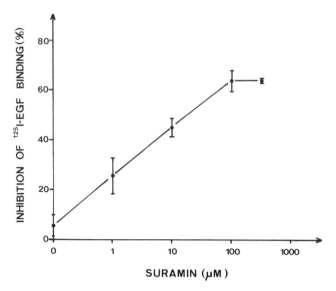

FIGURE 2. Emergence of CRGF inhibitory effect on HT-29 cells after suramin pretreatment. Results are expressed as in FIGURE 1.

EGF receptor affinity was decreased by the same order of magnitude as the one observed with noncolorectal cell lines. These findings are consistent with (i) a co-expression of CRGF and EGF receptors at the HT-29 cell surface and (ii) an inability of CRGF to bind to EGF receptors.

CONCLUSION

CRGF is a growth factor that appears to be specifically associated with some colorectal carcinoma cell lines. Immunologic and functional properties of CRGF distinguish it from other growth factors, such as TGF-alpha, TGF-beta, and PDGF.[3] Although immunologically related to EGF, CRGF represents a distinct molecule: more particularly, it binds to its own specific receptors, but is unable to do the same on EGF receptors. A further structural and biological characterization is warranted to evaluate the significance and generality of autocrine production of CRGF by colorectal carcinomas. However, there is the exciting possibility that CRGF might be used as a specific functional marker of colorectal tumor pathogenesis.

REFERENCES

1. GOUSTIN, A. S., E. B. LEOF, G. D. SHIPLEY & H. L. MOSES. 1986. Cancer Res. **46:** 1015–1029.
2. BETSHOLTZ, C., A. JOHNSSON, C. H. HELDIN & B. WESTERMARK. 1986. Proc. Natl. Acad. Sci. USA **83:** 6440–6444.
3. CULOUSCOU, J. M., M. REMACLE-BONNET, F. GARROUSTE, J. MARVALDI & G. POMMIER. 1987. Int. J. Cancer **40:** 646–652.

Lectins in Cancer Cells

REUBEN LOTAN[a,b] AND AVRAHAM RAZ[c]

[a]Department of Tumor Biology
The University of Texas
M. D. Anderson Cancer Center
Houston, Texas 77030

[c]The Program in Cancer Metastasis
Michigan Cancer Foundation
Detroit, Michigan 48201

INTRODUCTION

Normal development, cell growth, and differentiation involve various types of cellular interactions among cells and between cells and exogenous soluble or insoluble macromolecules. These interactions are mediated by specific cell surface components, some of which contain carbohydrates[1-18] and others of which are capable of binding carbohydrates.[19-26] Lectins are divalent or polyvalent carbohydrate-binding proteins that bind and precipitate glycoproteins and agglutinate cells.[27,28] The presence of lectin-like proteins in vertebrate tissue (liver) was observed 20 years ago and the first mammalian lectin was purified from liver.[29] Subsequently, different lectins were discovered in various other vertebrate tissues.[19-21,30] Such lectins are often referred to as endogenous lectins to distinguish them from plant lectins, which have been used extensively for vertebrate cell surface characterization.[27,28] The most prevalent endogenous lectins in various vertebrate tissues (e.g., spleen, thymus, heart, lung, bone marrow, skin, intestine, pancreas, liver, and muscle) bind galactosides.[19-21,30-35] Some of these lectins have been purified by affinity chromatography and were found to constitute two distinct classes of polypeptides; one exhibited an M_r of about 14,000 and the other ranged in M_r from 29,000 to 35,000.[19-21,35-38,39-49] Recently, a higher M_r lectin (M_r 67,000) has been discovered.[50] The galactoside-specific lectins have been detected in the cytoplasmic compartment of various normal cells as well as in the extracellular compartment, where they seem to be associated with the extracellular matrix (ECM) after being secreted by the cells.[20] Low levels of a 35,000 lectin have been detected on the surface of cultured 3T3 fibroblasts[51] as well as within the nuclei of these cells.[52] The levels of lectins in several vertebrate tissues are developmentally regulated.[19,20,25,39-41,53-56] Various functions have been proposed for the lectins, including mediation of intercellular recognition and adhesion,[19,21-26,31-35,39,56] as receptors for elastin,[50] in control of cellular proliferation,[33,46] and serving as constituents of the ECM.[20,36,57] Lectins isolated from different tissues and, often, from different species, exhibit

[b]Address for correspondance: Reuben Lotan, Ph.D., Department of Tumor Biology (Box 108), The University of Texas, M. D. Anderson Cancer Center, 1515 Holcombe Boulevard, Houston, Texas 77030.

immunologic crossreactivity, suggesting structural similarity.[37,40,58-60] Indeed, recently, lectins from different tissues and various species (e.g., electric eel, chick skin, mouse 3T3 fibroblasts and fibrosarcoma cells, bovine fibroblasts, human placenta, and human hepatoma) have been cloned, sequenced, and found to share considerable homology and some regions of identical peptide sequences.[41-49,61,62]

The presence of endogenous sugar-inhibitable hemagglutinating activity in tumor cells was observed in neuroblastoma[30] and embryonal carcinoma.[22] We have demonstrated the presence of lectin-like activity in various cultured tumor cell lines[23] and proposed that lectins on the surface of cancer cells might be involved in tumor cell aggregation, embolization in blood vessels, and adhesion to capillary endothelial cells, all of which are required for metastasis formation.[25,63] This paper describes the results of our studies on the properties and functions of cancer cell lectins.

INDICATIONS FOR THE PRESENCE OF CARBOHYDRATE-BINDING PROTEINS ON THE TUMOR CELL SURFACE

Asialofetuin was found to induce homotypic aggregation of various tumor cells including melanoma, fibrosarcoma, and carcinoma.[23] This property was lost when the asialofetuin was digested with pronase to yield glycopeptides. These results suggested that the intact asialofetuin molecule, which contains several oligosaccharide side chains, can serve as a bridge between carbohydrate-binding proteins on the surface of adjacent cells.[25,63] Of interest, the glycopeptides were effective inhibitors of aggregation induction by the intact asialofetuin molecule,[24] suggesting that they compete with asialofetuin for binding to the cell surface. Indeed, direct binding of fluorescently labeled asialofetuin to the surface of tumor cells could be demonstrated.[23] Syngeneic mouse serum was also an effective inducer of tumor cell aggregation.[31] Furthermore, B16 melanoma cells selected for reduced ability to aggregate in the presence of asialofetuin were also poorly aggregated with mouse serum, and exhibited a reduced lung colonization potential.[31] Since asialofetuin contains terminal nonreducing galactose residues, we presumed that it is bound by a galactoside-specific cell surface lectin.[23] Extracts of different murine and human tumor cells prepared by procedures that have been established for the extraction of endogenous lectins from normal tissue and cells[30,53] contain lectin activity detectable by hemagglutination assay and inhibitable by galactosides.[23]

PURIFICATION AND PROPERTIES OF ENDOGENOUS TUMOR LECTINS

The lectins mediating this hemagglutinating activity were purified from extracts of fibrosarcoma cells by affinity chromatography using immobilized asialofetuin.[35] The purified material consisted of two polypeptides of M_r 14,500 (L-14.5) and 34,000 (L-34), respectively. Similar lectins were purified from B16 and K-1735 murine melanoma, and carcinoma HeLa-S3 (unpublished data). The higher M_r polypeptide could not be dissociated by boiling in 9 M urea, 2% sodium dodecylsulfate, and 4 mM β-mercaptoethanol to the lower M_r polypeptide, and the peptide maps of the two polypeptides demonstrated only a partial overlap,[35] suggesting that L-34 is not a dimer of L-14.5. Separation by gel filtration of the two affinity-purified polypeptides revealed that both possess hemagglutinating capacity. Biopsies of several human tumors including those of the colon, lung, breast, and kidney were found to contain lactose-inhibitable hemagglutinating activity (TABLE 1). All of the tumors tested contained the 14.5-kD lectins, and a few also contained the higher M_r lectins (TABLE 1). ,xtensive studies carried out by Gabius *et al.* (reviewed in Ref. 64) on endogenous lectins extracted from different tumors revealed the presence of lectins with different sugar specificities, molecular weight, and requirements for calcium ions for activity.

TABLE 1. Analysis of Extracts of Fresh Human Tumor Biopsy Specimens for the Presence of Hemagglutinating Activity (HA), Its Inhibition by Lactose (LI), and Identification of Molecular Species of Affinity-Purified Lectins

Tumor Type	Specific HA[a]	LI (mM)[b]	M_r of Purified Lectins (kD)[c]	
Bladder carcinoma	5.7	0.58	14.5	
Breast carcinoma	2.7	1.2	14.5;	31
Breast carcinoma	10.0	1.2	14.5; 17.5	34
Breast carcinoma	7.6	1.2	14.6	
Colon carcinoma	4.9	1.2	14.5	
Renal carcinoma	9.6	18	14.5	
Lung carcinoma	13.5	9.4	14.5	
Squamous carcinoma of tongue	17.9	1.2	14.5	34

[a]The tumor samples were suspended in three volumes of Ca^{+2}-Mg^{+2}-free phosphate-buffered saline containing 4 mM β-mercaptoethanol, 4 mM EDTA (MEPBS), 0.3 M lactose, and 1 mM phenylmethylsulfonyl fluoride. The suspensions were frozen and thawed twice and then homogenized using initially a Polytron homogenizer and then a Dounce homogenizer. The homogenates were centrifuged at 100,000 g for 1 hour at 4°C and the supernatant was collected and dialyzed against MEPBS to remove the lactose. The concentration of protein was determined by a dye binding assay and the HA was analyzed using trypsinized fixed rabbit red blood cell as described earlier.[23] HA is expressed as titer^{-1} of the highest extract dilution to effect hemagglutination when a 25-μl sample is serially diluted, and the specific HA is the titer^{-1}/ml/mg protein.
[b]Concentration of lactose required to inhibit HA effected by a dilution of extract that agglutinates red blood cells up to a 1:4 dilution.
[c]M_r was estimated from the electrophoretic migration of protein bands on SDS-PAGE after staining with a silver stain, relative to the migration of standards.

APPLICATIONS OF MONOCLONAL ANTILECTIN ANTIBODIES AND POLYCLONAL ANTIBODIES FOR LECTIN LOCALIZATION IN TUMOR CELLS

To prepare monoclonal antilectin antibodies, syngeneic mice were immunized with crude extracts of B16-F1 melanoma cells. Several hybridomas were isolated, which produced antibodies capable of inhibiting the hemagglutinating activity of the lectin present in B16-F1 extracts. Lectin activity in extracts of tumor cell, including fibrosarcoma, lymphosarcoma, carcinoma, and melanoma cells, was also inhibited by these antibodies.[32] One of the monoclonal antibodies, designated 5D7, was shown by immunoblotting to bind specifically to both purified 14.5 and 34 kD lectins.[34] The mAb were found capable of binding to lectins of different tumor cell lines, indicating that these lectins share some antigenic determinants. The endogenous lectins have been localized, by indirect immunofluorescence staining, at the surface of different viable cultured tumor cells and, after fixation and permeabilization, in the intracellular compartment.[32] The presence of both lectins on the surface of various tumor and transformed cells was further demonstrated by labeling the proteins exposed on the surface of viable cells with ^{125}I by lactoperoxidase-catalyzed iodination followed by cell solubilization and immunoprecipitation with polyclonal antilectin antibodies prepared in rabbits by immunization with affinity-purified lectins.[35] Polyclonal rabbit antibodies against the 14.5-kD lectin isolated from rat lungs[37] (obtained from Dr. Samuel Barondes, the University of California at San Francisco) and an antiserum prepared in rabbits against a 34-kD lectin isolated from cultured 3T3 cells[38] (obtained from Dr. John Wang, Michigan State University, East Lansing, Michigan) recognized the lectins of the K-1735 melanoma cells (FIG. 1 A' and B'). These findings suggested that the K-1735 melanoma, 3T3 fibroblasts, and the normal rat lungs contain lectins, which share antigenic sites. Both anti-14.5 and anti-35 KD lectins were capable of binding to the surface of viable K-1735 cells (FIG. 1 A and B), indicating that both lectins are associated with the cell surface. Immunoblotting with antibodies that recognize both lectins (prepared against bovine lung 14- and 29-kD lectins[36]) in total cell extracts prepared from different tumor cell lines, established from human tumors in nude mice,

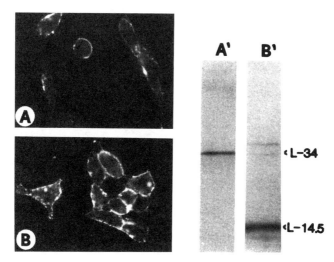

FIGURE 1. Binding of antilectin antibodies to the cell surface and to cell extracts of K-1735 murine melanoma cells. Cells were seeded on glass cover slips and cultured for 24 hours before they were incubated with antilectin antibodies [(**A**) rabbit anti-35-kD lectin from 3T3 fibroblasts[38]; (**B**) anti-14 kD-lectin from rat lung[37]] followed by fluorescein isothiocyanate-labeled goat anti-rabbit antibodies. The cells were observed and photographed using a Nikon fluorescence microscope (magnification ×238). For immunoblotting analysis the cells were solubilized with a mixture of ionic and nonionic detergents[40] and subjected to polyacrylamide gel electrophoresis followed by electrophoretic transfer onto nitrocellulose filters. The filters were then incubated with anti-35-kD antibodies (**A'**) or anti-14-kD antibodies (**B'**) followed by ^{125}I-labeled goat anti-rabbit IgG. Autoradiography revealed the nature of the proteins with which the antibodies reacted.

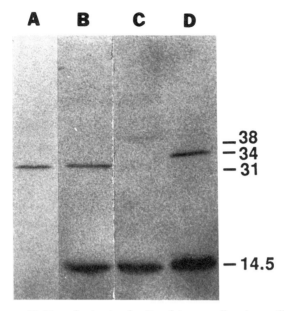

FIGURE 2. Immunoblotting of extracts of cultured tumor cells using antibodies against a mixture of 14- and 29-kD lectins from bovine lungs.[36] (**A**) KM12-L4 human colon carcinoma cells derived from liver metastasis in a nude mouse; (**B**) A375 MM human melanoma cell line; (**C**) SN12-L1 human renal carcinoma established from liver metastasis in a nude mouse; (**D**) murine lymphosarcoma RAW117 cell line. (Cells in A–C were obtained from Dr. I. J. Fidler of the University of Texas M. D. Anderson Cancer Center.)

revealed qualitative differences in lectin (or crossreactive protein) expression (FIG. 2). For example, human colon carcinoma cell line KM12 subline L4 (established from a nude mouse liver metastasis) was found to contain only a higher M_r lectin (31kD) whereas the human melanoma cell line A375 MM contains both a 14.5- and a 31-kD lectin. A renal carcinoma cell line SN12 contains a 14.5- and a 38-kD lectin and a murine lymphosarcoma RAW117-Pal 10 contains a 14.5- and a 34-kD lectin.

INCREASED EXPRESSION OF CELL SURFACE LECTINS AFTER TRANSFORMATION AND PROGRESSION TO A HIGHLY METASTATIC PHENOTYPE

Quantitative comparison of lectin expression on the surface of murine melanomas B16 and K-1735 and UV-2237 fibrosarcoma, performed by indirect immunofluorescence labeling with mAb 5D7 and flow microfluorimetry, revealed that in these three tumor systems, cell variants or clones that exhibit a higher metastatic potential *in vivo* bind more antibody at their surface then do the less metastatic ones.[34] The progression of BALB/c 3T3/A31 angiosarcoma cell variants from untransformed parental cell line through spontaneously transformed, tumorigenic, and metastatic cells was accompanied by a significant increase in the amount of ^{125}I-labeled cell surface lectins precipitable by a polyclonal antilectin serum.[35] Transformation also results in increased lectin expression on the cell surface; virally transformed cells and *ras*-transfected cells express more lectin on their surface than do their untransformed counterparts.[34] Further, immunoprecipitation of [^{35}S]methionine-labeled lectins from secondary rat embryo cells, which are considered normal, and from their transformed derivatives, obtained by transfection with various oncogenes (p53, *myc,* c-Ha-*ras*), demonstrated that the L-14.5 lectin was present in normal and transformed cells, but that the L-34 lectin was expressed only in the transformed cells.[35] Another indication for the relationship between the expression of lectins on the cell surface and the transformed phenotype was the increased binding of antilectin antibodies to the

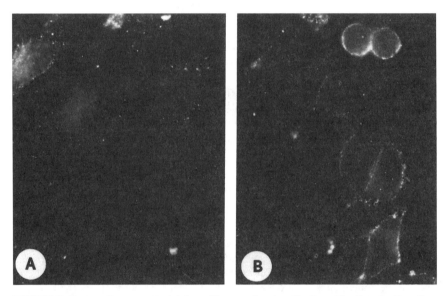

FIGURE 3. Immunofluorescence staining with monoclonal antilectin antibodies of cell surface lectins on normal rat kidney cells infected with a temperature-sensitive mutant Rous sarcoma virus (LA23). Cells were grown at 39°C (nonpermissive temperature) (**A**) or at 33°C (permissive temperature) (**B**) before they were incubated with the mAbs, followed by FITC-labeled rabbit anti-mouse Ig. Magnification ×950.

surface of normal rat kidney cells infected with a temperature-sensitive Rous sarcoma virus (LA23) when the cells are shifted from growth at the nonpermissive temperature (39°C, FIG. 3A) to the permissive temperature (33°C, FIG. 3B). Studies in other laboratories have demonstrated increased amounts of lectins (L-34 and higher M_r species) after transformation.[40,65]

PRELIMINARY REPORTS ON LECTIN GENE EXPRESSION

The genes coding for lectins of M_r of about 14 kD from chick embryo skin,[41,43] human placenta,[44,45] human lung,[42] human hepatoma cell line,[42] bovine heart,[47] rat lung,[48] and mouse fibrosarcoma,[61,62] were cloned or sequenced recently in several laboratories. The sequence homology ranged between 34 and 70% and this variability was not due to species differences.[62] The gene coding for L-34 from mouse fibrosarcoma UV-2237 cells was cloned and the carboxy terminal part of its amino acid sequence has been deduced from the nucleotide sequence.[62] It was found to include regions of sequence homology to the L-14.5, suggesting that these regions might be involved in forming the galactoside-binding site. The 35-kD lectin from 3T3 mouse fibroblasts was cloned and its complete nucleotide sequence was elucidated.[35,49] The carboxy terminal part of the lectin exhibits strong homology with L-34 of the fibrosarcoma cells. Of interest, the amino terminal portion of the 35-kD 3T3 lectin contains repetitive sequences of nine amino acids each and is rich in proline and glycine. This portion of the lectin molecule shows some homology with polypeptides of the heterogeneous nuclear ribonucleoprotein complex.[49] This homology might explain the association of the 35-kD lectin with RNA in 3T3 nuclei.[66] Analyses of lectin gene expression at the mRNA level were made possible by the availability of cDNA probes, and it was demonstrated that growth stimulation of quiescent 3T3 cells by serum leads to an increased expression of mRNA (1.3 kbp) for the 35-kD lectin.[46] The cDNA probes prepared after cloning of the two lectin genes of UV-2237 fibrosarcoma cells[61,62] were used to demonstrate by Southern blot analysis that the lectin genes of normal mouse embryo cells and syngeneic mouse tumor cells and virally transformed fibroblasts are similar in structure (restriction pattern) and copy number; the level of the mRNA for the 14.5-kD lectin (0.75 kbp, detected with the cDNA probe pL3) in the normal, and untransformed 3T3 and NRK cells was similar to the level in the virally transformed counterparts and melanoma cells, whereas the level of mRNA for the 34-kD lectin (1.65 kbp, detected with probe pM5) is higher in the transformed and tumor cells.[62] The levels of the mRNA for the L-14.5 and L-34 detected by the cDNA probes correlate with the level of lectin protein detected by immunoblotting of normal mouse embryo lung fibroblasts and B16 melanoma cells (FIG. 4). These findings indicate that L-34 lectin gene expression may be altered during changes in growth and transformation.

IMPLICATION OF TUMOR CELL SURFACE LECTINS IN CELL AGGREGATION, ATTACHMENT TO SUBSTRATUM, ANCHORAGE-INDEPENDENT GROWTH, AND EXPERIMENTAL LUNG COLONIZATION

Asialofetuin glycopeptides as well as antilectin mAbs inhibited asialofetuin-induced homotypic aggregation of tumor cells and suppressed the adhesion of tumor cells to the substratum.[24] The antilectin mAbs also inhibited the ability of certain tumor cells to form colonies in semi-solid medium (anchorage-independent growth), which is considered a property of transformed cells,[33] suggesting that the surface lectins may play a role in the expression of the transformed phenotype. Metastatic cells were preincubated with the antilectin mAbs *in vitro* before they were injected in the tail vein of syngeneic mice to test the hypothesis that lectins present on the surface of metastatic tumor cells mediate cellular interactions *in vivo,* such as embolization with platelets and lymphocytes.[25,63] This preincubation, but not a preincubation with anti-H2 antibodies, abrogated the ability of the metastatic cells to form lung tumor colonies.[24] Since the antibodies caused internalization of their complexes with the surface lectins at $37°C^{63}$ it is plausible to suggest that the inhibition of cell adhesion and of lung colonization resulted from removal of the lectins from the surface of the tumor cells. These results provide a strong support for a role of lectins in adhesion, growth regulation, and metastasis. They also provide a rationale for using antilectin antibodies or carbohydrate lectin inhibitors to suppress tumor growth and metastasis.

MODULATION OF LECTIN EXPRESSION DURING SUPPRESSION OF THE TRANSFORMED PHENOTYPE BY DIFFERENTIATION INDUCERS

The exposure of K-1735 melanoma cells grown in monolayer cultures to retinoic acid or dibutyryl cyclic AMP (Bt_2cAMP) resulted in marked changes in cell morphology. The treated cells extended long, dendrite-like processes characteristic of

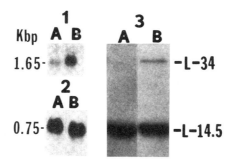

FIGURE 4. A comparison of lectin expression at the mRNA (1 and 2) and the protein (3) levels in mouse diploid fibroblasts (**A**) and B16-F1 melanoma cells (**B**). Total cellular RNA (10 μg) was analyzed by Northern blotting using cDNA probe pM5, which hybridizes to the L-34 lectin mRNA (1), and probe pL3, which hybridizes to the L-14.5 lectin mRNA (2).[62] Note that the level of the mRNA for the L-34 lectin is higher in the melanoma cells whereas the level of the L-14.5 lectin mRNA is similar in the normal and tumor cells. Extracts of normal and melanoma cells (3) were analyzed by immunoblotting (see FIGURE 2) using antibodies against both the L-14.5 and the L-34 lectins. Note that the normal cells do not contain detectable levels of the L-34 lectin.

differentiated neural-crest-derived cells (shown for Bt_2cAMP-treated cells in FIGURE 5A and B). The growth of the cells was inhibited by both agents in monolayer culture as well as in semi-solid agarose. Because the expression of endogenous lectins *in vivo* in normal rats, mice, and chickens is developmentally regulated,[6,39,40] we examined the effects of *in vitro* differentiation of melanoma cells on lectin expression. Immunoblot analyses revealed that differentiated cells express a significantly lower level of L-34 than do undifferentiated ones, while the levels of L-14.5 were not altered significantly (FIG. 5A' and B'). In addition, the binding of anti-L-34 antibodies to the surface of viable retinoic-acid- or Bt_2cAMP-treated cells was diminished markedly, whereas the binding of anti-L-14.5 antibodies was only reduced slightly (data not shown), suggesting that the decreased level of L-34 was not restricted to the major intracellular pool. These results demonstrate that L-34 lectin level is suppressed in differentiated melanoma cells. Since differentiation of these cells also leads to growth inhibition, it

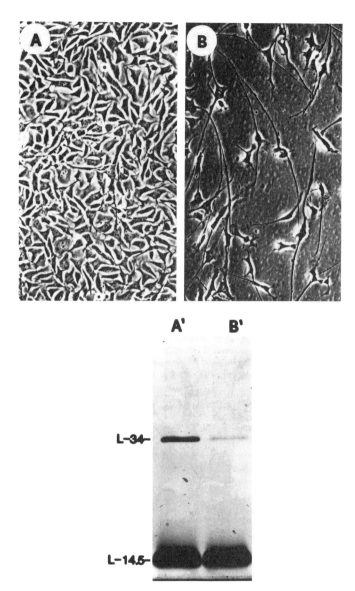

FIGURE 5. Effect of Bt$_2$cAMP on K-1735 melanoma differentiation and lectin levels. Cells were grown for 5 days in regular medium (**A**) or in medium supplemented with 1 mM Bt$_2$cAMP (**B**) and photographed under a phase-contrast microscope ($\times 108$). Note the extension of long dendrite-like processes by the treated cells. Untreated and treated cells were extracted and subjected to affinity chromatography as described elsewhere.[35] The purified lectin fractions were analyzed by polyacrylamide gel electrophoresis and silver staining of proteins. Note that the treatment resulted in a selective decrease in the level of the L-34 lectin.

was plausible that growth inhibition per se could be the reason for suppression of L-34 lectin production. However, this possibility was discounted since inhibition of the growth and DNA synthesis of the K-1735 melanoma cells by growing them to high density in serum-free medium did not alter the amount of the L-34 lectin. The modulation of lectin levels by treatment with differentiation inducers was not restricted to the K-1735 melanoma cells. Changes in lectin levels were found in several murine and human tumor cell lines whose differentiation is enhanced in culture by retinoic acid, Bt_2cAMP, or dimethylsulfoxide (DMSO). The expression of L-14.5 increased in F-9 embryonal carcinoma after differentiation to primitive endoderm, whereas it was suppressed after differentiation of neuroblastoma, leukemia, and breast carcinoma and was not altered in normal mouse lung cells.[67] The expression of L-34 is suppressed in all cells that express it prior to treatment, including those of melanomas and colon carcinoma. Human KM12 colon carcinoma cells, which express only a 31-kD lectin (FIG. 2), showed a marked reduction in lectin level after retinoic acid treatment. These results demonstrate that the production of the lectins is modulated by suppression of the transformed phenotype.

SUMMARY AND CONCLUSIONS

Studies carried out over the last few years have demonstrated that tumor cells and malignant tissues contain lectins that are similar in sugar-binding specificity, molecular size, and antigenicity to the lectins found in normal cells and tissues. Lectins from tumor cells also share marked sequence homology with lectins from normal tissues. Lectins were purified from various tumor cells by affinity chromatography and monoclonal and polyclonal antilectin antibodies were prepared against them. These enabled us to establish the following: (1) Lectins are present on the surface of all the tumor cells that were examined, albeit at varying levels. (2) The level of cell surface lectins increases after normal cells are transformed by transfection with certain oncogenes or by retroviruses, or when cells transformed with a temperature-sensitive viral mutant are switched from growth at the nonpermissive to the permissive temperature. (3) Among tumor cells differing in metastatic propensity, those exhibiting a higher potential express higher levels of surface lectins. (4) Tumor cell surface lectins might be involved in cell-cell adhesion, cell attachment to substratum, the expression of the transformed phenotype (anchorage-independent growth), and blood-borne metastasis. (5) The levels of the lectins in tumor cells are modulated by agents that suppress the transformed phenotype (as represented by anchorage-independence) or enhance differentiation. Numerous studies by others have shown that cell surface carbohydrate-containing molecules are modified after transformation, and our findings demonstrate that the expression of cell surface carbohydrate-binding proteins is also altered by transformation. Obviously, any of these changes may result in alterations in cellular interactions.

All the above findings implicate tumor cell lectins in cellular interactions (adhesion, attachment, possible binding of exogenous soluble glycoconjugates), cell growth and anchorage-independent growth, malignant transformation, tumor cell differentiation, and metastasis. It is clear that even if these lectins are involved in only a few of these fundamental processes, it is important to elucidate their functions and the mechanisms by which their expression is regulated during neoplastic transformation and tumor progression and the suppression of the transformed phenotype.

ACKNOWLEDGEMENTS

We thank Drs. S. H. Barondes and J. L. Wang for the antilectin antibodies and Mrs. Eleanor Felonia for assistance in manuscript preparation.

REFERENCES

1. SUBTELNY, S. & N. K. WESSELLS, Eds. 1980. The Cell Surface: Mediator of Developmental Processes. Academic Press. New York.
2. EDELMAN, G. M. 1983. Science 219: 450–457.
3. HAKOMORI, S.-I. 1981. Annu. Rev. Biochem. 50: 733–764.
4. OPPENHEIMER, S. B., M. EDIDIN, C. W. ORR & S. ROSEMAN. 1969. Proc. Natl. Acad. Sci. USA 63: 1395–1402.
5. ROSEMAN, S. 1970. Chem. Phys. Lipids 5: 270–297.
6. OPPENHEIMER, S. B. 1975. Exp. Cell Res. 92: 122–126.
7. SHUR, B. D. 1983. Dev. Biol. 99: 360–372.
8. RAUVALA, H., J.-P. PREELS & J. FINNE. 1983. Proc. Natl. Acad. Sci. USA 80: 3991–3997.
9. RAUVALA, H., W. G. CARTER & S.-I. HAKOMORI. J. Cell Biol. 88: 127–137.
10. RAUVALA, H. & S.-I. HAKOMORI. 1981. J. Cell Biol. 88: 149–159.
11. CHIPOWSKY, S., Y. C. LEE & S. ROSEMAN. 1973. Proc. Natl. Acad. Sci. USA 79: 2309–2312.
12. VICKER, M. G. & J. C. EDWARDS. 1972. J. Cell Sci. 10: 759–768.
13. LLOYD, C. W. & G. M. W. COOK. 1974. J. Cell Sci. 15: 575–590.
14. VICKER, M. G. 1976. J. Cell Sci. 21: 161–173.
15. EDWARDS, J. G., J. MCK. DYSART & C. HUGHES. 1976. Nature 264: 66–68.
16. BUSSIAN, R. W. & J. C. WRISTON. 1977. Biochim. Biophys. Acta 471: 336–340.
17. HUANG, R. T. 1978. Nature 276: 624–626.
18. FEIZI, T. & R. A. CHILDS. 1981. Trends Biochem Sci. 10: 24.
19. HARRISON, F. L. & C. J. CHESTERTON. 1980. FEBS Lett. 122: 157–165.
20. BARONDES, S. H. 1984. Science 2: 1259–1264.
21. MONSIGNY, M., C. KIEDA & A. C. ROCHE. 1983. Biol. Cell 47: 95–110.
22. GRABEL, L. B., S. D. ROSEN & G. R. MARTIN. 1979. Cell 17: 477–480.
23. RAZ, A. & R. LOTAN. 1981. Cancer Res. 41: 3642–3647.
24. MEROMSKY, L., R. LOTAN & A. RAZ. 1986. Cancer Res. 46: 5270–5272.
25. LOTAN, R. & A. RAZ. 1988. J. Cell. Biochem. 37: 107–117.
26. STOJANOVIC, D. & R. C. HUGHES. 1984. Biol. Cell. 51: 197–206.
27. GOLDSTEIN, I. J., R. C. HUGHES, M. MONSIGNY, T. OSAWA & N. SHARON. 1980. Nature 285: 66.
28. LIS, H. & N. SHARON. 1984. In Biology of Carbohydrates. V. Ginsburg & P. Robbins, Eds. Vol. 2: 1–86. Wiley. New York.
29. ASHWELL, G. & J. HARFORD. 1982. Annu. Rev. Biochem. 51: 531–554.
30. TEICHBERG, V., I. SILMAN, D. BEITSCH & G. RESHEFF. 1975. Proc. Natl. Acad. Sci. USA 72: 1383–1387.
31. LOTAN, R. & A. RAZ. 1981. Cancer Res. 43: 2088–2093.
32. RAZ, A., L. MEROMSKY, P. CARMI, R. KARAKASH, D. LOTAN & R. LOTAN. 1984. EMBO J. 3: 2979–2983.
33. LOTAN, R., D. LOTAN & A. RAZ. 1985. Cancer Res. 45: 4349–4353.
34. RAZ, A., L. MEROMSKY & R. LOTAN. 1986. Cancer Res. 46: 3667–3672.
35. RAZ, A., L. MEROMSKY, I. ZVIBEL & R. LOTAN. 1987. Int. J. Cancer 39: 353–360.
36. CERRA, R. F., M. A. GITT & S. H. BARONDES. 1985. J. Biol. Chem. 260: 10474–10477.

37. CERRA, R. F., P. L. HAYWOOD-REID & S. H. BARONDES. 1984. J. Cell Biol. **98:** 1580–1589.
38. ROFF, C. F. & J. L. WANG. 1983. J. Biol. Chem. **258:** 10657–10663.
39. REGAN, L. J., J. DODD, S. H. BARONDES & T. M. JESSEL. 1986. Proc. Natl. Acad. Sci. USA **83:** 2248–2252.
40. CRITTENDEN, S. L., C. F. ROFF & J. L. WANG. 1984. Mol. Cell. Biol. **4:** 1252–1259.
41. OHYAMA, Y., J. HIRABAYASHI, Y. ODA, S. OHNO, H. KAWASAKI, K. SUZUKI & K.-I. KASAI. 1986. Biochem. Biophys. Res. Commun. **134:** 51–56.
42. GITT, M. A. & S. H. BARONDES. 1986. Proc. Natl. Acad. Sci. USA **83:** 7603–7607.
43. HIRABAYASHI, J., H. KAWASAKI, K. SUZUKI & K.-I. KASAI. 1987. Biochemistry **101:** 775–787.
44. HIRABAYASHI, J., H. KAWASAKI, K. SUZUKI & K.-I. KASAI. 1987. Biochemistry **101:** 987–995.
45. PAROUTAUD, P., G. LEVI, V. I. TEICHBERG & A. D. STROSBERG. 1987. Proc. Natl. Acad. Sci. USA **84:** 6345–6348.
46. JIA, S., R. MEE, G. MORFORD, N. AGRWAL, P. VOSS, I. MOUTSATSOS & J. L. WANG. 1987. Gene **60:** 197–204.
47. SOUTHAN, C., A. AITKEN, R. A. CHILDS, W. M. ABBOTT & T. FEIZI. 1987. FEBS Lett. **214:** 301–304.
48. CLERCH, L. B., P. WHITNEY, M. HASS, K. BREW, T. MILLER, R. WERNER & D. MASSARO. 1988. Biochemistry **27:** 692–699.
49. JIA, S. & J. L. WANG. 1988. J. Biol. Chem. **263:** 6009–6011.
50. HINEK, A., D. S. WRENN, R. P. MECHAM & S. H. BARONDES. 1988. Science **239:** 1539–1541.
51. MOUTSATSOS, I. K., J. M. DAVIS & J. L. WANG. 1986. J. Cell Biol **102:** 477–483.
52. MOUTSATSOS, I., M. WADE, M. SCHINDLER & J. L. WANG. 1987. Proc. Natl. Acad. Sci. USA **84:** 6452–6456.
53. NOWAK, T. P., D. KOBILER, L. E. ROEL & S. H. BARONDES. 1977. J. Biol. Chem. **252:** 6026–6030.
54. NOWAK, T. P., P. L. HAYWOOD & S. H. BARONDES. Biochem. Biophys. Res. Commun. **68:** 650–657.
55. KOBILER, D. & S. H. BARONDES. 1977. Dev. Biol. **60:** 326–330.
56. COOK, G. M. W., S. E. ZALIK, N. MILOS & V. SCOTT. 1979. J. Cell Sci. **38:** 293–304.
57. ZALIK, S. E., L. W. THOMPSON & I. M. LEDSHAM. 1987. J. Cell Sci. **88:** 483–493.
58. BRILES, E. B., W. GREGORY, P. FLETCHER & S. KORNFELD. 1979. J. Cell Biol. **81:** 528–537.
59. ODA, Y. & K.-I. KASAI. 1983. Biochim. Biophys. Acta **761:** 237–245.
60. CARDING, S. R., R. A. CHILDS, R. THORPE, M. SPITZ & T. FEIZI. 1985. Biochem. J. **228:** 147–153.
61. RAZ, A., A. AVIVI, G. PAZERINI & P. CARMI. 1987. Exp. Cell Res. **173:** 109–116.
62. RAZ, A., P. CARMI & G. PAZERINI. 1988. Cancer Res. **48:** 645–649.
63. RAZ, A. & R. LOTAN. 1987. Cancer Metastasis Rev. **6:** 433–452.
64. GABIUS, H. J. 1987. Cancer Invest. **5:** 39–46.
65. CARDING, S. R., S. J. THORPE, R. THORPE & T. FEIZI. 1985. Biochem. Biophys. Res. Commun. **127:** 680–686.
66. LAING, J. G. & J. L. WANG. 1988. Biochemistry. In press.
67. LOTAN, R. *et al.* Unpublished data.

DISCUSSION

CHRISTOPHER FOSTER (*Hammersmith Hospital, London, England*): First, is there evidence that the endogenous lectins are glycoproteins? If so, does loss of expression of the lectins from the cell surface and into the cytosol with differentiation represent a change in glycosylation and hence intracellular targeting? Second, the lectins you report recognize galactose. This structure is the common terminal sugar of type I and type II oligosaccharides. If the lectins express any specificity, what is/are the preferred linkages of galactose recognized and are these oligosaccharides expressed on *specific* cell-surface recognition molecules?

R. LOTAN (*The University of Texas, Houston, Texas*): In answer to your first question, there is no evidence that any of the galactoside-specific lectins is a glycoprotein. To answer your second: Leffler and Barondes have determined the specificity of the endogenous lectins in rat lung and found that lactose is a preferred ligand. Derivatives of lactose with substitutions at C^4 or C^6 of galactose or at C^3 of the glucose are bound with lower affinity. Mammalian cell surface glycoconjugates contain N-acetyl-lactosamine sequences and this disaccharide is recognized by the endogenous lectins with an affinity slightly lower than that of lactose.

A. IMAM (*University of Southern California, Los Angeles, California*): In the earliest part of your presentation, a slide was projected showing inhibition of cell growth when incubated with antibody to a lectin. Have you studied the DNA synthesis of those cells?

LOTAN: Yes, in a 3-day treatment there was no decrease in DNA synthesis, suggesting no nonspecific toxicity.

M. MONSIGNY (*CNRS and Université d'Orleans, France*): According to the hypothesis on metastasis formation, do you have any evidence on interactions between the galactose-specific lectin and platelets or endothelial cells of capillaries?

LOTAN: Not yet; but I expect that membrane lectins of normal cells are involved in cell recognition according to the data you published showing that lectins of endothelial cells and of LL cells are both involved in cell adhesion.

A. CAPUTO (*Regina Elena Cancer Institute, Italy*): Is there any difference in endogenous lectins between primary and metastatic melanoma cells?

LOTAN: We found no qualitative differences between primary and metastatic cells. The only difference that we did find is that lectin levels were higher on the cell surface of metastatic cells than they were on nonmetastatic cells.

G. NERI (*Università D'Annunzio, Chieti, Italy*): Have you studied the effect of retinoic acid on the expression of endogenous lectins in normal cells? It would be interesting to try retinoic acid on the cranial cells of the neural crest, which seems to be the target of the teratogenic effect of retinoic acid in men and mice.

LOTAN: We have tried retinoic acid on normal mouse embryo lung fibroblasts (second passage *in vitro*) and found no effect on cell growth and no effect on the 14.5-kD lectin, which is the only lectin that we detected in these cells. We have not tried the experiment in cranial neural crest cells. However, in mouse embryonal carcinoma cells (F-9 cell line) retinoic acid induces differentiation to primitive endoderm and increases the expression of a 14.5-kD lectin.

CLAUDIO FRANCESCHI (*Università di Modena, Italy*): I agree with you that one of the physiological roles of endogenous lectins is likely the control of cell adhesion and cell proliferation. For example, I recently published a paper in which I demonstrated that only few selected monosaccharides were able to inhibit allogenic and autologous mixed lymphocyte cultures in humans.

P. A. RILEY (*University College, London, England*): In the experiments in which B16 melanoma metastasis was suppressed by retinoic acid, were the cells pretreated with retionic acid or were the animals receiving retinoic acid during the assay?

LOTAN: The cells were pretreated with RA because in this experimental metastatic system we wanted to determine the effect of changes in the cell surface lectins on lung colonization. In experiments not reported here we have been able to suppress the development of melanoma by injecting retinoic acid intraperitoneally after injection of tumor cells subcutaneously.

Endogenous Lectins and Drug Targeting

M. MONSIGNY, A. C. ROCHE, AND P. MIDOUX

Department of Biochemistry
Centre de Biophysique Moléculaire
Centre National de la Recherche Scientifique
45071 Orléans Cedex 2, France

INTRODUCTION

During the last decade, various cell surface receptors have been evidenced, including carbohydrate-binding proteins. Cell surface carbohydrate-specific receptors may be lectins,[1] ectoglycosyltransferases, or glycosidases. Glycolipids, glycoproteins, and proteoglycans have been shown to interact with lectins on the surface of a large number of animal cells (for reviews, see Refs. 2–14). The first membrane lectin of animal cells was characterized about 20 years ago by Ashwell, Morell and their colleagues.[6] Working on the metabolism of ceruleoplasmin, these authors treated this glycoprotein with neuraminidase, labeled the nonreducing terminal galactose residue, and found that this asialo-glycoprotein had a clearance drastically shorter than that of the native glycoprotein. Subsequently, they showed that asialo-glycoproteins were very efficiently taken up by hepatocytes. Rogers and Kornfeld[15] showed that ^{131}I-labeled lysozyme and serum albumin substituted with fetuin asialo-glycopeptides were taken up into rat liver *in vivo,* while the corresponding sugar-free proteins and the intact (sialylated) glycopeptide-protein conjugates were not, confirming the primary role of the nonreducing terminal galactose in the clearance of asialoglycoproteins.

Tumor cells also contain membrane lectins.[4,5,16–19] The specificity and the expression of tumor cell lectins depend on the cell type and/or the metastatic capacity of tumor cells. Roche *et al.*[20] identified, on the surface of 3LL Lewis lung carcinoma cells, a lectin that binds and takes up neoglycoproteins bearing α-D-glucose and α-D-mannose residues; Monsigny *et al.*[21] showed that L1210 murine leukemic cells internalize glycoconjugates containing α-L-fucose or β-D-galactose residues. Raz and Lotan have identified a β-galactose-specific lectin on B16 melanoma cells and other tumor cells[19,22] and have shown that the expression of this cell surface lectin is associated with transformation and metastasis.[18,23–25] Gabius[16] has identified β-galactose-specific lectins as well as lectins with other specificities on a large number of tumor cells of various origins. Paietta[26] showed evidence of a lectin on Hodgkin's disease cells.

The aim of this paper is to define the rationale of drug targeting based on specific sugar receptors called membrane lectins on the surface of certain type of cells. The first part deals with the characterization of membrane lectins and their ability to endocytose and to deliver defined ligands to lysosomes. The second part discusses various basic aspects of drug targeting. The third and forth parts summarize some interesting results in targeting toxic drugs and macrophage activators, respectively.

CHARACTERIZATION OF CELL SURFACE LECTINS
BY THE MEAN OF FLUORESCEINYLATED NEOGLYCOPROTEINS
AND GLYCOSYLATED POLYMERS

Among the endogenous lectins (which are mainly associated with the plasma membranes, the endoplasmic reticulum, the Golgi apparatus[6,27] and with the nucleus and the nucleolus),[28–30] those that are present at the cell surface are of primary interest with regard to recognition phenomena and to drug targeting. For this reason, we concentrated our efforts on techniques allowing the study of endogenous plasma membrane lectins.

Membrane lectins may be directly demonstrated by measuring the binding of labeled glycoconjugates to the cell surface. The glycoconjugates used are either native glycoproteins of glycosidase-treated glycoproteins or synthetic glycoproteins (called *neoglycoproteins;* see Refs 3, 31–33). Neoglycoproteins are obtained by adding an activated sugar, oligosaccharide, or glycopeptide to a protein such as serum albumin. The activated carbohydrate is usually either a glycosyl-alkylimidate, a glycosylphenyl-diazonium, or a glycosylphenyl-isothiocyanate.

The first neoglycoprotein was prepared by Goebel and Avery,[34] who synthesized diazophenylglycosides, coupled them to serum albumin, and showed that these conjugated carbohydrate-proteins were sugar-specific antigens.[35] Later, the same authors prepared new neoglycoproteins starting with oligosaccharides from type III *Pneumococcus* which were derivatized by using *p*-nitrobenzylbromide.[36]

Such glycosylphenylazo-protein conjugates were first used to study the properties of plant lectins[37,38]: concanavalin A and wheat germ lectin, respectively. Subsequently,

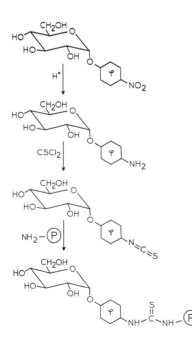

FIGURE 1. Scheme of synthesis of neoglycoproteins. A *p*-nitrophenylglycoside is reduced in the presence of palladium on charcoal; the aromatic amine is transformed into phenylisothiocyanate upon treatment with thiophosgene; and finally the activated sugar is added to a solution of protein at pH 8.5. Under optimal conditions, each step is roughly quantitative. According to the glycosylphenylisothiocyanate/protein ratio used, it is possible to obtain between a few and 50 sugar units per serum albumin molecule. P: serum albumin.

TABLE 1. Inhibition of the Hemagglutination Activity of Concanavalin A by α-Methyl Mannopyranoside, Mannosylated Serum Albumin, and Mannosylated Polymer

Compounds	Concentration (μM) Required for 50% Inhibiton		Ratio of α-Met Man/ Conjugate-Bound Man Concentrations
	a	b	
α-Met Man	2600	2600	1
Man_{20}-BSA^c	0.8	16	160
$PHA_{180}Man_{15}$ PL^d	6	100	25
$PHA_{165}Man_{30}$ PL	0.6	20	125
$PHA_{135}Man_{60}PL$	0.2	12	200

^{a}Expressed as conjugate concentration.
^{b}Expressed as free or bound mannose concentration.
^{c}BSA: bovine serum albumin.
^{d}PL: polylysine with an average polymerization degree of 190; PHA: polyhydroxyalcanoyl (in this case gluconoyl).

glycosylated cytochemical markers (i.e., glycosylated ferritin and glycosylated peroxidase) were prepared[39] and shown to be suitable to visualize lectins bound to cell-surface-associated glycoconjugates.[40–42]

Neoglycoproteins are also easily obtained by using glycosylphenylisothiocyanate (FIG. 1).[20,21,43,44] This type of neoglycoprotein is well adapted to prepare fluoresceinylated derivatives because the phenylthiocarbamyl group does not impair the emitted fluorescence.[20] Complex oligosaccharides are easily substituted with amino-phenylethylamine[45] by reductive amination in the presence of cyanoborohydride anion at neutral pH[46] and then transformed into phenylisothiocyanate derivatives.[47] Alternatively, a reducing oligosaccharide can be directly linked to the amino groups of a protein by reductive amination with cyanoborohydride anion[48] or can be oxidized and coupled to a protein.[49]

Lee et al.[50] developed neoglycoproteins based on the reaction of 2-imino-2-methoxyethyl-1-thioglycosides with the amino groups of a protein, leading to glycosyl-amidinated proteins. These neoglycoproteins were shown to react selectively with the liver lectin when the relevant specific sugar (galactose) was borne by the neoglycoprotein.[51]

Recently, we developed glycosylated polymers as specific biodegradable carriers of immunomodulators. These polymers are prepared by sequentially substituting poly-L-lysine with a sugar derivative such as hydroxysuccinimidyl α-D mannopyranosylphenylacetate ester and then with a biological response-modifier such as an active ester of N-acetyl muramyl dipeptide and finally with an acylating agent such as gluconolactone. Such a glycosylated polymer has a molecular mass of about 80,000, is strictly neutral, and displays a high specificity towards membrane lectins.[52] Neoglycoproteins, such as glycosylated serum albumin, containing about 20 sugar residues, or glycosylated polymers, containing 40 sugar residues, are quite useful for studying the specificity and the properties of endogenous membrane lectins because their apparent binding constant is much higher than that of the related free sugars or glycosides. Indeed, the concentrations required to inhibit the binding of a lectin to red blood cells are usually two or three orders of magnitude lower when neoglycoproteins or glycosylated polymers, instead of free glycosides, are used as agglutination inhibitors (TABLE 1).

Furthermore, neoglycoproteins can be easily labeled by either radioactive[15] or fluorescent probes.[20] The binding of neoglycoproteins to lectins increased relative to the number of bound sugars on the neoglycoprotein, but the binding is more specific when the number of sugars bound to a protein molecule is close to 20.[20]

Fluorescent neoglycoproteins[20,53] were used to visualize cell surface lectins and to study their binding, their internalization, and their intracellular degradation.[21] In order to select the more specific neoglycoproteins able to bind membrane lectins, cells are separately incubated in the presence of various fluoresceinylated neoglycoproteins either at 4°C or at 37°C for few minutes up to 2 hours. The cells are washed and the quantity of cell-associated neoglycoproteins is determined by using either a spectrofluorometric or a flow cytofluorometric method.

The spectrofluorometric method is based on a quantitative determination of the

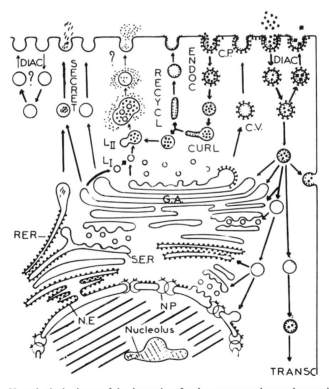

FIGURE 2. Hypothetical scheme of the dynamics of endogenous membrane glycoconjugates and lectins. CP = coated pit; CV = coated vesicle; CURL = compartment of uncoupling receptor ligand; DIAC = diacytosis; ENDOC = endocytosis; LI = primary lysosome; LII = secondary lysosome; NE = nuclear envelope; NP = nuclear pore; RECYCL = recycling; RER = rough endothelium reticulum; SECRET = secretion; SER = smooth endothelium reticulum; TRANSC = transcytosis.

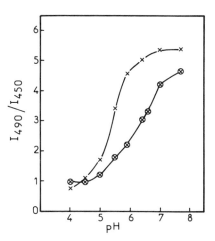

FIGURE 3. Relative fluorescence intensity ratio (I_{490}/I_{450}) at 520 mm emission wavelength of fluorescein bound to a neoglycoprotein before (⊗) and after (X) pronase treatment, upon excitation at 490 mm, and at 450 mm with regards to the pH of the solution at the time of spectrofluorimetric measurements.

component fluorescence upon solubilization of the cells in a borate buffer (pH 8.5) containing a surfactant (MAC 19s : $CH_2OH - (CHOH)_5 - CO - NH - (CH_2)_{12} - H$) which does not quench the fluorescein fluorescence.[20] Using this method, we showed that the neoglycoprotein bearing α-glucose is the best ligand of 3LL Lewis lung carcinoma cell lectin[20] and that mouse peritoneal macrophages internalize neoglycoproteins bearing α-mannose or β-N-acetyl-glucosamine, but not a protein substituted with N-acetyl-muramyl dipeptide.[54]

The flow cytofluorometric method is based on a quantitative determination of the fluorescence associated with single cells. Knowing (1) that endocytosed ligands are usually in an acidic environment (FIG. 2), (2) that the fluorescein quantum yield is maximum at neutral pH and decreases correlative with a decrease of pH from 7 to 4, and (3) that (*a*) the fluorescence quantum yield of fluorescein bound to a protein is lower than that of free fluorescein, (*b*) is dependent upon the number of fluorescein molecules linked to a protein, and (*c*) does not change upon acidification in the way free fluorescein does (FIG 3), we set up two quantitative methods allowing us to determine the binding, uptake, and intracellular degradation of neoglycoproteins. These methods require a standardization of the flow cytofluorometer[21] and either (1) the use of leupeptin (a permeant protease inhibitor) and of monensin (a permeant H^+/Na^+ ionophore), allowing the neutralization of intracellular microsomes[55] (TABLE 2), or (2) the use of two neoglycoproteins containing different numbers of fluorescein molecules.[56]

Using these methods, we showed that L1210 leukemic cells bind and internalize neoglycoproteins bearing α-L-fucose and to a lesser extent those bearing α-L-rhamnose or β-galactose but not that bearing α-galactose.[21] It was also shown that freshly isolated human monocytes only bind and internalize neoglycoproteins bearing 6-phosphomannose, but do not degrade the internalized neoglycoproteins, and that after three days in culture human monocytes bind, internalize and degrade neoglycoproteins bearing either α-mannose or 6-phosphomannose.[57] Recently, we showed that human colon adenocarcinoma cells primarily bind and internalize neoglycoproteins bearing either α-galactose, α-glucose or N-acetyl β-glucosamine.[58]

Fluoresceinylated neoglycoproteins may also be used to visualize endogenous lectins on tissue sections. For instance, a N-acetyl-β-glucosamine-specific lectin has been localized in the thyroid glands of various species.[59] An α-mannose-specific lectin and a β-galactose-specific endogenous lectin have been visualized in two distinct parts of a mesenteric lymph node.[5]

DRUG TARGETING: RATIONALE

The general aim in drug targeting is to increase the efficiency of a given drug by increasing the local drug concentration and/or decreasing its clearance rate. This statement applies to highly toxic drugs not usable as free drugs in clinics (such as diphtheria toxin and ricin) as well as for clinically used toxic drugs such as antitumor drugs (e.g., doxorubicin and methothrexate), antiviral drugs (e.g., arabinosyl cytosine

TABLE 2. Fluorescence Emissions at 520 mm of Cell-Associated Fluorescence upon Excitation at 450 nm and at 490 nm after a Postincubation in the Absence or Presence of Monensin

Incubation Temperature	Monensin	I_{490}	I_{450}	I_{490}/I_{450}
4°C	−	56	11	5.1
37°C	−	97	29	3.3
4°C	+	56	11	5.1
37°C	+	232	45	5.1

NOTE: I_{490} and I_{450} are the fluorescence emission intensities at 520 mm of Lewis lung carcinoma cells incubated for 2 hr in the presence of 100 μg/ml α-glucosylated fluorescein-labeled serum albumin minus the fluorescence intensities at 520 mm of cells incubated in the absence of fluorescein-labeled neoglycoprotein upon excitation at 490 mm and at 450 mm, respectively. In all cases, the cells were postincubated at 4°C in the absence or in the presence of 50 μM monensin.

[Ara C], arabinosyl adenosine [Ara A], 2'-deoxy 5-fluorouridine, and 2',3'-dideoxy 3'-azido thymidine) or antiparasite drugs (e.g., chloroquine and pyrimethamine).

For targeting biological response modifiers, a specific carrier is mainly needed to limit the clearance rate and to increase the uptake of small molecular weight modulators by specific cells. This is the case of macrophage activators such as N-acetyl muramyl dipeptide or lauroyltripeptide because macrophages have no receptor for such activators, and of anti-sens nucleotides because cells do not usually efficiently take up such compounds.

Various carriers have been proposed (for reviews see Refs. 60–62) such as (*a*) resealed red blood cells or ghosts; (*b*) microspheres or nanoparticles; (*c*) neutral or negatively charged liposomes; (*d*) genuine proteins or enzymatically/chemically modified serum albumin, glycoproteins, hormones, antibodies, and so forth; (*e*) polymers which are not easily biodegradable by mammalian cells (such as polyacrylamide and derivatives and polyvinylpyrrolidone); and (*f*) biodegradable polymers (such as polyesters, poly L-amino acids or polysaccharides).

The carriers may be directly substituted by the selected drug or are substituted by a spacer arm conjugated to the drug. Drugs containing a carboxylic group (e.g., methotrexate or indomethacine) may be conjugated to amino groups of the carriers; conversely, drugs containing an amino group may be linked to a carboxylic group of the carrier; in both cases, drug and carriers are associated through an amide linkage.

An hydroxyl group may be involved instead of an amino group, leading to ester linkages: carboxylic esters or phosphodiesters.

However, in many cases, this direct linkage leads to inefficient conjugates because the release of the drug is too slow or even does not occur. Consequently, it may be more appropriate to link the drug to a carrier through a spacer. Along that line, various spacers have been proposed. Among them, three classes of spacers are quite suitable because the spacer-drug linkage can be cleaved either by membrane-associated or lysosomal hydrolases, by an acidic environment, or by the reducing properties of the cytosol. The first class includes *peptides,* acting as substrates of various cathepsins or more generally intracellular proteases, *saccharides* as substrates of glycosidases, and *alkyl esters* as substrates of intracellular esterases. The second class includes acid-labile spacer-drug linkages: *cis*-aconityl, C-ribofuranosyl maleyl, hydrazone or thio-carbamyl acidic amino acyl derivatives. The third class includes spacer containing a disulfide bridge.

DRUG TARGETING: TARGETING TOXIC DRUG TO MEMBRANE LECTINS OF SPECIFIC CELLS

Site-specific drug delivery has been one of the main goals in therapeutics, since Paul Ehrlich's vision of targeted drugs as "magic bullets" for the eradication of diseases. Research in this field has expanded during the last several years thanks to a better understanding of the surface properties of different cell types and to the availability of monoclonal antibodies and other ligands to target drugs to specific cells bearing unique surface determinants. Toxic drug-antibody conjugates and immunotoxins have been extensively used. Along this line, we started to develop anthracycline carrier conjugates: DAC conjugate.[63] Even using selected monoclonal antibodies that specifically localized into the tumor,[64] *in vivo* experiments led to disappointing results.[65]

Membrane lectins of certain cells induce the internalization of their ligands[6,27] and therefore glycoconjugates specifically recognized by these lectins could be used as carriers of metabolite inhibitors and toxic drugs. Indeed, galactose-terminated glyco-proteins and neoglycoproteins have been used to carry antiparasitic drugs[66,67] or antiviral drugs[68–70] to parenchymal liver cells. Various antitumor drugs have been linked to glycoproteins, neoglycoproteins and to glycosylated polymers (Ref. 3 and references therein and Refs. 65, 69, and 71).

A neoglycoprotein bearing α-glucose was linked through a disulfide bridge, to gelonin, a toxic plant protein that inhibits protein synthesis in a cell-free system. This conjugate was one hundred times more toxic than free gelonin for 3LL Lewis lung carcinoma cells in culture.[20] The way the drug is bound to a carrier is quite important. Indeed, when gelonin is bound to a carrier by a noncleavable bridge, such as thioether linkage, the activity is usually dramatically reduced.

The efficacy of carrier-bound toxic drugs such as adriamycin, daunomycin, and primaquine is also quite dependent on the nature of the spacer. If the carrier is endocytosed and delivered to lysosomes, peptidic spacers are well adapted. As previously shown, daunomycin bound to wheat germ agglutinin through an Arg-Leu spacer (drug arm-carrier system) is even more efficient than free drug.[63] Other peptides such as Ala-Leu-Ala-Leu[66,67] or Gly-Phe-Leu-Gly[72] are also suitable. Daunorubicin bound to a neoglycoprotein bearing α-glucose through a dipeptide Tyr-Leu was

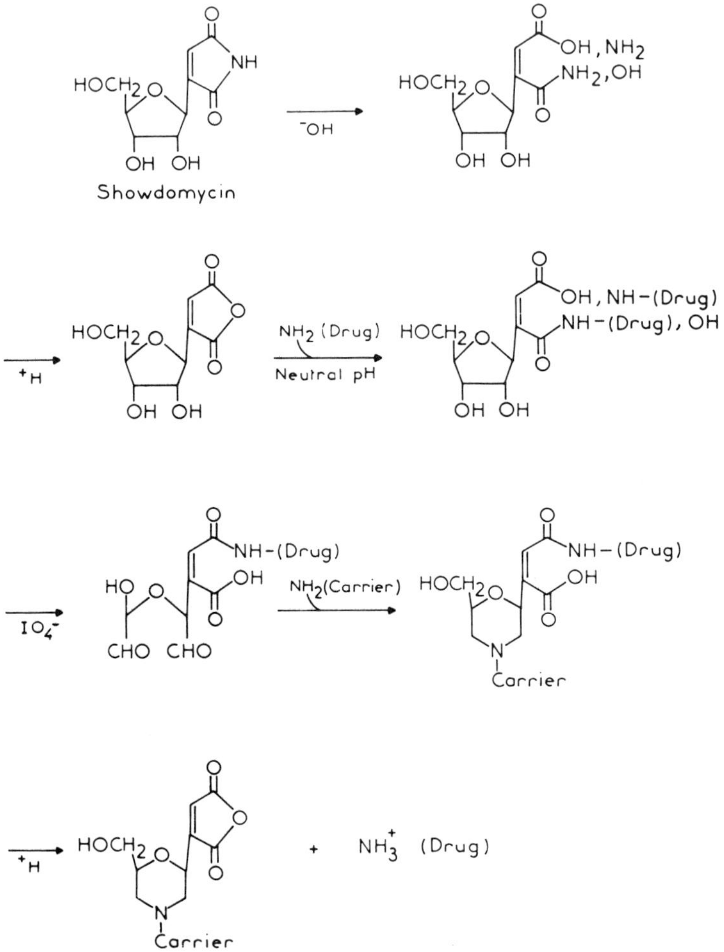

FIGURE 4. C-ribofuranosyl maleic anhydride as an acid-labile spacer: Showdomycin is hydrolyzed in alkaline medium and transformed to C-ribofuranosyl maleic anhydride in acidic medium. This compound is coupled with an amino group of a drug. The ribofuranosyl moiety is selectively oxidized by action of periodic acid and the aldehyde reacts with amino groups of a carrier. The drug acid-labile arm carrier conjugate is stable at neutral pH, and the drug is selectively released in slightly acidic medium.

three times more toxic towards cultured 3LL cells than was free Tyr-Leu-daunorubi-cin.[65]

The peptidyl drug can be linked to the carrier by using a heterobifunctional reagent such as ribosylthioethanoate.[73] However, carriers recognized by specific receptors and internalized may not be delivered to lysosomes. In such cases, an acid-labile spacer should be used because usually the pH inside endosomes is acidic. Several acid-labile spacers have been described including the maleyl derivatives *cis*-aconitic acid[74] or C-ribofuranosyl maleic acid (FIG. 4) as well as phenylthiocarbamyl derivatives.[75] Daunorubicin bound to a neoglycoprotein through the derivative of C-ribofuranosyl was shown to be stable at neutral pH and to release free and active daunorubicin at lower pHs (6 and below).

Methotrexate, bound to a neoglycoprotein bearing α-L-fucose was ten times more toxic for L1210 leukemia cells than was methotrexate bound to sugar-free serum albumin.[21] FUdr (5-fluorodeoxyuridine) bound to neoglycoproteins bearing α-glucose inhibited the growth of human colon carcinoma, whereas FUdr bound to a neoglyco-protein bearing α-mannose did not.[58] Daunomycin bound to glycosylated hydrophilic polymers (N-2-hydroxypropyl-methacrylamide copolymers) was more effective than that bound to nonglycosylated hydrophilic polymers.[72]

DRUG TARGETING: TARGETING OF IMMUNOMODULATORS TO MEMBRANE LECTINS OF MACROPHAGES AND MONOCYTES

Resident macrophages of mouse peritoneum, thioglycolate-elicited peritoneal macrophages, and macrophages derived from bone marrow in culture have specific receptors for mannose/fucose-terminated glycoproteins.[76] Mammalian lung macro-phages also bind, internalize, and degrade such glycoproteins as well as mannose-terminated macromolecules or neoglycoproteins.[77]

Uptake and transport of mannosylated ligands have been extensively studied[27]; mannosylated ligands are quickly transferred from endocytic vesicles to compartments of uncoupling receptor ligand (CURL), where receptor-ligand dissociation occurs, and to lysosomes. This suggests the potential efficiency of such a macromolecule to target drugs to macrophages which express mannose-specific receptors.

When macrophages were incubated with fluoresceinylated and mannosylated bovine serum albumin (F-Man-BSA), the fluorescence intensity of single cells was higher when the cells were incubated at 37°C to allow lysosomal degradation upon endocytosis than at 4°C, where only binding to the cell surface occurs.

When the cells incubated at 37°C were washed and further incubated at 4°C in the presence of monensin to equilibrate internal and external pH, the fluorescence intensity increases further, indicating that the endocytosed ligand was in acidic compartments.

Specificity of the binding was assessed by showing that binding and endocytosis of fluoresceinylated mannosylated serum albumin were inhibited by incubation in the presence of an excess of fluorescein-free mannosylated conjugate.

While mannose-specific receptors have been found on various macrophages,[78] human blood monocytes lack mannose-specific receptors; this receptor, however, appears after three days in culture.[79] Because circulating monocytes are precursors of tissue macrophages, it should be interesting to target immunomodulators to them; for this purpose, monocytes have been screened with a panel of neoglycoproteins to detect a putative membrane lectin.

Adherent cells obtained after mononuclear cell isolation of human blood through Ficoll-Paque, identified by using anti-monocyte monoclonal antibodies and corresponding to cells with a volume larger than lymphocytes, were labeled with F-(Man-6-P)-BSA, but not with F-Man-BSA.[57] Internalized F-(Man-6-P)-BSA is associated with endosomes (the pH of which is close to 5) since postincubation with monensin increased the fluorescence intensity; the extent of digestion of internalized neoglycoproteins can be assessed by comparing the cell-associated fluorescence of cells incubated in the absence and in the presence of a lysosomal protease inhibitor such as leupeptin. Freshly isolated monocytes have a low capacity to degrade endocytosed F-(Man-6-P)-BSA.[57] Therefore the drug should be linked by an acid-labile arm such as those cited above. The amount of F-(Man-6-P)-BSA endocytosed by matured monocytes is higher than that by freshly isolated monocytes and the capacity of the matured monocytes to degrade the ligand is also considerably increased.

Shepherd et al.[79] showed that rabbit alveolar macrophages express mannose-6-phosphate receptors. We have looked for the presence of such receptors on rat alveolar macrophages. Even at 4°C, F-(Man-6-P)-BSA gives a strong labeling compared to F-Man-BSA, which leads to a dull fluorescence. In order to know the effect of negative charges, we have used succinylated-F-Man-BSA; in this case the intensity of labeling is higher than with F-Man-BSA, but much lower than with F-(Man-6-P)-BSA.

Carriers able to specifically bind monocytes and macrophages receptors and to be internalized by adsorptive endocytosis could be suitable to selectively deliver immuno-stimulating drugs to these cells in order to induce their activation to a bactericidal or a tumoricidal state. As previously shown by Fidler[81,82] and Sone and Fidler,[83] MDP (N-acetyl muramyldipeptide, see Ref. 84) encapsulated within liposomes can be very efficient in rendering macrophages tumoricidal both in vitro and in vivo. The biological effect of free MDP depends on an internalization process based upon a fluid-phase nonspecific pinocytic phenomenon; the in vivo activation of tumor cytotoxic properties by liposome-encapsulated MDP does not appear to require participation of a putative cell surface receptor. As suggested by us[54] and confirmed by Fogler and Fidler[85] internalization of MDP is a prerequisite to the activation of macrophages. Furthermore, because the increased interleukin-1 secretion by peritoneal macrophages[54] induced by MDP is a saturable and concentration-dependent phenomenon, it is likely that specific receptors are localized inside the cell.

Because Man- and Man-6-P-specific receptors are present on most macrophages and mediate an active endocytosis of glycoconjugates containing mannose or mannose-6-phosphate in a terminal nonreducing position, muramyldipeptide bound to such glycoconjugates is internalized and released inside the macrophages with a high efficiency, leading to a very effective macrophage activation.

Experiments were conducted either with mouse thioglycolate-elicited peritoneal macrophages or with rat alveolar macrophages, which are known to be activated in vitro by free MDP.

MDP bound to Man-BSA rendered macrophages highly cytostatic and cytotoxic for tumor cells (3LL, L1210, P815); in comparison with free MDP, bound MDP was 100 times more efficient.[86,87] Mouse thioglycolate-elicited peritoneal macrophages incubated with 10 μg/ml free MDP led to a maximum of 30% cytotoxicity after a 24-hr stimulation and 72 hours of co-culture. MDP bound to BSA was slightly more active than free MDP because of its internalization by pinocytosis, but a similar concentration of MDP bound to Man-BSA led to 90% cytotoxicity and MDP bound to Man-BSA was still active at 0.1 μg/ml.

Interleukin-1 production increased upon incubation in the presence of Man-BSA or (Man-6-P)-BSA substituted with MDP. MDP bound to (Man-6-P)-BSA is much more active than MDP bound to Man-BSA in inducing the release of cytotoxic factors (unpublished data). In a similar way, rat alveolar macrophages were activated with a very low concentration of MDP bound to either Man-BSA or (Man-6-P)-BSA. All conjugates have about 20–25 sugar residues and 7 to 10 MDP residues bound per BSA molecule. The tumoricidal activation of macrophages is not LPS-dependent because experiments conducted in the presence or in the absence of polymyxin B gave identical results.

MDP[88] is known to increase hematopoiesis, so we have looked at the effect of the various MDP conjugates in comparison to free MDP. The effect of the drug was assessed by measuring the number of colonies in spleen after irradiation with cobalt-60. Mice received one injection (100 μg of conjugate corresponding to 5 μg of bound MDP or 5 μg MDP) before irradiation (day -7, -4, -1, or 0) or 1 day after; after ten days, mice were killed and the number of colonies was determined after spleen fixation (E-CFU). The neoglycoprotein without MDP was used as control. MDP bound to (Man-6-P)-BSA gave a significant increase of the E-CFU (15 compared to 3 in controls) and was more efficient than MDP bound to Man-BSA (in collaboration with M.L. Patchen, Bethesda, MD).

Although free MDP injected in mice has an adjuvant effect, no tumoricidal effect was detected, even when high concentration of MDP (200–500 μg per mouse) was injected.[81] On the contrary, the intravenous or intraperitoneal injection of MDP bound to Man-BSA, corresponding to 5–10 μg MDP, activated alveolar macrophages.[86,87]

This immunoactivator was also shown to reverse the development of lung metastases in more than two-thirds of mice upon intravenous injection in C57BL/6 mice in which a Lewis lung carcinoma primary tumor had been excised.[87] Because a similar result was obtained with beige mice, NK cells were not responsible for this improvement. We also showed that intraperitoneal injections of MDP-substituted mannosylated serum albumin led to the death of 100% of the mice bearing metastases, while the intravenous injection allowed long-term remission (more than 100 days) in 70% of the mice with metastases.[89]

More recently, using derivatives of poly L-lysine as a carrier, we found that the activity of the drug-arm-carrier conjugate was not only dependent on the nature of the sugar as signal recognition and of the drug as active molecule but also on the nature of the carrier. For instance, the kinetics of activation of macrophage by poly-L-lysine substituted with 40 mannose residues, 20 *N*-acetyl muramyl dipeptides, and 130 gluconoyl moieties was much faster than with MDP conjugated to mannosylated serum albumin. Furthermore, the secretion of tumor necrosis factor and/or interleukin-1 depends also on the nature of the carrier.[52]

CONCLUDING REMARKS

By targeting drugs, one may expect to specifically and directly act, on the one hand, on cells that are affected by a pathogenic agent or are pathogenic themselves, or to specifically act on cells such as macrophages, monocytes, and leukocytes in order to activate and induce them to become therapeutic agents.

In the case of targeting toxic drugs, the carrier should be as specific as possible. Glycosylated carriers may become efficient if they contain highly specific oligosaccharides able to bind selectively membrane lectins of tumor cells. To fulfill such a requirement, the structure of the actual high-affinity ligands of tumor cell membrane lectins has to be determined, and if such oligosaccharides are effectively found, it has yet to be demonstrated that normal cells have no membrane lectins able to bind such oligosaccharides. When such specific and selective carriers become available, the targeting of toxic drugs will be efficient only if all pathologic cells elicite a sufficient density of membrane lectins, if the membrane lectins are efficiently able to internalize the drug-carrier conjugate, and if the linkage between drug and carrier is able to release the toxic drug as an active drug in one of the intracellular compartments where the drug-carrier conjugate is transported.

Conversely, targeting of biological response-modifiers is more easily *attainable,* especially because biological response-modifiers may be not toxic at all or may have a very limited toxicity. Therefore, even when a part of the drug-carrier conjugate is taken up by cells that are not those that were expected, no deleterious effect will impair the efficacy of the drug. However, further work is required to achieve the controlled activation of monocytes and of macrophages and to extend this approach to the modulation of the activity of T-cytotoxic cells, T-suppressor cells, and NK cells.

In conclusion, targeting biological response-modifiers by means of glycosylated carriers is feasible and could become a new approach in the future therapy of certain diseases.

REFERENCES

1. GOLDSTEIN, I. J., R. C. HUGHES, M. MONSIGNY, T. OSAWA & N. SHARON. 1980. Nature **285:** 66.
2. MONSIGNY, M., C. KIEDA & A. C. ROCHE. 1979. Biol. Cell **86:** 289–300.
3. MONSIGNY, M., C. KIEDA & A. C. ROCHE. 1983. Biol. Cell **47:** 95–110.
4. MONSIGNY, M., C. KIEDA & A. C. ROCHE. 1984. *In* Cellular and Pathological Aspects of Glycoconjugate Metabolism. Colloq. INSERM-CNRS **126:** 357–372.
5. MONSIGNY, M., A. C. ROCHE, C. KIEDA, P. MIDOUX & A. OBRENOVITCH. 1988. Biochimie. In press.
6. ASHWELL, G. & J. HARFORD. 1982. Annu. Rev. Biochem. **51:** 531–534.
7. MONSIGNY, M., Ed. 1984. Biol. Cell **51:** 113–294.
8. SHARON, N. 1984. Biol. Cell **51:** 239–245.
9. OLDEN, K. & J. B. PARENT. 1987. Vertebrate Lectins, Van Nostrand. New York
10. LIENER, I. E., N. SHARON & I. J. GOLDSTEIN, Eds. 1986. The Lectins: Properties, Functions and Applications in Biological Medicine. Academic Press. New York.
11. BRANDLEY, B. K. & R. L. SCHNAAR. 1986. J. Leuk. Biol. **40:** 97–111.
12. LIS, H. & N. SHARON. 1986. Annu. Rev. Biochem. **55:** 35–67.
13. ROCHE, A. C. 1987. *In* Reconnaissance Moléculaire en Biologie INSERM. Colloque INSERM. In press.
14. DENNIS, J. W. & S. LAFERTE. 1987. Cancer Metast. Rev. **5:** 185–204.

15. ROGERS, J. C. & S. KORNFELD. 1971. Biochem. Biophys. Res. Commun. **45:** 622–627.
16. GABIUS, H. J. 1987. *In Vivo* **1:** 75–84.
17. GABIUS, H. J., K. VEHMEYER, R. ENGELHARDT, G. A. NAGEL & F. CRAMER. 1987. Cell Tiss. Res. **241:** 9–15.
18. LOTAN, R. & A. RAZ. 1988. J. Cell. Biochem. **37:** 107–117.
19. RAZ, A. & R. LOTAN. 1987. Cancer Metast. Rev. **6:** 433–452.
20. ROCHE, A. C., M. BARZILAY, P. MIDOUX, S. JUNQUA, N. SHARON & M. MONSIGNY. 1983. J. Cell Biochem. **22:** 131–140.
21. MONSIGNY, M., A. C. ROCHE & P. MIDOUX. 1984. Biol. Cell **51:** 187–196.
22. LOTAN, R., D. LOTAN & A. RAZ. 1985. Cancer Res. **45:** 4349–4353.
23. MEROMSKY, L., R. LOTAN & A. RAZ. 1986. Cancer Res. **46:** 5270–5275.
24. RAZ, A., L. MEROMSKY & R. LOTAN. 1986. Cancer Res. **46:** 3667–3672.
25. RAZ, A., L. MEROMSKY, I. ZVIBEL & R. LOTAN. 1987. Int. J. Cancer **39:** 353–360.
26. PAIETTA, E., R. J. STOCKERT, A. G. MORELL, V. DIEHL & P. H. WIERNIK. 1986. Proc. Natl. Acad. Sci. USA **83:** 3451–3456.
27. WILEMAN, T., C. CHARDING & P. STAHL. 1985. Biochem. J. **232:** 1–14.
28. SÈVE, A. P., J. HUBERT, D. BOUVIER, M. BOUTEILLE, C. MAINTIER & M. MONSIGNY. 1985. Exp. Cell Res. **157:** 533–538.
29. SÈVE, A. P., J. HUBERT, D. BOUVIER, C. BOURGEOIS, P. MIDOUX, A. C. ROCHE & M. MONSIGNY. 1986. Proc. Natl. Acad. Sci. USA **83:** 5997–6001.
30. BOURGEOIS, C. A., A. P. SÈVE, M. MONSIGNY & J. HUBERT. 1987. Exp. Cell Res. **172:** 365–376.
31. STOWELL, C. P. & Y. C. LEE. 1980. Adv. Carbohyd. Chem. Biochem. **37:** 225–281.
32. APLIN, J. D. & J. C. WRISTON, JR. 1981. Crit. Rev. Biochem. **10:** 259–306.
33. LEE, Y. C. & R. T. LEE. 1982. *In* The Glycoconjugates M. I. Horowitz Ed. Vol. 4, Academic Press, New York, pp 57–83.
34. GOEBEL, W. F. & O. T. AVERY. 1929. J. Exp. Med. **50:** 521–531.
35. AVERY, O. T. & W. F. GOEBEL. 1929. J. Exp. Med. **50:** 533–550.
36. GOEBEL, W. F. & O. T. AVERY. 1931. J. Exp. Med. **54:** 431–436.
37. IYER, R. N. & I. J. GOLDSTEIN. 1973. Immunochemistry **10:** 313–322.
38. PRIVAT, J. P., F. DELMOTTE & M. MONSIGNY. 1974. FEBS Lett **46:** 224–228.
39. KIEDA, C., F. DELMOTTE & M. MONSIGNY. 1977. FEBS Lett **76:** 257–261.
40. MONSIGNY, M., C. KIEDA, D. GROS & J. SCHRÉVEL. 1976. *In* VIth European Congress on Electron Microscopy, Jerusalem, Vol. **2:** 39–40. (Y. Ben-Schaul, Ed) TAL International Publishing Co. Israel.
41. MONSIGNY, M., C. KIEDA, A. OBRENOVITCH & F. DELMOTTE. 1976. *In* Protides of the Biological Fluids. H. Peeters, Ed.: 815–818. Pergamon Press, Oxford.
42. SCHRÉVEL, J., C. KIEDA, E. CAIGNEAUX, D. GROS, F. DELMOTTE & M. MONSIGNY. 1979. Biol. Cell **36:** 259–266.
43. BUSS, D. H. & I. J. GOLDSTEIN. 1968. J. Chem. Soc. C:1457–1461.
44. MACBROOM, C. R., C. H. SAMANEN & I. J. GOLDSTEIN. 1972. Meth. Enzymol. **28:** 212–219.
45. JEFFREY, A. M., D. A. ZOPF & V. GINSBURG. 1975. Biochem. Biophys. Res. Commun. **62:** 608–613.
46. BORCH, R. F., M. D. BERNSTEIN & H. D. DURST. 1971. J. Am. Chem. Soc. **93:** 2897–2904.
47. SMITH, D. F., D. A. ZOPF & V. GINSBURG. 1978. Meth. Enzymol. **83:** 169–171.
48. GRAY, G. 1974. Arch. Biochem. Biophys. **163:** 426–428.
49. ARAKATSU, Y., G. ASHWELL & E. A. KABAT. 1966. J. Immunol. **97:** 858–866.
50. LEE, Y. C., C. P. STOWELL & M. J. KRANTZ. 1976. Biochemistry **15:** 3956–3963.
51. KRANTZ, M. J., N. A. HOLTZMAN, C. P. STOWELL & Y. C. LEE. 1976. Biochemistry **15:** 3963–3968.
52. DERRIEN, D., P. MIDOUX, C. PETIT, V. PIMPANEAU, R. MAYER, M. MONSIGNY & A. C. ROCHE. 1988. Submitted for publication.
53. KIEDA, C., A. C. ROCHE, F. DELMOTTE & M. MONSIGNY. 1979. FEBS Lett **99:** 329–332.
54. TENU, J. P., A. C. ROCHE, A. YAPO, C. KIEDA, M. MONSIGNY & J. F. PETIT. 1982. Biol. Cell **44:** 157–164.
55. MIDOUX, P., A. C. ROCHE & M. MONSIGNY. 1987. Cytometry **8:** 327–334.

56. MIDOUX, P., A. C. ROCHE & M. MONSIGNY. 1986. Biol. Cell **58**: 221–226.
57. ROCHE, A. C., P. MIDOUX, P. BOUCHARD & M. MONSIGNY. 1985. FEBS Lett. **193**: 63–68.
58. GABIUS, H. J., R. ENGELHARDT, T. HELLMANN, P. MIDOUX, M. MONSIGNY, G. A. NAGEL & K. VEHMEYER. 1987. Anticancer Res. **7**: 109–112.
59. ALQUIER, C., R. MIQUELIS & M. MONSIGNY. 1988. Histochemistry. **89**: 171–176.
60. G. GREGORIADIS, J. SENIOR & G. POSTE, Eds. 1986. Targeting of Drugs with Synthetic Systems. Nato ASI Series. Plenum. New York.
61. DUNCAN, R. 1985. CRC Crit. Rev. Ther. Drug. Carrier Syst. **1**: 281–310.
62. VERT, V. 1986. CRC Crit. Rev. Ther. Drug. Carrier Syst. **2**: 291–327.
63. MONSIGNY, M., C. KIEDA, A. C. ROCHE & F. DELMOTTE. 1980. FEBS Lett. **109**: 181–186.
64. MIDOUX, P., T. MAILLET, F. THERAIN, M. MONSIGNY & A. C. ROCHE. 1984. Cancer Immunol. Immunother. **18**: 19–23.
65. DELMOTTE, F., P. J. LESCANNE, F. DAUSSIN, A. C. ROCHE, P. MIDOUX & M. MONSIGNY. 1985. Actual Chim. Ther. **12**: 129–140.
66. TROUET, A., R. BAURAIN, D. DEPREZ-DE-CAMPENEERE, M. MASQUELIER & P. PIRSON. 1982. *In* Targeting of Drugs. G. Gregoriadis, J. Senior & A. Trouet, Eds. :19–30. Plenum. New York.
67. TROUET, A., M. MASQUELIER, R. BAURAIN & D. DEPREZ-DE-CAMPENEERE. 1982. Proc. Natl. Acad. Sci. USA **79**: 626–629.
68. FIUME, L., C. BUSI, A. MATTIOLI, P. G. BALBONI & G. BARBANTI-BRODANO. 1981. FEBS Lett. **129**: 261–264.
69. FIUME L., C. BUSI, A. MATTIOLI, P. G. BALBONI, G. BARBANTI-BRODANO & Th. WIELAND. 1982. *In* Targeting of Drugs. G. Gregoriadis, J. Senior & A. Trouet, Eds. :1–17. Plenum. New York.
70. FIUME, L., B. BASSI, C. BUSI, A. MATTIOLI & G. SPINOSA. 1986. Biochem. Pharmacol. **35**: 967–972.
71. DELMOTTE, F., M. MONSIGNY, P. J. LESCANNE & F. DAUSSIN. 1985. French Patent No. 8508648.
72. DUNCAN, R., P. KOPECKOVA-REJMANOVA, J. STROHALM, I. HUME, H. C. CABLE, J. POHL, J. B. LLOYD & J. KOPECEK. 1987. Br. J. Cancer. **55**: 165–174.
73. MONSIGNY, M., F. DELMOTTE & A. C. ROCHE. 1983. French Patent No. 8314179; U.S. Patent No. 06646157.
74. SHEN, W. C. & H. T. P. RYSER. 1981. Biochem. Biophys. Res. Commun. **102**: 1048–1054.
75. DAUSSIN, F., E. BOSCHETTI, F. DELMOTTE & M. MONSIGNY. 1988. Eur. J. Biochem. **176**: 625–628.
76. STAHL, P. D. & S. GORDON 1982. J. Cell Biol. **93**: 49–56.
77. STAHL, P. D., J. S. RODMAN, M. J. MILLER & P. H. SCHLESINGER. 1978. Proc. Natl. Acad. Sci. USA **75**: 1399–1403.
78. STAHL, P. D., T. E. WILEMAN, S. DIMENT & V. L. SHEPHERD. 1984. Biol. Cell **51**: 215–218.
79. SHEPHERD, V. L., E. J. CAMPBELL, R. M. SENIOR & P. D. STAHL. 1982. J. Reticuloendothel. Soc. **32**: 423–431.
80. SHUR, B. D. 1982. Dev. Biol. **91**: 149–162.
81. FIDLER, I. J., S. SONE, W. E. FOGLER & Z. BARNES. 1981. Proc. Natl. Acad. Sci. USA **78**: 1680–1684.
82. FIDLER, I. J., Z. BARNES, W. E. FOGLER, R. KIRSH, P. BUGELSKI & G. POSTE. 1982. Cancer Res. **42**: 496–501.
83. SONE, S. & I. J. FIDLER. 1981. Cell Immunol. **57**: 42–50.
84. ADAM, A., J. F. PETIT, P. LEFRANCIER & E. LEDERER. 1981. Mol. Cell Biochem. **41**: 27–47.
85. FOGLER, V. E. & I. J. FIDLER. 1986. J. Immunol **136**: 2311–2317.
86. MONSIGNY, M., A. C. ROCHE & P. BAILLY. 1984. Biochem. Biophys. Res. Commun. **121**: 579–584.
87. ROCHE, A. C., P. BAILLY & M. MONSIGNY. 1985. Invasion Metastasis **5**: 218–232.
88. WUEST, B. & E. D. WACHSMUTH. 1982. Infection and Immunity **37**: 452–462.

89. ROCHE, A. C., P. MIDOUX, D. DERRIEN, C. PETIT, R. MAYER & M. MONSIGNY. 1988. *In* Advances in Biosciences, Vol. 68: 217–235. K. N. Mashi & W. Lange, Eds. Pergamon. Oxford.

DISCUSSION

R. LOTAN (*University of Texas, Houston, Texas*): Neoglycoproteins with Gal/ GalNAc might be removed from the circulation by hepatocyte asialoglycoprotein receptor. Can you use such neoglycoproteins as drug carriers?

MONSIGNY (*CNRS and Université d'Orléans, France*): Galactosylated (and *N*-acetylgalactosaminylated) polymers are rapidly taken up by parenchymal cells of the liver, upon intravenous injection. However, it is possible to inject toxic drug– neoglycoprotein conjugates inside the tumor itself; doing this greatly enhanced the efficiency of the conjugate. The glyco-conjugate helps the drug to be internalized and to be released, and lowers the rate of diffusion.

K. OLDEN (*Howard University Cancer Center, Washington, DC*): A useful neoglycoprotein would be one containing complex glycopeptides. Are you preparing such agents?

MONSIGNY: Endogenous lectins recognize oligosaccharides much better than monosaccharides. However, when an oligosaccharide has not exactly the right struc- ture, its affinity is very low. So the problem is to select the acutal oligosaccharide and to find a way to prepare it in large amounts, which is not an easy task. Furthermore, the actual ligand of an endogenous lectin may have an affinity high enough to be used as such to carry a drug. No doubt, in the future more elaborate neoglycoproteins will be shown to be more efficient than the simple neoglycoprotein we used.

C. FRANCESCHI (*University of Modena, Italy*): You can activate macrophages much better with your MDP-conjugates. Do you think that the effect of these components may be due, at least in part, to the fact that they, by being polymers, can alter receptors?

MONSIGNY: When we started working with MDP we looked for receptors on the surface of macrophages. Knowing that a small ligand bound to a macromolecule has an apparent affinity much higher than the free ligand (see TABLE 1 about the affinity of neoglycoproteins), we prepared a conjugate containing about 15 MDP molecules bound to serum albumin; this conjugate does not bind to macrophages and its ability to activate macrophages was not higher than that of free MDP. So far there are only two ways to make MDP very efficient to activate macrophages. One is by encapsulating MDP in liposomes, as done by Fidler and co-workers, in which liposomes fuse with the macrophage membrane and MDP is released in the cytosol. The second method is to prepare MDP-ligand conjugates such as MDP-neoglycoprotein. Mannosylated or 6-phosphomannosylated serum albumin or polymers bind and are endocytosed by macrophages, and MDP bound to such glycosylated macromolecules are up to 1000-fold more efficient than are free in *in vitro* experiments. Along the same line, we found that MDP bound to antitumor monoclonal antibodies are also very efficient in activating macrophages. This activation is quite specific because macrophages are activated by MDP-antibody conjugate-immune complexes not by the free MDP-antibody conjugate. Therefore MDP-antitumor antibody allows activation of macrophages that are mainly close to tumor cells.

A. CAPUTO (*Regina Elena Cancer Institute, Italy*): Endogenous lectins are transferred from the cell surface to the nucleus: what is the role of lectins in the nucleus?

MONSIGNY: We have no direct evidence that cell surface lectins are transferred to the nucleus. Nuclear lectins have been shown to be associated with DNA duplication on the one hand (see Olins *et al.* 1988. Biol. Cell), and with the synthesis of hnRNP on the other. The last result has been recently confirmed by Dr. J. Wang in Ann Arbor. Furthermore, S. Basu at Notre Dame, Indiana, has evidence that DNA polymerase is a glycoprotein. It is also known that several nonhistone proteins are glycoproteins. So nuclear lectins may be involved as regulatory molecules of the synthesis of DNA and RNA. Furthermore, we showed that the number of neoglycoproteins bound to the nucleus of dividing cells is about 10 times higher than that of neoglyproteins bound to the nucleus of contact arrested cells (see Seve *et al.* 1986. Proc. Natl. Acad. Sci. USA). This result has also been confirmed and extended by J. Wang using antibodies against a 34000 M_r galactose-specific lectin.

Therapeutic Strategies for Cancer Chemotherapy Based on Metabolic Consequences of DNA Damage[a]

NATHAN A. BERGER AND NORA HIRSCHLER

Hematology Oncology Division
Departments of Medicine and Biochemistry
R. L. Ireland Cancer Center
University Hospitals of Cleveland
Case Western Reserve University School of Medicine
Cleveland, Ohio 44106

Many chemotherapeutic agents cause DNA damage as their primary effect, yet the resultant lesions may not be the direct cause of tumor cell death. Studies in our laboratory suggest that in many cases chemotherapy-induced cell death may be the consequence of the metabolic effects of DNA damage.[1]

A major consequence of DNA damage in most eukaryotic cells is the utilization of NAD by the chromosomal enzyme, poly(adenosine diphosphoribose) polymerase (ADPRP).[1,2] The enzyme is chromatin-bound and usually exists in an inactive form.[3] It is activated by DNA strand breaks to cleave NAD at the glycosylic bond between the nicotinamide and adenosine diphosphoribose moieties.[3–5] The latter moieties are joined by the same enzyme into linear or branched chain polymers of ADP-ribose. Polymers have been identified in excess of 100 residues linked together by phosphodiester linkages alternating with ribose-ribose linkages.[3] The polymers are usually covalently attached to nuclear proteins such as the enzyme itself.[6] While its exact function in the DNA repair process is unknown, ADPRP is clearly activated by DNA strand breaks and the extent of its activation is proportional to the number of breaks.[6] The turnover of poly ADP-ribose is rapid and its half-life is brief—usually measured in minutes. We have shown that alkylating-agent-induced DNA damage can sufficiently activate the enzyme to deplete cellular NAD pools in 30–60 minutes.[1,2] Because NAD is required for enzymatic reactions involved in the generation of ATP, consumption of cellular NAD pools results in loss of ability to synthesize ATP, with consequent depletion of cellular ATP pools and loss of all energy-dependent functions. As outlined in FIGURE 1, this results in an inability of the cell to phosphorylate and use glucose.[7] Nucleoside- and deoxynucleoside-triphosphate pools decrease.[8] There is loss of the ability to conduct DNA, RNA, or protein synthesis.[1] Energy-dependent membrane pumps stop and membrane integrity cannot be maintained. Cells undergo marked swelling and lysis leading to cell death.

This mechanism appears to represent a final common pathway of cell death that is

[a]This research was supported by Grants CA-36983 and CA-43703 by the National Institutes of Health.

415

initiated by many diverse agents. It is actually a suicide pathway since the cell's own enzymes are active in causing its death. We have termed this pathway "type I cell death" to suggest that other biochemical pathways must exist by which chemotherapeutic agents kill cells.[1,9] These pathways are important to define and understand because they could provide a new basis for developing strategies for combination chemotherapy and/or biochemical modulation of chemotherapy. Thus, agents that interfere with pyridine nucleotide synthesis should sensitize cells to the NAD-lowering and cytotoxic effects of drugs that act by DNA strand disruptions and activation of ADPRP.

6-Aminonicotinamide (6-AN) and tiazofurin (TAZ) are each incorporated into analogues of NAD, and each interferes with the generation and maintenance of normal cellular pyridine nucleotide pools.[10–13] As false analogues of NAD, TAZ adenine dinucleotide and 6-aminonicotinamide adenine dinuleotide also function as inhibitors of pyridine-nucleotide-dependent dehydrogenases.[11–14] Their activities as competitive inhibitors are probably further enhanced by relative depletion of pyridine nucleotide pools. Each of these agents can be used to synergistically potentiate the tumoricidal activity of chemotherapeutic agents whose mechanism of action involves the production of DNA damage.[11,15]

As an example of the above pathway, we have shown that ADPRP is involved in steroid-induced cytotoxicity in steroid-sensitive tumor cells[16] by a mechanism that appears to involve the induction of a DNA endonuclease. Thus steroid treatment of sensitive cells leads to induction of DNA strand breaks,[17] activation of ADPRP, and

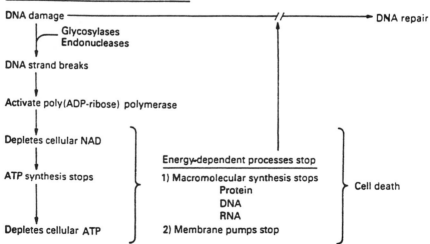

FIGURE 1. Schematic diagram of biochemistry of cell death initiated by DNA-damage-induced activation of poly(ADP-ribose) polymerase.

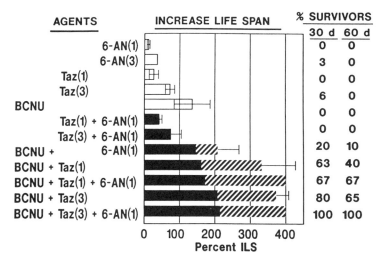

FIGURE 2. Mice were treated 25 hours after the intravenous administration of L1210 leukemia. The numbers in parentheses indicate the number of days each drug was given. Drugs given as single agents were all administered on the first day of therapy. Doses used were: TAZ, 12 mg/mouse; 6-AN, 0.4 mg/mouse; and BCNU, 0.4 mg/mouse. The *bar graphs* represent percent increase in life span (ILS). Values are expressed as the mean ± SD of at least three separate experiments with five mice in each group. *Open bars* indicate ILS for single agents; *solid bars* indicate additive effect of ILS expected from each combination determined by adding ILS produced by each individual agent. *Striped bars* represent synergistic ILS above that expected from the additive effects. The total bars (*solid and striped*) represent experimentally derived values. Long-term survivors of 30 and 60 days are indicated at right for each drug regimen.

depletion of NAD and ATP.[16] As predicted by the model of type I cell death, TAZ synergistically potentiates steroid-induced tumor cell killing.[16]

Another example using this strategy to enhance cytotoxicity for tumor cells *in vivo* is shown in FIGURE 2. In mice bearing the L1210 leukemia, administration of either TAZ, 6-AN, or BCNU as a single agent results in small increases in life span without significant increases in long-term survivors. Various combinations of TAZ and 6-AN are additive. However, combinations of BCNU with 6-AN produce synergistic increases in life span and an increase in long-term survivors at 30 and 60 days. Similarly, the combination of BCNU with TAZ results in a marked increase in life span and long-term survivors. Addition of 6-AN to the combination of BCNU and TAZ results in an even further increase in long-term survivors.[9]

The pathway of DNA strand breaks leading to activation of ADPRP, nucleotide depletion, loss of membrane integrity, and consequent cell death provides a mechanism to explain the interphase death that occurs in nonreplicating cells treated with a variety of chemotherapeutic agents. It may also contribute to the programmed cell death that occurs in the lymphoid system.[18] Recognition of this pathway as a mechanism of cell death suggests that different agents activating ADPRP or interfering with pyridine nucleotide metabolism may have synergistic, cytotoxic effects.[1,16] Thus, understanding biochemical pathways leading to cell death can provide the basis for developing new strategies for combination chemotherapy of malignant disease.

REFERENCES

1. BERGER, N. A. 1985. Poly(ADP-ribose) in the cellular response to DNA damage. Radiat. Res. **101:** 4–15.
2. SIMS, J. L., S. J. BERGER, & N. A. BERGER. 1983. Poly(ADP-ribose) polymerase inhibitors preserve NAD$^+$ and ATP pools in DNA-damaged cells: Mechanism of stimulation of unscheduled DNA synthesis. Biochemistry **22:** 5188–5194.
3. HAYAISHI, O. & K. UEDA. 1977. Poly(ADP-ribose) and ADP ribosylation of proteins. Annu. Rev. Biochem. **46:** 95–116.
4. BERGER, N. A., G. W. SIKORSKI, S. J. PETZOLD, & K. K. KUROHARA. 1979. Association of poly(adenosine diphosphoribose) synthesis with DNA damage and repair in normal human lymphocytes. J. Clin. Invest. **63:** 1164–1171.
5. COHEN, J. J., D. M. CATINO, S. J. PETZOLD, & N. A. BERGER. 1982. Activation of poly(adenosine diphosphate ribose) polymerase by SV 40 minichromosomes: Effects of DNA damage and histone H1. Biochemistry **21:** 4931–4940.
6. CARTER, S. G. & N. A. BERGER. 1982. Purification and characterization of human lymphoid poly(adenosine diphosphate ribose) polymerase. Biochemistry **21:** 5475–5481.
7. BERGER, S. J., D. C. SUDAR, & N. A. BERGER. 1986. Metabolic consequences of DNA damage: DNA damage induces alterations in glucose metabolism by activation of poly(ADP-ribose) polymerase. Biochem. Biophys. Res. Commun. **134:** 227–232.
8. DAS, S. K. & N. A. BERGER. 1986. Alterations in deoxynucleoside triphosphate metabolism in DNA damaged cells: Identification and consequences of poly(ADP-ribose) polymerase dependent and independent processes. Biochem. Biophys. Res. Commun. **137:** 1153–1158.
9. BERGER, N. A., S. J. BERGER, & S. L. GERSON. 1987. DNA repair, ADP-ribosylation and pyridine nucleotide metabolism as targets for cancer chemotherapy. Anti-Cancer Drug Design. **2:** 203–210.
10. DIETRICH, L. S., I. M. FRIEDLAND, & L. A. KAPLAN. 1958. Pyridine nucleotide metabolism: Mechanism of action of the niacin antagonist, 6-aminonicotinamide. J. Biol. Chem. **233:** 964–968.
11. BERGER, N. A., S. J. BERGER, D. M. CATINO, S. J. PETZOLD, & R. K. ROBINS. 1985. Modulation of nicotinamide adenine dinucleotide and poly(adenosine diphosphoribose) metabolism by the synthetic 'C' nucleoside analogs, Tiazofurin and Selenazofurin. A new strategy for cancer chemotherapy. J. Clin. Invest. **75:** 702–709.
12. COONEY, D. A., H. N. JAYARAM, G. GEBEYEHU *et al.* 1982. The conversion of 2-β-D-ribofuranosylthiazole-4-carboxamide to an analogue of NAD with potent IMP dehydrogenase-inhibitory properties. Biochem. Pharmacol. **31:** 2133–2136.
13. BERGER, S. J., I. MANORY, D. C. SUDAR, D. KROTHAPALLI, & N. A. BERGER. 1987. Pyridine nucleotide analog interference with metabolic processes in mitogen stimulated, human T lymphocytes. Exp. Cell Res. **173:** 379–387.
14. KOLBE, H., K. KELLER, K. LANGE, & H. HERKEN. 1976. Metabolic consequences of drug-induced inhibition of the pentose phosphate pathway in neuroblastoma and glioma cells. Biochem. Biophys. Res. Commun. **73:** 378–382.
15. BERGER, N. A., D. M. CATINO, & T. J. VIETTI. 1982. Synergistic antileukemic effect of 6-aminonicotinamide and 1,3-bis(2-chloroethyl)-1-nitrosourea on L1210 cells *in vitro* and *in vivo.* Cancer Res. **42:** 4382–4386.
16. BERGER, N. A., S. J. BERGER, D. C. SUDAR, & C. W. DISTELHORST. 1987. Role of nicotinamide adenine dinucleotide and adenosine triphosphate in glucorticoid-induced cytotoxicity in susceptible lymphoid cells. J. Clin. Invest. **79:** 1558–1563.
17. COMPTON, M. M. & J. A. CIDLOWSKI. 1986. Rapid *in vivo* effects of glucocorticoids on the integrity of rat lymphocyte genomic deoxyribonucleic acid. Endocrinology **118:** 38–45.
18. CARSON, D. A., S. SETO, D. B. WASSON, & C. J. CARRERA. 1986. DNA strand breaks, NAD metabolism and programmed cell death. Exp. Cell Res. **164:** 273–281.

DISCUSSION

C. S. FOSTER (*Hammersmith Hospital, London, England*): Several years ago, Otto Warburg indicated compromised oxidative phosphorylation in tumor cells. Do you think your observations represent the underlying mechanism for Warburg's observations or are there additional metabolic abnormalities in tumor mitochondria?

N. A. BERGER (*Case Western Reserve University, Cleveland, Ohio*): The altered oxidative phosphorylation in tumor cells may provide an advantage in trying to selectively interfere with energy metabolism in tumor cells as opposed to normal cells.

B. CHANCE (*University of Pennsylvania, Philadelphia, Pennsylvania*): Please explain how nuclear and cytosolic depletion of NAD will immediately affect the mitochondrial pool of NAD? We find that nicotinamide supplement to the diet affects the nuclear/cytosolic pool and not detectably the mitochondrial pool. Can you use NADH fluorescence to resolve this dilemma?

BERGER: Our studies have focused on total cellular NAD and ATP pools and we have shown that these pools can be rapidly and drastically depleted by the activation of poly (ADP-ribose) polymerase in response to DNA damage. We have no direct information on the time required for equilibration of the mitochondrial pools with the alterations in the cytoplasm. These, of course, would be interesting experiments to do.

C. FRANCESCHI (*University of Modena, Italy*): What happens to DNA breaks when cells are protected by NAM? How do you explain the synergistic effect of tiazofurin and 6-aminonicotinamide?

BERGER: At low levels of DNA damage nicotinamide inhibition of poly (ADP-ribose) polymerase tends to slow down the rate of repair of DNA strand breaks. At high levels of DNA damage, where activation of poly (ADP-ribose) polymerase can deplete pools of NAD + ATP, the use of nicotinamide to inhibit enzyme activity tends to preserve NAD + ATP pools and thereby facilitate DNA strand break repair. The answer to your second question is that tiazofurin and 6-aminonicotinamide are not synergistic with each other. In our system, they showed synergistic therapeutic effects when combined with a DNA-damaging agent such as BCNU. The latter synergy is explained by the ability of these compounds to interfere with synthesis of NAD. The lowered NAD pools interfere with the cell's ability to repair the BCNU-induced DNA damage. It also contributes to more rapid depletion of NAD + ATP pools.

H. C. BIRNBOIM (*Ottawa Regional Cancer Center, Ottawa, Canada*): Does nicotinamide protect cells by serving as substrate for NAD^+ synthesis or by inhibiting poly-ADPR polymerase and preventing the drastic drop in cellular NAD^+? There is human experience with high doses of nicotinic acid (for lowering cholesterol levels). Would this experience be of use to you in designing therapeutic strategies based upon perturbing NAD^+ metabolism?

BERGER: Nicotinamide has both effects. For short time periods it is probably the enzyme inhibition that is most important since 3-amino benzamide, which inhibits the enzyme but does not serve as precursor for NAD synthesis, is even more effective than nicotinamide. Nicotinic acid does not inhibit the enzyme and is not a very effective protector against rapid depletion of NAD. In more long-term experiments, nicotinic acid may be protective by increasing initial levels of NAD.

P. A. RILEY (*University College, London, England*): Is there any difference between single- and double-strand breaks in activating the poly(ADP-ribose) polymerase system?

BERGER: Double-stranded breaks appear to be more effective than single-stranded breaks in activating poly(ADP-ribose) polymerase and flush ends of a double-stranded break are more effective than staggered ends.

M. MIRANDA (*University of L'Aquila, Italy*): Have you any data about DNA-ligase activity in your experimental systems? What are the actions on this enzyme of NAD^+, ATP and ADP ribosylation?

BERGER: We have no direct data on DNA-ligase from our experiments. This is a somewhat controversial area, with some laboratories reporting that DNA damage stimulates the ADP-ribosylation and activation of DNA ligase, and other laboratories reporting that ADP-ribosylated proteins interfere with DNA-ligase function.

Use of Antiadhesive Peptide and Swainsonine to Inhibit Metastasis[a]

KENNETH OLDEN,[b] SURESH MOHLA,[b]
SHEILA A. NEWTON,[b] SANDRA L. WHITE,[b,c]
AND MARTIN J. HUMPHRIES[b]

[b]*Howard University Cancer Center, and*
[c]*Department of Microbiology*
Howard University College of Medicine
Washington, D.C. 20060

INTRODUCTION

Metastasis, the spread of tumor cells from a primary to a secondary site within the body, is the principal cause of morbidity and death in cancer patients. Consequently, there is great interest in understanding the biology and biochemistry of the metastatic process and in developing therapeutic strategies for the prevention and/or elimination of metastatic lesions. The process of tumor metastasis comprises a sequential series of migratory and invasive events that necessitate the passage of cells through both connective and vascular environments.[1-4] Thus, malignant cells must possess a phenotype that allows them to perform a multitude of functions.

In this chapter, we will review our preclinical findings which suggest that the employment of antimetastatic agents, developed on the basis of knowledge of the cell biology and biochemistry of the metastatic process, may be clinically efficacious in the management of human malignancies.[5-13]

POTENTIAL UTILITY OF ANTIMETASTATIC AGENTS IN MANAGEMENT OF MALIGNANCIES

A frequent criticism of research on the development of antimetastatic agents is that the effect of such drugs is likely to be marginal, limited to the rare cases where metastasis has not yet occurred. However, there are clinical situations where antagonists of inter-organ tumor cell spread could be of utility. Perhaps the most prominent example would be the administration of an antimetastatic agent in a prophylactic or adjuvant setting in patients undergoing surgical resection of their primary tumor. Dissemination of tumor cells due to mechanical manipulation and failure of inexperienced surgeons to remove all of the primary tumor probably accounts for a significant number of cancer deaths resulting from secondary metastasis.[14] Consistent with this

[a]This work was supported by Grants CA-14718 and CA-34918 from the U.S. Public Health Service, Grant PDT-312 from the American Chemical Society, and by a Howard University Faculty Research Support Grant.

421

view, large numbers of tumor cells are found in the venous blood during the manipulation of tumors at surgery, and a significant fraction of these circulating tumor cells subsequently lodge in the arterioles or capillaries of the first organ that they encounter. In fact, clinical trials involving the use of an antimetastatic agent, as an adjuvant to surgery for colorectal cancer, have shown that the survival or disease-free interval is significantly increased relative to that of untreated control patients.[15] Additionally, the chronic or long-term administration of a nontoxic antimetastatic agent would be useful in the management of patients presumed to be cured of their primary neoplasm, especially if these patients had no disseminated disease at the time of resection of their primary tumor. If malignant cells are left in the tumor bed that can proliferate and give rise to subsequent metastasis, an antimetastatic agent would limit the residual or recurring disease to the primary site. Another example of the practical application for antimetastatic agents would be in the area of chemoprevention for high-risk populations. While the chronic administration of such an agent would not prevent the formation of more readily treatable primary tumors, it would dramatically increase cancer survival statistics by eliminating death from disseminated disease. For example, such a preventive regimen might be recommended for women with a family history of breast cancer. Similar chemopreventive measures are now widely used for decreasing mortality from cardiovascular-system-related diseases.

The control of metastasis by means of antimetastatic drugs is a realistic objective since some of the biochemical events crucial for metastasis formation are likely to be similar for the various kinds of cancer. Also, it is time that more innovative and rational strategies, based on a detailed understanding of the biochemical events involved in the metastatic process, be pursued since their potential for therapeutic success is likely to be superior to that achieved by the random, semi-empirical screening approaches used in the past.

APPROACHES USED IN OUR LABORATORY TO INHIBIT METASTASIS

We have used three experimental approaches to inhibit experimental metastasis:

(1) co-injection of tumor cells with an antiadhesive peptide (Gly-Arg-Gly-Asp-Ser) previously reported to inhibit fibronectin-mediated cellular interactions[3,5–7,9,12,13];

(2) *in vitro* treatment of tumor cells with swainsonine or castanospermine prior to injection into recipient mice; and

(3) systemic administration of swainsonine or castanospermine to recipient mice.

The logic for employment of these approaches is based on findings which suggest that surface molecules, such as fibronectin and N-linked oligosaccharides, are involved in metastasis. For example, enzymatic modification of surface macromolecules alters the implantation and colonization potential of metastatic cells,[2] and transfer of plasma membrane vesicles from high- to low-metastatic variant cell lines significantly enhanced the colonizing activity of the low-metastatic lines.[92]

THE METASTATIC CASCADE

The dissemination of malignant cells from a primary tumor to secondary sites involves a complex series of events, referred to as the metastatic cascade, that must be successfully negotiated.[1-4,15] As the solid primary tumor becomes an invasive, malignant neoplasm, it becomes more highly vascularized, and often begins to grow more rapidly, invading surrounding tissue. Malignant tumor cells spread to distant sites (i) by direct extension into the body cavities, where the released tumor cells can seed onto tissue surfaces and develop new growth, or (ii) by invasion into blood vessels or lymphatic channels, where individual cells or clumps of cells may be released. After entry into the circulation, tumor cells have a tendency to form emboli by interaction with other tumor cells and blood components, especially platelets and fibrin.[16-19] The emboli lodge, by nonspecific entrapment or specific adhesion, in the capillary beds of distant organs or lymph nodes, where they penetrate the capillary endothelium/ basement membrane of susceptible organs and proliferate to form metastatic colonies.

From the above description, it is apparent that several requirements must be met in order for metastasis to occur. The first requirement is that tumor cells must detach from the primary tumor mass and enter the circulatory system. While the biochemical details of detachment and intravasation are not well understood, these processes probably require a combination of degradative and active migratory and invasive events.[20-23] The second requirement is that the tumor cells survive in circulation. In fact, the vast majority of cells that enter the circulatory system are destroyed by a combination of hemodynamic forces during the passage of blood through narrow capillary beds[24] and by host immune effector cells, such as the natural killer (NK) lymphocytes.[25,26] Survival in circulation and organ colonization potential of tumor cells are enhanced by the tumor cells' adherence both to other tumor cells and to blood cells, especially platelets.[18,19] The larger the embolus, the more likely it is that it will become lodged in capillary beds, by both specific and nonspecific mechanisms, where extravasation or invasion can occur.[16,17,27] Also, the aggregation of tumor cells to form emboli renders them more resistant to immune surveillance.[25,26]

The successful completion of the metastatic cascade, culminating in the establishment of metastases, is dependent on membrane properties or interactions unique to metastatic progenitor cells. These interactions are mediated by specific cell surface receptors, the identity of which is beginning to be determined, primarily through the use of molecular probes that perturb cell adhesion and tumor spread.[5-13,18-35] However, the exact function and mechanism of action of these cell surface molecules are yet to be determined.

Since each step of the complex metastatic cascade must be successfully negotiated for metastasis to occur, it is not surprising that the process is so inefficient. For example, of approximately 10^5–10^7 cells shed from a primary mouse tumor per day,[17,36] fewer than 1 percent of these give rise to metastases.[17,28,37] The inefficiency of the process can be exploited for therapeutic intervention; it is plausible that by blocking a single crucial step (e.g., the interaction of tumor cells with the extracellular matrix or basement membrane) the entire metastatic process would be unsuccessful.

FIBRONECTIN IN CELL RECOGNITION AND ADHESION

While the primary mechanism of tumor cell arrest in the target organ may be facilitated by nonspecific trapping of emboli, it is likely that metastasizing cells interact with specific adhesive proteins in the endothelial basement membrane layer or especially in the subendothelial extracellular matrix. In fact, an examination of the steps involved in the metastatic cascade (TABLE 1) reveals that there are multiple interactions of tumor cells with extracellular matrices. Over the past 15 years, a number of prominent connective tissue and basement membrane proteins have been identified whose localization *in vivo* would be consistent with a role in tumor cell adhesion. These include fibronectin, laminin, vitronectin/serum-spreading factor, fibrinogen, type IV and V collagen, and heparan sulfate proteoglycans. The majority of research on these proteins has been concerned with elucidation of the biochemical mechanisms by which they promote cell attachment *in vitro,* and subsequent development of antiadhesive agents that could be used in probing the functions of the individual factors *in vivo.*[3,4,12,13,21,34,35,41–43]

Fibronectin, possibly the best characterized extracellular matrix molecule,[3,4,12,13,21,34,45,41–47] has been shown to promote cell adhesion and migration during embryonic development, wound healing, and extracellular matrix assembly.[46,48,49] It is a large glycoprotein composed of two similar, disulfide-linked subunits of molecular weight 250–280 kDa (FIG. 1). The various binding activities of fibronectin (FIG. 1), located in distinct domains of the protein, were identified by limited proteolysis and subsequent bioassays or affinity chromatography.[3,4,12,13,21,34,35,41–58] An 11.5-kDa proteolytic fragment was isolated and sequenced, and overlapping peptides were synthesized to identify the major cell-binding amino acid sequence consisting of Gly-Arg-Gly-Asp-Ser (GRGDS).[50–58] The GRGDS pentapeptide was found to be an effective inhibitor of fibronectin-mediated adhesion.[5,6,44–58] Studies with radiolabeled fibronectin demonstrated that synthetic GRGDS inhibited fibronectin binding in a reversible, competitive manner.[56–58] While a second cell-binding domain, referred to as the type III connecting segment, has been shown to exist in the spliced region of the molecule, most cell types, including tumor cells, adhere to the central cell-binding domain.[49] The type III CS sequence is specific for cells derived from neural crest cells.

Recently, the receptor-protein complex recognizing the RGDS cell-binding sequence of fibronectin has been identified, purified, and sequenced.[59–67] The avian species exists as a polypeptide complex with subunit molecular weights of 155–160 kDa, 135 kDa, and 110–120 kDa bands,[63–65] whereas the mammalian form exists as a complex of two dissimilar, noncovalently linked subunits of molecular weight 140–160 kDa and 115–135 kDa.[68–73] Microinjection of the GRGDS peptide, which competitively inhibits receptor function, interrupts gastrulation and migration of neural crest cells in amphibians and in *Drosophila.*[74,75] Similarly, antibodies to the fibronectin receptor block gastrulation[76] and cell migration.[77–79]

The first indication that fibronectin was influencing tumorigenicity or malignancy was derived from indirect or correlative data. For example, it was observed that there is

TABLE 1. Events Involved in the Metastatic Cascade

- Invasion of normal tissue
- Detachment from primary tumor
- Intravasation into microvasculature or lymphatics
- Hematogenous dissemination with formation of emboli with host platelets and tumor cells
- Extravasation through the vessel wall into tissue of secondary arrest
- Proliferation or growth into a secondary metastatic tumor
- Angiogenesis or formation of blood supply

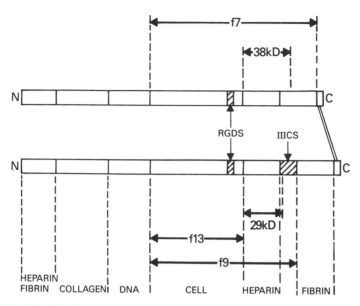

FIGURE 1. Map of the human plasma fibronectin dimer showing the derivation of the fragments used in this study. The sites of the RGDS tetrapeptide and the IIICS region are highlighted in the hatched areas. The binding specificities of the different fibronectin domains are listed under the diagram. The amino- and carboxyl-termini of the fibronectin chains are indicated by N and C, respectively.

a quantitative decrease in the the level of surface fibronectin on several virally transformed fibroblasts,[80-89] that fibronectin is present in and around tumors, and that it promotes the adhesion of both normal and transformed cells in culture.[34-43] With the recent development of specific peptides and monoclonal antibodies that block the adhesive activity of fibronectin, the hypothesized involvement in metastasis and invasion can now be examined.

THE EXPERIMENTAL METASTASIS MODEL

To investigate the biochemical events taking place during the seeding of metastatic cells and the effectiveness of new therapeutic agents, one must select an animal model that is fairly representative of the metastatic process in cancer patients. Such models have proven to be invaluable in the elucidation of various host factors and tumor properties involved in the pathogenesis of metastasis. Attempts to mimic the complete metastatic cascade from growth of the primary tumor, through vascular and/or lymphatic dissemination, to growth of secondary colonies, have met with limited success. The technique of subcutaneous implantation of cultured cells or tumor fragments into mice and other rodents has a number of associated problems that are not easy to resolve. These include (i) the inadvertent introduction of tumor cells into blood vessels of the host during the implantation process that may lead to false metastasis, and (ii) the administration of tumor cells into areas of the body where tumors of the same lineage would not normally develop. In the latter situation, the tumor cells may encounter microenvironments that could alter their propensity to metastasize.

In the present study, we have used the B16-F10 murine melanoma model developed by Fidler.[90,91] The cells were propagated in tissue culture, detached, and injected into the lateral tail vein of syngenic mice (C57BL/6) with or without the pentapeptide GRGDS. Two to three weeks later the animals were sacrificed and examined for evidence of metastasis. Metastasis was quantitated by determining the number of melanotic lesions in the lungs. By selection of the appropriate tumor cell inoculum, the assay was also used to examine animal survival and the potential clinical efficacy of the particular treatment. Intravenous administration of tumor cells has the disadvantage of bypassing the initial intravasation phases of the metastatic cascade from development of the primary malignant neoplasm to escape of metastatic cells into the blood vascular system. However, once in the bloodstream, the experimental metastasis model employed here reproduces the events occurring during authentic metastasis, including evasion of host defenses, interaction with blood cells and protein components, arrest in the microvasculature, extravasation, and proliferation to form metastatic foci.

EFFECT OF THE PENTAPEPTIDE GRGDS ON METASTASIS

Premixing and co-injection of GRGDS, the principal receptor-recognition site in the central cell-binding domain of fibronectin, with 50,000–100,000 B16-F10 melanoma cells elicited a dose-dependent inhibition of pulmonary metastasis.[5] The formation of visible tumor nodules was consistently inhibited by >90% at a peptide concentration of 3 mg per mouse. The injection of fewer cells resulted in complete inhibition of colonization and normal survival of recipient mice, demonstrating the potential clinical efficacy of GRGDS. The suppression of tumor formation in peptide-treated mouse lungs was confirmed by histologic sectioning and staining, thereby ruling out the possibility that GRGDS only blocks expression of the melanotic phenotype. Also, the inhibition of lung colonization did not result in metastasis to other organs, indicating that GRGDS did not merely alter target-organ specificity.

The specificity of GRGDS inhibition of metastasis was examined by comparison with other homologous peptides with respect to their antiadhesive and antimetastatic activities. These included GRGES, the prototype control peptide in which glutamic acid substitutes for the aspartic acid residue at position 4; GRDGS, in which the central glycine and aspartic acid residues are transposed; GRGD, which lacks the COOH-terminal serine residue; RGDS, which lacks the NH_2-terminal glycine residue; and SDGR, in which the sequence of the amino acids is reversed relative to RGDS. We found that there was a direct correlation between the capacity of peptides to disrupt cell adhesion *in vitro* and their relative inhibiting activities in the experimental metastasis assay.[5,6] For example, RGDS and SDGR were active in both assays, but less so than GRGDS; conversely GRGES, GRDGS and GRGD were either minimally active or completely inactive in inhibiting cell adhesion and metastasis. We therefore concluded that the antimetastatic activity of GRGDS is highly specific, since conservative amino acid substitutions or sequence alterations resulted in a dramatic loss of activity.

MECHANISM OF GRGDS ACTION

Because it was not possible to dissociate the antiadhesive and antimetastatic activities of GRGDS, we strongly suspected that the peptide blocked experimental metastasis through its ability to disrupt cell adhesive interactions. There are potentially many steps in the metastatic cascade at which GRGDS-dependent adhesion could be involved; for example, during homotypic and heterotypic cell-cell aggregation to form emboli, during specific arrest in the microvasculature, during adhesion and migration through basement membranes, and during invasion of subendothelial connective tissue.

As an initial step to investigate the mechanism involved in GRGDS-mediated inhibition of pulmonary colonization, the retention of [^{125}I]iododeoxyuridine-labeled B16-F10 cells in the lungs of C57BL/6 mice was monitored as a function of time after injection. In these studies, GRGDS rapidly promoted the loss of tumor cells from the lung[5,6]; the rate of loss was approximately twice as fast in mice administered the peptide. The enhanced clearance of cells from the lungs in the presence of GRGDS could account for almost all of the decrease in colony number caused by this peptide (e.g., 7 hours after injection of the tumor cells, the level of radioactivity in the lungs of peptide-treated mice was five-fold lower than in untreated mice[5]).

Recognizing that GRGDS might also affect crucial molecular events other than cell adhesion, we examined the possibility that the antimetastatic activity is due to inhibition of platelet aggregation and/or enhancement of NK cell activity by comparison of peptide activity in normal mice versus mice defective in either platelet or NK cell function. Thrombocytopenia was induced by injection of rabbit anti-mouse platelet antiserum, whereas mutant beige mice, which lack factors required for NK killing, were used as models for NK deficiency. The antimetastatic activity of GRGDS was fully retained in both types of animals,[6] indicating that neither platelets nor NK cells are required for expression of the antimetastatic activity of GRGDS.

EFFECT OF GRGDS ON METASTASIS OF HUMAN TUMORS

To determine whether GRGDS will also inhibit the metastasis of human tumors, we used the highly malignant, estrogen-receptor-negative MDA-MB-231 human breast cancer cell line in the above cell adhesion and experimental metastasis assays. Both adhesion and lung colonization after intravenous injection into athymic nude mice were dramatically inhibited.

OVERVIEW

The effectiveness of GRGDS treatment indicates an important role for fibronectin in tumor metastasis, and implies that adhesive interactions involving this sequence occur at least once during metastasis and are crucial for its successful completion. Inhibition of metastasis by GRGDS was not an artifact caused by cytotoxicity or suppression of tumorigenicity,[5] and the peptide possessed almost equal inhibiting activity when cells and peptide were injected into different tail veins without premixing.

Although GRGDS appears to block metastasis through its ability to interfere with tumor cell adhesion, the actual site(s) of action of the peptide awaits identification. Although the involvement of fibronectin in the adhesion of neoplastic cells *in vivo* had been hypothesized for some time, the most direct evidence is provided by our studies. However, the definitive assignment of function to a particular extracellular matrix molecule is uncertain since other glycoproteins (collagens, fibrinogen, vitronectin, thrombospondin, and von Willebrand factor[45,47]) contain an RGD sequence. This is of particular concern since such small peptides may not always possess complete specificity for the adhesive ligand from which they are derived. In any event, metastasis is unlikely to be due to adhesion to a single basement membrane glycoprotein, and inhibition of a single interaction may not completely abrogate metastasis. Studies focusing on a single molecule are important for understanding the cell biology of metastasis; however, interruption of several adhesive interactions will probably be required to prevent clinical metastasis. We predict that "cocktails" containing several antiadhesive molecules will be clinically efficacious in the management of human malignancies.

The most important effort for the future is the development of clinically useful agents. This technology will have limited clinical application unless the *in vivo* activity of the peptide can be substantially prolonged by chemical modification or by conjugation of GRGDS to inert carriers that confer an extended vascular half-life. However, there are situations, in addition to those mentioned earlier, where the technology could have immediate application, such as during autologous bone marrow transplantation, in which case the threat of metastasis is limited to the time of infusion. Malignant cells contaminating the bone marrow are often incompletely purged by current approaches and are unfortunately reinfused back into the patient. The co-infusion of an antiadhesive peptide, such as GRGDS, may prevent the inopportune seeding of metastatic cells; in this situation, the short circulation half-life would not be a problem.

A final concern is that GRGDS might interfere with normal cell adhesive processes, resulting in disruption of homeostasis. For example, it might be anticipated that chronic infusion of GRGDS would inhibit platelet aggregation or destroy the structural organization of microcapillaries through interference with endothelial cell adhesion.

SURFACE OLIGOSACCHARIDES IN CELL ADHESION AND METASTASIS

It is now well established that specific interactions between sugars and proteins play a major role in biological recognition and mediate many biological reactions.[98–100] This interaction requires both sugar moieties as well as protein molecules capable of binding specifically to these sugar residues. Such carbohydrate binding proteins have now been described in both plant and mammalian species.[101]

Alterations in the size and structure of the Asn-linked carbohydrate moieties of rodent and human tumor cell surface glycoproteins are now well documented.[93,94] The oligosaccharides are generally larger, more highly branched, and oversialylated at their chain termini. These alterations result from a quantitative increase in oligosaccharide branching or antennary structure.[95] Tumor cells are characterized by an increase in triantennary and tetra-antennary oligosaccharides containing the GlcNAC (β1,6) Man (α1,6) branch; these alterations are associated with a decrease in high mannose oligosaccharides. Consistent with these findings, parallel studies have demonstrated altered lectin-binding properties of normal and transformed cells and fluctuations in cellular levels of glycosidases and glycosyl transferases.[93,96,97] Also, a recent study demonstrated a direct correlation between the surface expression of β1,6 branched oligosaccharides and metastasis.[124]

Of interest, there is now increasing evidence from a number of different sources that tumor cell surface oligosaccharides play significant roles in metastasis and invasion: (i) The quantity of exposed sialic acid, and in particular the degree of sialylation of subterminal galactose and N-acetylgalactosamine residues, displays good correlation with metastatic potential;[102-104] consequently, sialogalactoproteins have been suggested as markers for metastatic activity.[105-107] (ii) Cell lines selected for their resistance to toxic concentrations of plant lectins, particularly wheat germ agglutinin, are often less metastatic than their parental lines.[108] Several of these mutants possessed altered surface carbohydrate structures and exhibited lesions in the enzymatic pathway of carbohydrate metabolism. (iii) Drug- or enzyme-induced modification of tumor cell surface carbohydrates alters metastatic potential in various experimental model systems.[7-9,109-114] Treatment with either tunicamycin,[7,113] an inhibitor of the synthesis of N-acetylglucosaminylpyrophosphatylpolyisoprenol, an intermediate in Asn-linked protein glycosylation,[115] swainsonine,[8,116] a potent inhibitor of Golgi α-mannosidase II (117), or castanospermine,[9] an inhibitor of glucosidase I,[118] has been reported to inhibit the experimental metastasis of B16 melanoma cells. Similar agents also inhibited invasion of malignant mouse MO4 cells into the chick heart tissue.[114]

The finding that many tumor cells contain endogenous lectins on their surface raised the idea that these carbohydrate-binding proteins might play a role in cell adhesion and metastasis.[119,120] For example, lectins on the surface of tumor cells can recognize and bind complementary glycoconjugates on the surface of other tumor cells and mediate homotypic aggregation, or they can bind to oligosaccharide moieties on the surface of host cells to mediate heterotypic aggregation or attachment to endothelial cells/basement membrane or extracellular matrix. These adhesive interactions are specific in that they are inhibited by specific sugars.[121] Additionally, pretreatment of malignant cells with antiendogenous lectin monoclonal antibody inhibited experimental metastasis.[122,123]

Whereas neoplastic transformation has been associated with a variety of structural changes in cell surface carbohydrates, it has been difficult to establish that a specific permutation is relevant to the metastatic process and not just an associated change. This difficulty has arisen because several features of the complex carbohydrates could contribute to the preferential recognition of highly branched oligosaccharides; some examples include (i) the distribution or density of terminal sugars such as galactose and sialic acid, (ii) the tertiary conformation of the oligosaccharide, and (iii) the primary structure or specific sequences of the various sugar residues and branching. The present approach of modifying a single characteristic using inhibitors of specific enzymes of the processing pathway may lead to identification of specific oligosaccharide structural features crucial for successful completion of the metastatic cascade.

Treatment of B16-F10 melanoma cells, in monolayer culture, for 24 hours with either swainsonine[8,9,116] or castanospermine[9] substantially inhibited ($\geq$80%) their pulmonary colonization capacity after intravenous injection into C57BL/6 mice. Swainsonine-mediated inhibition of pulmonary colonization was dose-dependent; half-maximal inhibition was observed at a swainsonine concentration of 100 ng/ml, and approximately 80% inhibition was obtained at 3 μg/ml. In other experiments with lower inocula, as much as 95% inhibition of colonization was obtained. Treatment with the drugs did not increase the frequency of extrapulmonary metastases by redistribution of cells to other sites. Also, the inhibition of pulmonary colonization could not be explained by cytotoxicity or suppression of tumorigenicity.[8,9]

One possible mechanism of action of tumor cell oligosaccharides in metastasis could be in the mediation of tumor cell-extracellular matrix interactions since rapid binding to either exposed basement membrane or endothelial cell surface macromolecules is crucial for arrest of metastatic cells.[125] Several lines of evidence provide a strong precedent for a partial involvement of cell surface carbohydrates in such adhesive processes: (i) modification of cell surfaces with either glycosidases, lectins, or blood group antibodies has been found to disrupt adhesion[128]; (ii) some specificity in the ability of purified glycopeptides to inhibit attachment to either laminin or type IV collagen has been reported[33]; (iii) lectin-resistant mutants exhibit altered adhesion to fibronectin and collagens,[126,127] while an unrelated mutant selected for detachability was found to possess a lesion in N-acetylglucosamine transferase[133]; (iv) treatment of cells with tunicamycin causes a weakening of adhesive forces concomitant with cellular rounding and detachment.[7,9,107,129,134–136] Tunicamycin-treated cells have also been reported to bind poorly to endothelial cell monolayers[113,136]; and (v) recently characterized hepatocyte surface proteins with lectin-like specificity for galactose residues may determine the organ selectivity of hepatic metastasizing tumor cells.[130–132] However, the kinetics of attachment and spreading of swainsonine- or castanospermine-treated B16 melanoma cells were comparable to those of untreated cells on both fibronectin- and laminin-coated dishes.

The fact that swainsonine and castanospermine had no significant effect on B16-F10 cell adhesion may mean that size rather than composition is important and the carbohydrate chain may function sterically to maintain protein conformation. Alternatively, the high mannose core may be the segment of the oligosaccharide chain important for mediation of the adhesive reaction.

In summary, several precedents exist for carbohydrate-mediated cellular functions that may be involved at different stages of the metastatic cascade and that may be inhibited by swainsonine. In general, alteration of surface carbohydrates appears to interfere with the adhesiveness of tumor cells together with their recognition of target organs. Lectin-resistant mutant cell lines often possess altered adhesive properties,[104,126,127] and some specificity in the ability of cellular glycopeptides to inhibit adhesion to extracellular matrix molecules has been demonstrated.[33] Although treatment with swainsonine or castanospermine did not affect attachment and spreading of the B16 melanoma cells, treatment of cells with either glycosidases,[128] lectins,[128] tunicamycin,[9,107,129] or neuraminidase[33] modulates their attachment and spreading *in vitro*. Recently, cell surface lectins on normal hepatocytes that aggregate liver-metastasizing tumor cells and that may be involved in the organ selectivity of these cells have been identified.[130–132]

The fact that the colonization capacity of B16-F10 melanoma cells was substantially impaired by swainsonine or castanospermine treatment suggests that the particular orientation and/or sequence of glycoprotein sugar residues is critical for the cells to be able successfully to colonize target organs.[8,9] The novel requirement for strict carbohydrate specificity that is apparent from our studies appears strongly to implicate an oligosaccharide-lectin-type of recognition event in metastatic colonization.

INHIBITION OF PULMONARY COLONIZATION BY SYSTEMIC ADMINISTRATION OF SWAINSONINE

The studies described above indicate that pretreatment of highly metastatic B16-F10 murine melanoma cells with inhibitors of the processing of Asn-linked oligosaccharides causes a dramatic decrease in their pulmonary colonization potential after intravenous injection into mice. Furthermore, the inhibition arising from pretreatment of cells with swainsonine has been found to be enhanced by simultaneous systemic administration of free drug to mice.[116] More recently we have shown that swainsonine has potent antimetastatic activity even when administered alone without pretreatment of tumor cells, and that the antimetastatic effect is maintained for up to five days after withdrawal of the drug.[10,11] The latter finding, together with pharmokinetic data which indicate that swainsonine is cleared rapidly from the body (Ref. 137, and our unpublished data), suggests that the drug initiates some kind of biological cascade which in turn leads to increased host resistance to metastasis formation.

Considering the time course over which swainsonine acts, one potential mechanism of action of the drug could be stimulation of immune system function. Interestingly, there is now substantial experimental evidence implicating carbohydrate moieties in immune cell recognition processes.[138,139] In particular, the relationship of cell surface oligosaccharides to NK cell function has been examined in some detail.[140-143] In a related approach to our own, suppression of murine lymphocyte proliferation caused by exposure to an endogenous immunosuppressive factor has been reported to be restored by swainsonine treatment.[144] Furthermore, the time course over which swainsonine inhibits experimental metastasis parallels previous reports of the kinetic profile of NK cell activation.[145-148] It is highly unlikely that the antimetastatic activity of swainsonine, under these conditions, is due to alteration of the tumor cell surface oligosaccharides since the drug is effective long after it has been excreted from the body (Ref. 137, and our unpublished results).

In our *in vivo* studies,[10] we have shown that swainsonine is able to interfere with pulmonary colonization of B16-F10 after only a short, systemic administration to mice, irrespective of the route of administration. The kinetics of inhibition of both experimental metastasis and pulmonary retention of injected tumor cells was consistent with swainsonine mediating its biological effects 1 to 3 days after administration. Also, NK cells appear to be required for swainsonine to function *in vivo,* since mice with depleted NK cell activity are unable to respond to swainsonine. The finding that the number of spleen cells per mouse is increased by three- to four-fold suggests that swainsonine acts as a mitogenic stimulus for spleen cell proliferation *in vivo.* The large increase in spleen cell number suggests that cell populations other than NK cells are also stimulated to proliferate as a result of swainsonine treatment.

MITOGENIC PROPERTIES OF SWAINSONINE

Direct evidence that the systemic administration of swainsonine promoted lympho-proliferation was determined by incorporation of [³H]thymidine into DNA of freshly isolated splenocytes.[11] Incorporation stimulation indices of four- to six-fold were routinely obtained using splenocytes isolated from animals treated with swainsonine for 48–72 hours. DNA synthesis remained elevated for up to three days after removal of the drug from the drinking water; this is consistent with our earlier finding that resistance to metastatic formation is retained for several days after withdrawal of the drug.[10] The absolute number of spleen cells obtained after a 48-hr administration was increased by as much as 64% in some experiments. Spleen cells isolated from untreated mice were also found to be mitogenically responsive to swainsonine *in vitro;* in fact, it was nearly as effective as PHA, a potent T-cell mitogen.

Several findings suggest that T cells are the primary targets of swainsonine. First, the mitogenic effect of swainsonine was drastically decreased in T-cell-deficient nude mice; second, T-cell-enriched mouse spleen cell cultures, prepared by depletion of NK activity with anti-asialo GM1, exhibited increased mitogenic stimulation; third, swainsonine stimulated the synthesis and secretion of interleukin-2 (IL-2) by two- to six-fold, and directly stimulated mitogenesis of T-helper cell clones.[155] IL-2 is a glycoprotein produced by T-helper cells which promotes the clonal expansion of activated T lymphocytes.[149] Preliminary studies, using flow cytometric analysis, suggest that swainsonine treatment also enhances the expression of IL-2 receptor. However, there is no effect on interferon (IFN) production within the limits of detection of the assay system used.

EFFECT OF SYSTEMIC ADMINISTRATION OF SWAINSONINE ON SPONTANEOUS METASTASIS

Although the studies described above, using the experimental metastasis model, may be extremely important in elucidation of many of the molecular events associated with metastasis, it has a major limitation in that several important differences exist between this model and authentic or spontaneous metastasis from a primary tumor.[150] To test swainsonine in clinically more relevant situations, we have set up spontaneous metastasis models using M5076 and B16-BL6 cell lines. The former is a highly invasive reticulosarcoma cell line which metastasizes to visceral organs, including liver and spleen, from dorsal scapular subcutaneous implants,[151–153] whereas the latter is a melanoma which can spontaneously metastasize from an intramuscular footpad injection to colonize the lung. B16-BL6 has been used extensively by other investigators, and has been found to be an excellent spontaneous metastasis model.[2] Another reason for using the B16-BL6 model was that work in our laboratory on experimental metastasis by tail vein injection had been done using the related B16-F10 melanoma.

In the case of M5076, the systemic administration of swainsonine decreased the mean colony number of controls from ≥300 to 16 in swainsonine-treated animals, an inhibition of approximately 95%.[156] We obtained similar results with the B16-BL6 cell line; for example, lungs of untreated mice had a median colony number of 121, whereas swainsonine-treated mice had a median number of 19. These findings indicate that swainsonine is effective in inhibiting both experimental and spontaneous metastasis.

EFFECT OF SWAINSONINE ON GROWTH RATE OF HUMAN BREAST CANCER CELLS

Swainsonine appears to induce changes in tumor cell proliferation and/or differentiation. For example, transformed NIH 3T3 cells, grown in the presence of swainsonine for 4–10 days, have been shown to lose their ability to grow in soft agar and acquire a morphology more characteristic of nontransformed cells.[154] Furthermore, a combination of swainsonine and the interferon-inducer polyriboinosinic-polyribocytidylic acid (poly I:C) inhibited growth of solid MDAY-D2 murine tumors, while neither agent alone had any significant effect.[116]

To determine whether swainsonine had tumor antiproliferative activity *in vivo,* MDA-MB-231 human breast carcinoma cells, negative for both estrogen and progesterone receptors, were injected subcutaneously into athymic nude mice. Tumors in control animals without swainsonine had a mean volume of 361 mm^3 at day 60 versus a mean tumor volume of 108 mm^3 in swainsonine-treated animals, an inhibition of approximately 70%. Furthermore, when swainsonine was withdrawn, the slower-growing tumor implants continued to grow at the slower rate, suggesting that they had become permanently altered.[157] Thus it appears that the systemic administration of swainsonine to tumor-bearing mice can lead to significant inhibition of tumor growth. Whether the change in tumor growth rate is due to differentiation is under investigation.

CONCLUSIONS

The results summarized above provide evidence that tolerable doses of GRGDS and swainsonine are effective in inhibiting tumor cell metastasis. Whether the antimetastatic effects of these agents will be seen with other tumor types remains to be seen. Mechanisms of metastasis may be quite different from one tumor to another, so we have avoided generalization because a sufficient number of tumor models has not been tested. However, in the case of swainsonine, we already have evidence that it is effective against murine melanoma (B16-F10 and B16-BL6), murine reticulum sarcoma (M5076), human breast carcinoma (estrogen-receptor-positive and -negative), and human osteosarcoma (Dr. Dvorit Samid, personal communication). We have no evidence that GRGDS will selectively destroy or inhibit the proliferation of established metastases; its role is essentially one of prevention or hindrance of tumor dissemination. But swainsonine has both antimetastatic and antiproliferative effects, as well as immunomodulatory activity, and hence may be used with curative intent.

In the last ten years considerable progress has been made in elucidating the mechanisms by which tumor cells metastasize. However, more emphasis should be paid to the rational development of new therapeutic approaches; such effort should enhance our still meager arsenal of agents effective in cancer therapy. Studies such as ours bring into question the cardinal tenet of adjuvant chemotherapy that only those drugs can be considered for adjuvant trials that exhibit antitumor activity, consistent with a previous suggestion.[14] For example, considerable therapeutic benefit could be derived from administration of adjuvant antimetastatic drugs which produce no tumor regression. Such agents would maintain the status quo in that they could change the malignant state to a quasi-benign one.[14] Agents such as those described here could have immediate application in inhibiting the spread of certain tumor cells during surgical treatment of malignant disease. Finally, these studies again emphasize the point that cell proliferation is not the only malignant characteristic of tumor cells that can be exploited in the development of cancer therapeutic agents. Thousands of drugs have been screened solely in terms of their ability to prevent or interfere with cell division. Since metastasis and invasion, unlike cell proliferation, are activities that are not shared by normal adult cells, the development of agents that interfere with these processes might be a more manageable assignment.

ACKNOWLEDGMENTS

We gratefully acknowledge Dr. Kenneth M. Yamada and Dr. Steven K. Akiyama for valuable collaboration, and Mrs. Kazue Matsumoto and Mrs. Karen Schweitzer for expert technical assistance.

REFERENCES

1. FIDLER, I. J., D. M. GERSTEN & I. R. HART. 1978. The biology of cancer invasion and metastasis. Adv. Cancer Res. **28:** 149–250.
2. NICOLSON, G. L. 1982. Cancer metastasis, organ colonization and the cell-surface properties of malignant cells. Biochem. Biophys. Acta **695:** 113–176.
3. MCCARTHY, J. B., M. L. BASARA, S. L. PALM, D. F. SAS & L. T. FURCHT. 1985. Cancer Metastasis Rev. **4:** 125–152.
4. LIOTTA, L. A., C. N. RAO & U. M. WEBER. 1986. Biochemical interactions of tumor cells with the basement membrane. Annu. Rev. Biochem. **55:** 1037–1057.
5. HUMPHRIES, M. J., K. OLDEN & K. M. YAMADA. 1986. A synthetic peptide from fibronectin inhibits experimental metastasis of murine melanoma cells. Science **233:** 467–470.
6. HUMPHRIES, M. J., K. M. YAMADA & K. OLDEN. 1988. Investigation of the biological effects of anti-cell adhesive synthetic peptides that inhibit experimental metastasis of B16-F10 murine melanoma cells. J. Clin. Invest. **81:** 782–790.
7. OLDEN, K., M. J. HUMPHRIES & S. L. WHITE. 1985. Biochemical effects and cancer therapeutic potential of tunicamycin. In Monoclonal Antibodies and Cancer Therapy. R. Reisfeld & S. Sell, eds.: 443–472. Alan R. Liss. New York.
8. HUMPHRIES, M. J., K. MATSUMOTO, S. L. WHITE & K. OLDEN. 1986. Oligosaccharide modification by swainsonine treatment inhibits pulmonary colonization by B16-F10 murine melanoma cells. Proc. Natl. Acad. Sci. USA **83:** 1752–1756.
9. HUMPHRIES, M. J., K. MATSUMOTO, S. L. WHITE & K. OLDEN. 1986. Inhibition of experimental metastasis by castanospermine in mice: Blockage of two distinct stages of tumor colonization by oligosaccharide processing inhibitors. Cancer Res. **46:** 5215–5222.
10. HUMPHRIES, M. J., K. MATSUMOTO, S. L. WHITE, R. J. MOLYNEUX & K. OLDEN. 1988. Augmentation of murine natural killer cell activity by swainsonine, a new immuno-modulator. Cancer Res. **48:** 1410–1415.
11. WHITE, S. L., K. SCHWEITZER, M. J. HUMPHRIES & K. OLDEN. 1988. Stimulation of DNA

synthesis in murine lymphocytes by the drug swainsonine: Immunomodulatory properties. Biochem. Biophys. Res. Commun. **150:** 615–625.

12. HUMPHRIES, M. J., M. OBARA, K. OLDEN & K. M. YAMADA. 1988. Role of fibronectin in adhesion, migration, and metastasis. Cancer Invest. In press.

13. HUMPHRIES, M. J., K. OLDEN & K. M. YAMADA. 1989. Fibronectin and cancer: Implications of cell adhesion to fibronectin for tumor metastasis. *In* Fibronectin and Health and Disease. S. Carsons, ed. Critical Reviews. CRC Press, Boca Raton, FL. In press.

14. HELLMAN, R. 1984. Antimetastatic drugs: laboratory to clinic. Clin. Exp. Metastasis **2:** 1–4.

15. NICOLSON, G. L. 1987. Tumor cell instability, diversification, and progression to the metastatic phenotype: From oncogene to oncofetal expression. Cancer Res. **47:** 1473.

16. FIDLER, I. J. 1973. The relationship of embolic homogeneity, number, size, and viability to the incidence of experimental metastasis. Eur. J. Cancer **9:** 233.

17. LIOTTA, L. A., J. KLEINERMAN & G. M. SAIDEL. 1984. The significance of hematogenous tumor cell clumps in the metastatic process. Cancer Res. **36:** 889.

18. GASIC, G. J. 1984. Role of plasma, platelets, and endothelial cells in tumor metastasis. Cancer Metastasis Rev. **3:** 99.

19. GASIC, G. J., G. P. TUSZYNSKI & E. GORELIK. 1986. Interaction of the hemostatic and immune systems in the metastatic spread of tumor cells. Int. Rev. Exp. Pathol. **29:** 173.

20. REICH, E. 1978. Activation of plasminogen: A general mechanism for producing localized extracellular proteolysis. *In* Molecular Basis of Biological Degradative Processes. R. D. Berlin, H. Herrmann, I. H. Lepow & M. J. Tanzer, Eds.:155. Academic Press. New York.

21. LIOTTA, L. A., V. P. THORGEIRSSON & S. GARBISA. 1982. Role of collagenases in tumor cell invasion. Cancer Metastasis Rev. **1:** 277.

22. GOLDFARB, R. H. & L. A. LIOTTA. 1986. Proteolytic enzymes in cancer invasion and metastasis. Semin. Thromb. Hemost. **12:** 294.

23. NAKAJIMA, M., D. R. WELCH, T. IRIMURA & G. L. NICOLSON. 1986. Basement membrane degradative enzymes as possible markers of tumor metastasis. Prog. Clin. Biol. Res. **212:** 113.

24. FIDLER, I. J. 1970. Metastasis: Quantitative analysis of distribution and fate of tumor emboli labeled with ^{125}I-5-iodo-2′-deoxyuridine. J. Natl. Cancer Inst. **45:** 773.

25. HANNA, N. 1985. The role of natural killer cells in control of tumor growth and metastasis. Biochem. Biophys. Acta **780:** 213.

26. HERBERMAN, R. B. 1986. Natural killer cells. Annu. Rev. Med. **37:** 347.

27. KNISELY, W. H. & M. S. MAHALEY. 1958. Relationship between size and distribution of "spontaneous" metastases and three sizes of intravenously injected VX2 carcinoma. Cancer Res. **18:** 900.

28. EDELMAN, G. M. 1983. Cell adhesion molecules. Science **219:** 450–457.

29. TURLEY, E. A. 1985. Proteoglycans and cell adhesion: Their putative role during tumorigenesis. Cancer Met. Rev. **3:** 325–340.

30. ROOS, E. 1984. Cellular adhesion, invasion and metastasis. Biochem. Biophys. Acta **738:** 263–284.

31. NICOLSON, G. L. 1984. Cell surface molecules and tumor metastasis. Exp. Cell Res. **150:** 3–22.

32. SCHIRRMACHER, V. 1985. Experimental approaches, theoretical concepts, and impacts for treatment strategies. Adv. Cancer Res. **43:** 1–73.

33. DENNIS, J. W., C. A. WALLER & V. SCHIRRMACHER. 1984. Identification of asparagine-linked oligosaccharides involved in tumor cell adhesion to laminin and type IV collagen. J. Cell Biol. **99:** 1416–1423.

34. RUOSLAHTI, E. 1985. Fibronectin in cell adhesion and invasion. Cancer Metastasis Rev. **3:** 43–51.

35. BRACKENBURY, R. 1985. Molecular mechanisms of cell adhesion in normal and transformed cells. Cancer Metastasis Rev. **4:** 450–457.

36. BUTLER, T. & P. GULLINE. 1975. Quantitation of cell-shedding into efferent blood of mammary adenocarcinoma. Cancer Res. **35:** 512.

37. WEISS, L., E. MAYHEW, D. GLAVES-RAPP & J. C. HOLMES. 1982. Metastatic inefficiency in mice bearing B16 melanomas. Br. J. Cancer 45: 44.
38. HYNES, R. O. 1985. Molecular biology of fibronectin. Annu. Rev. Cell Biol. 1: 67–90.
39. LINDER, E., A. VAHERI, E. RUOSLAHTI & T. WARTIOVAARA. 1975. J. Exp. Med. 142: 41.
40. STENMAN, S. & A. VAHERI. 1978. Distribution of a major connective tissue protein, fibronectin, in normal human tissue. J. Exp. Med. 147: 1054.
41. HYNES, R. O. & K. M. YAMADA. 1982. Fibronectins: Multifunctional modular glycoproteins. J. Cell Biol. 95: 369–377.
42. YAMADA, K. M. 1983. Cell surface interactions with extracellular materials. Annu. Rev. Biochem. 52: 761–799.
43. HYNES, R. O. 1987. Integrins: A family of cell surface receptors. Cell 48: 549–554.
44. HAYMAN, E. G., M. D. PIERSCHBACHER, Y. OHGREN, & E. RUOSLAHTI. 1983. Serum spreading factor (vitronectin) is present at the cell surface and in tissues. Proc. Natl. Acad. Sci. USA 80: 4003.
45. RUOSLAHTI, E. & M. D. PIERSCHBACHER. 1986. Arg-Gly-Asp: A versatile cell recognition signal. Cell 44: 517.
46. AKIYAMA, S. K. & K. M. YAMADA. 1987. Fibronectin. In Advances in Enzymology and Related Areas of Molecular Biology. A. Meister, Ed.: 1. Wiley. New York.
47. RUOSLAHTI, E. & M. D. PIERSCHBACHER. 1987. New perspectives in cell adhesion: RGD and integrins. Science 238: 491.
48. THIERY, J. P., J. L. DUBAND & G. C. TUCKER. 1985. Cell migration in the vertebrate embryo: Role of cell adhesion and tissue environment in pattern formation. Annu. Rev. Cell Biol. 1: 91.
49. YAMADA, K. M. 1988. Fibronectin domains and receptors. In Fibronectin. D. F. Mosher, Ed. Academic Press. New York. In press.
50. PIERSCHBACHER, M. D., E. G. HAYMAN & E. RUOSLAHTI. 1981. Location of the cell attachment site in fibronectin with monoclonal antibodies and proteolytic fragments of the molecule. Cell 26: 289.
51. PIERSCHBACHER, M. D., E. RUOSLAHTI, J. SUNDELIN, P. LIND & P. A. PETERSON. 1982. The cell attachment domain of fibronectin. Determination of the primary structure. J. Biol. Chem. 257: 9593.
52. PIERSCHBACHER, M. D., E. G. HAYMAN & E. RUOSLAHTI. 1983. Synthetic peptide with cell attachment activity of fibronectin. Proc. Natl. Acad. Sci. USA 80: 1224.
53. PIERSCHBACHER, M. D. & E. RUOSLAHTI. 1983. Cell attachment activity of fibronectin can be duplicated by small synthetic fragments of the molecule. Nature 309: 30.
54. YAMADA, K. M. & D. W. KENNEDY. 1984. Dualistic nature of adhesive protein function: Fibronectin and its biologically active peptide fragments can autoinhibit fibronectin function. J. Cell Biol. 99: 29.
55. YAMADA, K. M. & D. W. KENNEDY. 1985. Amino acid sequence specificities of an adhesive recognition signal. J. Cell. Biochem. 28: 99.
56. AKIYAMA, S. K. & K. M. YAMADA. 1985. The interaction of plasma fibronectin with fibroblastic cells in suspension. J. Biol. Chem. 260: 4492.
57. AKIYAMA, S. K. & K. M. YAMADA. 1985. Synthetic peptides competitively inhibit both direct binding to fibroblasts and functional biological assays for the purified cell-binding domain of fibronectin. J. Biol. Chem. 260: 10402.
58. AKIYAMA, S. K., E. HASEGAWA, T. HASEGAWA & K. M. YAMADA. 1985. The interaction of fibronectin fragments with fibroblastic cells. J. Biol. Chem. 260: 13256.
59. GREVE, J. M. & D. I. GOTTLIEB. 1982. Monoclonal antibodies which alter the morphology of cultured chick myogenic cells. J. Cell Biochem. 18: 221.
60. NEFF, N. T., C. LOWREY, C. DECKER, A. TOVAR, C. DAMSKY, C. BUCK & A. F. HORWITZ. 1982. A monoclonal antibody detaches embryonic skeletal muscle from extracellular matrices. J. Cell Biol. 95: 654.
61. CHAPMAN, A. E. 1984. Characterization of a 140 KD cell surface glycoprotein involved in myoblast adhesion. J. Cell Biochem. 25: 109.
62. DECKER, C., R. GREGGS, K. DUGGAN, J. STUBBS & A. HORWITZ. 1984. Adhesive multiplicity in the interaction of embryonic fibroblasts and myoblasts with extracellular matrices. J. Cell Biol. 99: 1398.

63. CHEN, W.-T., E. HASEGAWA, T. HASEGAWA, C. WEINSTOCK & K. M. YAMADA. 1985. Development of cell surface linkage complexes in cultured fibroblasts. J. Cell Biol. **100:** 1103.

64. HASEGAWA, T., E. HASEGAWA, W.-T. CHEN & K. M. YAMADA. 1985. Characterization of a membrane-associated glycoprotein complex implicated in cell adhesion to fibronectin. J. Cell Biochem. **28:** 307.

65. KNUDSEN, K. A., A. HORWITZ & C. A. BUCK. 1985. A monoclonal antibody identified a glycoprotein complex involved in cell-substratum adhesion. Exp. Cell Res. **157:** 218.

66. HORWITZ, A., K. DUGGAN, R. GREGGS, C. DECKER & C. A. BUCK. 1985. J. Cell Biol. **103:** 2134.

67. AKIYAMA, S. K., S. S. YAMADA & K. M. YAMADA. 1986. Characterization of a 140 KD avian cell surface antigen as a fibronectin-binding molecule. J. Cell Biol. **102:** 442.

68. BROWN, P. & R. L. JULIANA. 1985. Selective inhibiting of fibronectin-mediated cell adhesion by monoclonal antibodies to a cell surface glycoprotein. Science **228:** 1448.

69. PYTELA, R., M. D. PIERSCHBACHER & E. RUOSLAHTI. 1985. Identification and isolation of a 140 Kd cell surface glycoprotein with properties expected of a fibronectin receptor. Cell **40:** 191.

70. PATEL, V. P. & H. F. LADISH. 1986. The fibronectin receptor on mammalian erythroid precursor cells; characterization and developmental regulation. J. Cell Biol. **102:** 449.

71. GIANCOTTI, F. G., P. M. COMOGLIO & G. TARONE. 1986. J. Cell Biol. **103:** 429.

72. GIANCOTTI, F. G., P. M. COMOGLIO & G. TARONE. 1986. A 135,000 molecular weight membrane glycoprotein involved in fibronectin-mediated cell adhesion. Exp. Cell Res. **163:** 47.

73. CARDARELLI, P. M. & M. D. PIERSCHBACHER. 1987. Identification of fibronectin receptors on T lymphocytes. J. Cell Biol. **105:** 499.

74. BOUCAUT, J. C., T. DARRIBERE, T. J. POOLE, H. ASYAMA, K. M. YAMADA & J. P. THIERY. 1984. J. Cell Biol. **99:** 1822.

75. NAIDET, C., M. SEMERIVA, K. M. YAMADA & J. P. THIERY. 1987. Peptides containing the cell-attachment recognition signal Arg-Gly-Asp prevent gastrulation in *Drosophila* embryos. Nature **325:** 348.

76. DARRIBERE, T., K. M. YAMADA, K. E. JOHNSON & J. C. BOUCAUT. 1988. The 140-KDa fibronectin receptor complex is required for mesodermal cell adhesion during gastrulation in the amphibian *Pleurodeles* walthlii. Dev. Biol. In press.

77. SAGAGNER, P., B. A. IMHOF, K. M. YAMADA & J. P. THIERY. 1986. Homing of hemopoietic precursor cells to the embryonic thymus: Characterization of an invasive mechanism induced by chemotactic peptides. J. Cell Biol. **103:** 2715.

78. BONNER-FRASER, M. 1986. An antibody to a receptor for fibronectin and laminin perturbs cranial neural crest development *in vivo*. Dev. Biol. **117:** 528.

79. DONALDSON, D. J., J. T. MAHAN & G. N. SMITH. 1987. Newt epidermal cell migration *in vitro* and *in vivo* appears to involve Arg-Gly-Asp-Ser receptors. J. Cell Science **87:** 525.

80. HYNES, R. 1973. Alteration of cell-surface proteins by viral transformation and by proteolysis. Proc. Natl. Acad. Sci. USA **70:** 3170.

81. GAHMBERG, C. G. & S. HAKOMORI. 1973. Altered growth behavior of malignant cells associated with changes in externally labeled glycoprotein and glycolipids. Proc. Natl. Acad. Sci. USA **70:** 3329.

82. CRITCHLEY, D. R. 1974. Cell surface proteins of NIL1 hamster fibroblasts labeled by a galactose oxidase, tritiated barohydride method.

83. GAHMBERG, C. G., D. KIEHN & S. HAKOMORI. 1974. Changes in a surface-labeled galactoprotein and in glycolipid concentrations in cells transformed by a temperature-sensitive polyoma virus mutant. Nature **248:** 413.

84. HOGG, N. M. 1974. A comparison of membrane proteins of normal and transformed cells by lactoperoxidase labeling. Proc. Natl. Acad. Sci. USA **71:** 489.

85. PEARLSTEIN, E. & M. D. WATERFIELD. 1974. Metabolic studies on ^{125}I-labeled baby hamster kidney cell plasma membranes. Biochem. Biophys. Acta **362:** 1.

86. ROBBINS, P. W., G. G. WICKUS, P. E. BRANTON, B. J. GAFFNEY, C. B. HIRSCHBERG, P. FUCHS & P. M. BLUMBERG. 1984. The chick fibroblast cell surface after transformation by Rous sarcoma virus. Cold Spring Harbor Symp. Quant. Biol. **39:** 1173.

87. STONE, K. R., R. E. SMITH & W. K. JOKLIK. 1974. Changes in membrane polypeptides that occur when chick embryo fibroblasts and NRK cells are transformed with avian sarcoma viruses. Virology **58**: 86.

88. VAHERI, A. & E. RUOSLAHTI. 1974. Disappearance of a major cell-type specific surface glycoprotein antigen (SF) after transformation of fibroblasts by Rous sarcoma virus. Int. J. Cancer **13**: 579.

89. WARTIONAARA, J., E. LINDER, E. RUOSLAHTI & A. VAHERI. 1974. Distribution of fibroblast surface antigen. Association with fibrillar structures of normal cells and loss upon viral transformation. J. Exp. Med. **140**: 1522.

90. FIDLER, I. J. 1978. General considerations for studies of experimental metastasis. Methods Cancer Res. **15**: 399.

91. FIDLER, I. J. 1973. Selection of successive tumor lines for metastasis. Nature New Biol. **242**: 148.

92. POSTE, G. & NICOLSON, G. L. 1980. Blood-borne tumor cell arrest and metastasis modified by fusion of plasma membrane vesicles from highly metastatic cells. Proc. Natl. Acad. Sci. USA **77**: 399–403.

93. RAPIN, A. M. C. & M. M. BURGER. 1974. Tumor cell surfaces: General alterations detected by agglutinins. Adv. Cancer Res. **20**: 1–91.

94. WARREN, L., J. P. FUHRER & C. A. BUCK. 1973. Surface glycoproteins of cells before and after transformation by oncogenic viruses. Fed. Proc **32**: 80–85.

95. OGATA, S. I., T. MURAMATSU & A. KOBATA. 1976. New structural characteristics of the large glycopeptides from transformed cells. Nature **259**: 580–582.

96. ROBBINS, J. C. & G. L. NICOLSON. 1975. Surfaces of normal and transformed cells. *In* Cancer, A Comprehensive Treatise F. F. Becker, Ed. Vol. **4**: 3–54. Plenum Press. New York.

97. YOGEESWARAN, G. 1983. Cell surface glycolipids and glycoproteins in malignant transformation. Adv. Cancer Res. **38**: 289–350.

98. OLDEN, K. & J. B. PARENT, Eds. 1987. Vertebrate Lectins. Van Nostrand Reinhold. New York.

99. OLDEN, K., J. B. PARENT & S. L. WHITE. 1982. Carbohydrate moieties of glycoproteins: A re-evaluation of their function. Biochim. Biophys. Acta **650**: 209–232.

100. OLDEN, K., B. A. BERNARD, M. J. HUMPHRIES, T. K. YEO, K. T. YEO, S. L. WHITE, S. A. NEWTON, H. C. BAUER & J. B. PARENT. 1985. Function of glycoprotein glycans. Trends Biochem. Sci. **10**: 78–82.

101. SHARON, N. 1987. Lectins: An overview. *In* Vertebrate Lectins. K. Olden and J. B. Parent, Eds.: 27–45. Van Nostrand Reinhold. New York.

102. BOSMANN, T. W. & A. LIONE. 1973. Biochemical parameters correlated with tumor cell implantation. Nature **246**: 487–489.

103. YOGEESWARAN, G. & P. L. SALK. 1981. Metastatic potential is positively correlated with cell surface sialylation of cultured murine tumor cell lines. Science **212**: 1514–1516.

104. ALTEVOGT, P., M. FOGEL, R. CHEINGSONG, J. DENNIS, P. ROBINSON & V. SCHIRRMACHER. 1983. Cancer Res. **43**: 5138–5144.

105. YOGEESWARAN, G., B. S. STEIN & H. SEBASTIAN. 1978. Altered cell surface organization of gangliosides and sialylglycoproteins of mouse metastatic melanoma variant lines selected *in vivo* for enhanced lung implantation. Cancer Res. **38**: 1336–1344.

106. RAZ, A., W. L. MCLELLAN, I. R. HART C. D. BUCANA, L. C. HOYER, B. A. SELA, P. DRAGSTEN & I. J. FIDLER. 1980. Cell surface properties of B16 melanoma variants with differing metastatic potential. Cancer Res. **40**: 1645–1651.

107. IRIMURA, T. & G. L. NICOLSON. 1984. Carbohydrate chain analysis by lectin binding to electrophoretically separated glycoproteins from murine B16 melanoma sublines of various metastatic properties. Cancer Res. **44**: 791–798.

108. KERBEL, R. S., J. W. DENNIS, A. E. LARGARDE & P. FROST. 1982. Tumor progression in metastasis: an experimental approach using lectin resistant tumor variants. Cancer Metastasis Rev. **1**: 99–140.

109. GASIC, G. J., T. B. GASIC & C. C. STEWART. 1968. Antimetastatic effects associated with platelet reduction. Proc. Natl. Acad. Sci USA **61**: 46–52.

110. HAGMAR, B. & K. NORRBY. 1973. Influence of cultivation, trypsinization and aggregation on the transplantibility of melanoma B16 cells. Int. J. Cancer **11**: 663–675.

111. SINHA, B. K. & G. J. GOLDENBERG. 1974. The effect of trypsin and neuraminidase on the circulation and organ distribution of tumor cells. Cancer **34:** 1956–1961.

112. KIJIMA-SUDA, I., Y. MIUAMOTO, S. TOYOSHIMA, M. HOH & T. OSAWA. 1986. Inhibition of experimental pulmonary metastasis of mouse colon adenocarcinoma 26 sublines by a sialic acid: nucleoside conjugate having sialyltransferase inhibiting activity. Cancer Res. **46:** 858–862.

113. IRIMURA, T., R. GONZALEZ & G. L. NICOLSON. 1981. Effects of tunicamycin on B16 metastatic melanoma cell surface glycoproteins and blood-borne arrest and survival properties. Cancer Res. **41:** 3411–3418.

114. MAREEL, M. M., C. H. DRAGONETTI, R. J. HOOGHE & E. A. BRUYNEEL. 1985. Effect of inhibitors of glycosylation and carbohydrate processing on invasion of malignant mouse MO4 cells in organ culture. Clin. Exp. Metastasis **3:** 197–207.

115. TAKATSUKI, A., K. KOHNO & G. TAMURA. 1975. Inhibition of biosynthesis of polyisoprenal sugars in chick embryo microsomes by tunicamycin. Agr. Biol. Chem. **39:** 2089–2091.

116. DENNIS, J. W. 1986. Effects of swainsonine and polyinosinic: polycytidylic acid on murine tumor cell growth and metastasis. Cancer Res. **46:** 5131–5136.

117. ELBEIN, A. D., R. SOLF, P. R. DORLING & K. VOSBECK. 1981. Swainsonine: An inhibitor of glycoprotein processing. Proc. Natl. Acad. Sci. USA **78:** 7393–7397.

118. SAUL, R., J. P. CHAMBERS, R. J. MOLYNEUX & A. D. ELBEIN. 1983. Castanospermine, a tetrahydroxylated alkaloid that inhibits beta-glucosidase and beta glucoceribrosidase. Arch. Biochem. Biophys. **221:** 593–597.

119. HARRISON, F. L. & C. J. CHESTERTON. 1980. Factors mediating cell-cell recognition and adhesion. FEBS Lett. **122:** 157–165.

120. MONSIGNEY, J., C. KIEDA & A. C. ROCHE. 1983. Membrane glycoproteins, glycolipids, and membrane lectins as recognition signals in normal and malignant cells. Biol. Cell **47:** 95–110.

121. STOJANOVIC, D. & R. C. HUGHES. 1984. An endogenous carbohydrate-binding agglutinin of BHK cells. Purification, specificity and interaction with normal and ricin-resistant cell lines. Biol. Cell **51:** 197–206.

122. RAZ, A. & R. LOTAN. 1987. Endogenous galactoside-binding lectins: a new class of functional tumor cell surface molecules related to metastasis. Cancer Metastasis Rev. **6:** 433–452.

123. MEROMSKY, L., R. LOTAN & A. RAZ. 1986. Implications of endogenous tumor cell surface lectins as mediators of cellular interactions and lung colonization. Cancer Res. **46:** 5270–5275.

124. DENNIS, J. W., S. LAFERTE, C. WAGHORNE, M. L. BREITMAN & R. S. KERBEL. 1987. $\beta 1$–6 branching of Asn-linked oligosaccharides is directly associated with metastasis. Science **236:** 582–585.

125. LIOTTA, L. A., C. N. RAO & S. H. BARSKY. 1983. Tumor invasion and the extracellular matrix. Lab. Invest **49:** 636–649.

126. PENA, S. D. J. & R. C. HUGHES. 1978. Fibronectin-plasma membrane interactions in the adhesion and spreading of human fibroblasts. Nature **276:** 80–83.

127. DENNIS, J. W., C. WALLER, R. TIMPL & V. SCHIRRMACHER. 1982. Surface sialic acid reduces attachment of metastatic tumor cells to collagen type IV and fibronectin. Nature. **300:** 274–276.

128. GRIMSTAD, I. A., J. VARANI & J. P. MCCOY. 1984. Contribution of α-D-galactopyranosyl end groups to attachment of highly and low metastatic murine fibrosarcoma cells to various substrates. Exp. Cell Res. **155:** 345–358.

129. BUTLERS, T. D., V. DEVALIA, J. A. APLIN & R. C. HUGHES. 1980. Inhibition of fibronectin-mediated adhesion of human fibroblasts to substratum: effects of tunicamycin and some cell surface modifying agents. J. Cell Sci. **44:** 33–58.

130. SCHIRRMACHER, V., R. CHEINGSONG-POPOV & H. ARNHEITER. 1980. Hepatocyte-tumor cell interaction *in vitro*. I. Conditions for formation and inhibition by anti-H-2 antibody. J. Exp. Med. **151:** 984–989.

131. CHEINGSONG-POPOV, R., S. ROBINSON, P. ALTEVOGT & V. SCHIRRMACHER. 1983. Mouse hepatocyte carbohydrate-specific receptor and its interaction with liver-metastasizing tumor cells. Int. J. Cancer **32:** 356–366.

132. SPRINGER, G. F., R. CHEINGSONG-POPOV, V. SCHIRRMACHER, P. R. DESAI & H. TEGTMEYER. 1983. Proposed molecular basis of murine tumor cell-hepatocyte interaction. J. Biol. Chem. **258:** 5702–5706.

133. POUYSSÉGUR, J., M. WILLINGHAM & I. PASTAN. 1977. Role of cell surface carbohydrates and proteins in cell behavior: Studies on the biochemical reversion of an N-acetylglucosamine-deficient fibroblast mutant. Proc. Natl. Acad. Sci. USA **74:** 243–247.

134. DUKSIN, D. & P. BORNSTEIN. 1977. Changes in surface properties of normal and transformed cells caused by tunicamycin, an inhibitor of protein glycosylation. Proc. Natl. Acad. Sci. USA **74:** 3433–3437.

135. PRATT, R. M., K. OLDEN, K. M. YAMADA, S. H. OHANIAN & V. C. HASCOLL. 1979. Tunicamycin-induced alterations in the synthesis of sulfated proteoglycans and cell surface morphology in the chick embryo fibroblast. Exp. Cell Res. **118:** 245–252.

136. NICOLSON, G. L., T. IRIMURA, R. GONZALEZ & E. RUOSLAHTI. 1981. The role of fibronectin in adhesion of metastatic melanoma cells to endothelial cells and their basal lamina. Exp. Cell Res. **135:** 461–465.

137. BROQUIST, H. P., P. S. MASON, W. M. HAGLER & W. J. CROOM. 1985. Transmission of swainsonine into milk. Fed. Proc. **44:** 1860.

138. READING, C. L. & J. T. HUTCHINS. 1985. Carbohydrate structure in tumor immunity. Cancer Metastasis Rev **4:** 221–260.

139. WHITE, S. L. 1987. Role of lectins in immune recognition. *In* Vertebrate Lectins. K. Olden & J. B. Parent, Eds.: 182–194. Van Nostrand Reinhold. New York.

140. STUTMAN, O., P. DIEN, R. E. WISUN & E. C. LATTIME. 1980. Natural Cytotoxic cells against solid tumors in mice: Blocking of Cytotoxicity by D-mannose. Proc Natl. Acad. Sci. USA **77:** 2895–2898.

141. FORBES J. T., R. K. BRETTHAUER & T. N. OELTMANN. 1981. Mannose 6-, fructose 1-, and fructose 6-phosphates inhibit human natural cell-mediated cytotoxicity. Proc. Natl. Acad. Sci. USA **78:** 5797–5801.

142. WERKMEISTER, J. A., H. F. PROSS & J. C. RODER. 1983. Modulation of K562 cells with sodium butyrate. Association of impaired NK susceptibility with sialic acid and analysis of other parameters. Int. J. Cancer **32:** 71–78.

143. DENNIS, J. W. & S. LAFERTE. 1985. Recognition of asparagine-linked oligosaccharides on murine tumor cells by natural killer cells. Cancer Res. **45:** 6034–6040.

144. HINO, M., O. NAKAYAMA, Y. TSURUMI, K. ADACHI, T. SHIBATA, H. TERANO, M. KOHSAKA, H. AOKI & H. IMANAKA. 1985. Studies of an immunomodulator, swainsonine. I. Enhancement of immune response by swainsonine in vitro. J. Antibiol. **38:** 926–935.

145. HERBERMAN, R. B., M. E. NUNN, H. T. HOLDEN, S. STAAL & J. Y. DJEU. 1977. Augmentation of natural cytotoxicity reactivity of mouse lymphoid cells against syngeneic and allogeneic target cells. Int. J. Cancer **19:** 555–564.

146. WELSH, R. M. & R. M. ZINKERNAGEL. 1977. Heterospecific cytotoxic cell activity induced during the first three days of acute lymphocyte choriomeningitis virus infection in mice. Nature **268:** 646–658.

147. HERBERMAN, R. B., J. Y. DJEU, J. R. ORTALDO, H. T. HOLDEN, W. H. WEST & G. D. BONNARD. 1978. Role of interferon in augmentation of natural and antibody-dependent cell-mediated cytotoxicity. Cancer Treat. Rep. **62:** 1893–1896.

148. DJEU, J. Y., Y. A. HEINBAUGH, H. T. HOLDEN & R. B. HERBERMAN. 1979. Augmentation of mouse natural killer cell activity by interferon and interferon inducers. J. Immunol. **122:** 175–181.

149. MORGAN, O., F. RUSCETTI & R. GALLO. 1976. Selective *in vitro* growth of T lymphocytes from normal human bone marrows. Science **193:** 1007–1009.

150. STACKPOOLE, C. W. 1981. Distinct lung-colonizing and lung-metastasizing cell populations in B16 mouse melanoma. Nature **289:** 798–800.

151. POSTE, G., J. DOLL, A. E. BROWN, J. TZENG & I. ZEIDMAN. 1982. Comparison of the metastatic properties of B16 melanoma clones isolated from cultured cell lines, subcutaneous tumors, and individual lung metastases. Cancer Res. **42:** 2770–2778.

152. STACKPOOLE, C. W., D. M. FORNABAIO & A. M. ALTERMAN. 1985. Phenotypic interconversion of B16 melanoma clonal cell populations between metastasis and tumor growth rate. Int. J. Cancer **35:** 667–694.

153. HART, I. R. 1979. The selection and characterization of an invasive variant of the B16 melanoma. Am. J. Pathol. **97:** 587–600.
154. DeSANTIS, R., U. V. SANTER & M. C. GLICK. 1987. NIH 3T3 cells transfected with human tumor DNA lose the transformed phenotype when treated with swainsonine. Biochem. Biophys. Res. Commun. **142:** 348–353.
155. WHITE, S. L. Manuscript in preparation.
156. NEWTON, S. A. *et al.* Manuscript in preparation.
157. MOHLA, S. *et al.* Manuscript in preparation.

DISCUSSION

A. IMAM (*University of Southern California, Los Angeles, California*): What is the fate of those metastatic cells that do not form metastatic foci? How are they being killed?

K. OLDEN (*Howard University, Washington, D.C.*): Metastatic cells cleared from the lungs in the presence of GRGDS do not colonize other organs, so we are not simply altering organ specificity. I assume that they are killed by the same mechanisms that eliminate tumor cells in general, that is, hemodynamic stress that deforms and disrupts or breaks cells, leading to their death. Many of the cells cleared in the presence of GRGDS will also be killed by host immune surveillance mechanisms.

R. LOTAN: What do you know about the fate of the GRGDS-fibronectin receptor complex? Is it internalized?

OLDEN: The binding of GRGDS is so weak that we cannot do the experiment to find out. Your question is interesting: We would be interested in knowing to what degree the peptide can mimic the native molecule in terms of microfilament bundle organization and so forth. Apparently, such an experiment can be done with YIGSR, the laminin peptide. We are attempting to isolate or prepare fragments of fibronectin that bind more tightly to the receptors.

M. F. MARTELLI (*University of Perugia, Italy*): Concerning the future application of this compound in autologous bone marrow transplantation, what will be the effects of GRGDS on hemopoietic cell homing?

OLDEN: We have no data on the effects of GRGDS on the homing of bone marrow cells. But since we believe that GRGDS may be effective in inhibiting organ colonization of tumor cells existing in autologous bone marrow implants of cancer patients, that is an important concern of ours. There is now good evidence that mannosyl and galactosyl receptors are involved in homing to the bone marrow, but not to the spleen, so we may be lucky that GRGDS is not involved.

H. C. BIRNBOIM (*Ottawa Regional Cancer Center, Ontario, Canada*): Is GRGDS effective in preventing lung metastases if GRGDS is added after the intravenous injection of tumor cells?

OLDEN: It is most effective when added at the same time, but is partially active when added immediately before but not immediately after injection of cells. This is an affinity problem; once the surface receptor has already bound to native fibronectin, the cells cannot be displaced with the weakly binding peptide.

P. A. RILEY (*Middlesex School of Medicine, London, England*): Is it possible that the dramatic effects that you demonstrated in the B16 lung colony assay when melanoma cells were co-injected with GRGDS are due to the generation of a peptide-receptor complex with an immunogenic potential?

OLDEN: First, it is clear from our lung clearance data that a large portion of the GRGDS effect is expressed early, too early, to be mediated by humoral immunomechanisms. Also, the immune effector cells, that is, the NK cells, known to be involved in the early phase of cancer immune surveillance, were shown not to be involved in mediation of the GRGDS effect by using NK-deficient mice. Additionally, the receptor-peptide complex is very unstable. So, I believe that the GRGDS effect cannot be explained by immunogenic mechanisms, but, of course, anything is possible.

F. CUTTITTA (*National Institutes of Health, Bethesda, Maryland*): What effects do "D" forms of GRGDS have on metastasis experiments?

OLDEN: Substitution of D-amino acids is not effective in inhibition of cell attachment, spreading, and metastasis. This kind of experiment was done to demonstrate specificity of GRGDS.

M. MONSIGNY (*CNRS and Université d'Orleans, France*): First, How many RGDS molecules were linked to IgG molecules? Second, What is the required concentration of the bound peptide to inhibit cell adhesion in comparison with free peptide.

OLDEN: First, about 8 to 10 molecules were linked. Second, the bound peptide is about 10 times more active than the free peptide when assayed *in vitro*. Therefore, generating an effective or therapeutic level in circulation probably will not be a problem. However, we are exploring coupling to other polymers and different mechanisms of delivery of the peptide.

R. LOTAN (*University of Texas, Houston, Texas*): Given your findings that GRGDS is active in inhibition of metastasis in thrombocytopenic mice, and the fact that the luminal surface of capillary endothelial cells does not have any fibronectin, what type of intercellular adhesion do you think the GRGDS peptide inhibits?

OLDEN: I suspect that cancer cells are binding to fibronectin in exposed basement membrane and extracellular matrix stroma. However, in this experiment the platelet number is only reduced to 45% of normal, so there is still a possibility of platelet involvement.

The Surface Charge of Membranes Modulates the Interaction with the Anthracycline Daunomycin[a]

J. A. FERRAGUT, J. M. GONZALEZ-ROS,
A. V. FERRER-MONTIEL, AND P. V. ESCRIBA

Department of Neurochemistry
University of Alicante
03080 Alicante, Spain

The cytotoxic effects observed for nonpenetrating, polymerimmobilized anthracycline[1] strongly suggest that the interaction of the drugs with the plasma membrane of tumor cells may be important in the molecular mechanisms of drug cytotoxicity. In the present study, we have used positively (PCL) and negatively (NCL) charged liposomes (Avanti Polar Lipids) as model systems to determine how the surface charge of the membrane influences its interaction with ionized drugs such as daunomycin (DNM), an anthracycline antibiotic that contains an ionizable amino group (pK $\simeq$ 7.6–8.2) at the daunosamine sugar moiety.

Equilibrium binding of DNM to liposomes was carried out by using ultracentrifugation and fluorescence anisotropy techniques to distinguish between free and liposome-bound drug (FIG. 1). pH values ranging from 6 to 8.3 were chosen for the studies to produce different ionization states of the drug without changing the net charge of the stearylamine and the dicetyl phosphate groups present in the PCL and NCL, respectively. Binding parameters were obtained from Klotz plots, which gave straight lines with a slope equal to $1/n$, where n is the maximum number of binding sites per phospholipid molecule, and a y-intercept of $1/(n \cdot K_{app})$, where K_{app} is the apparent binding constant and $(n \cdot K_{app})$ equals the overall binding constant, K_s.

FIGURE 1 shows direct binding data corresponding to the interaction between DNM and NCL (A) or PCL (B) at different pH. It should be noted that the abscissa in B needs to include much higher liposome concentrations (in terms of phospholipid contents) to begin to evidence saturation.

The results clearly suggest that the interaction between the drug and the vesicles is mainly governed by the presence of negative charges at the liposome surface, while the ionization state of the drug, as evidenced from the results obtained at different pH, is less important in altering drug binding. Nonetheless, drug-binding parameters, as determined from Klotz plots, were sufficiently sensitive to the pH as to discriminate between binding of neutral and cationic species of the drug (TABLE 1). For either NCL or PCL, high pH values favoring the presence of un-ionized drug species result in larger K_s values due to an increased binding stoichiometry. Furthermore, fluorescence resonance energy transfer studies similar to those reported previously[2] indicate that the

[a]This work is supported by Grants 950/84 from CAICYT and 87/1634 from FISS.

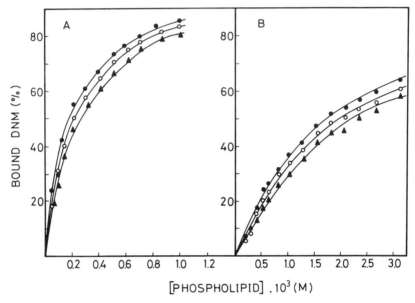

$[$PHOSPHOLIPID$] . 10^3$ (M)

FIGURE 1. Binding of DNM to negatively (**A**) and positively (**B**) charged liposomes at 25°C. The steady-state fluorescence anisotropy of a 2.5-μM DNM solution was determined in the presence of increasing amounts of liposomes by using a Perkin-Elmer LS-5 spectrofluorimeter equipped with total emission and automatic polarization accessories. Excitation wavelength was 480 nm and a Corning 3-68 filter was used to eliminate scattered light. The concentrations of free and liposome-bound DNM were determined at each phospholipid concentration as described by Burke and Tritton.[5] Buffers used in the experiments contained 100 mM NaCl in addition to either (●) 10 mM EPPS, pH 8.3; (○) 10 mM HEPES, pH 7.6; 10 mM HEPES, pH 7.0 (not shown), or (▲) 10 mM PIPES, pH 6.1.

TABLE 1. Binding Constants and Stoichiometries for the Interaction between DNM and Positively (PCL) or Negatively (NCL) Charged Liposomes at Different pH[a]

	pH			
	6.1	7.0	7.6	8.3
NCL				
n[b]	0.032	0.037	0.050	0.080
$K_s (M^{-1})$	4764	4975	5180	5515
$K_{app} \cdot 10^5 (M^{-1})$	1.50	1.34	1.04	0.70
PCL				
n	0.0048	0.0055	0.0071	0.0089
$K_s (M^{-1})$	485	500	570	675
$K_{app} \cdot 10^5 (M^{-1})$	1.0	0.91	0.78	0.74

[a]Binding parameters are averages from 2–3 measurements and were determined from Klotz plots, as indicated in the text. Linear correlation coefficients were better than 0.9 in all cases.
[b]n: moles of DNM/mol of phospholipid.

un-ionized form of the drug has access to hydrophobic domains located deep within the lipid bilayer, which are less accessible to ionized DNM (data not shown).

The observations reported here might have implications in cellular drug resistance. Since some drug-resistant cells have less negatively charged phospholipids than their parental drug-sensitive lines,[3] it is possible that a decreased drug binding to the resistant cell membrane could partly account for the observed decreased intracellular drug accumulation exhibited by most drug-resistant cells.[4]

REFERENCES

1. TRITTON, T. R. & G. YEE. 1982. The anticancer agent adriamycin can be actively cytotoxic without entering cells. Science **217:** 248–250.
2. FERRER-MONTIEL, A. V., J. M. GONZALEZ-ROS & J. A. FERRAGUT. 1988. Association of daunomycin to membrane domains studied by fluorescence resonance energy transfer. Biochim. Biophys. Acta **937:** 379–386.
3. FERRAGUT, J. A. Submitted for publication.
4. MYERS, C., K. COWAN, B. SINHA & B. CHABNER. 1987. The phenomenon of pleiotropic drug resistance. *In* Important Advances in Oncology. V. T. DeVita, S. Hellman & S. A. Rosenberg, Eds. Lippincott. Philadelphia, PA.
5. BURKE, T. G. & T. R. TRITTON. 1985. Structural basis of anthracycline selectivity for unilamellar phosphatidylcholine vesicles: An equilibrium binding study. Biochemistry **24:** 1768–1776.

Effect of ADP-Ribosyl Transferase Inhibitors on the Survival of Human Lymphocytes after Exposure to Different DNA-Damaging Agents[a]

C. FRANCESCHI,[b] M. MARINI,[c] G. ZUNICA,[c] D. MONTI,[b]
A. COSSARIZZA,[b] A. BOLOGNI,[b] C. GATTI,[b]
AND M. A. BRUNELLI[c]

[b]Institute of General Pathology
University of Modena
Modena, Italy

[c]Institute of Histology and General Embryology
University of Bologna
Bologna, Italy

ADP-ribosyl transferase (ADPRT) is a nuclear enzyme activated by single (SSB) and double (DSB) DNA strand breaks which synthesizes ADP-ribose homopolymers bound to nuclear proteins. The physiological role of ADPRT is still poorly understood, although it has been suggested that it plays a role in DNA repair and cell proliferation and differentiation.[1] Excessive activation of the enzyme can exhaust intracellular NAD^+ and ATP pools, thereby contributing to the so-called suicide response of extensively damaged cells.[2]

We asked whether ADPRT inhibitors, such as 3-aminobenzamide (3-ABA) and nicotinamide (NAM), would prevent the suicide response of human lymphocyte damaged by gamma rays in the same way as in lymphocytes damaged by oxygen radicals,[3] taking into account that cells damaged by both agents show increased ADPRT activity and DNA strand breaks.

MATERIALS AND METHODS

Human peripheral blood mononuclear cells were obtained and cultured as described.[4] Cells (1×10^6/ml) were exposed to gamma rays (10 Gy from a ^{60}Co source) or to 100 mU/ml xanthine oxidase (XOD) plus 100 μM hypoxanthine (HYP) in RPMI 1640 medium containing 10% human AB serum and incubated for 1 hour either in the absence or in the presence of 5 mM 3-ABA or 10 mM NAM. Cells were then resuspended in fresh medium, stimulated with phytohemagglutinin (PHA-P Difco, 1 μl/ml), and cultured for 3, 4 or 5 days, and [^{3}H]thymidine incorporation was evaluated as described.[4]

[a]This work was supported by CNR Grants No. 87.0129.44 (to C.F.) and No. 87.1489 (to M.A.B.) and by Minister of Education Grants 40% and 60%.

RESULTS AND DISCUSSION

TABLE 1 shows that—in time course studies—3-ABA and, to a lesser extent, NAM protect lymphocytes from oxygen radical damage, but not from gamma-ray-induced damage. The effect of 3-ABA and NAM does not seem to depend on the extent of damage since similar results were obtained when lymphocytes were exposed to lower concentrations of oxygen radicals by decreasing the concentration of hypoxanthine, or to lower gamma ray dose (5 Gy). The activation of ADPRT-mediated suicide mechanism does not seem to play a crucial role in the survival of gamma-ray-damaged lymphocytes, in accordance with previous observations in human fibroblasts.[5] We can speculate that toxic oxygen radicals and cell damages produced by an enzymatic system such as XOD/HYP and those produced by gamma rays are likely to differ (different radical species?[6] different amount of SSB versus DSB?). In conclusion, our

TABLE 1. Lymphocyte Survival after Damage with Gamma Rays (10 Gy) or Xanthine Oxidase (XOD, 100 mU/ml) plus Hypoxanthine (HYP, 100 μM) in the Presence or Absence of 3-aminobenzamide (3-ABA, 5 mM) or Nicotinamide (NAM 10 mM) during the Damaging Period

	Lymphocyte Survival		
Treatment	Day 3	Day 4	Day 5
Control + 3-ABA	106 ± 4	109 ± 10	97 ± 2
Control + NAM	110 ± 7	98 ± 7	73 ± 11
Gamma rays	33 ± 5	22 ± 1	17 ± 2
Gamma rays + 3-ABA	31 ± 7	22 ± 1	18 ± 2
Gamma rays + NAM	34 ± 5	23 ± 2	21 ± 3
XOD + HYP	5 ± 2	8 ± 5	36 ± 17
XOD + HYP + 3-ABA	51 ± 13	86 ± 20	117 ± 8
XOD + HYP + NAM	23 ± 9	56 ± 27	73 ± 22

NOTE: Data are means ±SEM of three experiments and are expressed as percentage of [³H] thymidine incorporation of undamaged untreated cultures.

data suggest that ADPRT inhibition affects the survival of damaged cells depending on the damaging agent and, according to preliminary data, on the donor age.

REFERENCES

1. CLEAVER, J. E., C. BOREK, K. MILAM & W. F. MORGAN. 1987. Pharmacol. Ther. **31:** 269–293.
2. CARSON, D. A., S. SETO, D. B. WASSON & C. J. CARRERA. 1986. Exp. Cell Res. **164:** 273–281.
3. CARSON, D. A., S. SETO & D. B. WASSON. 1986. J. Exp. Med. **163:** 746–751.
4. MARINI, M., G. ZUNICA & C. FRANCESCHI. 1985. Proc. Soc. Exp. Biol. Med. **180:** 246–257.
5. JAMES, M. R. & A. R. LEHMAN. 1982. Biochemistry **21:** 4007–4013.
6. EMERIT, I., S. H. KHAN & P. A. CERUTTI. 1985. J. Free Rad Biol. Med. **1:** 51–57.

Morphofunctional Aspects of Lipo-Melanosome as a Model for the Organelle Phenotype of Normal and Tumor Pigment Cells

M. MIRANDA, F. AMICARELLI, A. POMA,
A. M. RAGNELLI, A. BONFIGLI, O. ZARIVI,
P. AIMOLA, AND D. BOTTI

Department of Cell Biology and Physiology
University of L'Aquila
I-67100 L'Aquila, Italy

Melanosomes, the organelles where the enzyme tyrosinase (EC.1.14.18.1, monophenol, L-3,4-dihydroxyphenylalanine oxygen oxidoreductase) catalyzing the major steps of melanin synthesis is "compartmented," are expressed by both normal and abnormal pigment cells.[1] While a lot of information is available about free tyrosinase kinetics, very little is known about the effect of compartmentation on it; moreover, tyrosinase is known to be inhibited by reaction products[2] that are cytotoxic.[3] Two types of melanosomes are paradigmatic for both morphology and melanin synthesis, those of B16 and Harding-Passey mouse melanomas.[4] Melanogenesis is highly expressed in melanomas, where, according to the stage of progression, deranged diagnostic melanosomes are observed.[5] The present work investigates how lipo-melanosome (liposome-entrapped tyrosinase) morphology and L-3,4-dihydroxyphenylalanine (Dopa) oxidase activity change with respect to time of incubation with Dopa in order to obtain both a minimal model melanosome and to correlate the tyrosinase activity with the model organelle morphology.

For the experiments we used lipo-melanosomes with or without 4% cholesterol with respect to egg lecithin, prepared as described previously[6]; the lipo-melanosomes were incubated for various time points with 5 mM Dopa in 0.1 M phosphate buffer at pH 6.8 and 25°C, and samples were monitored at A_{475} in the presence of 5 mM Dopa in 0.1 M phosphate buffer at pH 6.8 and 25°C for Dopa oxidase activity[7] that was released from lipo-melanosome by adding 0.05% Triton X-100; morphology was controlled by electron microscopy of K-ammonium-molybdate-stained lipo-melanosomes.

The main results were: (*a*) Three different courses of Dopa oxidase inactivation relative to time of incubation with Dopa were linear and fast in the case of vesicles without cholesterol (FIGS. 1 A and B) but nonlinear and slower when lipo-melanosomes contained cholesterol (FIG. 1 C). (*b*) Three different morphologic patterns of the vesicles with or without cholesterol were seen after incubation with Dopa (FIGS. 2 A-A', B-B', C-C'); A' and B' show deranged membranes, while C' shows lipo-melanosomes more ordered, as expected in the presence of cholesterol. The possible conclusion is that the Dopa oxidase activity of cholesterol lipo-melanosomes, which are the most ordered after incubation with Dopa, is protected from the conditions in

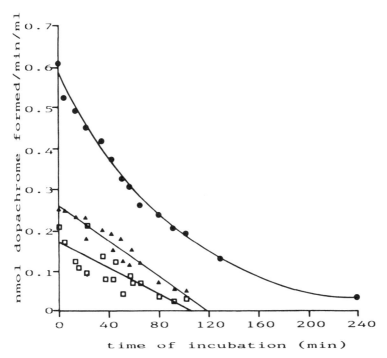

FIGURE 1. Decrease of lipo-melanosomes DOPA oxidase activity relative to time of incubation with L-Dopa. □-Lecithin lipo-melanosomes seen in FIGURE 2 A,A'; ▲-lecithin lipo-melanosomes seen in FIGURE 2 B,B'; ●-cholesterol-lecithin lipo-melanosomes seen in FIGURE 2 C,C'. Each point is the mean of 3 to 12 values. The equations and variances are: (□) $= -1.60 \ 10^{-3} x + 0.17$; $V = 1.0 \ 10^{-3}$; (▲) $y = -2.20 \ 10^{-3} x + 0.26$; $V = 0.30 \ 10^{-3}$; (●) $y = 0.62 \ x^4 + 1.70 \ x^3 + 2.10 \ x^2 + 1.58 \ x + 0.59$; $V = 0.19 \ 10^{-3}$.

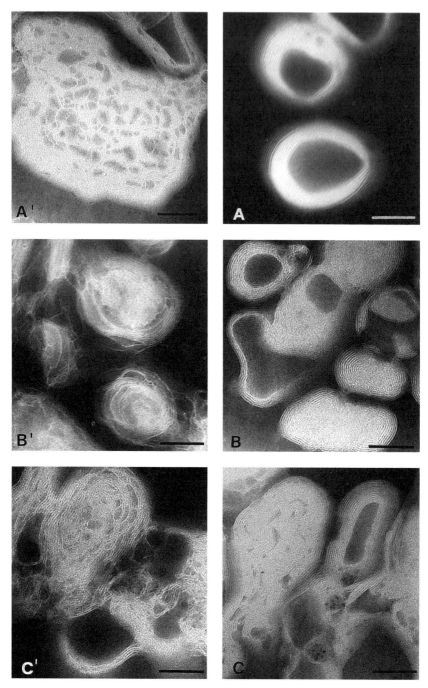

FIGURE 2. Electron micrographs of lipo-melanosomes incubated with L-Dopa. (A) Perturbed lipo-melanosome preparation at 0 min, **A'** at 240 min; (**B**) normal lipo-melanosome preparation at 0 min, **B'** at 240 min; (**C**) cholesterol-containing lipo-melanosomes at 0 min, **C'** at 240 min. Bar = 0.1 μm.

FIGURES 1 A and B, where enzyme activity decreases rapidly and membranes are deranged. It is well known that membrane lipid composition of cells changes in tumor,[4] and this may affect melanosome structure and function as observed in our model and in melanoma.[5]

REFERENCES

1. MASON, H. S. 1965. Annu. Rev. Biochem. **84:** 595–634.
2. MIRANDA, M. *et al.* 1979. Biochim. Biophys. Acta **585:** 398–404.
3. MIRANDA, M. *et al.* 1984. Mol. Gen. Genet. **193:** 395–399.
4. JIMBOW, M. *et al.* 1982. J. Invest. Dermatol. **79** (2):97–102.
5. JIMBOW, K. *et al.* 1988. Adv. Pigment Cell Res.: 169–182.
6. MIRANDA, M. *et al.* 1988. Colloids and Surfaces. Elsevier. Amsterdam.
7. MIRANDA, M. *et al.* 1987. Biochim. Biophys. Acta **913:** 386–394.

Index of Contributors

(Numbers in italics refer to comments in Discussion.)